# MOLECULAR THERMODYNAMICS
# OF NONIDEAL FLUIDS

# BUTTERWORTHS SERIES IN CHEMICAL ENGINEERING

**SERIES TITLES**

Chemical Process Equipment: Selection and Design   *Stanley M. Walas*

Constitutive Equations for Polymer Melts and Solutions   *Ronald G. Larson*

Gas Separation by Adsorption Processes   *Ralph T. Yang*

Heterogeneous Reactor Design   *Hong H. Lee*

Molecular Thermodynamics of Nonideal Fluids   *Lloyd L. Lee*

Phase Equilibria in Chemical Engineering   *Stanley M. Walas*

Transport Processes in Chemically Reacting Flow Systems   *Daniel E. Rosner*

**RELATED TITLES**

Catalyst Supports and Supported Catalysts   *Alvin B. Stiles*

Enlargement and Compaction of Particulate Solids   *Nayland Stanley-Wood*

Fundamentals of Fluidized Beds   *John G. Yates*

Liquid and Liquid Mixtures   *J.S. Rowlinson and F.L. Swinton*

Mixing in Process Industries   *N. Harnby, M. Edwards, and A.W. Nienow*

Solid Liquid Separation   *Ladislav Svarovsky*

Supercritical Fluid Extraction   *Mark A. McHugh and Val J. Krukonis*

# MOLECULAR THERMODYNAMICS

## OF NONIDEAL FLUIDS

## Lloyd L. Lee
*School of Chemical Engineering and Materials Science*
*University of Oklahoma*

**Butterworths**
Boston  London  Durban  Singapore  Sydney  Toronto  Wellington

**Library of Congress Cataloging-in-Publication Data**

*Lee,    Lloyd L.*
Molecular thermodynamics of nonideal fluids.

(Butterworths series in chemical engineering)
Bibliography: p.
Includes index.

| | | |
|---|---|---|
| | 1. Thermodynamics. | 2. Molecular dynamics. |
| | 3. Fluid mechanics. | I. Title.    II. Series. |
| QD504.L43 1987 | 536'.7 | 87-6384 |
| ISBN 0-409-90088-5 | | |

**British Library Cataloguing in Publication Data**

*Lee,    Lloyd L.*
Molecular thermodynamics of nonideal fluids.
1.  Fluids ——Thermal properties
2. Statistical thermodynamics
I. Title

| | |
|---|---|
| 536'.7 | QD504 |
| ISBN 0-409-90088-5 | |

Butterworth Publishers
80 Montvale Avenue
Stoneham, MA 02180

10 9 8 7 6 5 4 3 2 1

Printed in the United States of America

# CONTENTS

# CHAPTER V. MICROTHERMODYNAMICS

# CHAPTER VI. INTEGRAL EQUATION THEORIES

# CHAPTER VII. THEORIES FOR POLAR FLUIDS

# CHAPTER VIII. HARD SPHERES AND HARD-CORE FLUIDS

## CHAPTER XV. ADSORPTION: THE SOLID-FLUID INTERFACE

## APPENDIX A. *INTERMOLECULAR POTENTIALS*

## APPENDIX B. *GILLAN'S METHOD OF SOLUTION FOR INTEGRAL EQUATIONS*

## APPENDIX C. *MOLECULAR DYNAMICS PROGRAM IN THE N-V-E ENSEMBLE*

## APPENDIX D. *BIBLIOGRAPHY*

## INDEX

# PREFACE

The theoretical study of the physical properties of fluids (liquids and gases) has made significant advances in the last 30 years. The progress has been due chiefly to a three-prong development in (a) molecular theories based on the methods of statistical mechanics, (b) numerical simulations of molecular movements on fast electronic computers, and (c) scattering experiments with x-ray, neutrons, and other probes. Earlier, simple physical models were used to represent spherical molecules, such as argon. Gases like these are not of interest to practicing engineers. This state of affairs persisted into the 1950s and 1960s. In the 1970s, the situation improved considerably: fluids with polar forces (water, ammonia, and alcohols) and nonspherical shapes (polyatomics and hydrocarbons) were studied. In addition, charged particles (plasmas and ionic solutions), long-chain molecules (polymer solutions), and liquid metals were investigated. Studies of these "realistic" fluids were made possible by a combination of factors: development of the probability distribution function theories in statistical mechanics (e.g., the perturbation theories) and refinement of computer simulation methods (Monte Carlo random walks and molecular dynamics solution of the equations of motion). Currently, the new methods of investigation are expanding rapidly into fluid materials such as liquid crystals, colloidal solutions, biological fluids, polyelectrolytes, amphiphilic molecules, and polydisperse systems. Diverse phenomena found in the fluid state are being examined, such as phase transition, chemical reactions, adsorption at interfaces, and transport phenomena.

These new developments have made the field of molecular theories not only interesting but also "useful" to engineers. However, due to the interdisciplinary preparation and the requisite mathematical sophistication, the new information remained inaccessible to many segments of the engineering community. The situation has improved markedly since the 1970s and the uptake continued into the 1980s, as witnessed by the increased number of courses offered in engineering colleges on molecular thermodynamics and statistical mechanics, as well as seminars held at professional meetings on molecular-based studies of fluids. At the same time, there are needs for textbooks written for an engineering audience. Books in statistical mechanics were conventionally written for theoretical physicists and chemists. Although many are definitive works in the field (see Appendix D. Bibliography), few are suited as introductions to the subject. Others were devoted to specialized fields not of interest to engineers.

With this in mind, we have written this book to address a different need, the need to bring the molecular methods to engineers. Our presentation is introductory. Thus many subjects of prime interest to researchers were eliminated. It is a difficult, or even impossible, task to keep molecular theories on an elementary level. Compromises will have to be made. We are at the same time application-minded. Whenever possible, applications of interest to chemical engineers are included. We have presented additional material at the end of chapters on matters such as calculation of the ideal-gas heat capacities and mixture vapor-liquid equilibria for polar fluids. Due to the gap between theory and practice, some compromise of rigor will be inevitable, and we are aware of the risks involved. It is hoped that future developments will make these sacrifices unnecessary.

The materials in the book were the outgrowth of lectures given for the past seven years in a graduate course— the Modern Thermodynamics Seminar. The topics varied from year to year, reflecting the changes in interests and student needs. The main themes, however, have always revolved around the *molecular distribution functions*; i.e., treating the structure of matter in terms of the probabilistic distributions of molecules in space and time. This is the subject of statistical mechanics. The historical developments in the statistical mechanics of liquids could be summarized as (a) the partition function approach, (b) combinatorial studies, and (c) distribution function methods. Recent work has gravitated toward the last option, although the other approaches are being vigorously pursued at the same time. Partition functions are known explicitly only for very simple systems, such as the ideal gas and Ising models. For more complex systems, the approach is ineffective. Combinatorial studies were carried out, e.g., for mixture and adsorption problems (such as arrangements of polymer molecules on a lattice). The presence of sophisticated interaction forces soon taxes the combinatorial method to the limit. Distribution functions imply and implicate all these methods and more, since they are defined for any systems through a probability distribution. In principle, at least, the functions could be used for structural studies. A crucial connection is the probing of the molecular structure of matter by the scattering of x-ray and neutrons. They yield the probability distributions. These experiments lend physical support to the distribution functions. On the other hand, thermodynamic properties can be easily calculated from the probabilities. Three routes are available for this task: the energy equation, the virial equation, and the compressibility equation. These relations make the distribution functions highly useful in properties studies.

The first six chapters lead up to the integral equations. These equations are used to produce the distribution functions. Some model potentials— hard spheres and Lennard-Jones molecules— are examined. We then proceed to polar fluids and electrolyte solutions. The successful liquid state theory— the perturbation theory is introduced. For polyatomics, we offer one of the most promising approaches— the interaction site model. Interfacial behavior is investigated in a chapter on adsorption. Selected homeworks are included in the chapters. Some of these are "drill" exercises; the others are projects. To supplement the in-class experience, reading of current literature and possibly a research project applying what has been learned is recommended. In my seminar course, projects on behavior of fluids such as liquid crystals, biofluids, electrolyte solutions, and molecular sieves were assigned.

During the course of preparation, I have benefited by discussions with colleagues and students. Valuable suggestions were offered by P.T. Cummings, C.K. Hall, H.D. Cochran, and D. Henderson. Proofreading was done by J.L. Savidge, R.L. McFall, and W.M. Coleman. I particularly want to thank J.M. Haile and F.T.H. Chung for careful reading of the drafts and offering valuable suggestions. Some of the materials presented in this book are fruits of work by J.M. Haile, F.T.H. Chung, L.H. Landis, M.R. Brulé, S. Watanasiri, M.H. Li, and S.H. Brown. I thank them for their generous consent. Part of a sabbatical leave was devoted to the writing of this book. For all the help, I alone am responsible for any errors or omissions that remain. The choice of subjects is by necessity dictated by the author's familiarity with the topic material. One must be constantly vigilant in keeping current on the new developments in molecular theory. However, if a book is to be written at all, it has to start and to stop somewhere. The author has since developed a deeper sense of appreciation for the sayings of Jhwangjoe (*circa 275* B.C.)

> "*To pursue knowledge with one's lifetime is to span the boundless by the limited... Therefore, by following the middle way, you may maintain your health, replenish your life, care for the ones you love, and live out your years.*" — Inner Chapter: "The Regimen of Life"

*Norman, Oklahoma*  
*Fall 1987*

*Lloyd L. Lee*  
*D.H.D.*

# CHAPTER I

# INTRODUCTION

*Democritus of Abdera:* " ...οὐδὲν χρῆμα μάτην γίγνεται, ἀλλὰ
πάντα ἐκ λόγου τε καὶ ὑπ᾽ ἀνάγκης"*

In this chapter, we set forth the microscopic view of matter. A material system is composed of an enormous number of particles, for example, the molecules. We call this an N-body system. The molecules interact with one another by intermolecular forces. The basic question in molecular physics is the relation between molecular interactions and the bulk properties of the system. The answer is to be found in *statistical mechanics*. Statistical mechanics is a method of accounting whereby collective microscopic states are *averaged*, resulting in macroscopically measurable quantities. As the name implies, *mechanics* deals with forces, and *statistics* is a procedure of *averaging*. Thus statistical mechanics offers a prescription for averaging the molecular forces in order to obtain values for the gross properties. The method is applicable to the study of properties of fluids, solutions, plasmas, and crystals alike, In this chapter, we shall establish the basic terminology. The mechanical variables are called dynamic variables. They are functions of the positions and motions of many bodies. The velocities and configuration constitute the *phase space*. Hamilton's equations of motion are derived that describe the time evolution of the system in phase space. The description is based on Newtonian mechanics. However, a summary of quantum mechanics is presented in Section I.5 for comparison and later use. Other statistical concepts, such as the Gibbs ensembles and distribution functions, will be introduced in following chapters.

## I.1. *The N-Body System*

An N-body system is a collection of N material particles. These particles could be molecules, charged ions, or colloidal particles. In our approach, material is considered as composed of spatially discrete units. This view is in contrast to the continuum view where matter is continuous and homogeneous. The latter view is called macroscopic. The continuum view of matter is useful in a number of engineering studies such as fluid mechanics and heat transfer. However, the microscopic view is used for the study of properties because it is necessary to understand the specific forces of interaction among the particles that compose the material. When Newton's laws of

---

*Things do not happen fortuitously, but out of reason and necessity.

*1*

motion are applied at the molecular level, the treatment is called *classical*. When necessary, quantum mechanical corrections are incorporated. This happens at small de Broglie wavelengths (see below). For the fluids we study, a classical description is adopted whenever permissible.

The selection of the particles, whose collection constitutes the $N$-body system, is dictated by the particular physical process taking place. In general, we select a system of particles with maximum unity so that their internal structure and degrees of freedom have negligible effects on the physical process in progress. We illustrate this point by several simple examples.

### Example 1. *Motion of a Pendulum in a Gravitational Field*

For a small ball attached to a weightless cord, the particle chosen to represent the physical system is simply the ball itself. The internal structure of the ball (whether it is made of an alloy of metals, or an aggregate of $10^{23}$ molecules) has no direct bearing on the motion of the pendulum.

### Example 2. *The Harmonic Oscillator*

For a spring-and-block assembly, the particle is identified as the block. The spring is replaced in the physical representation by its Hookesian force.

$$F = - kx \tag{1.1}$$

i.e.

*Force = – (Spring constant)(Displacement)*

### Example 3. *Migration of Charged Particles in an Electric Field*

For the system depicted in its simplest form, the particles chosen to represent the system are the bodies carrying the electric charges. They may be dust particles, plasma, or colloidal particles in a solution.

### Example 4. *Compression of Gas in a Cylinder*

The particles chosen in this case are the gas molecules inside the cylinder.

## I.2. *The Hamiltonian and the Pair Potentials*

The particles in an $N$-body system are in constant motion undertaking movements in all directions. The total energy of the system, $TE$, is calculated as the sum of the kinetic contribution, $KE$, and the potential contribution, $PE$.

$$TE = KE + PE \tag{2.1}$$

The Hamiltonian, $H_N$, of the system is defined to be its total energy, $TE$, and is a function of the kinetic variables (i.e., the linear momenta $p_i$) and spatial variables (e.g., the center-of-mass positions $r_i$ of the particles)

$$H_N(p_1, \ldots, p_N, r_1, \ldots, r_N) = KE + PE \tag{2.2}$$

This is the Hamiltonian for an *isolated* system, i.e., there is no interaction with surroundings, such as an isothermal bath.

**KINETIC ENERGY**

Kinetic energy is associated with the motion of particles. It is separated into translational, rotational, and vibrational modes.

*Translational Energy*

$$KE_t = \frac{1}{2}\sum_{i=1}^{N} \frac{p_i^2}{m_i} \qquad (2.3)$$

*Rotational Energy*

$$KE_r = \frac{1}{2}\sum_{i=1}^{N} \alpha_i \cdot \mathbf{I}_i \cdot \alpha_i \qquad (2.4)$$

*Vibrational Energy*

$$KE_v = \frac{1}{2}\sum_{i=1}^{N}\sum_{j=1}^{n'} m_i\omega_{ij}^2 a_{ij}^2 \qquad (2.5)$$

where $p_i$ is the momentum, $m_i$ the mass of particle $i$, $I_i$ the moment of inertia, $\alpha_i$ the angular velocity, $\omega_{ij}$ the frequency of vibration of mode $j$ in molecule $i$, $a_{ij}$ the amplitude of vibration, and $n' = 3n - 5$ for linear molecules and $3n - 6$ for nonlinear molecules. These classical expressions also have their quantum counterparts (see below).

**POTENTIAL ENERGY**

Potential energy arises due to interactions among particles and between particles and surroundings. The strength of interaction is dependent on the distances that separate these particles. The total potential energy, $V_N$, is a function of the spatial configuration, $\mathbf{r}^N = \{\mathbf{r}_1, \ldots, \mathbf{r}_N\}$, of the particles. The position of particle $i$, as measured from the origin, is a vector $\mathbf{r}_i = (x_i, y_i, z_i)$. Therefore the configuration $\mathbf{r}^N$ is a vector of $3N$ dimensions. For nonspherical molecules, one must specify the angles of orientation for all $N$ particles, $\{\omega_1, \ldots, \omega_N\}$, where $\omega_i$ is the set of Euler angles $(\theta_i, \phi_i, \chi_i)$ of molecule $i$; $\theta$ is the polar angle, $\phi$ the azimuthal angle, and $\chi$ the rotational angle. For structured polyatomics such as the hydrocarbon *n*-butane, additional coordinates are needed to uniquely determine their conformations, e.g. the cis, trans, gauche, and cyclic arrangements. In contrast to the kinetic energy, which depends on motion, the potential energy is dependent on the spatial variables of the molecules. To account for the potential energy of the $N$ bodies, one assembles the particles into clusters of singlets, pairs, triplets, etc., and considers the total *PE* to be a sum of one-body energies, $u^{(1)}$, two-body energies, $u^{(2)}$, three-body energies, $u^{(3)}$, etc.

$$V_N(\mathbf{r}^N,\omega^N) = \sum_{i=1}^{N} u^{(1)}(\mathbf{r}_i,\omega_i) + \sum_{1=i\,<j}^{N} u^{(2)}(\mathbf{r}_i,\omega_i,\mathbf{r}_j,\omega_j) \qquad (2.6)$$

$$+ \sum_{1=i\,<j\,<k}^{N} u^{(3)}(\mathbf{r}_i,\omega_i,\mathbf{r}_j,\omega_j,\mathbf{r}_k,\omega_k) + \cdots + u^{(N)}(\mathbf{r}^N,\omega^N)$$

We have exhibited explicitly the position and angle variables. In case there are other spatial variables, they should also be included. The one-body potentials $u^{(1)}$ arise due to external fields (for example, an electric field). The pair potentials $u^{(2)}$ are evaluated by taking one pair of molecules at a time and measuring their interaction energy while iso-lated from all other molecules -- i.e., the interaction energy between a pair of particles is assumed to be undisturbed by the presence of a third particle. For example, in a system of four particles ($N=4$), we count the singlet energies $u^{(1)}(1)$, $u^{(1)}(2)$, $u^{(1)}(3)$, and $u^{(1)}(4)$, and the pair energies $u^{(2)}(12)$, $u^{(2)}(13)$, $u^{(2)}(14)$, $u^{(2)}(23)$, $u^{(2)}(24)$, and $u^{(2)}(34)$. However, in dense fluids, the sum of all these terms is not sufficient to account for the total energy of interaction. There are additional energies associated with the *residual* three-body forces. These forces are in excess of the sum of one-body and two-body forces. Thus we should also include $u^{(3)}(123)$, $u^{(3)}(124)$,..., and $u^{(3)}(234)$. Note that these higher-order forces contribute energies above and beyond the sum of pairs. The remainder is carried over to the residual four-body energies, $u^{(4)}$, excesses over the sum of energies of tri-plets. Finally, $u^{(N)}$ is the *excess energy* over the sum of energies of ($N-1$)-tuplets and is not the same as $V_N$, the *total* potential of the $N$-body system.

## Pairwise Additivity

It is often assumed that the total *PE* is adequately given by the sum of two-body energies. This is called the *pairwise additivity* (PA) assumption

$$V_N(\mathbf{r}^N, \omega^N) \approx \sum_{i<}^{N} \sum_{j} u^{(2)}(\mathbf{r}_i, \omega_i, \mathbf{r}_j, \omega_j) \qquad (2.7)$$

This assumption is valid only for dilute fluids. For argon, an accurate interaction poten-tial valid in the liquid state is the Barker-Fisher-Watts [1] potential. It contains three-body forces. However, PA is extensively used in theoretical work to simplify treatment. Under this assumption, the Hamiltonian is

$$H_N(\mathbf{p}^N, \mathbf{r}^N, \omega^N) = \sum_{j=1}^{N} \frac{p_j^2}{2m_j} + \sum_{i<}^{N} \sum_{j} u^{(2)}(\mathbf{r}_i, \omega_i, \mathbf{r}_j, \omega_j) \qquad (2.8)$$

For simplicity, we shall write $u(r)$ for $u^{(2)}(r)$. In real molecules the interaction potential is usually quite complicated. Quantum mechanical calculations exist for the most part for relatively small molecules (e.g., hydrogen [2], helium [3], and water [4]). On the other hand, simplified models are used in theoretical work. Some commonly used potentials are listed below. A detailed discussion on pair potentials is given in Appendix A.

## The Ideal Gas

$$u(r) = 0 \qquad (2.9)$$

## The Hard Spheres (HS)

$$u(r) = +\infty, \qquad r \leq d \qquad (2.10)$$

$$= 0, \qquad r > d$$

where $d$= hard-sphere diameter.

*The Square-Well Potential (SW)*

$$u(r) = +\infty, \qquad r \leq d \tag{2.11}$$

$$= -\varepsilon, \qquad d < r < \lambda$$

$$= 0, \qquad r \geq \lambda$$

where $d$= repulsive diameter, $\lambda$= attractive diameter, and $\varepsilon$= well depth.

*The Inverse-12 Soft-Sphere Potential (SS12)*

$$u(r) = 4\varepsilon(\frac{\sigma}{r})^{12} \tag{2.12}$$

where $\sigma$= collision diameter and $\varepsilon$= energy parameter.

*The Lennard-Jones Potential (LJ)*

$$u(r) = 4\varepsilon\left[(\frac{\sigma}{r})^{12} - (\frac{\sigma}{r})^6\right] \tag{2.13}$$

*The Kihara Potential (KH)*

$$u(r) = +\infty, \qquad r \leq d \tag{2.14}$$

$$= 4\varepsilon\left[\left[\frac{\sigma-d}{r-d}\right]^{12} - \left[\frac{\sigma-d}{r-d}\right]^6\right], \qquad r > d$$

where $d$= the hard-core diameter.

## RELATION BETWEEN THE FORCE AND THE POTENTIAL ENERGY

For a conservative system, the force acting on the system is given by the negative gradient of the potential energy

$$\mathbf{F} = -\nabla u(r) \tag{2.15}$$

This relation is general.

*Example 1. The Gravitational Field*

The gravitational field of the earth is conserved. We know from mechanics that the potential energy is

$$PE = u(h) = mgh = (mass)(gravity)(height) \tag{2.16}$$

Therefore the force, according to (2.15), should be

$$F = -\nabla u(h) = -\frac{\partial}{\partial h}(mgh) = -mg \tag{2.17}$$

i.e., $F = -mg$; this is precisely the expected result. The negative sign indicates that the force is acting downward (opposite to the direction of $h$).

*Example 2. The Spring Force*

   A frictionless spring is a conservative system. Its potential energy is known to be

$$PE = u(x) = \frac{1}{2}kx^2 = \frac{1}{2}(spring\ constant)(displacement)^2 \tag{2.18}$$

Therefore the force on the spring is

$$F = -\frac{\partial}{\partial x}u(x) = -kx \tag{2.19}$$

This is the well-known Hookesian force.

## I.3. *The Phase Space*

   The dynamic state of a system of simple (structureless) particles is completely determined by specification of the positions, $r_i$, and momenta, $p_i$ ($i=1,...,N$) of the $N$ particles composing the system. For a three-dimensional system (with $x$-, $y$-, and $z$-coordinates), these $2N$ quantities ($N$ $r_i$'s and $N$ $p_i$'s) constitute a $6N$-dimensional space with coordinates

$$(r_{x1}, r_{y1}, r_{z1}, p_{x1}, p_{y1}, p_{z1},..., r_{zN}, p_{xN}, p_{yN}, p_{zN}) \tag{3.1}$$

This space is called by Gibbs [5] the *phase space* (or the $\Gamma$-space). Each point in the phase space corresponds to a particular dynamic state of the system with positions and momenta of the $N$ particles specified by the coordinates of this chosen phase point. Since the particles are in constant motion, the movements of $N$ particles at subsequent instants trace out, in the phase space, a continuous trajectory, which represents the history of the states of the system. All dynamic properties of the system can be inferred from its phase space trajectories. For example, the total energy, the total momentum, the collision rates, and the diffusion constant could all be obtained from the trajectories. The phase space is of great importance in statistical mechanics. We must familiarize ourselves with this concept. Let us examine some sample systems.

*Example 1. One dimensional Harmonic Oscillator*

   For the spring-and-block system described earlier, the movement is in the $x$-

direction. The number of particles is $N=1$. The phase space is then two dimensional with coordinates $(x, p_x)$. The motion of the block obeys the law of conservation of energy (assuming no frictional loss):

$$\frac{1}{2}kx^2 + \frac{1}{2}\frac{p_x^2}{m} = constant \tag{3.2}$$

The trajectory of this particle in the phase space $(x, p_x)$ is therefore described by an ellipse.

*Example 2. Pair of Charged Particles in Three-Dimensional* Space

Since $N=2$, the phase space is 12 dimensional, with three components in the $x$-, $y$-, and $z$-directions for each of $r_1$, $p_1$, $r_2$, and $p_2$. A geometrical representation of the locus of the phase points of this system is not possible. However, the physical idea is the same as in the above example.

**GENERALIZED PHASE SPACE**

For polyatomic molecules the kinetic energy now comprises, in addition to the translational mode, rotational and vibrational modes. Thus $KE$ contains

$$KE = \frac{1}{2}\sum_i^N \frac{p_i^2}{m_i} + \frac{1}{2}\sum_i^N \frac{J_i^2}{m_i} + \frac{1}{2}\sum_i^N\sum_j^{n'} m_i\omega_{ij}^2 a_{ij}^2 \tag{3.3}$$

To completely specify the dynamic state of the system, we need fix additional variables -- i.e., the angular momenta, $J_i$ ($3N$ variables)*, the vibrational frequencies and amplitudes, $\omega_{ij}a_{ij}$ ($n'N$ variables). Now the Hamiltonian is of the form

$$TE = KE + PE = H_N(p^N, J^N, (\omega a)^{Nn'}, r^N) \tag{3.4}$$

The dimensions of the generalized phase space are

$$3N + 3N + n'N + 3N = (n' + 9)N \tag{3.5}$$

For anisotropic molecules, the potential energy also depends on the $3N$ Euler angles of the molecules. The phase space will have $(n' + 12)N$ dimensions

$$H_N = H_N(p^N, J^N, (\omega a)^{Nn'}, r^N, \omega^N) \tag{3.6}$$

where $\omega^N = \{\omega_1, \omega_2, ..., \omega_N\}$ and $\omega_i = (\theta_i, \phi_i, \chi_i)$ are the Euler angles of the orientation of

---

*Note that the angular momentum, $J_i = \{J_{ix}, J_{iy}, J_{iz}\}$, has three projections in the x-, y-, and z-directions. $J_i$ is the total momentum of molecule $i$ irrespective of its symmetry. Internally, spherical top molecules have only one value for the moment of inertia (i.e., $I_{xx} = I_{yy} = I_{zz}$), symmetrical top molecules (e.g. $NH_3$) have two values ($I_{xx} = I_{yy}$ and $I_{zz}$), and asymmetric molecules (e.g., $H_2O$) have three values ($I_{xx}$, $I_{yy}$ and $I_{zz}$). They do not affect the vectorial nature (three-dimensionality) of the external total angular momentum unless the external coordinates coincide with the symmetry axes of the molecule (in which case one or two components are zero.)

molecule *i*. (We have used the symbol ω for both vibrational frequencies and Euler angles in deference to convention. The context will make the distinction clear.) In case other conformational distinctions are present, further augmentation of the dimensionality of the phase space will be required.

## I.4. *The Equations of Motion*

In this section we shall derive an equation of motion which describes the time evolution of a system of *N* simple particles. Since the particles are in motion, their positions, $\mathbf{r}_i$, and momenta, $\mathbf{p}_i$, are functions of time:

$$\mathbf{r}_i = \mathbf{r}_i(t) \tag{4.1}$$

$$\mathbf{p}_i = \mathbf{p}_i(t), \qquad i=1,2,..., N$$

As in the mechanics for macroscopic bodies, the motion of the individual particle *i* of mass $m_i$ is determined by Newton's second law:

$$m_i \frac{d^2\mathbf{r}_i}{dt^2} = \mathbf{F}_i \tag{4.2}$$

where $\mathbf{F}_i$ is the force acting on particle *i*. For conservative systems, the force is derived from a potential energy function. For simple systems with pair interactions only

$$\mathbf{F}_i = -\sum_{j\neq i}^{N} \nabla u(r_{ij}), \qquad r_{ij} = |\mathbf{r}_i - \mathbf{r}_j| \tag{4.3}$$

where *u* is the interaction potential, and the gradient $\nabla$ is with respect to $\mathbf{r}_i$. Since the momentum

$$\mathbf{p}_i = m_i \frac{d\mathbf{r}_i}{dt} \tag{4.4}$$

(4.2) can also be written as

$$\frac{d\mathbf{p}_i}{dt} = -\sum_{j\neq i}^{N} \nabla u(r_{ij}) \tag{4.5}$$

For structureless particles, $H_N$ is a function of $\mathbf{p}_i$ and $\mathbf{r}_i$: $H_N = H_N(\mathbf{p}^N, \mathbf{r}^N)$; we can form the partial derivatives

$$\frac{\partial H_N}{\partial \mathbf{p}_i} = \frac{\mathbf{p}_i}{m_i} \tag{4.6}$$

and

$$\frac{\partial H_N}{\partial \mathbf{r}_i} = \sum_{j \neq i}^{N} \nabla u(r_{ij}) \tag{4.7}$$

Comparing (4.6 & 7) with (4.4 & 5) gives

$$\dot{\mathbf{r}}_i = \frac{\partial H_N}{\partial \mathbf{p}_i} \tag{4.8}$$

$$\dot{\mathbf{p}}_i = -\frac{\partial H_N}{\partial \mathbf{r}_i}, \qquad i=1,...,N \tag{4.9}$$

where the dot denotes the time derivative $d/dt$. (4.8 & 9) are *Hamilton's equations of motion* and are valid for simple $N$-body systems. They govern the time evolution of the phase points, thus determining the dynamic states of the system. Every trajectory in the phase space represents a solution to eqs. (4.8 & 9) when the initial conditions are given

$$\mathbf{r}_i(0) = \mathbf{r}_i^0 \tag{4.10}$$

$$\mathbf{p}_i(0) = \mathbf{p}_i^0, \qquad i=1,..., N$$

*Example 1. The One-Dimensional Harmonic Oscillator*

Let the initial position be $r(0) = L$ and initial velocity (or momentum) $p(0) = 0$. Its total energy is

$$TE = H_1(p, r) = KE + PE = \frac{1}{2}\frac{p^2}{m} + \frac{1}{2}kr^2 \tag{4.11}$$

where $p$ and $r$ are functions of time: $p = p(t)$ and $r = r(t)$. Hamilton's equations are then

$$\frac{\partial H_1}{\partial p} = \frac{p}{m} = \frac{dr}{dt} \tag{4.12}$$

and

$$\frac{\partial H_1}{\partial r} = kr = -\frac{dp}{dt} \tag{4.13}$$

To solve (4.12) and (4.13), we differentiate (4.12) with respect to time to get

$$\frac{d^2r}{dt^2} = \frac{1}{m}\frac{dp}{dt} = -\frac{k}{m}r \tag{4.14}$$

where we have used (4.13). Let $\omega \equiv \sqrt{k/m}$. Then (4.14) can be solved as an ordinary differential equation

$$r(t) = A \, \sin(\omega t) + B \, \cos(\omega t) \tag{4.15}$$

Differentiating we get

$$\frac{dr}{dt} = A\omega \, \cos(\omega t) - B\omega \, \sin(\omega t) = \frac{p(t)}{m} \tag{4.16}$$

The initial conditions, $r(0) = L$ and $p(0) = 0$, imply

$$r(t) = L \, \cos(\omega t), \qquad p(t) = -L\omega m \, \sin(\omega t) \tag{4.17}$$

To see that this time dependence describes an ellipse in the $(r\text{-}p)$ space, we note that $\cos^2(\omega t) + \sin^2(\omega t) = 1$. Thus the time variable could be eliminated from (4.17) to give

$$\frac{1}{2}kr^2 + \frac{1}{2}\frac{p^2}{m} = \frac{kL^2}{2} \tag{4.18}$$

## I.5. *Quantum Mechanics*

As mentioned earlier, a physical system can be treated classically if its de Broglie thermal wavelength, $\Lambda$, is small. The de Broglie thermal wavelength is defined as

$$\Lambda \equiv \frac{h}{\sqrt{2\pi mkT}} \tag{5.1}$$

where $h = 6.624 \times 10^{-27}$ *erg·s* is the Planck constant, $m =$ mass of the particle, $k = 1.038 \times 10^{-16}$ *erg/K* is the Boltzmann constant, and $T =$ absolute temperature. $\Lambda$ is a measure of the magnitude of thermal fluctuations at the given temperature. When $\Lambda$ is large in comparison to the dimension characteristic of the material system, quantum calculations [6] must be used.

In heuristical terms, the difference between the classical approach and the quantum approach lies in the basic mathematical representation of the physical variables. In classical mechanics, the physical quantity, called *dynamic variable*, is a function of the phase space coordinates $(\mathbf{p}_1,..., \mathbf{p}_N; \mathbf{r}_1,..., \mathbf{r}_N)$. In contrast, quantum mechanics treats a physical quantity, or an *observable*, as an operator which operates on a wave function $\psi$, the latter being a function of position $\mathbf{r}$ and time $t$: $\psi = \psi(\mathbf{r}, t)$. The so-called wave function has a probabilistic interpretation; i.e., the product of $\psi$ with its complex conjugate, $\psi^*\psi$, gives the probability density that at time $t$ the system will be found at location $\mathbf{r}$. Therefore quantum mechanics requires a change of mathematical representation from the mathematics of functions to the mathematics of operators. And there is a one-to-one correspondence between the functions (dynamic variables) of classical mechanics and the operators (observables) in quantum mechanics. A few examples are given in Table I.1.

We now turn to a discussion of the time evolution of the physical system. In classical mechanics, the trajectory $\mathbf{r}^N(t)$, $\mathbf{p}^N(t)$ of the phase points obeys the Hamilton equations (4.8 & 4.9), as described in section I.4. A corresponding equation of motion exists in the quantum case for the wave function $\psi$, i.e., the Schrödinger equation

*Table I.1.  Correspondence of Physical Quantities in Classical and Quantum Mechanics*

| *Classical Mechanics* **Dynamic Variable** (a function) Arguments = coordinates | *Quantum Mechanics* **Observable** (an operator) Operand = wave function $\psi$ |
|---|---|
| $x$-coordinate | $x$ as multiplier |
| $y$-coordinate | $y$ as multiplier |
| $z$-coordinate | $z$ as multiplier |
| Momentum, **p** | Gradient, $-i\hbar\nabla$ |
| Angular momentum, $\mathbf{L} = \mathbf{r}\times\mathbf{p}$ | $-i\hbar\ \mathbf{r}\times\nabla$ |
| Energy, $E$ | $-i\hbar\ \partial/\partial t$ |

$$i\hbar\ \frac{\partial\psi(\mathbf{r},\ t)}{\partial t} = \mathbf{H}\psi(\mathbf{r},\ t) \tag{5.2}$$

where $\hbar = h/2\pi$.  The Hamiltonian operator, **H**, is obtained by replacing the classical functions by the corresponding quantum operators as listed in the table.  We show three cases

**TRANSLATIONAL MOTION OF A PARTICLE**

[Classical]

$$TE = KE + PE = \frac{1}{2}\frac{p^2}{m} + V(r) \tag{5.3}$$

[Quantum]

$$i\hbar\frac{\partial\psi}{\partial t} = \frac{-\hbar^2}{2m}\nabla^2\psi + V(r)\psi \tag{5.4}$$

**ROTATIONAL MOTION OF A PARTICLE**

[Classical]

$$TE = KE + PE = V(r) + \frac{1}{2}\alpha\cdot\mathbf{I}\cdot\alpha \tag{5.5}$$

[Quantum]

$$\psi = \rho(r)\alpha(\theta,\ \phi)\tau(t) \tag{5.6}$$

$$i\hbar\frac{\partial\tau}{\partial t} = E\tau(t) \tag{5.7}$$

$$\frac{1}{r^2}\frac{d}{dr}[r^2\frac{d}{dr}\rho(r)] - \frac{2m}{\hbar^2}[E - V(r)]\rho(r) - \frac{\lambda}{r^2}\rho(r) = 0 \tag{5.8}$$

$$\frac{1}{\sin\theta}\frac{\partial}{\partial\theta}[\sin\theta\frac{\partial}{\partial\theta}\alpha(\theta,\phi)] + \frac{1}{\sin^2\theta}\frac{\partial^2}{\partial\phi^2}\alpha(\theta,\phi) + \lambda\alpha(\theta,\phi) = 0 \tag{5.9}$$

## VIBRATIONAL MOTION OF A HARMONIC OSCILLATOR

[Classical]

$$TE = KE + PE = \frac{p^2}{2m} + \frac{1}{2}kx^2 \tag{5.10}$$

[Quantum]

$$i\hbar\frac{\partial\psi}{\partial t} = -\frac{1}{2}\frac{\hbar^2}{m}\frac{\partial^2}{\partial x^2}\psi + \frac{1}{2}kx^2\psi \tag{5.11}$$

These equations can be solved with appropriate boundary conditions. They give the time evolution of the wave function $\psi(r, t)$. The energy levels are the eigenvalues of the operator **H**. We show the most common cases below.

## TRANSLATIONAL ENERGY LEVELS

$$\varepsilon_t = \frac{n^2h^2}{8mL^2} \tag{5.12}$$

where $n$ =quantum number =0,1,2,..., $h$ =Planck's constant, $m$ =mass of the particle, $L$ =the linear dimension of the system. There are no degeneracies in the energy levels. The eigenfunctions for a free particle in a one-dimensional box are

$$\psi_n(x) = \sqrt{\frac{2}{L}}\sin\left[\frac{n\pi}{L}\right], \qquad n = 0,1,2,... \tag{5.13}$$

## ROTATIONAL ENERGY LEVELS

(For rigid rotators: i.e., two masses $m_1$ and $m_2$ at fixed distance $2R$ apart)

$$\varepsilon_r = \frac{J(J+1)\hbar^2}{2\bar{m}R^2} \tag{5.14}$$

where $J$ =quantum number = 0,1,2,..., $\bar{m}$= reduced mass =$m_1m_2/(m_1+m_2)$, $R$ =radius of rotation. The degrees of degeneracies are $2J+1$. The eigenfunctions are the so-called spherical harmonics

$$Y_{J,m}(\theta, \phi) = (2\pi)^{-1/2}\exp(im\phi)\left[\sqrt{\frac{2J+1}{2}\frac{(J-m)!}{(J+m)!}}\right]P_J^m(\cos\theta) \qquad (5.15)$$

where $J = 0,1,2,...$; $m = -J, -(J-1),...,0,1,..., J$. $P_J^m(x)$ is the associated Legendre function.

**VIBRATIONAL ENERGY LEVELS**

(For harmonic oscillators in one dimension $x$)

$$\varepsilon_v = h\nu(L+\frac{1}{2}) \qquad (5.16)$$

where $L$ = quantum number = $0,1,2,...$, $\nu$ = frequency of vibration. There is no degeneracy. The eigenfunctions are the Hermite polynomials

$$\psi_L(u) = \left[\frac{(2m\omega/h)^{1/2}}{2^L L!}\right]H_L(u)\exp\left[\frac{-u^2}{2}\right] \qquad (5.17)$$

where $u \equiv (2\pi m\omega/h)^{1/2}x$, and $H_L(.)$ is the Hermite polynomial.

Physics of *small* systems (on the scale of $\Lambda$) is described by quantum mechanics. Classical mechanics is asymptotically valid at small de Broglie thermal wavelength: $\Lambda \ll 1$. When we treat systems of heavy particles at high temperatures, a classical description is sufficient.

## References

[1]   J.A. Barker, R.A. Fisher, and R.O. Watts, Mol. Phys. **21**, 657 (1971).

[2]   V. Magnasc, G.F. Musso, and R. McWeeny, J. Chem. Phys. **46**, 4015; **47**, 1723, 4617, and 4629 (1967).

[3]   H.F. Schaefer III, D.R. McLaughlin, F.E. Harris, and B.J. Alder, Phys. Rev. Lett. **25**, 988 (1970).

[4]   O. Matsuoka, E. Clementi, and M. Yoshimine, J. Chem. Phys. **64**, 1351 (1976).

[5]   J.W. Gibbs, *Statistical Mechanics*, (Yale University Press, New Haven, Connecticut, 1902).

[6]   J.O Hirschfelder, C.F. Curtiss, and R.B. Bird, *Molecular Theory of Gases and Liquids*, (Wiley, New York, 1964).

## Exercises

1.   *One-dimensional harmonic oscillator.* For the harmonic oscillator described in Example I.4.1, solve for $x(t)$ and $p(t)$ using Hamilton's equations of motion. The initial conditions are $x(0) = 0$, $p(0) = 4$ kg m/s, and $k = 25$ N/m, $m = 2$ kg. Sketch the trajectory in the phase space $(x, p)$.

2.    *Vibrational energy.* For the one-dimensional harmonic oscillator, show that its total energy is given by $(1/2)m\omega^2a^2$ (where $\omega$= frequency, $a$= amplitude).

3.    *The pendulum.* Sketch the motion of the pendulum in its phase space $(p\text{-}\theta)$ representation. ($p$= momentum; $\theta$= angular displacement).

4.    *Intermolecular forces.* For three molecules of argon interacting with the Lennard-Jones potential,

$$u(r) = 4\varepsilon\left[(\frac{\sigma}{r})^{12} - (\frac{\sigma}{r})^6\right]$$

where $\varepsilon/k$ =119.8$K$, and $\sigma$ =3.405$A$, and located at $(0,0,0)$, $(5,1,4.2)$, and $(7,0,0)$ (units in angstroms), calculate (i) the force on each molecule and (ii) the position of each molecule $3\times10^{-14}$ s later (using approximate finite difference method). ($k$= Boltzmann constant).

# CHAPTER II

# THE STATISTICAL ENSEMBLES

## II.1. *Review of Thermodynamics*

Consider a physical system $K$ containing $c$ different chemical species, each with $N_j$ molecules ($j = 1,2,...,c$). The external deformation coordinates of the system $K$ (volume, surface area, altitude in a gravitational field, etc.) are denoted by $X_1, X_2,..., X_n$. The phase rule says that at equilibrium the extensive properties of the system are dependent on $c+n+1$ variables (e.g., temperature, $X_1, X_2,..., X_n, N_1,..., N_c$).

As the system $K$ undergoes changes from one equilibrium state to another, there are accompanying changes in system energy, pressure, enthalpy, etc. The rules governing these changes are formulated in classical thermodynamics as the three laws of thermodynamics. Further analysis differentiates the laws into two groups: those governing processes (the first and second laws) and those governing properties (the third law).

### LAWS OF THERMODYNAMICS

The first law states (for a closed system)

$$dU = \delta Q - \delta W \tag{1.1}$$

In words, the increase of the internal energy of the system is equal to the heat absorbed by the system minus the work done on the surroundings. Meanwhile, the system undergoes a change of state from one equilibrium state to another.

The second law is (for an isolated system)

$$dS \geq 0 \tag{1.2}$$

In words, the change in entropy for an isolated system is non-negative. This law affirms the directional change of spontaneous processes in an isolated system.

The third law establishes an absolute scale for the entropy. It acts like a boundary condition. For a pure and perfectly crystalline substance

$$\lim_{T \to 0} S \to 0 \tag{1.3}$$

(1.3) affirms the value of the entropy at absolute zero temperature.

## EXTERNAL DEFORMATION COORDINATES AND CONJUGATE FORCES

The work term $\delta W$ in (1.1) is generated by macroscopically deforming some extensive properties of the system. In the thermodynamics of fluids, the most common external deformation coordinate is the *volume* ($V$) of the container. For a differential change in volume, $dV$, there is associated a resisting force, in this case the pressure ($P$). The differential work produced, $\delta W$, is

$$\delta W = P \, dV \tag{1.4}$$

This concept can be generalized to other deformation coordinates. For example, if there is a free surface $\Sigma$, it can be deformed by stretching. The resisting force is the surface tension, $\gamma$.

$$\delta W = -\gamma \, d\Sigma \tag{1.5}$$

To each external coordinate $X_i$ there corresponds a resisting force that opposes deformation. This resistance is called the conjugate force $F_i$. The product of the pair gives the work

$$\delta W = F_i \, dX_i \tag{1.6}$$

which has the units of energy. Although $F_i$ is called a force, it does not always have the units of mechanical force (mass)(length)/(time)$^2$. It is a *generalized force*. For example, when the external coordinate is the electric charge ($Q_e$, coulomb), the conjugate force is the electric potential ($V_e$, volt), and

$$\delta W = V_e \, dQ_e \tag{1.7}$$

The commonly used deformation coordinates are volume, surface area, length, charges, and magnetization. The associated conjugate forces are pressure, surface tension, tensile force, electric potential, and magnetic field strength, respectively.

For systems where more than one type of deformation coordinate exists, all conjugate pairs should be taken into account. For example, for plasma in an electric field $E$ over gravity, the work term should be

$$\delta W = P \, dV - E \, dQ_e - mg \, dh \tag{1.8}$$

In general,

$$\delta W = \sum_i F_i \, dX_i \tag{1.9}$$

## II.2. *The Information Entropy*

In classical thermodynamics, the entropy is defined by heat and temperature:

$$dS = \frac{\delta Q_{rev}}{T} \tag{2.1}$$

In the following, we shall discuss an *information entropy*, which is defined in terms of probability distributions. This kind of entropy is more suited to our statistical discussions. Let $P=\{p_1, p_2,..., p_n\}$ be a probability distribution of $n$ events. In other words, $p_i$ is the probability of event $i$. Let all the events be mutually independent. We define a quantity $S$, called the information entropy, for the distribution $P$ as

$$S \equiv -k \sum_{i=1}^{n} p_i \ln p_i \tag{2.2}$$

where $k$ is a proportionality constant. We note that

1. The so-called information entropy is defined for the probability distribution $P$.
2. It is called information entropy because of its first introduction in communication theory [1]
3. It is a nonnegative quantity that is zero only when either $p_i = 0$ or $p_i = 1$.
4. $S$ is also a measure of *uncertainty*. The use of information is to eliminate a corresponding amount of uncertainty.

To gain a better understanding of this quantity, we look at some examples.

*Example 1. Tossing a Coin*

When a balanced coin is tossed, the probability, $p_1$, of getting a head (event 1) is 0.5; for a tail (event 2), $p_2 = 0.5$. The probability distribution $P =\{p_1, p_2\} =\{0.5, 0.5\}$. Therefore the information entropy for this distribution is

$$S = - k\, p_1 \ln(p_1) - k\, p_2 \ln(p_2) \tag{2.3}$$

$$= - k\, [0.5 \ln(0.5) + 0.5 \ln(0.5)]$$

$$= k \ln(2) = 0.69315k$$

*Example 2. At Crossroads*

A stranger arrives at a crossroads in town and wants to find the way to the library. What is his uncertainty, or lack of information in this case? Let us denote the four street directions at the crossroads as $E, W, N,$ and $S$. There are four choices (four events). The probability distribution is $P= \{P_E, P_N, P_S, P_W\} = \{1/4, 1/4, 1/4, 1/4\}$. The information entropy is

$$S = -k\left[\frac{1}{4}\ln(\frac{1}{4}) + \frac{1}{4}\ln(\frac{1}{4}) + \frac{1}{4}\ln(\frac{1}{4}) + \frac{1}{4}\ln(\frac{1}{4})\right] \qquad (2.4)$$

$$= k\ln(4) = 1.38629k$$

Now if some passerby points the way, the uncertainty is removed. Therefore, the quantity 1.38629k can be construed as the value of the information "go west." Upon comparing examples 1 and 2, we note that in the first case the probability $p_i = 0.5$, and $S = 0.69315k$; in the second case, $p_i = 0.25$, and $S$ is increased twofold to 1.38629k. Thus lower probability entails higher uncertainty, a proportionality conforming to our intuitive concept of information. In the following section, we shall return to the phase space and define thereupon an information entropy.

## II.3. A Distribution Game

As shown in Section I.2, the total energy, i.e., the Hamiltonian, of the $N$-particle system is a function of the phase point $(\mathbf{r}_1,..., \mathbf{r}_N, \mathbf{p}_1, \ldots, \mathbf{p}_N)$. We partition the phase space into $M$ cells (eventually we take the limit $M \rightarrow \infty$) with lengths $\Delta\mathbf{r}_1,...,\Delta\mathbf{r}_N, \ldots,$ $\Delta\mathbf{p}_1,..., \Delta\mathbf{p}_N$. As time elapses, the phase point $P$ travels from cell to cell. Since the Hamiltonian is a function of the phase points, the $N$-particle system will in general experience different energies, $\varepsilon_j$.

*Table II.1. Energies of Phase Points*

| 6N-Dimensional Phase Space | |
|---|---|
| Cell No. | Total Energy |
| 1 | $\varepsilon_1$ |
| 2 | $\varepsilon_2$ |
| ... | ... |
| M | $\varepsilon_M$ |

We pose the question: what is the probability, $p_j$, that the phase point $P$ will be found in cell $j$? This is a distribution problem. Before embarking on to a real physical system involving an enormous number ($10^{23}$) of molecules, we examine first a simple but meaningful distribution game.

Consider a special kind of roulette game. One is given 10 identical chips to play per game. There are four colors on the roulette table: red, black, yellow, and green. To play red, you need to pay $3 per chip; to play black, $2 per chip; to play yellow, $1 per chip; and to play green, $0 per chip. For each game, you must spend $8 while distributing the 10 chips among the four colors. *What is the probability that a given color will receive a chip?* This problem could be contrasted with a physical one, that of localized adsorption. There are four unequally energized sites on a solid surface with energy expenditures, 1 kT, 2 kT, 3 kT, and 0 kT, respectively. Ten molecules are adsorbed with a combined energy of 8 kT. We are interested in the probabilities of occupation of the sites. These two problems are clearly isomorphic. The information can be obtained by a brute-force tabulation (Table II.2) of all allowed combinations:

Table II.2. Tabulation of the Distributions of Chips

| Trial No. | Red ($3) | Black ($2) | Yellow ($1) | Green ($0) | Degeneracies |
|---|---|---|---|---|---|
| 1 | 2 | 1 | 0 | 7 | 360 |
| 2 | 1 | 2 | 1 | 6 | 2,520 |
| 3 | 0 | 4 | 0 | 6 | 210 |
| 4 | 0 | 3 | 2 | 5 | 2,520 |
| 5 | 0 | 0 | 8 | 2 | 45 |
| 6 | 0 | 2 | 4 | 4 | 3,150 |
| 7 | 0 | 1 | 6 | 3 | 840 |
| 8 | 1 | 1 | 3 | 5 | 5,040 |
| 9 | 2 | 0 | 2 | 6 | 1,260 |
| 10 | 1 | 0 | 5 | 4 | 1,260 |
| Sum of Chips | 12,060 | 25,980 | 49,500 | 84,510 | 172,050 |
| Probability, $p_i$ | 0.070 | 0.151 | 0.2878 | 0.4912 | |
| (cf. $p_i$ from Max S) | (0.079) | (0.146) | (0.271) | (0.504) | |

The degeneracies arise due to permutations of chips within each set of distribution. For example, in trial 1, there are $_{10}C_2$ ways of selecting 2 chips out of 10 playing on red; next there are $_8C_1$ ways to choose 1 chip from the remaining 8 on black; yellow requires no choice. Finally there are $_7C_7$ ways to play on green. The total number of combinations for trial 1 is

$$_{10}C_2 \cdot {}_8C_1 \cdot {}_7C_7 = \frac{10 \cdot 9}{1 \cdot 2} \frac{8}{1} \frac{7!}{7!} = 360 \tag{3.1}$$

Similarly, the total number of combinations for trial 2 is

$$_{10}C_1 \cdot {}_9C_2 \cdot {}_7C_1 \cdot {}_6C_6 = \frac{10}{1} \frac{9 \cdot 8}{1 \cdot 2} \frac{7}{1} \frac{6!}{6!} = 2520 \tag{3.2}$$

We found the probabilities by dividing the number of chips played on each color by the total number of chips played in 10 trials (i.e., 172,050) On red, there are totally 12,060 chips played. This number comes from

$$2(360)+1(2520)+1(5040)+2(1260)+1(1260) = 12,060 \tag{3.3}$$

Therefore probability(red) = 12,060/172,050 = 0.070. For black, probability(black) = 25,980/172,050 = 0.1510, etc.

This procedure can be applied, in principle, to all distribution problems. However, when the number of chips increases to millions and the number of colors is equally vast, the tabulation method becomes impractical. A method of *maximum entropy* is proposed.

## THE METHOD OF MAXIMUM ENTROPY

We use the same distribution game for illustration. Let the probability that a chip will fall on red be $p_1$, on black, $p_2$, on yellow $p_3$, and on green, $p_4$. The entropy associated with this distribution is

$$S = - k\sum_{i=1}^{4} p_i \ln p_i \tag{3.4}$$

with the constraints

$$\sum_{i}^{4} p_i = 1 \tag{3.5}$$

$$\sum_{i}^{4} p_i D_i = <D> \tag{3.6}$$

where $D_i$ is the dollars per chip for the $i$th color (e.g., $D_{red}$ = $3), and $<D>$ is the average dollars per chip per game. In the above example, $<D>$ = $0.8 (or 80¢, obtained from $8/10 chips per game).

To find the *equilibrium* distribution, we maximize the entropy (3.4) subject to the constraints (3.5 & 6). The Lagrange multipliers method can be used. The Lagrangian function $L$ with Lagrange multipliers $\lambda_1$ and $\lambda_2$ is constructed as

$$L(p_1,..., p_4, \lambda_1, \lambda_2) = - k\sum_{i}^{4} p_i \ln p_i + \lambda_1 \left[ \sum_{i}^{4} p_i - 1 \right] + \lambda_2 \left[ \sum_{i}^{4} p_i D_i - <D> \right] \tag{3.7}$$

The conditions of maximum $S$ with the given constraints are

$$\frac{\partial L}{\partial p_i} = 0, \qquad \frac{\partial L}{\partial \lambda_1} = 0, \qquad \frac{\partial L}{\partial \lambda_2} = 0 \tag{3.8}$$

where $i$ = 1, 2, 3, 4. The first of conditions (3.8) gives

$$\frac{\partial L}{\partial p_i} = 0 = - k(\ln p_i +1) + \lambda_1 + \lambda_2 D_i \tag{3.9}$$

or

$$\ln p_i = \frac{\lambda_1}{k} - 1 + \frac{\lambda_2}{k} D_i \tag{3.10}$$

Since $\lambda_1$ and $\lambda_2$ are constants to be determined, we can define new constants $Z$ and $\beta$:

$$\ln Z \equiv 1 - \frac{\lambda_1}{k}; \qquad \beta \equiv -\frac{\lambda_2}{k} \tag{3.11}$$

Then

$$p_i = \frac{e^{-\beta D_i}}{Z} \tag{3.12}$$

Substitute (3.12) into (3.5) to get

$$1 + e^{-\beta} + e^{-2\beta} + e^{-3\beta} = Z \tag{3.13}$$

Substituting (3.12) into (3.6) gives

$$e^{-\beta} + 2e^{-2\beta} + 3e^{-3\beta} = 0.8Z \tag{3.14}$$

Simultaneous solution of (3.13 & 14) gives $\beta$= 0.61849, and Z=1.98537. Therefore, the probabilities are $p$(red) = 0.079, $p$(black) = 0.146, $p$(yellow) = 0.271, and $p$(green) =0.504 (compared to 0.07, 0.15, 0.29, and 0.49, respectively, from direct tabulation). The agreement is remarkable. The differences occur because in the maximum entropy approach, differentiation with respect to continuous values of $p_i$ was used. For small number of chips (in this case, 10), $p_i$ makes quite large discrete jumps. When the number of cases becomes large, as for $10^{23}$ molecules, maximum entropy results should match the tabulation results.

Before proceeding further, we observe some general results of the solution. Substitution of (3.12) into (3.4) gives

$$S = - k\sum_i p_i[-\beta D_i - \ln Z] = k \ln Z + k\beta\sum_i p_iD_i \tag{3.15}$$

$$= k \ln Z + k\beta<D> = k[\ln 1.98537 + 0.61894(0.8)] = 1.1806k$$

Since $Z = \Sigma_i \exp(-\beta D_i)$,

$$\frac{\partial \ln Z}{\partial \beta} = \frac{1}{Z}\sum_i (-D_i)e^{-\beta D_i} = -\sum_i D_i\, p_i = -<D> \tag{3.16}$$

Therefore the average cost per chip is related to the partial derivative of the factor ln Z. Further differentiation gives

$$\frac{\partial^2 \ln Z}{\partial \beta^2} = \frac{1}{Z}\sum D_i^2\, e^{-\beta D_i} - \frac{1}{Z^2}\left[\sum_i D_i\, e^{-\beta D_i}\right]^2 = \sum_i p_i\, (D_i - <D>)^2 \tag{3.17}$$

The second partial derivative of ln Z with respect to $\beta$ is the variance of the cost per chip.

## II.4. *Gibbs Ensembles*

The definition of an ensemble was introduced by Gibbs [2] in 1902 as the conceptual foundation of statistical mechanics. An *ensemble* is a conceptual construction in which a given physical system is duplicated *ad infinitum*. The rationale for such a construction is to make a statistical count of all the states that the protosystem could possibly be in, thus arriving at an average property corresponding to the macroscopical state of the system.

Gibbs gives the following description: "...Let us imagine a great number of independent systems identical in nature, but differing in phase, that is, in their condition with respect to configuration and velocity...". Therefore given a protosystem, $K_0$, with a given configuration and velocity distribution, we can make $x$ images of $K_0$ (i.e., $K_1, K_2, \ldots, K_x$), each with a different configuration and velocity distribution. In the 6$N$-dimensional phase space originally constructed for $K_0$, these $x$ systems are represented by $x$ phase points. They form a *cloud* in the phase space. This collection of systems is called an *ensemble*.

The protosystem $K_0$ could be duplicated in different ways by varying the macroscopic constraints imposed. In some cases, we would like to keep the content, $N$ particles, the total volume, $V$, and the total energy, $E$, the same in the duplication process. The resulting ensemble is called a *microcanonical ensemble*. On the other hand, constant $N$, $V$, and $T$ (temperature) give a *canonical ensemble*. Constant $V$, $T$, and $\mu$ (chemical potential) produce the *grand canonical ensemble*. In addition, constant $N$, $P$ (pressure), and $T$ yield the *N-P-T* ensemble, or the *isothermal-isobaric ensemble*.

Suppose a certain kind of ensemble has been constructed, and a cloud of points is formed in the phase space. We wish to determine how the points are distributed. To facilitate consideration, we partition the phase space as before into M cells with volumes $\Delta r_1 \cdots \Delta r_N \Delta p_1 \cdots \Delta p_N$ and number them consecutively. The question can then be posed: "What is the probability, $p_i$, that the $i$th cell contains a phase point?" This question is reminiscent of our previous problem with the roulette game. A direct analogy can be drawn between the two. (We note that it is the phase points -- the dynamic states -- that are distributed, not the molecules themselves.)

*Table II.3.  Comparison of Distributions*

| Roulette | | Ensemble |
|----------|------|----------|
| 4 Colors | <=> | $M$ Cells |
| 10 Chips | <=> | $x$ Phase points |

Recall that we solved the roulette problem by two methods: direct tabulation and maximization of entropy. For a large ensemble ($M \to \infty$), practicality precludes the direct tabulation method, even though conceptually one could still carry out the "thought experiment." We shall use the maximum entropy instead.

## II.5. *The Canonical Ensemble*

Given a protosystem, $K_0$, the canonical ensemble is formed by duplicating $K_0$ at constant number of particles $N$, constant compositions, constant external deformation coordinates (e.g., constant volume), and constant temperature. This choice corresponds to *closed systems* in thermodynamics. We divide the phase space into $M$ cells, $M$ being a large number. Let $p_i$ be the probability that a phase point (representing a member of the ensemble) is found in cell $i$. The probability distribution is $\{p_1, p_2, \ldots, p_M\}$ for all $M$ cells. We define the entropy

$$S \equiv -k \sum_{i=1}^{M} p_i \ln p_i \qquad (5.1)$$

with the normalization of probabilities

II. Statistical Ensembles

$$\sum_{i=1}^{M} p_i = 1 \tag{5.2}$$

and the definition of average energy as

$$\sum_{i=1}^{M} p_i \varepsilon_i = <E> \tag{5.3}$$

where

$$\varepsilon_i = [\Delta \mathbf{r}^N \Delta \mathbf{p}^N]_i^{-1} \int_{[\Delta \mathbf{r}^N \Delta \mathbf{p}^N]_i} d\mathbf{r}^N \, d\mathbf{p}^N \, H_N(\mathbf{r}^N, \mathbf{p}^N) \tag{5.4}$$

is the energy associated with the $i$th cell. In the limit, $M \to \infty$ and $\Delta \mathbf{r}^N \Delta \mathbf{p}^N \to 0$, $\varepsilon_i$ is simply the value of the Hamiltonian at the $i$th point. We can maximize $S$ subject to the constraints (5.2 & 3) by using the Lagrangian method. The results are

*The Probability*

$$p_i = \frac{\exp(-\beta \varepsilon_i)}{Z_N} \tag{5.5}$$

*The Partition Function of the Canonical Ensemble*

$$Z_N = \sum_{i=1}^{M} \exp(-\beta \varepsilon_i) \tag{5.6}$$

*The Entropy*

$$S = k \ln Z_N + k\beta <E> \tag{5.7}$$

Now we identify $S$ as the thermodynamic entropy and $<E>$ as the internal energy. We have from thermodynamics

$$S = -\frac{A}{T} + \frac{<E>}{T} \tag{5.8}$$

Comparison of (5.7) with (5.8) shows that (see also the differential relations below)

$$\beta = \frac{1}{kT} \tag{5.9}$$

and

$$A = -kT \ln Z_N \tag{5.10}$$

We have identified $\beta$ as the reciprocal temperature multiplied by a constant $k$, and the partition function $Z_N$ as a function of the Helmholtz free energy. This relation is of great importance because it connects microscopic distributions to macroscopic thermodynamic functions. The probability (5.5) is called the Boltzmann (-Maxwell) distribution. The distribution maximizes the entropy of the ensemble. In thermodynamics, we know that when the entropy of a system is at a maximum, the system is at equilibrium. Therefore the Boltzmann distribution characterizes the equilibrium state of the system. In fact, in statistical mechanics, Boltzmann distribution is synonymous with *equilibrium*. In other words, the state of equilibrium is defined in statistical mechanics as the one where the Boltzmann distribution law prevails.

The formula for the partition function $Z_N$ in (5.6) corresponds to the quantum mechanical expression

$$Z_N = Tr \, \exp(-\beta \mathbf{H}) \tag{5.11}$$

or

$$Z_N = \sum_i e^{-\beta \varepsilon_i} \tag{5.12}$$

where $\mathbf{H}$ is the Hamiltonian operator with eigenvalues $\varepsilon_i$. $Tr$ stands for the trace operator. In the classical limit (i.e., small $\Lambda$), the discrete sum is replaced by a phase space integral. Let the classical Hamiltonian be of the form

$$H_N(\mathbf{p}^N, \mathbf{r}^N) = \frac{1}{2m} \sum p_i^2 + V_N(\mathbf{r}^N) \tag{5.13}$$

The elementary volume in classical phase space, $d\mathbf{p}^N \cdot d\mathbf{r}^N$, is associated with a single quantum state by

$$d\mathbf{p}^N \cdot d\mathbf{r}^N <=> N! h^{3N} (\text{quantum state}) \tag{5.14}$$

(A detailed discussion could be found in Hill [3]). Therefore (5.6) can be transcribed in the classical limit as

$$Z_N = \frac{1}{N! h^{3N}} \int d\mathbf{p}_1 \cdots d\mathbf{p}_N d\mathbf{r}_1 \cdots d\mathbf{r}_N \, \exp[-\beta H_N(\mathbf{p}^N, \mathbf{r}^N)] \tag{5.15}$$

The Hamiltonian above contains the translational energy only. For more general KE with rotational and vibrational energies, KE includes three parts:

$$KE = E_t(\mathbf{p}^N) + E_r(\mathbf{J}^N) + E_v(L) \tag{5.16}$$

where $\mathbf{J}$ is the angular momentum and $L$ is the quantum number of vibration. The translational and rotational contributions to the partition function can usually be treated classically as integrals (due to small rotational temperatures). However, the vibrational contribution must retain its quantum form (since the vibrational energy levels are widely separated). The partition function is consequently a hybrid of classical and quantum expressions*

$$Z_N = \frac{1}{N! h^{3N}} \left[ \int dp^N \exp\left[-\beta \sum_i^N \frac{p_i^2}{m}\right]\right] \tag{5.17}$$

$$\cdot \left[ \int dJ^N \left[\prod_i^N \left\{a_i^{-1}(2J_i+1)\exp\left[-\beta \frac{J_i(J_i+1)h^2}{2\overline{m}R_i^2}\right]\right\}\right]\right]$$

$$\cdot \left[\sum_{i=1}^N \sum_{j=1}^{\kappa_i} \sum_{L=0}^\infty \exp\left[-\beta(L+\frac{1}{2})h\nu_{ij}\right]\right]$$

$$\cdot \left[\int dr^N \exp(-\beta V_N(r^N))\right]$$

where, as before, $dJ^N = dJ_1\, dJ_2 \cdots dJ_N$, $a_i$ is a rotational symmetry number, $\overline{m}R_i^2$ is the moment of inertia of molecule $i$, and $\nu_{ij}$ is the vibrational frequency of the $j$th mode in the $i$th molecule. Explanations will be found in Chapter III, where a detailed discussion of the ideal gas ($V_N = 0$) is given.

As mentioned earlier, the energy $\varepsilon_i$ of the $i$th cell in the phase space is dependent on the coordinates $r_1,..., r_N, p_1,..., p_N$. We call these internal coordinates. On the other hand, when the $N$-body system is taken as a whole with respect to its surroundings, $\varepsilon_i$ is also dependent on the external coordinates (e.g., volume, altitude in a gravitational field):

$$\varepsilon_i = \varepsilon_i(r^N, p^N; X_1, \ldots, X_n) \tag{5.18}$$

where the external coordinates are given as $X_i$. The microscopic force $F_{j,i}$ conjugate to the external coordinate $X_j$ is obtained by taking the gradient of $\varepsilon_i$:

$$F_{j,i} \equiv -\frac{\partial \varepsilon_i}{\partial X_j} \tag{5.19}$$

The macroscopically measured force is actually an average of these microscopic forces:

$$\langle F_j \rangle = \sum_{i=1}^M p_i F_{j,i} \tag{5.20}$$

Taking the derivative of the partition function with respect to $X_j$ gives

$$\frac{\partial}{\partial X_j} \ln Z_N(\beta; X_1, \ldots, X_s)\bigg|_\beta = \frac{1}{Z_N} \frac{\partial}{\partial X_j} \sum_{i=1}^M e^{-\beta\varepsilon_i} = \frac{1}{Z_N} \sum_i e^{-\beta\varepsilon_i}\left[-\beta \frac{\partial \varepsilon_i}{\partial X_j}\right] \tag{5.21}$$

---

*In case $V_N$ involves angular coordinates, the configurational integral should contain angle integration.

$$= \frac{\beta}{Z_N} \sum_i e^{-\beta \varepsilon_i} F_{j,i} = \beta \sum_i \frac{e^{-\beta \varepsilon_i}}{Z_N} F_{j,i} = \beta \sum_i p_i F_{ji} = \beta <F_j>$$

We see that the derivative of $\ln Z_N$ with respect to an external coordinate $X_j$ returns the conjugate force $<F_j>$. In particular, when $X_j = V$, the conjugate force is the pressure

$$\frac{\partial \ln Z_N}{\partial V}\bigg|_{T,N} = \frac{P}{kT} \tag{5.22}$$

Recall that the Helmholtz free energy $A = -kT \ln Z$. Equation (5.22) is consistent with the thermodynamic relation

$$\frac{\partial A}{\partial V}\bigg|_{T,N} = -P \tag{5.23}$$

Similarly, we can get the other generalized forces from (5.21). Statistical mechanics furnishes us a general formalism for treating external deformation coordinates.

## II.6. *Comments on the First Law of Thermodynamics*

We examine the first law of thermodynamics from the probabilistic perspective. The first law says

$$dU = \delta Q - \delta W \tag{6.1}$$

In our formulation, $U = <E>$. Therefore,

$$dU = d<E> = d(\sum_i p_i \varepsilon_i) = \sum_i p_i \, d\varepsilon_i + \sum_i \varepsilon_i \, dp_i \tag{6.2}$$

$$= -\sum_i \sum_j p_i F_{ji} \, dX_j + \sum_i \varepsilon_i \, dp_i = \sum_i \varepsilon_i \, dp_i - \left[ \sum_j <F_j> \, dX_j \right]$$

We identify $\sum <F_j> dX_j$ as the work term $\delta W$. The term $\sum \varepsilon_i \, dp_i$ can be viewed as the definition of heat. In other words, heat is the energy mode associated with the disturbances of the probabilities $(dp_i)$. By this interpretation, it is clear that heat transfer is caused by movements on the microscopic scale. This explains the difficulty in converting heat into other useful forms of energy.

We also know that the heat capacity $C_v$ is related to the internal energy by

$$C_v = \frac{\partial <E>}{\partial T}\bigg|_{V,N} \tag{6.3}$$

Thus, from (5.24),

$$\frac{\partial^2 \ln Z_N}{\partial \beta^2} = kT^2 C_v = \sum_i (E_i - \langle E \rangle)^2 p_i \tag{6.4}$$

Heat capacity is then a measure of the magnitude of energy fluctuations about the mean, $\langle E \rangle$ (i.e., statistically the variance of energy fluctuations).

## II.7. *The Grand Canonical Ensemble*

The grand canonical ensemble is formed from the protosystem $K_0$ by a duplication process that keeps the temperature $T$, the volume $V$, and the chemical potential $\mu$ constant. In other words, every image of $K_0$ in the ensemble has the same $T$, $V$ and $\mu$ as the protosystem. Macroscopically this corresponds to an *open* system with fixed volume that allows heat and mass to transfer across the boundaries. In an open system, the number of molecules is not conserved. Therefore the images of $K_0$ may have more (or less) molecules than $K_0$. Let the maximum number be $N_{max}$ ($N_{max}$ being arbitrarily large and in the thermodynamic limit, $N_{max} \rightarrow \infty$). We represent the ensemble by a phase space of $6N_{max} + 1$ dimensions (i.e., with $3N_{max}$ values for positions, $3N_{max}$ values for momenta, and one number axis $n$ indicating the number of molecules for that subsystem.) We partition the phase space into $M$ (finally, $M \rightarrow \infty$) cells of size $\Delta r_{1x} \Delta r_{1y} \cdots \Delta p_{1x} \cdots \Delta p_{Nz} \Delta n$. With each cell is associated an energy $\varepsilon_j = \varepsilon_j(r_1,..., r_{N_j}, p_1,...,p_{N_j}, N_j)$ where $N_j$ is the number of molecules in the $j$th cell.

Since there are $x$ members in the grand canonical ensemble, each differing in positions and momenta, there is again a cloud of points in the phase space distributed among the $M$ cells. We ask how are these phase points distributed? Or what is the probability, $p_j$, that cell $j$ contains a phase point? This distribution is to be determined by the maximum entropy principle. The entropy for this distribution is

$$S = -k \sum_{j=1}^{M} p_j \ln p_j \tag{7.1}$$

which is to be maximized subject to the constraints

$$\sum_{j=1}^{M} p_j = 1 \tag{7.2}$$

$$\sum_{j=1}^{M} p_j \, \varepsilon_j = \langle E \rangle \tag{7.3}$$

where $\langle E \rangle$ is the average energy per subsystem of the ensemble, and

$$\sum_{j=1}^{M} p_j \, N_j = \langle N \rangle \tag{7.4}$$

Thus we maximize $S$ in (7.1) subject to (7.2, 3 & 4). As before, we form the Lagrangian function, $L$

$$L = S + \lambda_1(\sum p_j - 1) + \lambda_2(\sum p_j \varepsilon_j - <E>) + \lambda_3(\sum p_j N_j - <N>) \tag{7.5}$$

where $\lambda_1$, $\lambda_2$, and $\lambda_3$ are the multipliers introduced to account for the three constraints. The optimum values of $p_j$ are found by requiring

$$\frac{\partial L}{\partial p_j} = 0, \qquad j=1,...,M; \qquad \frac{\partial L}{\partial \lambda_k} = 0, \qquad k=1, 2, 3 \tag{7.6}$$

Carrying out the partial differentiation, we get

$$-k[\ln p_j + 1] + \lambda_1 + \lambda_2 \varepsilon_j + \lambda_3 N_j = 0, \tag{7.7}$$

$$\sum p_i - 1 = 0, \qquad \sum p_j \varepsilon_j - <E> = 0, \qquad \sum p_j N_j - <N> = 0$$

Rewriting the multipliers as $\ln \Xi = 1 - \lambda_1/k$, $\beta = -\lambda_2/k$, and $\alpha = -\lambda_3/k$, we have

$$p_j = \frac{e^{-\beta \varepsilon_j - \alpha N_j}}{\Xi} \tag{7.8}$$

From (7.2), we have

$$\Xi = \sum_j^M \exp(-\beta \varepsilon_j - \alpha N_j) \tag{7.9}$$

And from (7.1), we have

$$S = k \ln \Xi + k\beta<E> + k\alpha<N> \tag{7.10}$$

Note that earlier $\beta = (kT)^{-1}$. From thermodynamics, we also know that the Gibbs free energy, $G$, is

$$G = \mu<N> = <E> + PV - TS \tag{7.11}$$

Rearrangement gives

$$S = \frac{<E>}{T} + \frac{PV}{T} - \frac{\mu<N>}{T} \tag{7.12}$$

Comparison of (7.10) with (7.12) gives (see also the differential relations below)

$$k \ln \Xi = \frac{PV}{T} \tag{7.13}$$

$$\alpha = -\frac{\mu}{kT} \tag{7.14}$$

In the grand canonical ensemble, the partition function $\Xi$ is related to the thermodynamic product PV, and the Lagrange multiplier $\alpha$ is a measure of the chemical potential $\mu$. Similar derivatives can be obtained from the partition function $\Xi$; i.e.,

$$\frac{\partial \ln \Xi}{\partial \beta} = -\sum_j \varepsilon_j \frac{e^{-\beta \varepsilon_j - \alpha N_j}}{\Xi} = -\sum_j p_j \varepsilon_j = -<E> \tag{7.15}$$

$$\frac{\partial \ln \Xi}{\partial \alpha} = -kT \frac{\partial \ln \Xi}{\partial \mu} = \frac{1}{\Xi} \sum_j -N_j e^{-\beta \varepsilon_j - \alpha N_j} = -\sum_j p_j N_j = -<N> \tag{7.16}$$

For the variations of the external coordinate $X_k$, we have

$$\frac{\partial \ln \Xi}{\partial X_k} = \frac{1}{\Xi} \sum_j -\frac{\partial \varepsilon_j}{\partial X_k} \beta e^{-\beta \varepsilon_j - \alpha N_j} = \sum_j F_{k,j} \beta p_j = \beta <F_k> \tag{7.17}$$

where $<F_k>$ is the equilibrium (generalized) force corresponding to $X_k$. For $X_k =$ volume, we have,

$$\left. \frac{\partial \ln \Xi}{\partial V} \right|_{T,N} = \beta P = \frac{P}{kT} \tag{7.18}$$

This fact is easily seen from the identity $k \ln \Xi = PV/T$. Furthermore, the fluctuations (variance) in energy are given by

$$\left. \frac{\partial^2 \ln \Xi}{\partial \beta^2} \right|_{N,\mu} = \sum_j [\varepsilon_j - <E>]^2 p_j = kT^2 C_v \tag{7.19}$$

where $C_v$ is the constant-volume heat capacity. The number fluctuations (or density fluctuations) are given by

$$\left. \frac{\partial^2 \ln \Xi}{\partial \alpha^2} \right|_{T,N} = \sum_j [N_j - <N>]^2 p_j \tag{7.20}$$

In the following we shall examine a grand canonical ensemble for mixtures. Let the species be $\nu = a,b,...,c$. Thus the number of $\nu$-molecules in phase cell j is $N_{j,\nu}$. Equation (7.4) is now replaced by c equivalent equations for all c species

$$\sum_{j=1}^{M} p_{j,\nu} N_{j,\nu} = <N_\nu> \tag{7.21}$$

$\nu = a,b,...,c$. $<N_\nu>$ is the *average* number of $\nu$ molecules in the open system. The

Lagrange multipliers $\alpha_v$ now correspond to all c species. The partition function becomes

$$\Xi = \sum_j^M \exp(-\beta\varepsilon_j - \sum_v^c \alpha_v N_{j,v}) \tag{7.22}$$

Comparison with thermodynamic formulas gives

$$\alpha_v = -\frac{\mu_v}{kT} \tag{7.23}$$

i.e. $\alpha_v$ are chemical potentials for different species. As before, in the classical limit, the quantum states of N particles correspond to

$$d\mathbf{p}^N \cdot d\mathbf{r}^N = N_a! N_b! ... N_c! h^{3N}(\text{quantum state}) \tag{7.24}$$

where $\{N_a, N_b, ...,N_c\}$ is a *partition* of N (i.e. $\sum_v N_v = N$). The product of the factorials $N_v!$ results from *correct* Boltzmann counting for mixtures, since permutations of identical molecules within each species leave the quantum state invariant. Note also that for different species, the masses $m_v$ will also be different. Therefore, the kinetic energies $p_i^2/m_v$ are different. As a consequence, the de Broglie thermal wave lengths are different for different species

$$\Lambda_v^2 = \frac{h^2}{2\pi m_v kT} \tag{7.25}$$

In the classical limit, the grand canonical partition function becomes

$$\Xi = \sum_{N_a \geq 0}^\infty \sum_{N_b \geq 0}^\infty \cdots \sum_{N_c \geq 0}^\infty \frac{1}{N_a!}\left[\frac{e^{-\mu_a/kT}}{\Lambda_a^3}\right]^{N_a} \frac{1}{N_b!}\left[\frac{e^{-\mu_b/kT}}{\Lambda_b^3}\right]^{N_b} \cdots \frac{1}{N_c!}\left[\frac{e^{-\mu_c/kT}}{\Lambda_c^3}\right]^{N_c} \tag{7.26}$$

$$\cdot \int d\mathbf{r}^N \mathbf{p}^N \, exp(-\beta V_N)$$

where $N = N_a + N_b + \cdots + N_c$. The chemical potentials act as statistical weighting factors on the molecular numbers $N_v$. We have discovered a probabilistic interpretation of the chemical potentials. It is convenient to define a quantity, the *activity* as $z_v \equiv \exp(-\beta\mu_v)/\Lambda_v^3$. $z_v$ has the units of number density. Thus the factors in (7.26) could be simplified

$$\Xi = \sum_{N_a \geq 0}^\infty \sum_{N_b \geq 0}^\infty \cdots \sum_{N_c \geq 0}^\infty \frac{z_a^{N_a}}{N_a!} \frac{z_b^{N_b}}{N_b!} \cdots \frac{z_c^{N_c}}{N_c!} \int d\mathbf{r}^N \mathbf{p}^N \, exp(-\beta V_N) \tag{7.27}$$

This formula is for the case of simple spherical molecules. Essentially similar reasoning leads to formulas for more complicated molecules.

## II.8. *The Microcanonical Ensemble*

As mentioned in section II.4, the microcanonical ensemble is also called the N-V-E (constant $N$, constant $V$, and constant $E$) ensemble. To construct an N-V-E ensemble, we reproduce $x$ images of a given protosystem $K_0$ with fixed content ($N=\{N_1, N_2,..., N_c\}$), fixed volume ($V$), and constant energy ($E$). In the limit, we let $x\rightarrow\infty$. Macroscopically, this ensemble corresponds to an isolated system since neither $N$, $V$, nor $E$ can change. For a given value of $E$, we know from quantum mechanics that there are $\Omega$ degeneracies; i.e., there are $\Omega$ independent states (eigenfunctions or wave functions) corresponding to $E$. For example, in rotation there are $2J+1$ degeneracies corresponding to the angular energy $J(J+1)$ (see section I.5). The $x$ subsystems are distributed among these eigenstates with equal probability:

$$p_i = \frac{1}{\Omega} \tag{8.1}$$

Therefore the information entropy is given by

$$S = - k \sum_{i=1}^{\Omega} p_i \ln p_i = - k \ln(\frac{1}{\Omega}) = k \ln \Omega \tag{8.2}$$

$\Omega$ is the partition function of the microcanonical ensemble. It is a measure of entropy. To establish the relation with thermodynamics, we know from the first law that

$$dU = T \, dS - P \, dV + \sum_{v=1}^{c} \mu_v \, dN_v \tag{8.3}$$

Rearrangement gives

$$dS = \frac{1}{T} \, dU + \frac{P}{T} \, dV - \sum \frac{\mu_v}{T} \, dN_v \tag{8.4}$$

Equivalently

$$k \, d \ln(\Omega) = \frac{1}{T} \, dU + \frac{P}{T} \, dV - \sum \frac{\mu_v}{T} \, dN_v \tag{8.5}$$

Therefore

$$\left.\frac{\partial \ln \Omega}{\partial U}\right|_{V,N} = \frac{1}{kT}, \quad \left.\frac{\partial \ln \Omega}{\partial V}\right|_{U,N} = \frac{P}{kT}, \quad \left.\frac{\partial \ln \Omega}{\partial N_v}\right|_{UVN_\gamma} = - \frac{\mu_v}{kT} \tag{8.6}$$

## II.9. *The Isothermal-Isobaric Ensemble*

The constant content ($N=\{N_1, \ldots , N_c\}$), constant pressure ($P$), and constant temperature ($T$) ensemble has gained much interest in computer simulation recently. In this ensemble, the images of a protosystem are reproduced with fixed **N**, $P$ and $T$. Volume ($V$) and energy ($E$) are allowed to vary. The $x$ subsystems spread across the phase space $\Gamma$. As before, we partition $\Gamma$ into $M$ cells (finally, $M \rightarrow \infty$). The maximum entropy method is used to find the equilibrium distribution. The information entropy is

$$S = -k \sum_{i=1}^{M} p_i \ln p_i \tag{9.1}$$

subject to the following constraints:

$$\sum_{i=1}^{M} p_i = 1 \tag{9.2}$$

$$\sum_{i=1}^{M} p_i \varepsilon_i = <E> \tag{9.3}$$

$$\sum_{i=1}^{M} p_i V_i = <V> \tag{9.4}$$

where $\varepsilon_i$ is the energy of cell $i$, and $V_i$ is the volume corresponding to cell $i$. The Lagrange multiplier method gives the probability

$$p_i = Y^{-1} \exp(-\beta\varepsilon_i - \gamma V_i) \tag{9.5}$$

where $Y$ is the partition function

$$Y = \sum_{i=1}^{M} \exp(-\beta\varepsilon_i - \gamma V_i) \tag{9.6}$$

Properly speaking, integration over volume should be used in (9.6) since volume is a continuous variable. $\beta$ and $\gamma$ are Lagrange multipliers. Substitution of $p_i$ into (9.1) gives

$$S = k \ln Y + k\beta<E> + k\gamma<V> \tag{9.7}$$

The physical meaning of $Y$, $\beta$, and $\gamma$ could be obtained from the definition of Gibbs free energy G:

$$S = -\frac{G}{T} + \frac{<E>}{T} + \frac{P<V>}{T} \tag{9.8}$$

Comparison with (9.7) gives

$$G = -kT \ln Y \tag{9.9}$$

and

$$\gamma = \frac{P}{kT} \tag{9.10}$$

as well as $\beta = 1/kT$. The same type of differential relation holds:

$$\frac{\partial \ln Y}{\partial \beta} = - <E> \tag{9.11}$$

$$\frac{\partial \ln Y}{\partial \gamma} = - <V> \tag{9.12}$$

$$\frac{\partial^2 \ln Y}{\partial \beta^2} = kT^2 C_p \tag{9.13}$$

Note also the volume fluctuations

$$\frac{\partial^2 \ln Y}{\partial \gamma^2} = \sum p_i [V_i - <V>]^2 \tag{9.14}$$

In fact, the identification (9.9 and 10) could be confirmed by the differential relations (9.11 and 12). Physically, the *N-P-T* ensemble represents a *closed system* with a movable boundary (variable volume). The system is closed with respect to mass transfer. However, it could exchange energy with surroundings through heat transfer and mechanical work. Pressure is kept constant through volume change. An example of such a system could be a gas in a cylinder in contact with a constant-temperature bath and compressed by a movable piston at the boundaries.

## SUMMARY OF GIBBS ENSEMBLES

We have discussed several ensembles: the canonical (*N-V-T*) the grand canonical (μ-*V-T*), the microcanonical (*N-V-E*) and the *N-P-T* ensemble. Each ensemble corresponds to a macroscopic thermodynamic system. Recently, new ensembles have been proposed and tested in computer simulations [4]. For example, Andersen [5] has proposed an *N-P-H* ensemble. Enthalpy, instead of energy, is the controlled variable. Discussion of these specialized ensembles is beyond the scope of this book. Interested reader is referred to the literature.

To each ensemble there is associated a partition function ($Z_N$, $\Xi$, $\Omega$, or $Y$). The partition functions are related to well-known thermodynamic functions when the system is at equilibrium or when the distribution is given by the Boltzmann distribution. Thus $Z_N$, $\Xi$, $\Omega$, and $Y$ are related to the Helmholtz free energy, the product $PV$, the entropy, and the Gibbs free energy, respectively. We summarize the ensembles in Table II.4.

We have just seen the close relationship between the partition functions and thermodynamic functions. The partition function links quantities on the molecular level (such as interaction potentials and vibrational frequencies) to measurable macroscopic properties (such as $PV$ and energy). It serves as a bridge between the microscopic and the macroscopic. From the derivation, we observe that the partition function is a normalization factor for the probabilities. On the other hand, it is also a kind of average of the microscopic energies or the particle number. Unfortunately, only for very simple systems is it possible to evaluate the partition function in closed form. Ideal-gas molecules (which have no potential energy) and the one-dimensional Ising model

*Table II.4.  Summary of Ensembles*

| Ensemble Type | Fixed Variables | Partition Function | Thermodynamic Relation |
|---|---|---|---|
| Canonical | $N, V, T$ | $Z_N = \sum_i \exp(-\beta\varepsilon_i)$ | $A = -kT \ln Z_N$ |
| Microcanonical | $N, V, E$ | $\Omega = \sum_i 1$ | $-TS = -kT \ln \Omega$ |
| Grand canonical | $\mu, V, T$ | $\Xi = \sum_{i,N} \exp(\beta N\mu - \beta\varepsilon_i)$ | $-PV = -kT \ln \Xi$ |
| Isothermal -isobaric | $N, P, T$ | $Y = \sum_{i,V} \exp(-\beta PV - \beta\varepsilon_i)$ | $G = -kT \ln Y$ |

(particles arranged on a one-dimensional lattice with nearest-neighbor interactions) are examples of such systems.  For molecules with more realistic interactions, direct evaluation of the partition function from its definition is mathematically impossible.  In many cases, the partition function is not explicitly evaluated.  It is used as an exact formula, though implicit, to show the functional dependence on correct groups of variables.  Due to the difficulty of evaluating the partition function, an alternative in statistical mechanics is developed, i.e., the molecular distribution functions (see Chapter IV).  In the following chapter, we shall discuss applications of the partition function to ideal-gas molecules.  Afterwards, we shall introduce the molecular distribution functions.

## References

[1]   C. Shannon, *A Mathematical Theory of Communication*, (University of Illinois Press, Urbana, Illinois, 1949).

[2]   J. W. Gibbs, *Statistical Mechanics: Developed with Especial Reference to the Rational Foundation of Thermodynamics*, (Yale University Press, New Haven, Connecticut, 1902).

[3]   T.L. Hill, *Statistical Mechanics*, (McGraw, New York, 1956).

[4]   J.M. Haile and H.W. Graben, Mol. Phys. **40**, 1433 (1980); also J.R. Ray, H.W. Graben, and J.M. Haile, Nuovo Cimento **64**, 19 (1981).

[5]   H.C. Andersen, J. Chem. Phys. **72**, 2384 (1980).

## Exercises

1.   Calculate the entropy $S$ for the game of throwing a die.  The throwing of a die gives six possible outcomes.  If the die is balanced, the probability of the appearance of each face is 1/6.  Also calculate the expected value of the game.

2.   For a roulette of five colors: one will play 10 chips with $9 per game.  (The rules are the same as described earlier.)  Find the entropy, $S$, Helmholtz free energy, internal energy (or average dollars), and temperature ($1/\beta$) of this game.  Use the tabulation and the maximum entropy methods.

| Red | Black | Yellow | Green | Blue |
|-----|-------|--------|-------|------|
| $4  | $3    | $2     | $1    | $0   |

3.  Given the partition function of a system at equilibrium, we can find the equation of state (or, equivalently, the pressure) of the system. The reverse is also true provided the system follows the ideal gas behavior at low pressure. For the van der Waals equation of state

$$P = \frac{RT}{v-b} - \frac{a}{v^2}, \qquad v = \frac{V}{N}$$

Find the partition function and the internal energy for this gas. Note that for an ideal gas

$$Z_N = \frac{V^N}{\Lambda^{3N}N!}, \qquad U = \frac{3}{2} NkT$$

4.  The conjugate force to the surface area in a system with a free surface (e.g., water in a shallow container) is the surface tension. Find the expression for the change in internal energy $dU$ by taking into account the surface forces. Namely, what is the expression for the first law for a system with surface effects?

5.  *One-dimensional Ising Model.* In the study of ferromagnetism, binary alloys, and adsorption, the Ising model representation has proven to be useful. In the example of ferromagnetism, the particles are aligned on the lattice. The spin of the particle can be in the up (+1) or down (−1) directions. Assuming nearest-neighbor interaction on a one-dimensional lattice, the partition function is

$$Q_c = \left[ e^K \cosh C + (e^{2K}\sinh C + e^{-2K})^{1/2} \right]^N$$

where $K = J/kT$, $J$ is the interaction energy between neighbors, T is the absolute temperature, k is the Boltzmann constant, $C = - uH/kT$, $u$ is the magnetic moment per spin, and $H$ is the magnetic field strength. Find the internal energy, the Helmholtz free energy, and the magnetization, $M$ (the magnetization is related to $Q_c$ by $M = (u/N)(\partial \ln Q_c/\partial C)$).

# CHAPTER III
# THE IDEAL GAS

Historically, an ideal gas is considered to fulfill the following conditions:

(1)   Its $P$-$V$-$T$ relation is governed by the ideal-gas equation of state

$$PV = \tilde{n}RT \tag{1}$$

where $P$ is the pressure, $V$ is the volume of the gas, $\tilde{n}$ is the number of moles present, R is the gas constant, and T is the absolute temperature.

(2)   The internal energy of an ideal gas is a function of temperature only. It does not depend on pressure.

In molecular theory this definition can be made precise. Ideal gas is composed of molecules lacking mutual interaction (the potential energy is zero). However, the molecules possess kinetic energies, $KE_t$, $KE_r$, and $KE_v$, derived from the translational, rotational and vibrational motions, respectively. Given $N$ such molecules, the Hamiltonian is

$$H_N = KE = KE_t + KE_r + KE_v \tag{2}$$

Note that $H_N$ is not a function of the configuration $\mathbf{r}^N$ (i.e., the spatial arrangements) of the gas molecules. From the definition, ideal-gas molecules have no excluded volume (volumeless) and exert no attractive forces (in other words, they are free particles). A simple example of this ideal behavior is monatomic molecules at infinite dilution.

## III.1. *Monatomic Molecules*

In a monatomic gas, only the translational motion of the particles contributes to the kinetic energy

$$\frac{p^2}{2m} \tag{1.1}$$

where $\mathbf{p}$ is the momentum and $m$ is the mass of the molecule. Thus the Hamiltonian (total energy) for $N$ molecules is given by

$$H_N = \sum_i^N \frac{p_i^2}{2m_i} \tag{1.2}$$

where $i$ is a subscript for the $i$th molecule. For identical particles, the mass $m$ is the same. In three dimensions, the momentum $\mathbf{p}_i$ is a vector quantity, and

$$p_i^2 = p_{ix}^2 + p_{iy}^2 + p_{iz}^2 \tag{1.3}$$

According to section II.5 the partition function is given by the $6N$-fold integral

$$Z_N = \frac{1}{N!h^{3N}} \int_0^L dx_1 \int_0^L dy_1 \int_0^L dz_1 \cdots \int_0^L dz_N \int_{-\infty}^\infty dp_{1x} \int_{-\infty}^\infty dp_{1y} \int_{-\infty}^\infty dp_{1z} \tag{1.4}$$

$$\cdots \int_{-\infty}^\infty dp_{Nz} \exp[-\frac{\beta}{2m} \sum_{i=1}^N (p_{ix}^2 + p_{iy}^2 + p_{iz}^2)]$$

where $L$ is the length of the sides of the volume, $V = L^3$. Since the integrand is independent of positions $(x_1, y_1, z_1, \ldots, z_N)$, spatial integration gives a factor $L^{3N}$ or $V^N$. The integration over the momenta is given by the well-known definite integral

$$\int_{-\infty}^\infty dt \, \exp(-at^2) = \sqrt{\frac{\pi}{a}} \tag{1.5}$$

Thus

$$Z_N = \frac{V^N (2\pi mkT)^{3N/2}}{N!h^{3N}} \tag{1.6}$$

Introducing the de Broglie thermal wavelength

$$\Lambda^2 \equiv \frac{h^2}{2\pi mkT} \tag{1.7}$$

we have

$$Z_N = \frac{V^N}{N!\Lambda^{3N}} \tag{1.8}$$

As indicated earlier, this partition function can be used to derive the thermodynamic properties.  The Helmholtz free energy is given by

$$A = - kT \ln Z_N \tag{1.9}$$

$$= - kT(N \ln V - \ln N! - N \ln \Lambda^3)$$

$$= - kT(N \ln V - N \ln N + N - N \ln \Lambda^3)$$

$$= NkT[\ln(\rho\Lambda^3) - 1]$$

where we have used the Stirling approximation for the factorials at large $N$, i.e.,

$$\ln N! \approx N \ln N - N \tag{1.10}$$

We note that since $\rho = P/kT$ (see below), we could write

$$\ln \rho = \ln P - \ln T - \ln k \tag{1.11}$$

and

$$\ln \Lambda^3 = - \frac{3}{2} \ln T + \ln(kC) \tag{1.12}$$

where

$$C \equiv \frac{(h/\sqrt{2\pi mk})^3}{k} \tag{1.13}$$

Thus

$$A = NkT[\ln P - \frac{5}{2} \ln T - 1 + \ln C] \tag{1.14}$$

The pressure and internal energy can be obtained from the derivatives of $Z_N$:

$$P = kT \left[ \frac{\partial \ln Z_N}{\partial V} \right]_{N,T} = \frac{NkT}{V} \tag{1.15}$$

and

$$U = - \left[ \frac{\partial \ln Z_N}{\partial \beta} \right]_{N,V} = \frac{3NkT}{2} \tag{1.16}$$

Relation (1.15), the ideal-gas equation of state, is a consequence of our molecular definition.  Equation (1.16) gives the internal energy as a function of temperature only. When rotational and vibrational energies are present, the internal energy equation includes more terms.  However, the pressure equation remains the same.  This will be discussed in the sections for polyatomic molecules.  Other thermodynamic quantities are derived as

*Enthalpy*

$$H = U + PV = \frac{5NkT}{2} \tag{1.17}$$

*Heat Capacity at Constant Volume*

$$C_v = \frac{3k}{2} \tag{1.18}$$

*Heat Capacity at Constant Pressure*

$$C_p = \frac{5k}{2} \tag{1.19}$$

*Gibbs Free Energy*

$$G = A + PV = NkT \ln(\rho\Lambda^3) \tag{1.20}$$

$$= NkT \left[ \ln P + \ln \left[ \frac{\Lambda^3}{kT} \right] \right]$$

$$= NkT \left[ \ln P - \frac{5}{2} \ln T + \ln C \right]$$

*Entropy*

$$S = \frac{U-A}{T} = Nk[\frac{5}{2} - \ln(\rho\Lambda^3)] \tag{1.21}$$

Equation (1.21) is also known as the Sackur-Tetrode equation. The chemical potential is

$$\mu = kT \ln(\rho\Lambda^3) \tag{1.22}$$

We have given a complete description of the thermodynamic properties of ideal monatomic gas. This information is relevant to dilute noble gases, e.g., argon, krypton, and xenon. At low pressures, the specific volume of gas is large, and the molecules, on the average, are separated by large distances. As a result, the interaction energies which are short-ranged become negligible ($PE = 0$). This fulfills the conditions for an ideal gas.

## III.2. *Alternative Derivation*

We shall demonstrate that a semiclassical treatment of the quantum-mechanical expression for the translational energy results in the same partition function for an ideal gas. Recall that in I.2. the energy levels for translation are

$$\varepsilon_{ni} = \frac{n_i^2 h^2}{8mL^2}, \qquad n_i = 0,1,2,... \tag{2.1}$$

where $n$ is the quantum number and $i$ denotes the $i$th molecule. The quantum partition function is

$$Z_N = \prod_i^{3N} \sum_{n_i} \exp(-\beta \varepsilon_{n_i}) \tag{2.2}$$

$$= \prod_i^{3N} \sum_{n_i} \exp[-\beta n_i^2 h^2/(8mL^2)]$$

At high temperatures, the quantum levels $h^2/8mkTL^2$ are closely spaced. The sum over $n_i$ can be approximated by the integral

$$Z_N = \prod_i^{3N} \int_0^\infty dn_i \, \exp\left[-\frac{\beta n_i^2 h^2}{8mL^2}\right] \tag{2.3}$$

$$= \frac{L^{3N}(2\pi mkT)^{3N/2}}{h^{3N}} = \frac{V^N}{\Lambda^{3N}}$$

We next introduce a factor $1/N!$ to the quantum partition function in order to account for the indistinguishability of $N$ identical molecules. $N!$ refers to the number of permutations of $N$ particles. Since the quantum states are unaffected by such permutations, we divide (2.3) by $N!$. This procedure is known as *correct Boltzmann counting* [1]. Introducing this correction, we have

$$Z_N = \frac{V^N}{N!\Lambda^{3N}} \tag{2.4}$$

The result is identical to (1.8), derived from purely classical considerations.

## III.3. *Diatomic Molecules: Rotation*

In diatomics, the two constituent atoms can revolve around each other as well as vibrate along the connecting axis. Thus the rotational and vibrational motions contribute to the kinetic energy. The Hamiltonian is given by

$$H_N = KE_t + KE_r + KE_v + \cdots \tag{3.1}$$

Other terms that contribute to the kinetic energy come from electron motion, nuclear spin, bond rotation, etc. We consider $KE_r$, the rotational contribution, first. As noted earlier, the quantum energy levels of a rigid rotator are

$$\varepsilon_J = \frac{J(J+1)h^2}{2\overline{m}R^2} = J(J+1)k\Theta_r \tag{3.2}$$

where we have introduced the characteristic rotational temperature $\Theta_r$ as

$$\Theta_r \equiv \frac{h^2}{2\overline{m}R^2k} \tag{3.3}$$

$J$ is the quantum number for rotation (i.e., the total momentum), $\overline{m} = m_1m_2/(m_1+m_2)$ the reduced mass, $R$= radius of rotation (or half-bond length), and $h = h/2\pi$. Table III.1 gives the characteristic rotational temperatures for a number of diatomic molecules.

*Table III.1. The Vibrational and Rotational Temperatures for Diatomic Molecules*

|          | $\Theta_v$, K | $\Theta_r$, K | $a^*$ |
|----------|--------|--------|-------|
| $H_2$    | 6210   | 85.4   | 2     |
| $N_2$    | 3340   | 2.86   | 2     |
| CO       | 3070   | 2.77   | 1.128 |
| HCl      | 4140   | 15.2   | 1     |
| $Cl_2$   | 810    | 0.346  | 2     |

*$a$ is a symmetry number (see below).

For each quantum energy $\varepsilon_J$, there are $2J+1$ independent eigenfunctions (or degeneracies). The source of the degeneracies comes from the solutions to the differential equation (I.5.9) for the rigid rotator. Since each of the $2J+1$ states is populated with equal probability $p_J$=exp($-\beta\varepsilon_J$)/$Z_r$, where $Z_r$ is the rotational partition function for one rotator, one needs to multiply the Boltzmann factor by $2J+1$, to account for the $2J+1$ states:

$$Z_{N,r} = \left\{ \frac{1}{a} \sum_{J=0}^{\infty} (2J+1)e^{-\beta J(J+1)k\Theta_r} \right\}^N \tag{3.4}$$

where $Z_{N,r}$ is the rotational partition function for $N$ rotators, and $a$ is a rotational symmetry number, which is 2 for homonuclear diatomics (e.g., $N_2$, $O_2$, $Cl_2$, etc.) and 1 for heteronuclear diatomics (e.g., HCl, CO). This factor is needed again for correct counting . For identical atoms, the interchange of the positions of the two gives back the same (quantum) state. To eliminate double counting, we divide the partition function by 2. To apply classical approximation, we replace the summation in (3.4) by the integration

$$Z_{N,r}(\text{classical}) = \left[ \frac{1}{a} \int_0^{\infty} dJ\ (2J+1)\ e^{-J(J+1)\Theta_r/T} \right]^N \tag{3.5}$$

This approximation is justified when the temperature is high (for example five times higher than the characteristic rotational temperature $T > 5\Theta_r$). The result is

$$Z_{N,r} = \left[\frac{T}{a\Theta_r}\right]^N \tag{3.6}$$

The complete partition function is

$$Z_N = Z_{N,t}Z_{N,r} = \frac{1}{N!}\left[\frac{V}{\Lambda^3}\right]^N\left[\frac{T}{a\Theta_r}\right]^N \tag{3.7}$$

The thermodynamic properties of ideal diatomics can be derived as

$$\frac{A}{NkT} = \ln \rho -1 - \frac{5}{2}\ln T + \ln a\Theta_r\left[\frac{h}{\sqrt{2\pi mk}}\right]^3 \tag{3.8}$$

$$= \ln P -1 - \frac{7}{2}\ln T + \ln \frac{a\Theta_r}{k}\left[\frac{h}{\sqrt{2\pi mk}}\right]^3 \tag{3.9}$$

$$P = \frac{NkT}{V} \tag{3.10}$$

$$U = \frac{5}{2}NkT \tag{3.11}$$

$$H = \frac{7}{2}NkT \tag{3.12}$$

$$C_v = \frac{5}{2}k \tag{3.13}$$

$$C_p = \frac{7}{2}k \tag{3.14}$$

$$\frac{G}{NkT} = \ln \rho - \frac{5}{2}\ln T + \ln a\Theta_r\left[\frac{h}{\sqrt{2\pi mk}}\right]^3 \tag{3.15}$$

$$= \ln P - \frac{7}{2}\ln T + \ln \frac{a\Theta_r}{k}\left[\frac{h}{\sqrt{2\pi mk}}\right]^3$$

$$\frac{\mu}{kT} = \ln \rho - \frac{5}{2}\ln T + \ln a\Theta_r\left[\frac{h}{\sqrt{2\pi mk}}\right]^3 \tag{3.16}$$

$$= \ln P - \frac{7}{2}\ln T + \ln \frac{a\Theta_r}{k}\left[\frac{h}{\sqrt{2\pi mk}}\right]^3$$

$$\frac{S}{Nk} = \frac{7}{2} - \ln \rho + \frac{5}{2}\ln T - \ln a\Theta_r\left[\frac{h}{\sqrt{2\pi mk}}\right]^3 \tag{3.17}$$

$$= \frac{7}{2} - \ln P + \frac{7}{2} \ln T - \ln \frac{a\Theta_r}{k} \left[ \frac{h}{\sqrt{2\pi mk}} \right]^3$$

## III.4  *Diatomic Molecules: Vibration*

For diatomic molecules with vibrational energy, the energy levels are again given by quantum mechanics as in (I.5.16):

$$\varepsilon_L = (L + \frac{1}{2})h\nu \tag{4.1}$$

We rewrite (4.1) as

$$\varepsilon_L = Lk\Theta_v' + \frac{1}{2}k\Theta_v' \tag{4.2}$$

where $\Theta_v' = h\nu/k$ is the characteristic vibrational temperature specific to the diatomic under study. Since for most substances the characteristic vibrational temperatures are quite large, we cannot treat the vibration mode classically. The quantum expression for the partition function will have to be used. Let us denote the translational partition function by $Z_t = (V/\lambda^3)^N/N!$, rotational partition function by $Z_r = (T/a\Theta_r)^N$, and vibrational partition function by $Z_v$. The total canonical partition function $Z_N$ is then

$$Z_N = Z_t Z_r Z_v = Z_t Z_r \left[ \sum_{L=0}^{\infty} \exp(-\beta(L+\frac{1}{2})k\Theta_v') \right]^N \tag{4.3}$$

$$= \frac{1}{N!} \left[ \frac{VT \exp(-\Theta_v'/2T)}{\Lambda^3 a\Theta_r[1 - \exp(-\Theta_v'/T)]} \right]^N$$

Knowing the partition function $Z_N$, we can derive the following thermodynamic properties:

$$\frac{A}{NkT} = \ln \rho - \frac{5}{2} \ln T + \ln a\Theta_r \left[ \frac{h}{\sqrt{2\pi mk}} \right]^3 + \ln(1 - e^{-\Theta_v'/T}) + \frac{1}{2} \frac{\Theta_v}{T} \tag{4.4}$$

$$U = \frac{5}{2}NkT + \frac{Nk\Theta_v'}{e^{\Theta_v'/T} - 1} + \frac{\Theta_v'NK}{2} \tag{4.5}$$

$$H = \frac{7}{2}NkT + \frac{Nk\Theta_v'}{e^{\Theta_v'/T} - 1} + \frac{\Theta_v'Nk}{2} \tag{4.6}$$

$$P = \frac{NkT}{V} \tag{4.7}$$

$$C_v = \frac{5}{2}k + \frac{k\Theta_v'^2 \exp(\Theta_v'/T)}{T^2[\exp(\Theta_v'/T) - 1]^2}$$ (4.8)

$$C_p = \frac{7}{2}k + \frac{k\Theta_v'^2 \exp(\Theta_v'/T)}{T^2[\exp(\Theta_v'/T) - 1]^2}$$ (4.9)

$$\frac{G}{NkT} = \ln \rho - \frac{5}{2}\ln T + \ln a\Theta_r \left[\frac{h}{\sqrt{2\pi mk}}\right]^3 + \ln\left[1 - \exp\left[\frac{-\Theta_v'}{T}\right]\right] + \frac{\Theta_v'}{2T}$$ (4.10)

$$\frac{\mu}{kT} = \ln \rho - \frac{5}{2}\ln T + \ln a\Theta_r \left[\frac{h}{\sqrt{2\pi mk}}\right]^3 + \ln\left[1 - \exp\left[\frac{-\Theta_v'}{T}\right]\right] + \frac{\Theta_v'}{2T}$$ (4.11)

$$\frac{S}{Nk} = \frac{7}{2} - \ln \rho + \frac{5}{2}\ln T - \ln a\Theta_r \left[\frac{h}{\sqrt{2\pi mk}}\right]^3$$ (4.12)

$$+ \frac{\Theta_v'}{T[\exp(\Theta'_v/T) - 1]} - \ln\left[1 - \exp\left[\frac{-\Theta_v'}{T}\right]\right]$$

Since eqs. (4.4–12) contain all three modes of kinetic motion, they should be used in actual calculations for diatomic ideal-gas properties.

### III.5. *Polyatomic Molecules*

Consider a polyatomic molecule with $n$ atoms. Each atom has a nucleus and a number of electrons. In an exact quantum-mechanical treatment, all these groupings and their interactions should be included in the Schrödinger equation. The mathematics becomes very complicated. In the Born-Oppenheimer approximation, the nuclei are considered fixed when the motions of the lightweight electrons are treated. We then obtain the electronic energy levels for fixed distances $r_1, r_2, \ldots, r_n$ of the nuclei. Next, we allow the distances $r_1, \ldots, r_n$ to vary. These motions are separated into the center-of-mass motion (translation), rotation, and vibration. Therefore the energy KE is given by the sum

$$KE = KE_t + KE_r + KE_v + KE_e$$ (5.1)

We have dealt with $KE_t$, $KE_r$, and $KE_v$ previously, and the same procedure is used here for polyatomics. The contribution $Q_e$ of the electronic energy levels to the partition function $Z_N$ is

$$Z_N = Q_t^N Q_r^N Q_v^N Q_e^N$$ (5.2)

where

$$Q_e = \omega_1 e^{-\beta a_1} + \omega_2 e^{-\beta a_2}$$ (5.3)

and $\omega_i$ is the degree of degeneracies for energy level $a_i$. We have exhibited only two terms in the electronic energy levels because of the large separation in $a_i$ compared to most kT. Therefore at ordinary temperatures, the molecule can rarely reach beyond $a_2$.

The vibrational motion of the $n$ atoms is analyzed in terms of the so-called normal coordinates. It turns out that only $3n-6$ (for nonlinear molecules) or $3n-5$ (for linear molecules) modes of vibration are nonzero. Therefore the vibrational partition function is given by

$$\ln Q_v = -\sum_{i=1}^{n'} \left\{ \frac{\Theta_i'}{2T} + \ln\left[1 - \exp\left(-\frac{\Theta_i'}{T}\right)\right] \right\} \tag{5.4}$$

where $n' = 3n-6$ (for nonlinear molecules, such as $NH_3$, and $CH_3OH$) and $n'=3n-5$ (for linear molecules, such as $O_2$, and $CO_2$). And $\Theta_i'$ are the characteristic vibrational temperatures of the polyatomic molecule. For example, for the molecule $BF_3$, there are $3(4)-6=6$ independent modes of vibration. The characteristic vibrational temperatures are known to be 2070, 2070, 1270, 995, 631, 631 K. (Note that $BF_3$ is a nonlinear polyatomic molecule.) For the linear molecule $CO_2$, there are $3(3)-5=4$, $\Theta_i'$, i.e., 3360, 1890, 954, and 954 K.

The rotational motion also distinguishes a nonlinear molecule (e.g., $CHCl_3$) from a linear molecule (e.g., $C_2H_2$). For a linear molecule, all three components of rotational modes are indistinguishable, as in the case of the diatomics. For a nonlinear molecule, we can distinguish mechanically three principal moments of inertia: $I_x$, $I_y$, and $I_z$. A molecule whose $I_x = I_y = I_z$ (e.g., $CH_4$) is called a spherical top; if only two $I$'s are equal, it is called a symmetric top (e.g., $C_6H_6$); if all three principal moments of inertia are different (e.g., $H_2O$), it is called an asymmetrical top. The rotational energy levels for the most general case (an asymmetrical top) are very complicated. A quantum mechanical description is difficult. Except for very light atoms (e.g., H) at low temperatures, a classical treatment is close enough for practical purposes. The results for the partition function are

$$\ln Q_r = \ln\frac{\sqrt{\pi}}{a} + \sum_{\alpha=x,y,z} \ln\left[\frac{2I_\alpha kT}{h}\right]^{1/2} \tag{5.5}$$

$$= \ln\frac{\sqrt{\pi}}{a} + \frac{1}{2}\sum_{\alpha=x,y,z} \ln\frac{T}{\Theta_\alpha}$$

where the characteristic rotational temperatures $\Theta_\alpha$ are defined as $h^2/2I_\alpha k$, and $a$ is again the symmetry number; e.g., for $CH_4$, $a=12$; for $H_2O$, $a=2$; and for $NH_3$, $a=3$.

The translational mode contributes simply

$$Q_t = \frac{V}{\Lambda^3} \tag{5.6}$$

where $\Lambda$ is the de Broglie thermal wavelength defined earlier. We obtain the Helmholtz free energy from the known relation

$$A = -kT \ln Z_N = -kT \ln(Q_t^N Q_r^N Q_v^N Q_e^N) \tag{5.7}$$

Thus,

$$\frac{A}{NkT} = \ln \rho - 1 - \left[\frac{3}{2}+J\right] \ln T + \sum_{i=1}^{n'} \ln \left[1 - \frac{e^{-\Theta_i'}}{T}\right] + \sum_{i=1}^{n'} \frac{\Theta_i'}{2T}$$ (5.8)

$$+\frac{1}{2} \ln \left[\frac{h^2 a^2 \Theta_x \Theta_y \Theta_z}{2\pi^2 mk}\right] - \ln \sum_{j=1}^{2} [\omega_j \exp(-\beta a_j)]$$

The other thermodynamic properties are obtained as

$$\frac{U}{NkT} = \frac{3}{2} + J + \sum_{i=1}^{n'} \left[\frac{\Theta_i'}{2T} + \frac{\Theta_i'/T}{\exp(\Theta_i'/T) - 1}\right]$$ (5.9)

$$+ \frac{\displaystyle\sum_{j=1}^{2} a_j \omega_j e^{-\beta a_j}}{kT \displaystyle\sum_{j=1}^{2} \omega_j e^{-\beta a_j}}$$

$$\frac{H}{NkT} = \frac{U}{NkT} + 1$$ (5.10)

where $J=3/2$ for nonlinear molecules and 1 for linear molecules. For linear molecules, $\Theta_x \Theta_y \Theta_z$ should be replaced by $\Theta^2$.

The heat capacities are

$$\frac{C_v}{k} = \frac{3}{2} + J + \sum_{i=1}^{n'} \left[\frac{\Theta_i'}{T}\right]^2 \frac{\exp(\Theta_i'/T)}{[\exp(\Theta_i'/T) - 1]^2}$$ (5.11)

$$+ \frac{1}{(kT)^2}\left[\frac{\displaystyle\sum_{j=1}^{2} a_j \omega_j \exp(-a_j/kT)}{\displaystyle\sum_{j=1}^{2} \omega_j \exp(-a_j/kT)} - \frac{\left[\displaystyle\sum_{j=1}^{2} a_j \omega_j \exp(-a_j/kT)\right]^2}{\left[\displaystyle\sum_{j=1}^{2} \omega_j \exp(-a_j/kT)\right]^2}\right]$$

$$C_p = C_v + k$$ (5.12)

The entropy is

$$\frac{S}{Nk} = -\ln \rho + \frac{5}{2} + J + (\frac{3}{2}+J) \ln T - \frac{1}{2} \ln\left[\frac{h^2 a^2 \Theta_x \Theta_y \Theta_z}{2\pi^2 mk}\right]$$ (5.13)

$$+ \sum_{i=1}^{n'} \left\{\frac{\Theta_i'/T}{\exp(\Theta_i'/T) - 1} - \ln\left[1 - \exp\left[\frac{\Theta_i'}{T}\right]\right]\right\} - \ln\left[\sum_{j=1}^{2} \omega_j \exp\left[\frac{-a_j}{kT}\right]\right]$$

$$+ \frac{\displaystyle\sum_{j=1}^{2} a_j \omega_j \exp(-a_j/kT)}{kT \displaystyle\sum_{j=1}^{2} \omega_j \exp(-a_j/kT)}$$

The Gibbs free energy is

$$\frac{G}{NkT} = \ln \rho - (\frac{3}{2}+J) \ln T + \frac{1}{2} \ln \left[ \frac{h^2 a^2 \Theta_x \Theta_y \Theta_z}{2\pi^2 mk} \right] \qquad (5.14)$$

$$+ \sum_{i=1}^{n'} \left\{ \frac{\Theta_i'}{2T} + \ln \left[ 1 - \exp \left( \frac{-\Theta_i'}{T} \right) \right] \right\}$$

$$- \ln \left[ \sum_{j=1}^{2} \omega_j \exp \left( \frac{-a_j}{kT} \right) \right]$$

The chemical potential for a pure fluid is

$$\frac{\mu}{kT} = \frac{G}{NkT} \qquad (5.15)$$

We note that, in this section, thermodynamic functions contain the electronic contribution, whereas it was absent in the discussion for diatomics. In fact, eqs. (5.7-15) are more general than eqs. (4.3-12) in that when the electronic energy contributions are set to zero, (5.7-15) reduce to (4.3-12). On the other hand, there are kinetic energy contributions in (5.7-15) worth further discussion. For example, the ethane molecule possesses torsional motion. This mode is neither Raman nor infrared active, so direct spectroscopic information about the frequency is unavailable. The potential energy is dependent on the dihedral angle $\Phi$ between the two methyl groups -$CH_3$. The energy exhibits periodical variations with amplitude $V_o$. The form is approximated by

$$u(\Phi) = \frac{1}{2}V_o (1 - \cos 3\Phi) \qquad (5.16)$$

At high temperatures, $kT \gg V_o$, the rotation is effectively free and contributions to $C_v$ approach the value $(1/2)k$. For small $T$, the motion is approximated by torsional harmonic oscillation. In this limit, $C_v$ approaches the low-temperature behavior of a one-dimensional Einstein solid. Other modes of kinetic energy include the nuclear spin contribution, anharmonicity, and nonrigidity corrections. Specialized texts are available for further study.

## III.6. *Calculation of Ideal-Gas Heat Capacity*

In principle, all the information needed in formulas (5.7-15) can be obtained from thermal, spectroscopic, and molecular measurements. It is then a simple matter to substitute the characteristic temperatures ($\Theta_i'$, or $\Theta_x$, etc.) into the proper equations for calculating the ideal-gas properties. However, approximations have been made:

(1) We assumed that the vibrational motion is harmonic. Correction of anharmonicity is required at high temperatures.

(2) We assumed the absence of coupling between rotational and vibrational motions. A correction is needed.

(3) For large molecules, such as benzene, there are 12 atoms. Therefore 30 vibrational characteristic temperatures, $\Theta_i'$, are needed in a calculation. This is very time consuming.

An alternative method is to examine the given formulas from (5.7) to (5.15) to see if some simplification can be effected. For example, for $CO_2$, the vibrational characteristic temperatures are 3360 K, 1890 K, 954 K, and 954 K. Evaluation of (5.11) shows that only the low characteristic temperatures (954 K) contribute significantly to the heat capacity. For practical purposes, one can approximate (5.11) by two terms from the vibrational contributions, and ignore the higher characteristic temperatures.

**SIMPLIFIED FORMULAS**

If we consider only the dominant vibrational characteristic temperature, we have the contribution

$$C_{p,1} = C\left[\frac{x}{\sinh}(x)\right]^2 \tag{6.1}$$

where $C$ is a constant to be determined, and $x = \Theta_{min}/2T = D/T$. $D$ can be treated as an empirical constant. For electronic contribution, let the ground state energy $a_1$ be taken as zero, and let $m=2$. We have, after recombinations,

$$C_{p,2} = E\left[\frac{y}{\cosh}(y)\right]^2 \tag{6.2}$$

where $E$ is a constant, $y = a_{max}/2kT = F/T$, and $F$ is a parameter. The energy level of nuclear spin is extremely large ($\Delta e_n$ is about 1 Mev) or, equivalently, in units of $kT$, $10^{10}K$); at most temperatures in practice, the molecule is at its nuclear ground state. There is no contribution from nuclear spin.

For molecules such as ethane, there is internal bond rotation (torsional vibrations). The energy barrier is periodical. At higher temperatures ($kT >> V_o$), the contribution to $C_p$ is $(1/2)k$. This number can be included with the $(3/2)k$ in (5.11). We call the combined constant $B$, i.e.,

$$C_{p,3} = B \tag{6.3}$$

Finally, we arrive at the simplified expression for the heat capacity

$$C_p^* = B + C\left[\frac{D/T}{\sinh(D/T)}\right]^2 + E\left[\frac{F/T}{\cosh(F/T)}\right]^2 \tag{6.4}$$

Equation (6.4) contains only five constants. Direct integration of (6.4) gives the enthalpy

$$H^* = BT + CT(\frac{D}{T})\coth(\frac{D}{T}) - ET(\frac{F}{T})\tanh(\frac{F}{T}) + A \qquad (6.5)$$

where A is a constant of integration. The entropy is obtained from the thermodynamic formula $\int dT\,(C_p/T)$:

$$S^* = B \ln T + C\left[\frac{D}{T}\coth(\frac{D}{T}) - \ln\sinh(\frac{D}{T})\right] \qquad (6.6)$$

$$- E\left[\frac{F}{T}\tanh(\frac{F}{T}) - \ln\cosh(\frac{F}{T})\right] + G$$

where $G$ is the integration constant. The constants $A, B, C, D, E, F,$ and $G$ are then determined from fitting eqs. (6.4-6) to reported experimental data. A number of experimental sources, such as the API project 44 [2], and JANAF Thermochemical Tables [3], could be utilized in this correlation effort. The constants, $A$ to $F$, were determined for 60 hydrocarbons and common gases found in natural gas and petroleum refining operations. [4] [5] [6] The results are presented in Table III.2. It is seen that the constant-pressure ideal heat capacity $C_p$ is fitted to within 0.05% (in average absolute deviations) for most substances over wide ranges of temperatures (from 298 K to 1000 K or 1500 K). The agreement in enthalpy and entropy predictions is equally satisfactory.

*Table III.2. Coefficients in the Simplified Formulas for Heat Capacity (cal/(mol K)), Enthalpy (cal/mol), and Entropy (cal/mol) of Ideal Gases*

| Compound | A cal/mol | B cal/(mol K) | C cal/(mol K) | D K | E cal/(mol K) | F K | G cal/(mol K) |
|---|---|---|---|---|---|---|---|
| Methane | -30787.5 | 8.00318 | 19.2633 | 2148.79 | 10.4423 | 1017.07 | -7.28270 |
| Ethane | -42778.2 | 10.3525 | 32.0989 | 1845.59 | 19.8704 | 846.785 | -13.1462 |
| Propane | -58099.3 | 13.7962 | 45.4301 | 1802.34 | 30.1245 | 815.820 | -25.4424 |
| n-Butane | -72674.8 | 18.6383 | 57.4178 | 1792.73 | 38.6599 | 814.151 | -46.1938 |
| Ethylene | -29634.1 | 8.51142 | 23.0661 | 1841.98 | 15.7134 | 825.296 | -1.54433 |
| Propylene | -45987.4 | 12.3781 | 35.9330 | 1808.60 | 23.8008 | 823.973 | -15.7445 |
| Benzene | -59270.3 | 13.0616 | 53.9027 | 1691.83 | 42.5244 | 771.023 | -19.4500 |
| Toluene | -75315.5 | 17.5653 | 66.1975 | 1713.33 | 50.2196 | 783.938 | -36.3540 |
| Oxygen | -3497.45 | 6.96302 | 2.40013 | 2522.05 | 2.21752 | 1154.15 | 9.19749 |
| Nitrogen | -119.39 | 6.95808 | 2.03952 | 1681.6 | 0.505863 | 6535.68 | 5.05495 |
| Carbon dioxide | -7460.81 | 7.54056 | 7.51625 | 1442.70 | 5.38023 | 647.502 | 6.25230 |

## III.7. *Ideal-Gas Mixtures*

Ideal-gas mixtures are, by definition, molecules whose components are free of interaction. Consequently, each component is individually an ideal gas. Ideal gases differ from one another by molecular mass. For simplicity, we restrict the discussions to monatomic gases. Generalization to polyatomics is straightforward. For C such gases with $N_A, N_B, \ldots, N_C$ and masses $m_A, m_B, \ldots, m_C$, the Hamiltonian is

$$H_N = \sum_i \sum_j \frac{p_{i(j)}^2}{2m_j} \tag{7.1}$$

where the subscripts $i(j)$ denote the $i$th molecule of the $j$th species. The partition function

$$Z_N = \frac{V^N}{N_A! N_B! ... N_C! h^{3N}} \int_{-\infty}^{\infty} dp_{1(1)x}\, dp_{1(1)y}\, dp_{1(1)z} \cdots dp_{N_C(C)z} \cdot \exp[-\beta H_N] \tag{7.2}$$

can be integrated to give

$$Z_N = \frac{V^N}{\prod N_j! \Lambda_j^{3N_j}} \tag{7.3}$$

Note that in (7.3) the correct Boltzmann counting is the product $\prod N_j!$. Each factor represents the number of permutations of $N_j$ identical molecules. These permutations do not give rise to a new (different) quantum state. The multiplicity is removed (or correctly counted) by division of $\prod N_j!$. Knowledge of the partition function leads to the thermodynamic functions

$$A = -kT \ln Z_N = -kT \left[ N \ln V - \sum \ln N_j! - \sum N_j \ln \Lambda_j^3 \right] \tag{7.4}$$

$$= -NkT \left[ \sum x_j \ln \left[ \rho_j \Lambda_j^3 \right] - 1 \right]$$

$$= -NkT \left[ \sum_j x_j \ln(x_j P) - (\frac{5}{2}) \ln T + \sum_j x_j \ln C_j \right]$$

Note that $\rho_j = x_j P/kT$ from the ideal-gas equation for the $j$th component. $C_j$ is the constant $C$ in (1.13) for species $j$. The equation of state is obtained from

$$\frac{P}{kT} = \frac{\partial \ln Z_N}{\partial V}\bigg|_{N,T} = \frac{N}{V} \tag{7.5}$$

Multiplication by the mole fraction $x_j$ gives

$$x_j PV = x_j NkT = N_j kT \tag{7.6}$$

Thus component $j$ individually obeys the ideal-gas equation of state. Let $P_j = x_j P$, the partial pressure. We have $P_j V = N_j kT$ (Dalton's law).

The internal energy and enthalpy are the same as given earlier for pure monatomics. The entropy is

$$\frac{S}{Nk} = \frac{5}{2} - \sum_j x_j \ln(\rho_j \Lambda_j^3) \tag{7.7}$$

$$= \frac{5}{2} - \sum_j x_j \ln(x_j P) + \frac{5}{2} \ln T - \sum_j x_j \ln C_j$$

The Gibbs free energy is

$$\frac{G}{NkT} = \sum x_j \ln(\rho_j \Lambda_j^3) \tag{7.8}$$

$$= \sum_j x_j \ln(x_j P) - \frac{5}{2} \ln T + \sum_j x_j \ln C_j$$

The chemical potential is

$$\frac{\mu_j}{kT} = \frac{\partial (G/kT)}{\partial N_j}\bigg|_{P,T,N_k} \tag{7.9}$$

$$= \ln(\rho_j \Lambda_j^3) = \ln(x_j P) - \frac{5}{2} \ln T + \ln C_j$$

## COMMENTS ON THE CHEMICAL POTENTIAL

An observation could be made regarding the definition of the chemical potential in classical thermodynamics.

$$\lim_{P \to 0} \frac{\mu_j}{RT} = \ln(x_j P) + B_j(T) \tag{7.10}$$

where $B_j$ was stated in textbooks to be a function of temperature only. However, an explicit expression for $B_j(T)$ could not be derived in the macroscopic approach. Comparison with (7.10) shows that for monatomics, $B_j$ ought to be

$$B_j(T) = -\frac{5}{2} \ln T + \ln\left[\frac{(h/\sqrt{2\pi m_j k})^3}{k}\right] \tag{7.11}$$

Similar expressions hold for polyatomics.

## III.8. *Properties of Mixing*

In thermodynamics, the property change upon mixing is measured at constant temperature and pressure. Thus the entropy change upon mixing $N_A$ molecules of component A with $N_B$ molecules component B ($N = N_A + N_B$) is

$$\Delta S_m = S(N \text{ molecules of mixture at } T, P) \tag{8.1}$$

$$-S_A^o(N_A \text{ molecules of pure A at } T, P)$$

$-S_B^0(N_B$ molecules of pure B at $T$, $P$)

where superscript 0 denotes the property for pure substances. For monatomics, we apply eq. (7.7) to get

$$\Delta S_m = \frac{5}{2}Nk - \sum_{j=A,B} N_j k \, \ln(x_j P) + \frac{5}{2}Nk \, \ln T - \sum_{j=A,B} N_j k \, \ln C_j \tag{8.2}$$

$$- \frac{5}{2}N_A k + N_A k \, \ln P - \frac{5}{2}N_A k \, \ln T + N_A k \, \ln C_A$$

$$- \frac{5}{2}N_B k + N_B k \, \ln P - \frac{5}{2}N_B k \, \ln T + N_B k \, \ln C_B$$

$$= - k(N_A \, \ln x_A + N_B \, \ln x_B)$$

or

$$\frac{\Delta S_m}{N} = -k \sum_j x_j \, \ln x_j \tag{8.3}$$

The same results are obtainable for polyatomics. Other mixing properties could be similarly derived:

$$\Delta U_m = 0 \tag{8.4}$$

$$\Delta H_m = 0 \tag{8.5}$$

$$\Delta A_m = NkT \sum_j x_j \, \ln x_j \tag{8.6}$$

$$\Delta G_m = NkT \sum_j x_j \, \ln x_j \tag{8.7}$$

Also from the fact that $V=NkT/P$ and $V_j = N_j kT/P$, we obtain

$$\Delta V_m = 0 \tag{8.8}$$

These relations satisfy the so-called ideal solution conditions.

As in thermodynamics, the partial molar quantities are defined as the changes of the properties at constant temperature and pressure upon the addition of a unit mole of substance $j$, e.g.,

$$\bar{S}_j = \frac{\partial S}{\partial N_j} \bigg|_{P,T,N_k} \tag{8.9}$$

Differentiation of appropriate expressions gives

$$\bar{U}_j = U_j^o \tag{8.10}$$

$$\bar{H}_j = H_j^o$$

$$\bar{V}_j = V_j^o$$

$$\bar{S}_j = S_j^o - k \ln x_j$$

$$\bar{A}_j = A_j^o + kT \ln x_j$$

$$\bar{G}_j = G_j^o + kT \ln x_j$$

These relations conform to conventional thermodynamics.

## References

[1]    K. Huang, *Statistical Mechanics*, (Wiley, New York, 1963) p.154

[2]    American Petroleum Institute, Project 44, *Selected Values of Properties of Hydrocarbons and Related Compounds*, (Thermodynamics Research Center, Texas A&M University, College Station, Texas, 1975).

[3]    D.R. Stull, *Joint Army Navy Air Force (JANAF) Thermochemical Tables*, (Dow Chemical Company, 1965), No. PB-168370.

[4]    F.A. Aly and L.L. Lee, Fluid Phase Equil. **6**, 169 (1981).

[5]    A. Fakeeha, A. Kache, Z.U. Rehman, Y. Shoup, and L.L. Lee, Fluid Phase Equil. **11**, 225 (1983).

[6]    Z.U. Rehman and L.L. Lee, Fluid Phase Equil. **22**, 21 (1985).

## Exercises

1.    Carbon dioxide is a linear molecule

$$O=C=O$$

Its vibrational characteristic temperatures are 954, 954, 1890, and 3360K. Calculate the heat capacity, $C_p$ for temperatures given in the following table, using the exact formula (5.11) and the approximate formula (6.4). Compare with the experimental data and discuss the differences.

| T (K) | $C^p$, (Cal/gmol K) |
|---|---|
| 50.0 | 6.955 |
| 100.0 | 6.981 |
| 150.0 | 7.228 |
| 200.0 | 7.733 |
| 273.15 | 8.594 |

2.  Discuss the correct Boltzmann counting after referring to reference [1]. What is the role of uncertainty (entropy S) in deciding the Boltzmann counting? How does one obtain the correction factor for a mixture of $N_A + N_B$ molecules of species A and B?

3.  What is the difference in entropies of mixing at constant pressure as compared to the value at constant volume for ideal gases? Find the volume and pressure changes, if any, during either process.

# CHAPTER IV
# THE STRUCTURE OF LIQUIDS

## IV.1. *A Probabilistic Description*

In statistical mechanics the structure of a liquid (i.e., how the molecules are arranged spatially) is expressed in terms of probabilistic quantities: the *molecular distribution functions*. As we shall see shortly, the usefulness of the partition function defined earlier is limited to simple physical systems, such as the ideal gas and one-dimensional Ising model. For general interaction potentials $u(r)$ the partition functions are difficult to evaluate. As an alternative, the distribution function is developed. The distribution functions give the time-averaged spatial configuration of the molecules in the liquid (i.e., the liquid structure), and are accessible experimentally as the *radial distribution functions* obtained in X-ray and neutron scattering. This type of representation proves to be fruitful, since the original many-body problem is now presented in terms of few-body coordinates. The statistical description is necessitated by the unusually large number of particles involved. For example, for one mole of gas, there are more than $10^{23}$ molecules. It is impractical to account for them individually. It is sufficient, for statistical purposes, to know how these molecules are *distributed* with respect to one another. In equilibrium statistics, only the time-averaged *spatial* distribution is of interest. For time-dependent processes, the dynamics of the system is determined by an additional factor, namely the momentum distribution function. Both static and dynamic correlations should be considered, and a full phase space description is required.

The functions we shall introduce include the pair correlation function, $g(r)$ (specifying the spatial distribution of a pair of molecules), the direct correlation function, $C(r)$, the total correlation function, $h(r)$, and the background correlation function, $y(r)$. Triplet correlations will also be discussed due to their importance in dense liquids. For all practical purposes, these distribution functions will give an adequate description of the structure of liquids. There are three ways these distribution functions could be obtained: (i) from scattering experiments, (ii) from computer simulation, and (iii) from integral equation theories. We shall discuss these methods in the following chapters. Molecular distribution functions could be used to obtain the thermodynamic properties of the system. A full discussion will be given in Chapter V.

Before introducing these distributions, it is instructive to review probability theory [1]. We shall make use of the concepts of joint probability and marginal probability. Joint probability is a function of several random variables, expressing the likelihood of their simultaneous occurrence. Marginal probability is one with a reduced set of variables. Marginal probabilities are obtainable from joint probabilities by integration or summation over the unspecified variables. This reduction is designed to exhibit some gross but meaningful features of the original distributions. The relation between the probabilities can best be illustrated by a simple example: the throwing of two dice. The random variable (i.e., the face value) of the first die is denoted by $x$, that of the second by $y$. The joint probability, $p^{(2)}(x,y)$, is a function of both $x$ and $y$. It is the probability that die 1 has value $x$ and die 2 has value $y$. The probability is normalized by

$$\sum_{x=1}^{6}\sum_{y=1}^{6} p^{(2)}(x,y) = 1 \tag{1.1}$$

In the present case, $p^{(2)}(x,y) = 1/36$ because die 1 and die 2 are independent of each other. From this joint probability, we could formulate many marginal probabilities. For example, the probability of $x = 3$, irrespective of the values of $y$, is given by summing $p^{(2)}$ over all unspecified $y$ values:

$$p^{(1)}(x=3) = \sum_{y=1}^{6} p^{(2)}(x=3,y) \tag{1.2}$$

$p^{(1)}(x)$ is called the marginal probability in relation to $p^{(2)}(x,y)$. Different combinations with respect to the joint probability could be formed. For example, the sum $x+y = 5$ has a probability of 1/9; and the inequality $x > y$ has a probability of 5/36, all involving summation of probabilities of allowed combinations of $x$ and $y$. In case the random variables $x$ and $y$ assume continuous values, such as the velocities of two colliding particles or the surface positions of two adsorbed molecules, the summation used earlier should be replaced by integration over continuous intervals. In addition, we are not restricted to only two random variables. It is entirely legitimate to formulate joint probabilities $p^{(n)}(x_1, x_2, \ldots, x_n)$ of $n$ random variables.

## IV.2. *The n-Body Distribution Functions in Canonical Ensemble: Monatomic Fluids*

We have shown that in the phase space of a canonical ensemble the probability that the $i$th cell contains a phase point (i.e., occurrence of a particular spatial configuration and momentum distribution) is given by

$$p_i = \frac{\exp(-\beta\varepsilon_i)}{Z_N} \tag{2.1}$$

or, classically,

$$dp^{(N)}(\mathbf{p}^N, \mathbf{r}^N) = \frac{\exp[-\beta H_N(\mathbf{p}^N, \mathbf{r}^N)]\, d\mathbf{p}^N\, d\mathbf{r}^N}{N! h^{3N} Z_N} \tag{2.2}$$

That is, the differential $dp^{(N)}(\mathbf{p}^N, \mathbf{r}^N)$ gives the joint probability that the state (phase point) of the system is to be found in the differential volume formed by $d\mathbf{p}_1$

$dp_2 \cdots dp_N \, dr_1 \, dr_2 \cdots dr_N$ centered at the coordinates $(\mathbf{p}_1,...,\mathbf{p}_N, \mathbf{r}_1,...,\mathbf{r}_N)$.

Various specialized probabilities can be obtained from this general expression by *marginalizing* over the phase space coordinates [2] [3]. For monatomic molecules with spherically symmetric potentials, the Hamiltonian $H_N$ is

$$H_N(\mathbf{p}^N, \mathbf{r}^N) = \sum_{i=1}^{N} \frac{p_i^2}{2m} + V_N(\mathbf{r}_1,..., \mathbf{r}_N) \tag{2.3}$$

The classical canonical partition function is then

$$Z_N = \frac{Q_N}{N! \Lambda^{3N}} \tag{2.4}$$

where the configurational integral, $Q_N$, is defined as

$$Q_N \equiv \int dr_1 \cdots dr_N \exp[-\beta V_N(\mathbf{r}_1, \ldots, \mathbf{r}_N)] \tag{2.5}$$

This quantity has the units of volume to the $N$th power, i.e., $V^N$. For an ideal gas, there is no energy of interaction: $Q_N = V^N$. The configurational marginal probability (i.e., the probability independent of momenta) is

$$dp^{(N)}(\mathbf{r}^N) = \frac{1}{Q_N} \exp[-\beta V_N(\mathbf{r}_1, \ldots, \mathbf{r}_N)] \, dr_1 \cdots dr_N \tag{2.6}$$

Note that the coordinates of the momenta $dp_1 \cdots dp_N$ have been integrated out. The term $dp^{(N)}(\mathbf{r}^N)$ gives the probability of a particular configuration (i.e., arrangement of molecules in space) irrespective of their momenta. Next, we single out a few of the molecules for consideration. Select $n$ ($n<N$) molecules and ask what the probability is for these $n$ molecules fixed at positions $\mathbf{r}_1, \mathbf{r}_2,...,\mathbf{r}_n$ irrespective of the positions of the remaining molecules. This is obtained by integration over the $N-n$ unspecified positions:

$$dp^{(n)}(\mathbf{r}^n) = \frac{1}{Q_N} dr_1 \cdots dr_n \left[ \int dr_{n+1} \, dr_{n+2} \cdots dr_N \exp(-\beta V_N(\mathbf{r}^N)) \right] \tag{2.7}$$

$dp^{(n)}(\mathbf{r}_1, \ldots, \mathbf{r}_n)$ is the differential probability that molecule 1 is found in the differential volume $dr_1$ centered at position $\mathbf{r}_1$, and at the same time molecule 2 in $dr_2$ at $\mathbf{r}_2,...$, up to $n$. On the other hand, if we are interested in the presence of a molecule, regardless of the labels, at position $\mathbf{r}_1$ (there are $N$ such molecules to choose from) and another molecule, also unlabeled, occupying $\mathbf{r}_2$ (there are $N-1$ such molecules), up to $n$, such a probability should be higher than the one defined in (2.7). In fact, it is $N(N-1)(N-2)...(N-n+1)$ times the value of (2.7)

$$dp_g^{(n)}(\mathbf{r}^n) \equiv \frac{N!}{(N-n)!} \, dp^{(n)}(\mathbf{r}^n) \tag{2.8}$$

$$= \frac{N!}{(N-n)!Q_N} \, d\mathbf{r}_1 \cdots d\mathbf{r}_n \left[ \int d\mathbf{r}_{n+1} \cdots d\mathbf{r}_N \, e^{-\beta V_N} \right]$$

This new probability $p_g^{(n)}$ is called the *generic probability*, since it counts molecules without reference to their labels. This picture is realistic, as with identical molecules permutations do not change the molecular state.

*DEFINITION:* **The n-Body Generic Density Function.** $\rho^{(n)}(\mathbf{r}_1, \ldots, \mathbf{r}_n)$ is defined as the probability density of $p_g^{(n)}(\mathbf{r}^n)$; i.e.,

$$\rho^{(n)}(\mathbf{r}_1, \ldots, \mathbf{r}_n) \equiv \frac{N!}{(N-n)!Q_N} \int d\mathbf{r}_{n+1} \cdots d\mathbf{r}_N \, e^{-\beta V_N(\mathbf{r}^N)} \tag{2.9}$$

The most commonly used density functions are the one-body, two-body, and three-body density functions. The one-body density function $\rho^{(1)}(\mathbf{r}_1)$ is also called the *singlet density*:

$$\rho^{(1)}(\mathbf{r}_1) = \frac{N}{Q_N} \int d\mathbf{r}_2 \cdots d\mathbf{r}_N \, \exp(-\beta V_N(\mathbf{r}_1, \ldots, \mathbf{r}_N)) \tag{2.10}$$

The two-body density function is called the *pair density*:

$$\rho^{(2)}(\mathbf{r}_1, \mathbf{r}_2) = \frac{N(N-1)}{Q_N} \int d\mathbf{r}_3 \cdots d\mathbf{r}_N \, \exp[-\beta V_N(\mathbf{r}_1, \ldots, \mathbf{r}_N)] \tag{2.11}$$

The three-body density function is called the *triplet density*:

$$\rho^{(3)}(\mathbf{r}_1, \mathbf{r}_2, \mathbf{r}_3) = \frac{N(N-1)(N-2)}{Q_N} \int d\mathbf{r}_4 \cdots d\mathbf{r}_N \, \exp[-\beta V_N(\mathbf{r}_1, \ldots, \mathbf{r}_N)] \tag{2.12}$$

Before discussing the properties of these $\rho^{(n)}$, we define another class of reduced probabilities: the *correlation functions*.

*DEFINITION:* **The n-Body Correlation Function.** $g^{(n)}$ is defined as the ratio

$$g^{(n)}(\mathbf{r}_1, \ldots, \mathbf{r}_n) \equiv \frac{\rho^{(n)}(\mathbf{r}_1, \ldots, \mathbf{r}_n)}{\rho^{(1)}(\mathbf{r}_1) \cdots \rho^{(1)}(\mathbf{r}_n)} \tag{2.13}$$

In particular, the *singlet correlation function* $g^{(1)}$ is simply given as

$$g^{(1)}(\mathbf{r}_1) = \frac{\rho^{(1)}(\mathbf{r}_1)}{\rho^{(1)}(\mathbf{r}_1)} = 1 \tag{2.14}$$

The *pair correlation function* (pcf) $g^{(2)}(\mathbf{r}_1, \mathbf{r}_2)$ is

$$g^{(2)}(\mathbf{r}_1,\mathbf{r}_2) = \frac{\rho^{(2)}(\mathbf{r}_1,\mathbf{r}_2)}{\rho^{(1)}(\mathbf{r}_1)\rho^{(1)}(\mathbf{r}_2)} \tag{2.15}$$

and the *triplet correlation function* is

$$g^{(3)}(\mathbf{r}_1,\mathbf{r}_2,\mathbf{r}_3) = \frac{\rho^{(3)}(\mathbf{r}_1,\mathbf{r}_2,\mathbf{r}_3)}{\rho^{(1)}(\mathbf{r}_1)\rho^{(1)}(\mathbf{r}_2)\rho^{(1)}(\mathbf{r}_3)} \tag{2.16}$$

These correlation functions, in contrast to the density functions defined earlier, are dimensionless. For example, the pair density has the units of density squared, whereas the pcf is dimensionless.

## IV.3. *Properties of Distribution Functions*

### IV.3.1. THE n-BODY DENSITY FUNCTIONS

From definition (2.9) we can derive the following properties for the n-body density functions:

*Recursion Relation*

Given $\rho^{(n+1)}$, we obtain $\rho^{(n)}$ by integrating over $dr_{n+1}$:

$$\int dr_{n+1}\, \rho^{(n+1)}(\mathbf{r}_1, \ldots ,\mathbf{r}_n,\mathbf{r}_{n+1}) \tag{3.1}$$

$$= \frac{N!}{(N-n-1)!Q_N}\int dr_{n+1} \cdots dr_N \exp(-\beta V_N)$$

$$= (N-n)\, \rho^{(n)}(\mathbf{r}_1, \ldots ,\mathbf{r}_n)$$

In particular, when $n=0$,

$$\int dr_1\, \rho^{(1)}(\mathbf{r}_1) = N \tag{3.2}$$

Namely, the integration of the singlet density over the whole volume gives the total number of molecules, $N$. The interpretation is like this: since $\rho^{(1)}(\mathbf{r})$ gives the probability that there is a molecule at $\mathbf{r}$, integration over all the volume simply counts the total number of molecules $N$. (Recall that $\rho^{(1)}$ is a generic probability.)

*Normalization Condition*

The integration of $\rho^{(n)}$ over $dr^n$ gives

$$\int dr_1 \cdots dr_n\, \rho^{(n)}(\mathbf{r}_1,...,\mathbf{r}_n) \tag{3.3}$$

$$= \frac{N!}{(N-n)!Q_N} \int d\mathbf{r}_1 \cdots d\mathbf{r}_{n+1} \cdots d\mathbf{r}_N \exp(-\beta V_N)$$

$$= N(N-1)\cdots(N-n+1) = \frac{N!}{(N-n)!}$$

This condition gives the number of ways that one can choose $n$ molecules from $N$. The above conditions are derived from definition (2.9) and are based on purely mathematical grounds. In the following, we introduce some physical arguments. For an ideal gas, $V_N = 0$; thus $Q_N = V^N$.

*The Ideal-Gas Limit*

$$\rho^{(n)}(\mathbf{r}_1,...,\mathbf{r}_n) = \frac{N!}{(N-n)!Q_N} \int d\mathbf{r}_{n+1} \cdots d\mathbf{r}_N \cdot 1 \tag{3.4}$$

$$= \frac{N! V^{N-n}}{(N-n)! V^N} = \frac{N(N-1)(N-2) \cdots (N-n-1)}{V^n}$$

$$= (\frac{N}{V})^n (1 - \frac{1}{N})(1 - \frac{2}{N}) \cdots (1 - \frac{N-n-1}{N}) = \rho^n (1 - O(\frac{1}{N}))$$

Therefore for an ideal gas, the $n$-body distribution function is simply the number density $\rho$ raised to the power $n$.

If we assume that the total potential energy $V_N(\mathbf{r}_1,\mathbf{r}_2, \ldots ,\mathbf{r}_N)$ is pairwise additive and the pair interaction potential $u(r_{ij})$ is short-ranged (i.e., $u(r_{ij}) = 0$ for $r_{ij}$ large, where $r_{ij}$ is the intermolecular distance), we have the

*Limit of Large Separation*

$$\lim_{\mathbf{r}_{n+1} \to \infty} \rho^{(n+1)}(\mathbf{r}_1, \ldots ,\mathbf{r}_n,\mathbf{r}_{n+1}) \tag{3.5}$$

$$= \frac{N!}{(N-n-1)!Q_N} \int d\mathbf{r}_{n+2} \cdots d\mathbf{r}_N \lim_{\mathbf{r}_{n+1} \to \infty} \exp[-\beta \sum_{i<j}^{N} \sum^{N} u(r_{ij})]$$

As the distance $\mathbf{r}_{n+1}$ of the $(n+1)$th molecule goes to infinity, this molecule for all practical purposes disappears from the system. All factors $u(r_{n+1,j})$ involving the position $\mathbf{r}_{n+1}$ vanish. To keep track of the number of molecules, we shall indicate the total number by a subscript.

$$\lim_{\mathbf{r}_{n+1} \to \infty} V_N(\mathbf{r}_1,...,\mathbf{r}_n,\mathbf{r}_{n+1},\mathbf{r}_{n+2}, \ldots ,\mathbf{r}_N) = V_{N-1}(\mathbf{r}_1,...,\mathbf{r}_n,\mathbf{r}_{n+2},...,\mathbf{r}_N) \tag{3.6}$$

$$\lim_{\mathbf{r}_{n+1} \to \infty} Q_N = VQ_{N-1} \tag{3.7}$$

The volume factor, $V$, has appeared due to integration over $d\mathbf{r}_{n+1}$. Equation (3.5) can now be written as

$$\lim_{r_{n+1} \to \infty} \rho^{(n+1)}(\mathbf{r}_1,...,\mathbf{r}_{n+1}) \tag{3.8}$$

$$= \frac{N}{V} \frac{(N-1)!}{(N-n-1)!Q_{N-1}} \int dr_{n+2} \cdots dr_N \, e^{-\beta V_{N-1}(\mathbf{r}_1, \ldots, \mathbf{r}_n, \mathbf{r}_{n+2},...,rr_N)}$$

$$= \rho \, \rho^{(n)}(\mathbf{r}_1,...,\mathbf{r}_n)$$

where we have made the assumption

$$\frac{(N-1)!}{(N-1-n)!Q_{N-1}} \int dr_{n+2} \cdots dr_N \exp(-\beta V_{N-1}) \tag{3.9}$$

$$= \frac{N!}{(N-n)!Q_N} \int dr_{n+1} \cdots dr_N \exp(-\beta V_N)$$

For $N$ large (on the order of $10^{23}$), this assumption is valid. As a molecule is removed from the system, the $(n+1)$th-order distribution function is reduced to the $n$th-order distribution function. A usual practice to render the results independent of the system size is to take the thermodynamic limit. The limit is defined as $\lim N \to \infty$, and the density, $\rho = N/V$, is held constant. Therefore (3.9) is valid in the thermodynamic limit.

In the absence of external potential, the $n$-body density functions are independent of the choice of origin, i.e., translation of the origin does not disturb the density functions. This property is called *translation-invariance* (or homogeneity). For a translation-invariant function $f(\mathbf{r})$

$$f(\mathbf{r}-\mathbf{a}) = f(\mathbf{r}) \tag{3.10}$$

where $\mathbf{a}$ is an arbitrary vector. Furthermore, if $f(\mathbf{r}) = f(r,\theta,\phi) = f(r)$, $f$ is called *isotropic*, since it ceases to depend on the angles $\theta$ and $\phi$. Note that $r$, $\theta$ and $\phi$ are the spherical coordinates of the vector $\mathbf{r}$. In a uniform fluid with no external forces, all distribution functions are translation invariant. Furthermore, they are isotropic, if the potentials of interaction are isotropic. For example, the pair density

$$\rho^{(2)}(\mathbf{r}_1, \mathbf{r}_2) = \rho^{(2)}(|\mathbf{r}_1 - \mathbf{r}_1|, |\mathbf{r}_2 - \mathbf{r}_1|) = \rho^{(2)}(0, |\mathbf{r}_2 - \mathbf{r}_1|) \tag{3.11}$$

$$= \rho^{(2)}(\mathbf{r}_{12}) = \rho^{(2)}(r_{12},\theta,\phi) = \rho^{(2)}(r_{12})$$

where $r_{12} = |\mathbf{r}_2 - \mathbf{r}_1|$. For the triplet density, a similar consideration leads to

$$\rho^{(3)}(\mathbf{r}_1,\mathbf{r}_2,\mathbf{r}_3) = \rho^{(3)}(r_{12},r_{23},r_{31}) \tag{3.12}$$

## IV.3.2. DISTRIBUTION FUNCTIONS FOR ISOTROPIC FLUIDS

For isotropic functions, the angle dependence disappears. Only radial distances remain. Simplifications of the density functions result.

### The Singlet Density

For isotropic, homogeneous fluids,

$$\rho^{(1)}(\mathbf{r}_1) = N \frac{\int d\mathbf{r}_2 \cdots d\mathbf{r}_N \, e^{-\beta V_N(\mathbf{r}_1,\dots,\mathbf{r}_N)}}{\int d\mathbf{r}_1 \, d\mathbf{r}_2 \cdots d\mathbf{r}_N \, e^{-\beta V_N(\mathbf{r}_1,\dots,\mathbf{r}_N)}} \tag{3.13}$$

Since $V_N(\mathbf{r}_1,\dots,\mathbf{r}_N)$ is translation invariant, we may change to relative coordinates by moving the origin from 0 to $\mathbf{r}_1$: i.e., $\mathbf{r}_1' = \mathbf{r}_1$, $\mathbf{r}_2' = \mathbf{r}_2-\mathbf{r}_1$, $\mathbf{r}_3' = \mathbf{r}_3 - \mathbf{r}_1$,..., $\mathbf{r}_N' = \mathbf{r}_N-\mathbf{r}_1$.

$$\rho^{(1)}(r_1) = N \frac{\int d\mathbf{r}_2' \cdots d\mathbf{r}_N' \, \exp(-\beta V_N(\mathbf{r}_2',\mathbf{r}_3',\dots,\mathbf{r}_N'))}{\int d\mathbf{r}_1' \, d\mathbf{r}_2' \cdots d\mathbf{r}_N' \, \exp(-\beta V_N(\mathbf{r}_2',\mathbf{r}_3',\dots,\mathbf{r}_N'))} \tag{3.14}$$

$$= \frac{N}{V} = \rho$$

Note that in the denominator the integration with respect to $d\mathbf{r}_1'$ has been carried out, since the integrand was independent of the coordinate $\mathbf{r}_1'$. The result is that the singlet density is equal to the number density $\rho$.

Another consequence of isotropy for the pcf is:

### The Pair Correlation Function

$$g^{(2)}(r_{12}) \equiv \frac{\rho^{(2)}(r_{12})}{\rho^2} \tag{3.15}$$

Since the pcf is a function only of the intermolecular distance $r_{12}$, we could write it as $g^{(2)}(r_{12})$. This quantity corresponds to the so-called radial distribution function (rdf), $g(r)$, obtained from X-ray and neutron scattering experiments. This identification endows $g^{(2)}(r)$ with physical significance. Figures IV.1 and 2 exhibit the rdf for monatomic argon and diatomic nitrogen. We see the general oscillatory behavior of $g(r)$. These oscillations correspond to successive shells of neighbors surrounding a central molecule: the first peak corresponds to the first layer of neighbors, the second peak to the next shell of neighbors. The ensuing peaks become weaker and less defined because diffusional motions in a liquid tend to randomize higher neighbors.

### The n-body Correlation Functions

As $\rho^{(1)}(r) = \rho$ for isotropic fluids, definition (2.13) for $n$-body correlation function becomes

$$g^{(n)}(\mathbf{r}_1,\dots,\mathbf{r}_n) = \rho^{-n}\rho^{(n)}(r_{12},r_{13}, \dots ,r_{1n},r_{23},\dots,r_{n-1,n}) \tag{3.16}$$

The same recursion relation for density functions is applicable to the correlation functions.

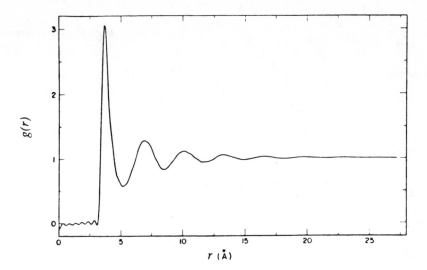

FIGURE IV.1. *The radial distribution function g(r) of argon at 85 K obtained from the neutron scattering experiments of Yarnell et al. (Phys. Rev. A7, 2130 (1973)).*

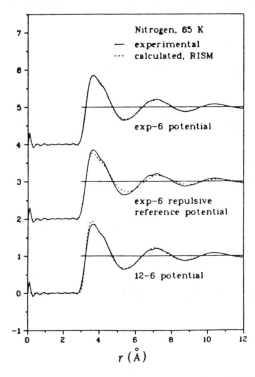

FIGURE IV.2. *Atom-atom pair correlation functions of nitrogen at 65 K obtained from neutron scattering (Narten et al. J. Chem. Phys. 73, 1248 (1980)). The lines are experimental results. The dotted curves are theoretical calculations from the reference interaction site model.*

*Recursion*

$$\int d\mathbf{r}_{n+1} g^{(n+1)}(\mathbf{r}_1, \ldots, \mathbf{r}_{n+1}) = V g^{(n)}(\mathbf{r}_1,\ldots,\mathbf{r}_n)(1 - O(\frac{1}{N})) \tag{3.17}$$

*Normalization*

$$\int d\mathbf{r}_1 \cdots d\mathbf{r}_n \, g^{(n)}(\mathbf{r}_1, \ldots, \mathbf{r}_n) = V^n(1 - O(\frac{1}{N})) \tag{3.18}$$

*Ideal Gas Limit*

$$g^{(n)}(\mathbf{r}_1, \ldots, \mathbf{r}_n) = 1 \tag{3.19}$$

*Limit of Large Separation*

$$\lim_{r_{n+1}\to\infty} g^{(n+1)}(\mathbf{r}_1, \ldots, \mathbf{r}_{n+1}) = g^{(n)}(\mathbf{r}_1, \ldots, \mathbf{r}_n) \tag{3.20}$$

Note that eqs. (3.17-20) are valid only for translation-invariant $V_N$ and for large $N$. Two other important limiting properties of $g^{(n)}$ are the high temperature and low density limits.

*High Temperature Limit*

As temperature $T \to \infty$, $\beta=1/kT \to 0$, therefore $\beta V_N \to 0$. From the definitions of $g^{(n)}$ and $\rho^{(n)}$, we have

$$\lim_{T\to\infty} g^{(n)}(\mathbf{r}_1, \ldots, \mathbf{r}_n) = \rho^{-n} \frac{N!}{(N-n)!} \frac{\int d\mathbf{r}_{n+1} \cdots d\mathbf{r}_N \exp(-\beta V_N)}{\int d\mathbf{r}_1\ldots d\mathbf{r}_N \exp(-\beta V_N)} \tag{3.21}$$

$$= 1 - O(\frac{1}{N})$$

$$\lim_{T\to\infty} \rho^{(n)}(\mathbf{r}_1, \ldots, \mathbf{r}_n) = \lim_{\beta\to 0} \frac{N!}{(N-n)!} \frac{\int d\mathbf{r}_{n+1} \cdots d\mathbf{r}_N \exp(-\beta V_N)}{\int d\mathbf{r}_1 \cdots d\mathbf{r}_N \exp(-\beta V_N)} \tag{3.22}$$

$$= \frac{N!}{(N-n)!} \frac{V^{N-n}}{V^N} = \left[\frac{N}{V}\right]^n (1 - (\frac{1}{N})) = \rho^n(1 - O(\frac{1}{N}))$$

In particular,

$$\lim_{T\to\infty} g^{(2)}(r) = 1 - O(\frac{1}{N}) \tag{3.23}$$

In the thermodynamic limit, $N \to \infty$ (at constant $\rho$), terms of order $O(1/N)$ vanish.

**Low Density Limit**

As $\rho \to 0$, it can be shown from the cluster expansion of $g^{(2)}$ (see Chapter V) that

$$\lim_{\rho \to 0} g^{(2)}(r) = \exp[-\beta u(r)] \tag{3.24}$$

where $u(r)$ is the pair potential.

## IV.4. *Other Correlation Functions*

In liquid theories, correlation functions other than the density functions are also used: notably, the total correlation function $h(r)$, giving the number fluctuations; the direct correlation function (dcf) $C(r)$, giving the isothermal compressibility; and the background correlation function (bcf) $y(r)$, exhibiting the indirect correlation between molecules. These functions are related, in one way or another, to the pair density. Their physical significance is not always transparent. The usefulness justifies their existence.

*DEFINITION:* **The Total Correlation Function $h$.** The total correlation function $h(\mathbf{r}_1,\mathbf{r}_2)$ is defined as

$$h(\mathbf{r}_1,\mathbf{r}_2) \equiv g^{(2)}(\mathbf{r}_1,\mathbf{r}_2) - 1 \tag{4.1}$$

For isotropic, homogeneous fluids with no external potential, $h(\mathbf{r}_1,\mathbf{r}_2) = h(|\mathbf{r}_2 - \mathbf{r}_1|)$ $= h(r_{12}) = g(r_{12}) - 1$. At low densities, $h(r) = \exp[-\beta u(r)] - 1$. For an ideal gas, $h(r) = 0$.

Another function $y(r)$ often appears in applications, e.g., in the perturbation theories of fluids. It is called the *y-correlation function*, the *cavity distribution function*, the *indirect correlation function*, or the *background correlation function*, since its cluster diagram [4] shows that $y(\mathbf{r}_1,\mathbf{r}_2)$ gives the indirect correlation of molecule 1 with molecule 2 via Mayer bonds that connect all molecules except 1 and 2.

*DEFINITION:* **The Background Correlation Function.** The background correlation function $y(\mathbf{r}_1,\mathbf{r}_2)$ is defined as

$$y(\mathbf{r}_1,\mathbf{r}_2) \equiv g^{(2)}(\mathbf{r}_1,\mathbf{r}_2) \exp[\beta u(\mathbf{r}_1,\mathbf{r}_2)] \tag{4.2}$$

At low densities, $y(r) = 1$. The y-function is continuous for all $r$ values (see Figure IV.3) even for hard-core fluids. Therefore it is much used in numerical work where "discontinuities" are avoided at all costs.

Another important correlation function $C(r)$ is defined below:

*DEFINITION:* **The Ornstein-Zernike Relation.** The direct correlation function, $C^{(2)}(\mathbf{r}_1,\mathbf{r}_2)$ or $C(\mathbf{r}_1,\mathbf{r}_2)$, is defined in terms of the total correlation function $h(\mathbf{r}_1,\mathbf{r}_2)$ through a convolution integral

$$h(\mathbf{r}_1,\mathbf{r}_2) - C(\mathbf{r}_1,\mathbf{r}_2) \equiv \rho \int d\mathbf{r}_3 \, h(\mathbf{r}_1,\mathbf{r}_3)C(\mathbf{r}_3,\mathbf{r}_2) \tag{4.3}$$

This relation is called the *Ornstein-Zernike* (OZ) relation. It defines the direct correlation function in terms of $h(\mathbf{r}_1,\mathbf{r}_2)$. For isotropic fluids with translation invariance, we obtain

$$h(r) - C(r) = \rho\!\int ds\; h(s)C(|\mathbf{r}{-}\mathbf{s}|) \tag{4.4}$$

or, equivalently,

$$h(r) - C(r) = \rho\!\int ds\; h(|\mathbf{r}{-}\mathbf{s}|)C(s) \tag{4.5}$$

The convolution integral can be solved by standard mathematical techniques (e.g., Fourier or Laplace transforms). Figure IV.4 shows some typical dcf for a continuous potential. This function does not have a simple physical meaning. Obviously, $C(\mathbf{r}_1,\mathbf{r}_2)$ is related to the compressibility derivatives $\partial\beta\mu/\partial\rho$ in thermodynamics (see below). It could be considered as an inhomogeneous counterpart of the derivative of the chemical potential. In addition, it serves as the matrix inverse to the total correlation function $h(\mathbf{r}_1,\mathbf{r}_2)$, which gives the number fluctuations in a grand canonical ensemble. For mixtures, these relations constitute the basis of the *Kirkwood-Buff solution theory*. The usefulness of the OZ relation resides in its correspondence to the compressibility-fluctuation reciprocity relation.

In nonuniform systems, one often employs the one-body dcf, $C^{(1)}(r)$.

*DEFINITION:* **The Singlet Direct Correlation Function**

$$C^{(1)}(\mathbf{r}) \equiv \ln[\rho^{(1)}(\mathbf{r})\Lambda^3] + \beta[w(\mathbf{r}) - \mu] \tag{4.6}$$

where $\rho^{(1)}$ is the singlet density, $w(r)$ is the one-body potential in an inhomogeneous system, and $\mu$ is the chemical potential. This quantity plays an important role in the theory of liquids. It is used to derive the integral equations (e.g., Percus-Yevick [5] and hypernetted chain equations [6]). It is related to the potential distribution theorem [7], and could be used in the polyatomic interaction site model for the derivation of a new OZ relation. Lebowitz and Percus [8] have shown that $C^{(1)}$ could be differentiated to give the two-body dcf

$$C^{(2)}(\mathbf{r}_1,\mathbf{r}_2) = \frac{\delta C^{(1)}(\mathbf{r}_1)}{\delta\rho^{(1)}(\mathbf{r}_2)} \tag{4.7}$$

where $\delta/\delta\rho$ is a derivative for functionals. A physical interpretation of $C^{(1)}$ can be made. It is the *inhomogeneous* counterpart of the chemical potential, i.e., the chemical potential of a system in an external field where there exist spatial density variations (inhomogeneities). Note that $\ln(\rho^{(1)}\Lambda^3)$ is the ideal gas chemical potential. For $w{=}0$, $C^{(1)}$ gives the configurational part of the chemical potential. Since any external field $w(\mathbf{r})$ would have the effect of counterbalancing the chemical potential, $C^{(1)}$ is the *remainder* of the external driving forces over the homogeneous nonideal chemical potential.

*DEFINITION:* **The Structure Factor $S(k)$.** The function obtained in scattering experiments is related to the total correlation function via a Fourier transform:

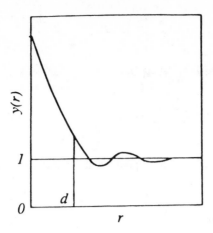

FIGURE IV.3. The background correlation function $y(r)$ for hard spheres. The radial distribution function $g(r)$ coincide with $y(r)$ for $r > d$ (= the hard sphere diameter).

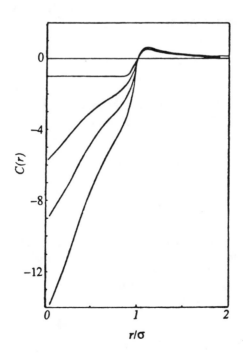

FIGURE IV.4. The direct correlation function $C(r)$ at four different densities. The lowest curve corresponds to the highest density.

$$\rho h(r) = \frac{1}{(2\pi)^3} \int d\mathbf{k} \ e^{i\mathbf{k} \cdot \mathbf{r}} [S(k) - 1] \tag{4.8}$$

$$= \frac{1}{2\pi^2 r} \int_0^\infty dk \ k \ \sin(kr)[S(k) - 1]$$

This simple relation connects the experimental structure factor to the probabilistic function, $g(r) = 1 + h(r)$.

## IV.5. *The Meaning of* $g^{(2)}(r)$

The rdf was first used in scattering experiments for the analysis of the fluid structure. We could explain this in probabilistic terms. Given a molecule in the differential volume $d\mathbf{r}_1$ at position $\mathbf{r}_1$, the probability of finding any one of the $N-1$ remaining molecules in $d\mathbf{r}_2$ at $\mathbf{r}_2$ is related to the two-body specific distribution function $p^{(2)}(\mathbf{r}_1, \mathbf{r}_2)$ by

$$(N-1)p^{(2)}(\mathbf{r}_1, \mathbf{r}_2) \ d\mathbf{r}_1 \ d\mathbf{r}_2 = \frac{\rho^2}{N} \ g^{(2)}(r_{12}) \ d\mathbf{r}_1 \ d\mathbf{r}_2 \tag{5.1}$$

Carrying out a transformation of variables for the position vectors $\mathbf{r}_1' = \mathbf{r}_1$ and $\mathbf{r}_{12}' = |\mathbf{r}_2 - \mathbf{r}_1|$ gives the probability that a specific molecule is in $d\mathbf{r}_1'$ at $\mathbf{r}_1'$ while any one of the remaining $N-1$ molecules is found in $d\mathbf{r}_2'$ at $\mathbf{r}_2'$:

$$\frac{\rho^2}{N} \ g^{(2)}(r_{12}) \ d\mathbf{r}_1' \ d\mathbf{r}_{12}' \tag{5.2}$$

We let the given molecule move over the entire volume (i.e., $\mathbf{r}_1'$ could be anywhere in $V$). Integration of (5.2) with respect to $d\mathbf{r}_1'$ gives

$$\int_V d\mathbf{r}_1' \frac{\rho^2}{N} \ d\mathbf{r}_{12}' g^{(2)}(r_{12}') = \frac{\rho^2}{N} V g^{(2)}(r_{12}') \ d\mathbf{r}_{12}' \tag{5.3}$$

$$= \rho g^{(2)}(r_{12}') \ d\mathbf{r}_{12}'$$

Equation (5.3) gives the probability that a second molecule is to be found in the volume $d\mathbf{r}_{12}'$ at a distance $r_{12}'$ from a central molecule. For $g^{(2)}$ independent of orientations, we could form a spherical shell of thickness $dr$ at a distance $r \ (= r_{12})$ from the central molecule and calculate the probability of finding a second molecule (among the remaining $N-1$ molecules) in the annular space $4\pi r^2 \ dr$, i.e.,

$$\int_0^\pi d\theta \sin\theta \int_0^{2\pi} d\phi \ r^2 \ \rho g^{(2)}(r) \ dr = \rho g^{(2)}(r) \ 4\pi r^2 \ dr \tag{5.4}$$

This expression gives the number of molecules in the spherical shell of thickness $dr$ at distance $r$ from a central molecule. Integrating (5.4) over a coordination distance $L$ gives the so-called coordination number $N(L)$:

$$N(L) \equiv \int_0^L \rho g(r)\, 4\pi r^2\, dr \tag{5.5}$$

$N(L)$ counts the number of molecules inside a sphere of radius $L$ surrounding the central molecule. Namely, it is a counter of neighbors. For a crystalline solid, $L$ is a well-defined quantity (i.e., equal to the lattice spacing). In a face-centered cubic lattice, each central molecule has 12 nearest neighbors: $N(L)=12$. In a liquid, on the other hand, constant diffusional motion "melts" the lattice structure, and the coordination distance $L$ is "smeared." The coordination number for the first neighbor shell, being dependent on the range $L$, is not well defined. However, $N(L)$ continues to be used in liquid theory, giving valuable statistical information. For liquid argon, the coordination number is depicted in Figure IV.5. When integrated to $L=\infty$, (5.5) gives the value $N-1$, i.e., the total number of neighbors.

We have shown that $g(r)$ gives the distribution of molecules at radial distances $r$ from a given center. Experimentally, it corresponds to the radial distribution function, $g(r)$. We shall thus use the notation $g^{(2)}(r) = g(r)$ interchangeably. The liquid structure thus exhibited has a *local* flavor; i.e., it is from the perspective of a central molecule. This picture coexists with the global bulk density as an indicator of distribution and

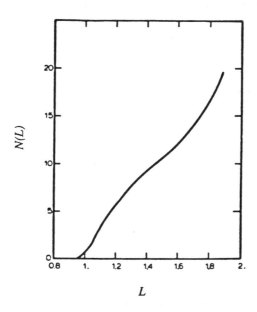

*FIGURE IV.5. Number of solvent molecules $N(L)$ present within a distance $L$ from a solute molecule obtained from the molecular dynamics simulation of Lennard-Jones particles.*

gives a microscopic view inside a macroscopic system. The functional form of $g(r)$ is dependent on the interaction potential. As an illustration, we inspect the qualitative behavior of $g(r)$ for some commonly used potentials. Figure IV.6 shows the rdf for HS, SW, and LJ potentials. Each one reflects some characteristics of its original potential. The rdf of HS gives a sharp peak at $r=d$ for the first neighbors. Longer-range oscillations represent layers of successive neighborhoods. For SW, two sharp cusps are in evidence, one arising from the repulsive discontinuity, and the other from the attractive discontinuity, in the pair potential. For LJ, the transition from repulsion to attraction is smooth. This is reflected in the rdf being continuous and smooth around the corners.

The pcf shows distinct structural differences for the solid, liquid, and gaseous states. For crystalline solids, molecules are regularly arranged on the lattice sites. They could vibrate with respect to the equilibrium positions, but they rarely execute large-scale translational motions. The pcf of the solid is long-ranged. It oscillates over great distances without approaching unity (see Figure IV.7(d)). Thus for crystalline solids, the rdf exhibits *long-range order.*

On the other hand, the pcf of dilute gas gives one prominent peak at the location $r_0$ corresponding to the minimum in the pair potential (see Figure IV.7(b)). This is the distance where a neighboring molecule is most likely to be found. Beyond $r_0$, the chances of finding another molecule are no better than that given by the bulk density $\rho$ (i.e., $g(r)=1$). For an ideal gas, $g(r)$ is featureless. Since the pair potential is zero, $g(r)$ is unity everywhere (Figure IV.7(a)). From a structural point of view, ideal gas has *no structure.* For real gases, there exists a shell of nearest neighbors.

In the liquid state (inclusive of dense gas), we encounter a situation intermediate between the solid and the gas. The pcf exhibits several peaks, indicating the presence of first-, second-, and third-neighbor shells (Figure IV.7(c)). The molecules are not as sparsely distributed as in the dilute gases. However, the structure is less regular than in a solid. For limited distances, liquids (and dense gases) possess short-range order. Structurally, they represent the state intermediate of orderliness (the crystalline state) and randomness (the gaseous state).

## IV.6. *The n-Body Distribution Functions in Grand Canonical Ensemble: Monatomic Fluids*

It is also possible to define the n-body distribution functions in a grand canonical ensemble GCE. The GCE is the ensemble for open systems. The correlation functions are well behaved in a GCE; e.g., it is possible to obtain the isothermal compressibility, $K_T$, from the total correlation function in GCE.

As discussed earlier, the number of molecules in a GCE is weighted by a factor involving the chemical potential

$$P_N = \frac{[e^{\beta\mu}/\Lambda^3]^N Q_N}{N! \; \Xi} \tag{6.1}$$

The quantity in brackets is called the *activity* and will be denoted by $z$ ( $\equiv \exp(\beta\mu)/\Lambda^3$). The $N$-body probability function is now (contrast this formula with $dp^{(N)}$ of (2.2) in CE)

$$dp^{(N)}(\mathbf{p}^N, \mathbf{r}^N) = \frac{e^{\beta\mu N}}{N! h^{3N} \Xi} e^{-\beta H_N(\mathbf{p}^N, \; \mathbf{r}^N)} \; d\mathbf{p}^N \; d\mathbf{r}^N \tag{6.2}$$

We note, however, that $N$ is a variable in GCE since the number of molecules fluctuates

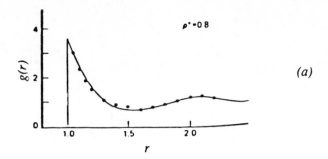

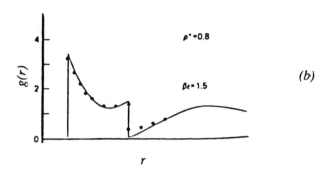

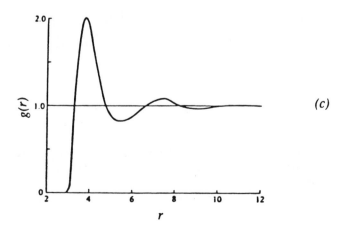

FIGURE IV.6. The pair correlation functions for different pair potentials. (a): g(r) for hard spheres; (b): g(r) for square-well potential; and (c): g(r) for Lennard-Jones potential.

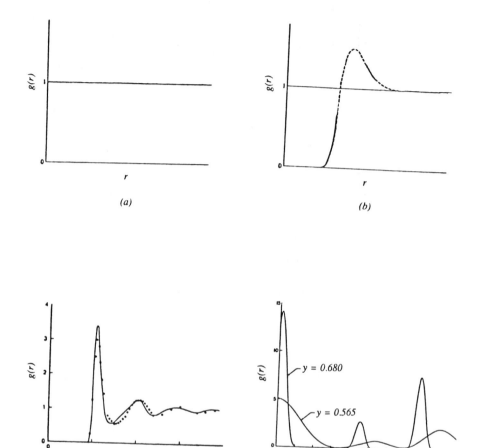

FIGURE IV.7. The effects of the fluid state on the radial distribution functions. (a): Ideal gas; (b): Dilute gas state; (c): Dense gas state (including liquids); and (d): Solid state (in this case, hard spheres near close packing). y = $\pi \rho d^3/6$ is the packing fraction. (Kincaid and Weis, Mol. Phys., 1977)

in an open system. The probability weight for such fluctuations is determined by the factor $\exp(\beta\mu N)$. If we write the $n$-body density in the canonical ensemble of $N$ molecules as $\rho_N^{(n)}$, the GCE $n$-body density is the weighted average

$$\rho^{(n)}(\mathbf{r}_1, \ldots, \mathbf{r}_n) = \sum_{N \geq n}^{\infty} P_N \rho_N^{(n)} = \langle \rho_N^{(n)} \rangle \tag{6.3}$$

$$= \frac{1}{\Xi} \sum_{N \geq n}^{\infty} \frac{z^N}{(N-n)!} \int d\{N-n\} \exp[-\beta V_N(\mathbf{r}^N)]$$

where $\Xi$ is the GCE partition function. We shall use the angular brackets $\langle \cdots \rangle$ to denote the average weighted by $P_N$. Note that the summation over the number of molecules starts at $N=n$, not at $N=0$, because the minimum number of molecules in the system is $n$ for $n$ particles.

The $n$-body correlation function is defined similarly:

$$g^{(n)}(\mathbf{r}^n) = \frac{\rho^{(n)}(\mathbf{r}^n)}{\rho^{(1)}(\mathbf{r}_1)\rho^{(1)}(\mathbf{r}_2) \cdots \rho^{(1)}(\mathbf{r}_n)} \tag{6.4}$$

The normalization condition is

$$\int dr_1 \cdots dr_n \, \rho^{(n)}(\mathbf{r}^n) = \frac{1}{\Xi} \frac{\sum z^N Q_N}{(N-n)!} \tag{6.5}$$

$$= \sum P_N \frac{N!}{(N-n)!} = \langle N^n \rangle \, [1 + O(\frac{1}{\langle N \rangle})]$$

In particular, for $n=2$, the normalization of the pair density gives

$$\int dr_1 \cdots dr_2 \, \rho^{(2)}(\mathbf{r}_1, \mathbf{r}_2) = \langle N^2 \rangle - \langle N \rangle \tag{6.6}$$

This is the well-known fluctuation formula giving the fluctuations of number of molecules in a GCE. It will be shown (Chapter V) that the isothermal compressibility is related to the number fluctuations by

$$\int dr_1 \, dr_2 \, [\rho^{(2)}(\mathbf{r}_1, \mathbf{r}_2) - \rho^{(1)}(\mathbf{r}_1)\rho^{(1)}(\mathbf{r}_2)] = \rho V[\rho k T K_T - 1] \tag{6.7}$$

where $K_T$ is the isothermal compressibility

$$K_T = -\frac{1}{V} \left[ \frac{\partial V}{\partial P} \right]_{N,T} \tag{6.8}$$

For isotropic, homogeneous fluids,

$$\rho^{(1)} = \frac{<N>}{V} = \rho \tag{6.9}$$

and

$$\rho^{(n)} = \rho^n g^{(n)} \tag{6.10}$$

In the ideal-gas limit ($V_N$=0)

$$\rho^{(n)} = \frac{1}{V^n} \sum P_N \frac{N!}{(N-n)!} = z^n = \rho^n \tag{6.11}$$

and

$$g^{(n)} = 1 \tag{6.12}$$

Also note that as all $r_{ij} \to \infty$ ($1=i<j=n$),

$$\lim_{\substack{\text{all } r_{ij} \to \infty}} g^{(n)} = 1 \tag{6.13}$$

We note that no term of order $O(1/N)$ appears in the GCE. This is in contrast to the CE case. At large separations, the GCE pcf approaches unity and is well behaved.

**MIXTURES**

As an example, we exhibit explicitly the density functions for mixtures. Without loss of generality, we consider a binary mixture of species A and B with $N_A$ and $N_B$, molecules each. There are three distinct pair densities, $\rho_{AA}^{(2)}$, $\rho_{AB}^{(2)}$, and $\rho_{BB}^{(2)}$. The chemical potentials $\mu_i$, ($i$=A,B), distinguish the species A and B. Thus (6.2) becomes

$$dp^{(N)}(\mathbf{p}^N, \mathbf{r}^N) = \frac{e^{\beta\mu_A N_A} e^{\beta\mu_B N_B}}{N_A! N_B! h^{3N} \Xi} \, e^{-\beta H_N(\mathbf{p}^N, \mathbf{r}^N)} \, d\mathbf{p}^N \, d\mathbf{r}^N \tag{6.14}$$

Therefore the pair density $\rho_{AA}^{(2)}$ is given by integration over the differentials $dr^{N-2} dp^N$ and multiplication by $N_A(N_A - 1)$

$$\rho_{AA}^{(2)}(12) = \frac{1}{\Xi} \sum_{N_A \geq 2}^{\infty} \sum_{N_B \geq 0}^{\infty} \frac{z_A^{N_A} z_B^{N_B}}{(N_A-2)! N_B!} \int d\,\{N-2_A\} \, \exp[-\beta V_N(\mathbf{r}^N)] \tag{6.15}$$

Similarly, $\rho_{AB}^{(2)}$ is given by

$$\rho_{AB}^{(2)}(12) = \frac{1}{\Xi} \sum_{N_A \geq 1}^{\infty} \sum_{N_B \geq 1}^{\infty} \frac{z_A^{N_A} z_B^{N_B}}{(N_A-1)!(N_B-1)!} \int d\,\{N-1_A-1_B\} \, \exp[-\beta V_N(\mathbf{r}^N)] \tag{6.16}$$

where $2_A$ denotes the vector positions of two molecules of species A.

SINGLET DIRECT CORRELATION FUNCTION

The singlet direct correlation function defined earlier also has a grand canonical ensemble representation. We consider first a nonuniform system (i.e. system with spatially variable densities) under the influence of an external potential $w(\cdot)$ (e.g., arising from interaction with a *wall*). The potential energy is given by

$$V_N(\{N\}) + \sum_{k=1}^{N} w(k) \equiv \sum_{i<j}^{N} \sum u(ij) + \sum_{k=1}^{N} w(k) \tag{6.17}$$

$$= V_{N-1}(\{N-1\}) + \sum_{i=1}^{N-1} u(i, N) + \sum_{k=1}^{N} w(k)$$

where we have singled the $N$th particle for special consideration. The sum $\sum_{i=1}^{N-1} u(i,N) \equiv \sum_{i=1}^{N-1} U(i)$ is the potential generated by a *test particle* "N" located at $\mathbf{r}_N$. We shall denote this source energy by $U(\cdot)$ (i.e., an external potential in addition to the wall potential $w(\cdot)$). The partition function under $w(\cdot)$ is

$$\Xi[w] \equiv \sum_{N\geq 0}^{\infty} \frac{z^N}{N!} \int d\{N\} \, \exp[-\beta V_N] \, \exp[-\beta \sum_{k=1}^{N} w(k)] \tag{6.18}$$

The singlet density is given by

$$\rho^{(1)}(\mathbf{r}_N;w) \equiv \frac{1}{\Xi[w]} \sum_{N\geq 1}^{\infty} \frac{z^N}{(N-1)!} \int d\{N-1\} \, \exp[-\beta V_N] \, \exp[-\beta \sum_{k=1}^{N} w(k)] \tag{6.19}$$

$$= z e^{-\beta w(\mathbf{r}_N)} \frac{1}{\Xi[w]} \sum_{N\geq 1}^{\infty} \frac{z^{N-1}}{(N-1)!} \int d\{N-1\} \, \exp[-\beta V_N] \, \exp[-\beta \sum_{k=1}^{N-1} w(k)]$$

Applying the definition of the singlet direct correlation function (4.6) and changing the index to $M=N-1$, we have

$$\exp[C^{(1)}(\mathbf{r}_N;w)] = \rho^{(1)}(\mathbf{r}_N;w)\Lambda^3 \exp[-\beta\mu + \beta w(\mathbf{r}_N)] \tag{6.20}$$

$$= \frac{1}{\Xi[w]} \sum_{M\geq 0}^{\infty} \frac{z^M}{M!} \int d\{M\} \, \exp[-\beta V_M] \, \exp[-\beta \sum_{j}^{M} u(jN)] \, \exp[-\beta \sum_{k=1}^{M} w(k)]$$

$$= \frac{1}{\Xi[w]} \sum_{M\geq 0}^{\infty} \frac{z^M}{M!} \int d\{M\} \, \exp[-\beta V_M] \, \exp[-\beta \sum_{k=1}^{M} [U(k) + w(k)]]$$

$$= \frac{\Xi[U+w]}{\Xi[w]}$$

Namely, $\Xi[U+w]$ is the partition function under the influence of the test particle plus the inhomogeneity $w(\cdot)$, whereas $\Xi[w]$ is the partition function under the influence of $w(\cdot)$ only. Simply stated, the singlet dcf accounts for the influence of a *test particle* located at $\mathbf{r}_N$. As we shall see, the chemical potential is related to the energy of inserting a test particle into a bath of fluid particles, i.e., the so-called Kirkwood charging process (see the following chapter). The singlet dcf appears naturally in this charging process.

## IV.7. *The Correlation Functions for Molecular Fluids:*
## *The Spherical Harmonic Expansions*

Most real molecules are polyatomic. The correlation functions introduced previously are suitable only for simple molecules, monatomic and nonpolar. For polyatomic and/or polar molecules, the nature of the intermolecular interaction changes. We note two prominent features of a molecular fluid:

1.  The interaction between a pair of polyatomic molecules depends not only on the positions but also on the relative orientation of the two molecules. For example, the interaction energy between two carbon dioxide molecules depends on the mutual orientation of the two molecules, in addition to the distance of separation. A number of landmark orientations or cross sections are noted in Table IV.1 – for example, parallel, antiparallel, end-to-end, $T$, and cross orientations. These orientations can be characterized by the values of the polar angle $\theta$, azimuthal angle $\phi$, and rotational angle $\chi$.

2.  The angle-dependent forces are caused by electrostatic interactions. These forces arise due to charge separation in the molecule: forming dipoles (HCl and CO), quadrupoles ($CO_2$, $Cl_2$, etc.), and octopoles (methane). Strong orientational forces arise in hydrogen-bonding fluids, such as water. In addition, quantum dispersion forces and induction forces are in operation. They all contribute to the interaction energy and eventually to the physical properties of the fluid.

To account for the orientational dependence of noncentral forces for nonspherical molecules, we must redefine the geometric representation. The angular coordinates are given by the Euler angles. For example, Figure IV.8 depicts the angles $\theta$ (the polar angle), $\phi$ (the azimuthal angle), and $\chi$ (the rotational angle) for a single dumbbell-like molecule.

*Table IV.1. Cross-Sections of Pairs of Linear Molecules*

| Configuration | Representation | $\theta_1$ | $\theta_2$ | $\phi$ |
|---|---|---|---|---|
| Head-to-end | →→ | 0 | 0 | – |
| Head-to-head | →← | 180 | 0 | – |
| Parallel | ↑ ↑ | 90 | 90 | – |
| Antiparallel | ↑ ↓ | 90 | 90 | 180 |
| Tee | ⊢ | 90 | 0 | – |
| Cross | + | 90 | 90 | 90 |

Since the two-body potential involves a pair of molecules, we need to describe the position and angle coordinates of two molecules simultaneously. This can be done by two-center body-fixed relative (or intermolecular) coordinates or by space-fixed (or laboratory) coordinates. We discuss the relative coordinates first.

The intermolecular axis $r_{12}$ is established by connecting two sites in the two molecules. This axis is taken as the z-axis for these two molecules (see Figure IV.9). The sites could be chosen to coincide with the atoms, centers of mass, or simply some geometric sites. Relative to this z-axis, molecule 1 forms a polar angle $\theta_1$, and molecules 2 forms $\theta_2$. The vectorial direction of molecule 1 is also chosen by convenience; in this case, it coincides with the bond vector L connecting the two atoms. The azimuthal angles $\phi_1$ and $\phi_2$ are determined by arbitrarily fixing a common x-axis. For

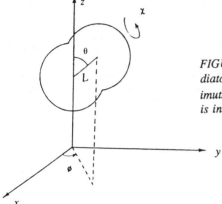

FIGURE IV.8. Definition of the Euler angles for a diatomic molecule. $\theta$ is the polar angle, $\phi$ is the azimuthal angle, and $\chi$ is the rotational angle (which is indistinguishable here for diatomics).

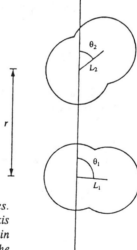

FIGURE IV.9. The two-body relative coordinates. The z-axis coincides with the intermolecular axis (from one site in molecule 1 to another site in molecule 2). $L_i$ is the bond length. and $r_{12}$ is the intermolecular distance. $\phi_2 - \phi_1$ is the difference between the azimuthal angles of the two molecules.

nonlinear molecules, additional angles of rotation, $\chi_1$ and $\chi_2$, need to be specified. Finally, the intermolecular axis $r_{12}$ is oriented with respect to the laboratory with components $r_{12}$, $\Theta$, and $\Phi$. These angles (capitalized) refer to the laboratory frame distinct from the intermolecular frame given above.

In general, to determine the relative orientation of two nonspherical but linear molecules, seven quantities: $\theta_1$, $\theta_2$, $\phi_1$, $\phi_2$, $r = (r, \Theta, \Phi)$ need to be specified. Simplification is possible by first taking the relative angle $\phi_{12} = \phi_2 - \phi_1$. We still have six variables $(r, \Theta, \Phi, \theta_1, \theta_2, \phi_{12})$. $\theta_1$, $\theta_2$, and $\phi_{12}$ refer to relative coordinates, whereas $\Theta$ and $\Phi$ refer to laboratory coordinates.

In space-fixed coordinates, the *x*-, *y*- and *z*-axes refer to the laboratory. All three vectors: the bond vectors $L_1$, $L_2$, and the intermolecular vector $r_{12}$, share a common reference framework (see Figure IV.10). The space variables $r_{12}$, $\Theta$, $\Phi$, $\Theta_1$, $\Theta_2$, $\Phi_1$ and $\Phi_2$ uniquely determine the mutual orientation of the molecules.

In terms of angular parameters, the pair potential is expressed as

$$u(12) = u(r_{12}, \Theta, \Phi, \theta_1, \theta_2, \phi_{12}) \tag{7.1}$$

or

$$u(12) = u(r_{12}, \omega_1, \omega_2) \tag{7.2}$$

To rewrite the statistical mechanical formulas to reflect the angle dependence, we consider an assembly of $N$ polyatomic molecules with Hamiltonian

$$H_N(p^N, J^N, r^N, \omega^N) = KE_t + KE_r + V_N(r^N, \omega^N) \tag{7.3}$$

where we have postulated separation of the kinetic energy from the potential energy. The total potential energy $V_N$ is

$$V_N(r^N, \omega^N) = \sum_{i=1}^{N} u^{(1)}(r_i, \omega_i) + \sum_{i<}^{N} \sum_{j} u^{(2)}(r_i, r_j, \omega_i, \omega_j) \tag{7.4}$$

$$+ \sum_{i<} \sum_{j<} \sum_{k}^{N} u^{(3)}(r_i, r_j, r_k, \omega_i, \omega_j, \omega_k) + \cdots$$

$$+ u^{(N)}(r^N, \omega^N)$$

For pairwise additive energy,

$$V_N(r^N, \omega^N) = \sum_{i<}^{N} \sum_{j} u^{(2)}(r_i, r_j, \omega_i, \omega_j), \tag{7.5}$$

Assuming that the rotational kinetic energy $KE_r$ is decoupled from the configurational energy (see, e.g., Gray and Gubbins [9]), the partition function in the canonical ensemble is

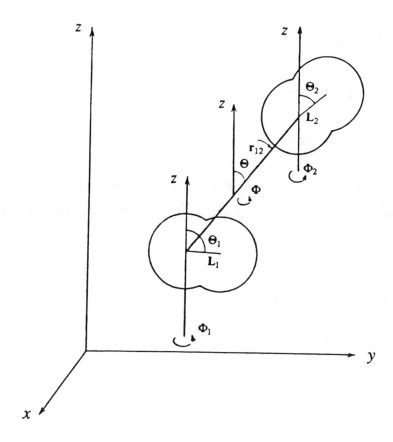

FIGURE IV.10. The space-fixed coordinates. All three vectors $L_1$, the bond length for molecule 1, $L_2$, the bond length for molecule 2, and $r_{12}$, the intermolecular axis, refer to the common x-, y-, and z-axes.

$$Z_N = \frac{Q_N}{N!\Lambda^{3N}} = \frac{1}{N!\Lambda^{3N}} \frac{1}{\Omega^N} \int dr^N \, d\omega^N \, \exp[-\beta V_N(\mathbf{r}^N,\omega^N)] \tag{7.6}$$

where the angle integration $\int d\omega$ is defined, for nonlinear molecules, as

$$\int d\omega(\cdot) = \int\limits_0^\pi d\theta \int\limits_0^{2\pi} d\phi \int\limits_0^{2\pi} d\chi \, \sin\theta(\cdot) \tag{7.7}$$

or, for linear molecules, as

$$\int d\omega(\cdot) = \int\limits_0^\pi d\theta \int\limits_0^{2\pi} d\phi \, \sin\theta(\cdot) \tag{7.8}$$

Consequently, the normalization factor $\Omega$ is $8\pi^2$ for nonlinear molecules and $4\pi$ for linear molecules.

The $N$-body *specific* probability distribution function $p^{(n)}$ is defined as

$$dp^{(N)}(\mathbf{r}^N,\omega^N) \equiv \frac{N!}{Q_N} \exp[-\beta V_N] \, dr^N \, d\omega^N \tag{7.9}$$

where $Q_N$ is the configurational integral

$$Q_N \equiv \frac{1}{\Omega^N} \int dr^N \, d\omega^N \, \exp[-\beta V_N(\mathbf{r}^N,\omega^N)] \tag{7.10}$$

The *generic* $n$-body density $\rho^{(n)}$ is defined as

$$\rho^{(n)}(\mathbf{r}^N,\omega^N) \equiv \frac{N! \int dr^{N-n} d\omega^{N-n} \exp[-\beta V_N]}{(N-n)!\Omega^{N-n}Q_N} \tag{7.11}$$

where $dr^{N-2}=dr_3 \cdots dr_N$ and, similarly, $d\omega^{N-2}=d\omega_3 \cdots d\omega_N$. In particular, the pair density is

$$\rho^{(2)}(\mathbf{r}_1,\mathbf{r}_2;\omega_1,\omega_2) = \frac{N!}{(N-2)!\Omega^{N-2}Q_N} \int dr^{N-2} \, d\omega^{N-2} \, \exp[-\beta V_N] \tag{7.12}$$

The angular pcf $g^{(2)}$ is

$$g^{(2)}(\mathbf{r}_1,\mathbf{r}_2;\omega_1,\omega_2) = \frac{\rho^{(2)}(\mathbf{r}_1,\mathbf{r}_2;\omega_1,\omega_2)}{\rho^{(1)}(\mathbf{r}_1,\omega_1)\rho^{(1)}(\mathbf{r}_2,\omega_2)} \tag{7.13}$$

In the absence of external potentials (as with homogeneous fluids),

$$g^{(2)}(\mathbf{r}_1,\mathbf{r}_2;\omega_1,\omega_2) = \rho^{-2}\rho^{(2)}(\mathbf{r}_1,\mathbf{r}_2;\omega_1,\omega_2) \tag{7.14}$$

Higher-order correlation functions, $g^{(n)}$, are similarly defined.

## SPHERICAL HARMONIC EXPANSIONS

We know from mathematics that a function $f(x,y,z;\Theta,\Phi,\chi)$ of angles $\Theta$, $\Phi$, and $\chi$ can be expanded in terms of the spherical harmonics $Y_{LM}(\Theta,\Phi)$ or, more generally, the rotation matrices $D^L_{MN}(\Phi\Theta\chi)$ (Rose [10]); thus

$$g^{(2)}(r_{12},\Theta,\Phi;\Theta_1,\Phi_1,\chi_1;\Theta_2,\Phi_2,\chi_2) = \sum_{(L)}\sum_{(M)}\sum_{(N)} G(L'L''L,N'N'';r_{12}) \tag{7.15}$$

$$\cdot C(L'L''L,M'M''M)D^{L'*}_{M'N'}(\Phi_1\Theta_1\chi_1)D^{L''*}_{M''N''}(\Phi_2\Theta_2\chi_2)\cdot Y_{LM}{}^*(\Theta,\Phi)$$

for nonlinear molecules, where $Y_{LM}{}^*$ is the complex conjugate of $Y_{LM}$. $C(L'L''L,M'M''M)$ is the Clebsch-Gordan coefficient as defined by Rose, and $G(L'L''L;N'N'';r_{12})$ are the expansion coefficients for the given function. The summation on $(L)$ is over all $L'$, $L''$, and $L$ values that obey the triangular inequality $\Delta(L'L''L)$; i.e., $|L'-L''|\leq L\leq L'+L''$ for all $L'$, $L''$ and $L$. The summation on $(M)$ is such that for any $M'$, $-L'\leq M'\leq L'$. The summation on $(N)$ is similar to the summation on $(M)$.

For linear molecules, there is no dependence on the rotational angle $\chi$. Thus the quantum number $N$ is set to zero, and

$$D^L_{MN}(\Phi\Theta\chi) = D^L_{M0}(\Phi,\Theta,0) = \left[\frac{4\pi}{2L+1}\right]^{1/2} Y_{LM}{}^*(\Theta,\Phi) \tag{7.16}$$

Therefore

$$g^{(2)}(r_{12},\Theta,\Phi;\Theta_1,\Phi_1;\Theta_2,\Phi_2) \equiv \sum_{(L)}\sum_{(M)} G'(L'L''L;r_{12})C(L'L''L;M'M''M) \tag{7.17}$$

$$\cdot Y_{L'M'}(\Theta_1,\Phi_1).Y_{L''M''}(\Theta_2,\Phi_2).Y_{LM}{}^*(\Theta,\Phi)$$

for linear molecules, where $G'(L'L''L;r_{12})$ are new expansion coefficients. Expansion (7.17) is for laboratory-fixed coordinates. For the two-center relative frame, the spherical harmonic expansion simplifies to

$$g^{(2)}(r_{12},\theta_1,\theta_2,\phi_{12}) \equiv 4\pi \sum_{(L)(M)} g_{LL'M}(r_{12})Y_{LM}(\theta_1,\phi_{12})Y_{L'\underline{M}}(\theta_2,0) \tag{7.18}$$

It is customary to introduce a factor $4\pi$ in the expansion. As will be seen later, the coefficient $g_{000}(r_{12})$ coincides with the angular average of the angular pcf: $<g^{(2)}(12)>_\omega$. The summation on $(L)$ is for all $L$ and $L'$ from zero to infinity, and the summation on $(M)$ is confined to $|M|\leq min(L,L')$. $\underline{M}$ is defined as the negative $M$, $(\underline{M}=-M)$.

Some examples of the angle-dependent correlation functions are given in the figures below. Figure IV.11 shows the stereograph of such a $g^{(2)}$ for a quadrupolar fluid.

Several of the spherical harmonic coefficients $g_{LL'M}(r)$ are given in Figure IV.12. They are oscillatory functions of $r$. The angular dependence of pair potentials in polar and multipolar fluids can be easily expressed in terms of spherical harmonics. One such potential (Lennard-Jones potential plus quadrupole-quadrupole interaction) is depicted in Figure IV.13. However, the expansion of apcfs in terms of spherical harmonics suffers from slow convergence: i.e., a large number of terms in the series have to be retained in order to have any reproducibility of the original apcfs. Haile and Gray [11] have tested the convergence of the relative frame spherical harmonic expansions for LJ+QQ apcf in the liquid state. At least 19 terms in the expansion had to be used in order to obtain convergence.

There is a simple correspondence between the space-fixed coefficients, $G'(L'L''L; r)$ and the two-center body-fixed coefficients $g_{L'L''M}(r)$; i.e.,

### From Two-Center Coordinates to Space-Fixed Coordinates

$$G'(L'L''L; r) = 4\pi \left[\frac{4\pi}{2L+1}\right]^{1/2} \sum_{(M)} g_{L'L''M}(r) C(L'L''L; M\underline{M}0) \tag{7.19}$$

where the summation on $(M)$ is from $-min(L',L'')$ to $+min(L',L'')$. $G'(L'L''L; r)$ and $g_{L'L''M}(r)$ have been defined in eqs. (7.17) and (7.18), respectively.

### From Space-Fixed Coordinates to Two-Center Coordinates

$$g_{L'L''M}(r) = \frac{1}{4\pi} \sum_{(L)} \left[\frac{2L+1}{4\pi}\right]^{1/2} G'(L'L''L; r) C(L'L''L, M\underline{M}0) \tag{7.20}$$

where the summation on $(L)$ is from $|L'-L''|$ to $L'+L''$.

We have listed a few spherical harmonics $Y_{LM}(\theta,\phi)$ from $L=0$ to $L=3$ in Table IV.2. To obtain the coefficients of expansion for a given function, the triple integration with respect to $\theta_1, \theta_2$, and $\phi_{12}$ is used in (7.18)

$$g_{LL'M}(r) = \frac{1}{2} \int_0^\pi d\theta_1 \sin \theta_1 \int_0^\pi d\theta_2 \sin \theta_2 \int_0^{2\pi} d\phi_{12} \tag{7.21}$$

$$\cdot g^{(2)}(r;\theta_1,\theta_2,\phi_{12}) Y_{LM}{}^*(\theta_1,\phi_{12}) Y_{L',-M}{}^*(\theta_2,0)$$

## IV.8.  *The Correlation Functions for Molecular Fluids: The Site-Site Correlation Functions*

The angular pcf $g^{(2)}(r_{12},\Theta,\Phi,\Theta_1,\Theta_2,\Phi_{12})$ of (7.17) depends on the six variables $r_{12}$, $\Theta$, $\Phi$, $\Theta_1$, $\Theta_2$, and $\Phi_{12}$. In practice, it is cumbersome to keep a variable list of so many dimensions. Spherical harmonic expansions greatly reduce the effort of bookkeeping. However, there remains the problem of slow convergence with the series of spherical harmonics [12]. An alternative, although not entirely equivalent, is the *site-site correlation functions*. This function gives the probabilistic distribution of sites in a molecule with respect to sites in another molecule. It is related to the structure factor obtained in scattering experiments.

For polyatomics, the interaction could be attributed to many force centers

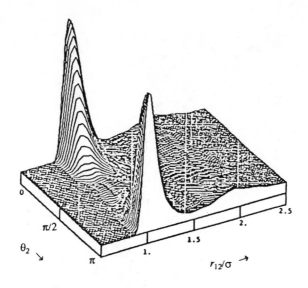

FIGURE IV.11. The angluar pair correlation function, $g(r_{12},\theta_1,\theta_2,\phi_{12})$ at the reduced quadrupolar moment $Q^{*2} = Q^2/\epsilon\sigma^5 = 1$. Here g is plotted as a function of $r_{12}$ and $\theta_2$. $\theta_1$ is fixed at $\pi/2$ and $\phi_{12}$ is left unspecified. (Haile, Thesis, 1976)

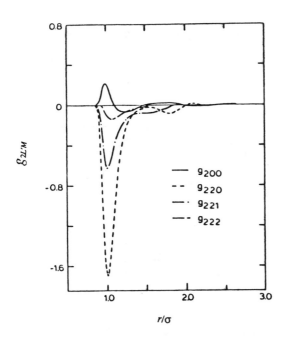

FIGURE IV.12. The expansion coefficients $g_{2L'M}$ for a Lennard-Jones plus quadrupole fluid. The condition is at $kT/\epsilon = 1.277$, $\rho\sigma^3 = 0.85$, and $Q^2/\epsilon\sigma^5 = 1$. (Haile, Thesis, 1976)

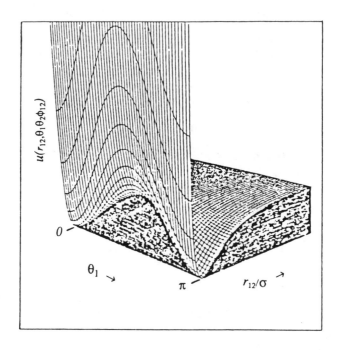

FIGURE IV.13. Surface of the intermolecular pair potential $u(r_{12}\theta_1\theta_2\phi_{12})$ for the Lennard-Jones plus quadrupole interaction. $Q^2/\varepsilon\sigma^5 = 1.0$, $\theta_2 = \pi/2$, and $\phi_{12}$ is undefined. (Haile, Thesis, 1976)

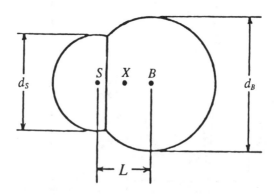

FIGURE IV.14. A hard dumbbell molecule composed of two fused hard spheres, B and S, with diameters $d_B$ and $d_S$. X is the center of mass. L is the bond length.

Table IV.2. The Spherical Harmonics, $Y_{LM}$.

$$Y_{LM}(\theta,\phi) = \left\{ \frac{2L+1}{4\pi} \frac{(L-M)!}{(L+M)!} \right\}^{1/2} \exp(iM\phi)\, P_L^M(x=\cos\theta)$$

$$P_L^0(x) = \frac{1}{2^L L!} \frac{d^L}{dx^L}(x^2-1)^L$$

$$P_L^M(x) = (-1)^M (1-x^2)^{m/2} \frac{d^m}{dx^m} P_L^0(x)$$

$$Y_{0,0}(\theta,\phi) = (\frac{1}{4\pi})^{1/2}$$

$$Y_{1,-1}(\theta,\phi) = (\frac{3}{8\pi})^{1/2} \exp(-i\phi)\, \sin\theta$$

$$Y_{1,0}(\theta,\phi) = (\frac{3}{4\pi})^{1/2} \cos\theta$$

$$Y_{1,1}(\theta,\phi) = (\frac{3}{8\pi})^{1/2} \exp(i\phi)\, (-\sin\theta)$$

$$Y_{2,-2}(\theta,\phi) = (\frac{15}{32\pi})^{1/2} \exp(-i2\phi)\, \sin^2\theta$$

$$Y_{2,-1}(\theta,\phi) = (\frac{15}{8\pi})^{1/2} \exp(-i\phi)\, \sin\theta \cos\theta$$

$$Y_{2,0}(\theta,\phi) = (\frac{5}{16\pi})^{1/2} (3\cos^2\theta - 1)$$

$$Y_{2,1}(\theta,\phi) = (\frac{15}{8\pi})^{1/2} \exp(i\phi)(-\sin\theta \cos\theta)$$

$$Y_{2,2}(\theta,\phi) = (\frac{15}{32\pi})^{1/2} \exp(i2\phi)\, \sin^2\theta$$

$$Y_{3,-3}(\theta,\phi) = (\frac{35}{64\pi})^{1/2} \exp(-i3\phi)\, \sin^3\theta$$

$$Y_{3,-2}(\theta,\phi) = (\frac{105}{32\pi})^{1/2} \exp(-i2\phi)\, \sin^2\theta \cos\theta$$

$$Y_{3,-1}(\theta,\phi) = (\frac{21}{64\pi})^{1/2} \exp(-i\phi)\, \sin\theta\, (5\cos^2\theta - 1)$$

$$Y_{3,0}(\theta,\phi) = (\frac{7}{16\pi})^{1/2} \cos\theta\, (5\cos^2\theta - 3)$$

*For negative quantum number $M$, use the conjugate relation, $Y_{LM} = (-1)^M Y^*_{LM}$.

distributed inside the molecule. Figure IV.14 shows the case for diatomics. Two interaction sites (force centers) are given for each molecule. In general, there exist $m$ sites in a polyatomic molecule. They are labeled by Greek letters, $\alpha,\gamma = 1,2,...,m$. In case these sites coincide with the atoms, we call the model atom-atom interaction model. In space-fixed coordinates, the position of the center of mass of the $i$th molecule is $R_i$. Let the position of site $\alpha$ in molecule $i$ relative to the center be $L_i^\alpha$ (i.e., the *bond radius*). The position of site $\alpha$ with respect to the space-fixed coordinates is

$$\mathbf{r}_i^\alpha = \mathbf{R}_i + \mathbf{L}_i^\alpha, \qquad \alpha=1,2,...,m; \quad i=1,2,...,N \tag{8.1}$$

In the interaction site model (ISM) the total potential energy is assumed to be the sum of interactions of all pairs of sites belonging to different molecules, i.e.,

$$V_N(\{N\}) = \sum_{i<}^{N}\sum_j u(ij) \tag{8.2}$$

where the site-site potentials $u(ij)$ are

$$u(ij) = \sum_\alpha^m\sum_\gamma^m u_{\alpha\gamma}(\mathbf{r}_i^\alpha,\mathbf{r}_j^\gamma), \qquad i \ne j \tag{8.3}$$

Namely, the interaction between molecule $i$ and molecule $j$ is equal to the sum of site-site interactions $u_{\alpha\gamma}$ of all the sites in $i$ with all the sites in $j$. The sites within the same molecule may also interact [13]. $u_{\alpha\gamma}$ is taken to be dependent only on the relative distance $r_{ij}^{\alpha\gamma} = |\mathbf{r}_i^\alpha - \mathbf{r}_j^\gamma|$, and not on the orientation of the sites.

The bond radius $\mathbf{L}_i^\alpha$ gives the location of site $\alpha$ in molecule $i$ relative to $\mathbf{R}_i$. For flexible molecules, $\mathbf{L}_i^\alpha$ is a variable. For rigid molecules, $\mathbf{L}_i^\alpha$ is a fixed quantity.

To define the site-site correlation function, we make use of the density operators

$$\rho_{op}(\mathbf{r}) \equiv \delta_D(\mathbf{r},\mathbf{r}_i) \tag{8.4}$$

where $\delta_D(\mathbf{r},\mathbf{r}_i)$ is the Dirac delta function in three dimensions. $\mathbf{r}_i$ is a field variable (i.e., it disappears after integration). $\rho_{op}$ operates on an equilibrium ensemble according to

$$<\rho_{op}(\mathbf{r})> = <\sum_{i=1}^N \delta_D(\mathbf{r},\mathbf{r}_i)> = \frac{1}{Q_N}N\int dr_2 \cdots dr_N \exp[-\beta V_N(\mathbf{r},\mathbf{r}_2,\ldots,\mathbf{r}_N)] \tag{8.5}$$

$$= \rho^{(1)}(\mathbf{r})$$

That is, the ensemble average of $\rho_{op}$ is the singlet density $\rho^{(1)}$. The product $\rho_{op}(r)\rho_{op}(r')$ can be separated into two parts

$$\rho_{op}(\mathbf{r})\rho_{op}(\mathbf{r}') = \sum_i^N \delta_D(\mathbf{r},\mathbf{r}_i)\sum_j^N \delta_D(\mathbf{r}',\mathbf{r}_j) \tag{8.6}$$

$$= 2\sum_{i<}\sum_j \delta_D(\mathbf{r},\mathbf{r}_i)\delta_D(\mathbf{r}',\mathbf{r}_j) + \sum_{k=1}^N \delta_D(\mathbf{r},\mathbf{r}_k)\delta_D(\mathbf{r}',\mathbf{r}_k)$$

Therefore the ensemble average is

$$<\rho_{op}(r)\rho_{op}(r')> = 2Q_N^{-1}\frac{N(N-1)}{2}\int dr_3 \cdots dr_N \exp[-\beta V_N(\mathbf{r},\mathbf{r}',\mathbf{r}_3,\ldots,\mathbf{r}_N)] \tag{8.7}$$

$$+ \delta(\mathbf{r},\mathbf{r}')Q_N^{-1}N\!\int d\mathbf{r}_2 \cdots d\mathbf{r}_N \exp[-\beta V_N(\mathbf{r},\mathbf{r}_2,\mathbf{r}_3,\ldots,\mathbf{r}_N)]$$

$$= \rho^{(2)}(\mathbf{r},\mathbf{r}') + \delta_D(\mathbf{r},\mathbf{r}')\rho^{(1)}(\mathbf{r})$$

where $\rho^{(2)}$ is the pair density and $\rho^{(1)}$ is the singlet density. It is also possible to define the Fourier transform (FT) of the density operator, denoted by $\rho_k$, as

$$\rho_k \equiv FT\{\,\rho_{op}(\mathbf{r})\} = \int d\mathbf{r}\ \exp(-i\mathbf{k}\cdot\mathbf{r})\sum_{j=1}^{N} \delta_D(\mathbf{r},\mathbf{r}_j) \tag{8.8}$$

$$= \sum_{j=1}^{N} \exp(-i\mathbf{k}\cdot\mathbf{r}_j)$$

where $i$ is the imaginary number $\sqrt{-1}$.

Using operator notation, we define the site-site pair density as the ensemble average

$$\rho_{\alpha\gamma}^{(2)}(\mathbf{r},\mathbf{r}') \equiv \Big\langle \sum_{i\neq j}^{N} \delta_D(\mathbf{r},\mathbf{r}_i^{\alpha})\delta_D(\mathbf{r}',\mathbf{r}_j^{\gamma})\Big\rangle \tag{8.9}$$

It gives the probability that site $\alpha$ in one molecule is situated at $\mathbf{r}$ while site $\gamma$ in a second molecule is situated at $\mathbf{r}'$. In an isotropic fluid, the site-site pair correlation function (ss pcf) is defined by

$$g_{\alpha\gamma}^{(2)}(\mathbf{r},\mathbf{r}') \equiv \rho^{-2}\rho_{\alpha\gamma}^{(2)}(\mathbf{r},\mathbf{r}') \tag{8.10}$$

The total correlation function is

$$h_{\alpha\gamma}(\mathbf{r},\mathbf{r}') = g_{\alpha\gamma}^{(2)}(\mathbf{r},\mathbf{r}') - 1 \tag{8.11}$$

Inside a polyatomic, we define an intramolecular pair operator

$$\overline{\rho}_{op}^{\alpha\gamma}(\mathbf{u}) = \delta_D(\mathbf{u},\ \mathbf{L}_j^{\alpha}-\mathbf{L}_j^{\gamma}) \tag{8.12}$$

where $\mathbf{u}$ is an internal position vector relative to $\mathbf{r}_j^{\gamma}$. The Fourier transform of $\rho_{op}^{\alpha\gamma}(\mathbf{u})$ is

$$FT\ \{\rho_{op}^{\alpha\gamma}(\mathbf{u})\} = \int d\mathbf{u}\ \exp(i\mathbf{k}\cdot\mathbf{u})\delta_D(\mathbf{u},\mathbf{L}) = \exp(i\mathbf{k}\cdot\mathbf{L}) \tag{8.13}$$

where we have written $\mathbf{L}$ for $\mathbf{L}_j^{\alpha}-\mathbf{L}_j^{\gamma}$. Next let the vector $\mathbf{k}$ coincide with the $z$-axis. The vector $\mathbf{L}$ forms an angle $\theta_1$ with respect to $\mathbf{k}$, and $\phi_1$ with respect to the $x$-axis. We then take the angular average of (8.13):

$$\hat{\omega}_{\alpha\gamma}(k) \equiv \frac{1}{4\pi} \int_0^\pi d\theta_1 \sin\theta_1 \int_0^{2\pi} d\phi_1 \exp(i\mathbf{k}\cdot\mathbf{L}) \tag{8.14}$$

Since $\mathbf{k}$ is the $z$-axis, $\mathbf{k}\cdot\mathbf{L} = kL \cos\theta_1$, where $k=|\mathbf{k}|$ and $L=|\mathbf{L}|$. Thus (8.14) can be written as

$$\hat{\omega}_{\alpha\gamma}(k) = \frac{1}{4\pi} \int_{-1}^1 d\cos\theta_1 \int_0^{2\pi} d\phi_1 \exp(ikL \cos \theta_1) \tag{8.15}$$

$$= \frac{1}{2} \int_{-1}^1 dx \exp(ikLx) = \frac{1}{2ikL} \exp(ikLx)\Big|_{-1}^1 = \frac{\sin(kL)}{kL}$$

The quantity $\hat{\omega}_{\alpha\gamma}(k)$ is called the *intramolecular structure*. Because $L_j^\alpha - L_j^\gamma$ has been kept constant in the derivation, (8.15) is valid for rigid molecules only. The inverse transform of (8.15) gives the angle-averaged intramolecular structure in real space

$$\omega_{\alpha\gamma}(r) = FT^{-1}\{\hat{\omega}_{\alpha\gamma}(k)\} = \frac{1}{2\pi^2 r} \int_0^\infty dk\, k \sin (kr)\, \hat{\omega}_{\alpha\gamma}(k) \tag{8.16}$$

$$= \frac{1}{2\pi^2 r} \int_0^\infty dk\, k \sin (kr)\, \frac{\sin (kL)}{kL} = \frac{1}{4\pi rL} \frac{2}{\pi} \int_0^\infty dk \sin (kL) \sin (kr)$$

$$= \frac{1}{4\pi rL} \delta_D(r,L)$$

Since the Dirac delta puts $r$ and $L$ on equal footing, $\omega_{\alpha\gamma}(r)$ is often written as $\delta_D(r, L)/4\pi L^2$. The same result could be obtained from the surface integral

$$\omega_{\alpha\gamma}(r) = \frac{1}{4\pi L^2} \int_{sph.surf.} dS\, \delta_D(\mathbf{r},\mathbf{L}) \tag{8.17}$$

where the limits of integration are the surface of a sphere of radius $L$.

The Fourier transform of the total correlation function $h_{\alpha\gamma}(r)$ is given by

$$\hat{h}_{\alpha\gamma}(\mathbf{k}) = \int d\mathbf{r} \exp(i\mathbf{k}\cdot\mathbf{r}) h_{\alpha\gamma}(r) \tag{8.18}$$

It is customary to define a site-site direct correlation function (ss dcf) $\hat{C}_{\alpha\gamma}(k)$ by the matrix relation

$$\hat{h}_{\alpha\gamma}(\mathbf{k}) \equiv \hat{\omega}_{\alpha\mu}(\mathbf{k})\hat{C}_{\mu\nu}(\mathbf{k})[I - \rho_\varepsilon \hat{\omega}_{\varepsilon\xi}\hat{C}_{\xi\eta}]_{\nu\zeta}^{-1}\hat{\omega}_{\zeta\gamma}(\mathbf{k}) \tag{8.19}$$

where $I$=unit matrix. Equation (8.19) is a generalization of the Ornstein-Zernike relation to molecular fluids. It was proposed without proof by Chandler and Andersen [14]. Other alternatives to (8.19) have since been proposed. We shall refer to (8.19) as the *reference interaction site model* (RISM) OZ relation. Note that $[\cdots]_{\nu\zeta}^{-1}$ is the $\nu\zeta$

element of the inverted matrix $(I - \rho\hat{\omega}\hat{c})$. We have adopted the convention that repeated indices are summed (e.g., $\hat{\omega}_{\alpha\mu}\hat{C}_{\mu\nu} = \sum_{\mu=1}^{m} \hat{\omega}_{\alpha\mu}\hat{C}_{\mu\nu}$). The inverse transform of $\hat{C}(k)$ gives the ss dcf in real space:

$$C_{\alpha\gamma}(r) = \frac{1}{8\pi^3} \int d\mathbf{k} \, \exp(-\mathbf{k}\cdot\mathbf{r}) \, \hat{C}_{\alpha\gamma}(\mathbf{k}) \qquad (8.20)$$

Equation (8.19) can be written in real space as

$$h_{\alpha\gamma}(12) = \sum_{\beta}^{m} \sum_{\delta}^{m} \int d3 \, d4 \, \omega_{\alpha\beta}(13) C_{\beta\delta}(34) \omega_{\delta\gamma}(42) \qquad (8.21)$$

$$+ \sum_{\beta}^{m} \sum_{\delta}^{m} \int d3 \, d4 \, \omega_{\alpha\beta}(13) C_{\beta\delta}(34) \rho_{\delta} h_{\delta\gamma}(42)$$

where $\int d3$, for example, is shorthand for $\int d\mathbf{r}_3^\beta$. This equation reduces to the atomic OZ equation for atomic fluids where the intramolecular $\omega_{\alpha\gamma}(12)$ is set equal to the Dirac delta.

The ss pcf $g_{\alpha\gamma}^{(2)}$ could be obtained from the angular pcf $g^{(2)}(r_{12},\omega_1,\omega_2)$ defined earlier by angle averaging, but not vice versa. The angular pcf is more general than the ss pcf and contains more structural information. The site-site correlation function can be obtained from approximate integral equations (such as the RISM Percus-Yevick equation) or from computer simulations. Klein et al. [15] reported simulation results for water and ammonia from Monte Carlo and molecular dynamics methods. The HH, NH, and NN (in $NH_3$) correlations are shown in Figure IV.15, and the HH, OH, and OO (in $H_2O$) correlations are shown in Figure IV.16. The OH correlation shows a peak at about 1.9 Å, corresponding to the hydrogen bonding distance.

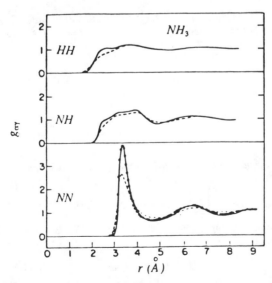

FIGURE IV.15. The site-site pair correlation functions for ammonia from molecular dynamics simulation of Klein et al. (ref. [13]). $T = 272\ K$, $v = 26.5$ cc/mol. (Klein et al., J. Chem. Phys., 1979)

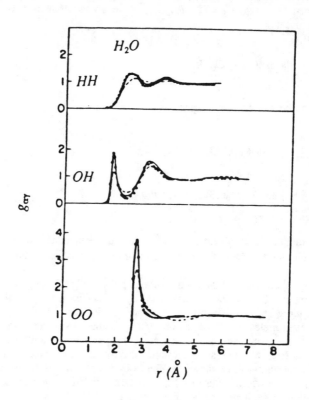

FIGURE IV.16. *The site-site pair correlation functions for water from molecular dynamic simulation of Klein et al. (ref. [13]). T = 277 K and v = 18.0 cc/mol. (Klein et al., J. Chem. Phys., 1979)*

## References

[1]   W. Feller, *An Introduction to Probability Theory and its Applications* (John Wiley, New York, 1950).

[2]   T. Hill, *Statistical Mechanics* (McGraw-Hill, New York, 1956).

[3]   C.G. Gray and K.E. Gubbins, *Theory of Molecular Fluids*, Volume 1 (Clarendon Press, Oxford, 1984).

[4]   R.O. Watts and I.J. McGee, *Liquid State Chemical Physics* (John Wiley, New York, 1976).

[5]   J.K. Percus and G.J. Yevick, Phys. Rev. **110**, 1 (1958).

[6]   T. Morita and K. Hiroike, Progr. Theor. Phys., (Kyoto) **23**, 1003 (1960).

[7]   B. Widom, J. Chem. Phys. 39, 2808 (1963).

[8]   J.L. Lebowitz and J.K. Percus, Phys. Rev. **122**, 1675 (1961).

[9]   C.G. Gray and K.E. Gubbins, *Ibid.*

[10]  M.E. Rose, *Elementary Theory of Angular Momentum* (John Wiley, New York, 1957).

[11]  J.M. Haile and C.G. Gray, Chem. Phys. Lett. **76**, 583 (1980).

[12]  J.M. Haile and C.G. Gray, *Ibid.* (1980).

[13]  T.A. Weber J. Chem. Phys. **69**, 2347 (1978).

[14]  D. Chandler and H.C. Andersen, J. Chem. Phys. **57**, 1930 (1972).

[15]  M.L. Klein, I.R. McDonald, B.J. Berne, M. Rao, D.L. Beveridge, and P.K. Mehrota, J. Chem. Phys. **71**, 3889 (1979).

## Exercises

1.   *Coordination Number.* In the study of the structure of materials, the concept of coordination number is very useful. The coordination number $N(L)$ is the average number of particles found within a radius of $L$ from any central molecule. It is related to the radial distribution function by

$$N(L) = \rho \int_0^L dr \; 4\pi r^2 g(r)$$

Find the coordination number for

(i)   a two-dimensional lattice (a square lattice with mesh 1 mm $\times$ 1 mm) with five values of $r$, i.e., $r$ =4 mm, 8 mm, 13 mm, 16 mm, and 25 mm. Plot a histogram of $N(r)$.

(ii) A three-dimensional face-centered cubic lattice of hard spheres of diameter $d$ at close packing. ($d$=1).

Also estimate the rdf $g(r)$ by taking the differences in N(r) (or using finite differences formulae). Set $\rho$ = 0.004 molecules/mm$^3$ in (i) and $\rho d^3$ = 1.4142 in (ii).

2.   Find the low density approximations of the rdf $g(r)$ for a Lennard-Jones potential at the reduced temperatures of $kT/\varepsilon$ =0.7, 1.0, 1.26, 2.0, and 10.0.

3.   Given the low density $g(r)$ for the Lennard-Jones potential at $kT/\varepsilon$ =1.0, find the direct correlation function $C(r)$ numerically by solving the Ornstein-Zernike equation ($\rho\sigma^3$ =0.001).

4.   The Clebsch-Gordan coefficients obey the orthogonality condition

$$\sum_{(L)} C(L'L''L,M,-M,O)C(L'L''L,N,-N,O) = \delta_{MN}$$

where $\delta_{MN}$ is the Kronecker delta. Use computer programming to show this for $L'=2$, and $L''=3$ at a series of values of $M$ and $N$.

# CHAPTER V
# MICROTHERMODYNAMICS

$B$y microthermodynamics we mean the interpretation of macroscopic thermodynamic properties in terms of molecular forces. There are three major routes to thermodynamics based on the molecular distribution functions: (i) the energy equation, (ii) the pressure equation, and (iii) the isothermal compressibility equation. All three equations express thermodynamic properties in terms of molecular correlation functions. The basic idea is that if we know, for example, the spatial distributions of molecules as well as the interaction energies among them, we could sum the energies between the molecules over the intermolecular distances, as given by the distribution, to obtain the total energy. For pressure, the quantity summed is the *virial*, and for isothermal compressibility, the number fluctuations. For the first two properties (energy and pressure), we could work in the canonical ensemble (CE), whereas for the number fluctuations, we must work in the grand canonical ensemble (GCE). As we shall see shortly, for all practical purposes, only the pair, and, sometimes the triplet correlation functions are needed for a full determination of the thermodynamic properties. Higher-order correlations are rarely required.

We shall give a detailed derivation of the three thermodynamic quantities in sections V.1, V.2, and V.3. A brief introduction to the cluster series is also given in order to derive the virial coefficients. Molecular fluids are investigated in the spherical harmonic representation as well as the site-site interaction representation. Evaluation of the triplet correlation functions poses numerical problems due to the high dimensionality of the independent variables. The superposition approximation of Kirkwood [1] is usually invoked. We shall also discuss some of its improvements. In this chapter, a major effort is made to establish the connections between the microscopic force laws (i.e., the distribution functions) and the macroscopic properties (i.e., thermodynamic quantities). Specific applications will be postponed to future chapters.

## V.1. *The Internal Energy*

The internal energy arises due to the interaction forces among the molecules, thus *internal* to the system. We have derived the differential relation in a CE

$$U = <E> = -\left.\frac{\partial \ln Z_N}{\partial \beta}\right|_{V,N} \tag{1.1}$$

For pairwise additive potentials

$$Z_N = \frac{1}{N!\Lambda^{3N}} \int dr_1 \cdots dr_N \exp\left[-\beta \sum_{i<j}^{N}\sum u(r_{ij})\right] \tag{1.2}$$

Carrying out the differentiation with respect to $\beta$, we get

$$-\left.\frac{\partial \ln Z_N}{\partial \beta}\right|_{V,N} = \frac{-1}{Z_N}\left\{\frac{1}{N!\Lambda^{3N}}\int dr^N\left[-\sum_i\sum_j^N u(r_{ij})\right]\exp(-\beta V_N) - \frac{3NZ_N}{\Lambda}\frac{\partial\Lambda}{\partial\beta}\right\} \tag{1.3}$$

For identical particles, $u(r_{ij}) = u(r_{km})$ for all $i,j,k$, and $m$ as long as $r_{ij} = r_{km}$. There are $N(N-1)/2$ distinct pairs in the summation. Thus we have

$$U = \frac{1}{Q_N}\frac{N(N-1)}{2}\int dr_1\,dr_2\cdots dr_N\,u(r_{12})e^{-\beta V_N} + \frac{3N}{\Lambda}\frac{\partial\Lambda}{\partial\beta} \tag{1.4}$$

Since $\Lambda = h\beta^{1/2}(2\pi m)^{-1/2}$, we have

$$\frac{\partial\Lambda}{\partial\beta} = \frac{\Lambda}{2\beta} \tag{1.5}$$

Also, noting that

$$\rho^{(2)}(\mathbf{r}_1, \mathbf{r}_2) = \frac{1}{Q_N}\frac{N(N-1)}{2}\int dr_3\,dr_4\cdots dr_N\,e^{-\beta V_N} \tag{1.6}$$

we have

$$U = \frac{3}{2}NkT + \frac{1}{2}\int_V dr_1\int_V dr_2\,u(r_{12})\rho^{(2)}(\mathbf{r}_1, \mathbf{r}_2) \tag{1.7}$$

For an isotropic and homogeneous fluid, $\rho^{(2)}(\mathbf{r}_1, \mathbf{r}_2) = \rho^2 g^{(2)}(\mathbf{r}_1, \mathbf{r}_2) = \rho^2 g^{(2)}(r_{12})$. The integration with respect to $\int dr_1\,dr_2$ can be transformed to $\int dr_1\,dr_{12} = V\int dr_{12}$:

$$U = \frac{3}{2}NkT + \frac{\rho^2 V}{2}\int dr_{12}\,u(r_{12})g^{(2)}(r_{12}) \tag{1.8}$$

or

$$\frac{U}{NkT} = \frac{3}{2} + \frac{\rho}{2kT}\int_0^\infty dr_{12}\,4\pi r_{12}^2\,u(r_{12})g^{(2)}(r_{12}) \tag{1.9}$$

The upper limit can be taken to be infinity, provided

1.    The intermolecular forces are short-ranged (e.g., under $\cong 10^{-9}m$) in comparison to the macroscopic dimension ($\cong 1$ $m$) of the container.

2.    Surface effects are negligible; e.g., in the thermodynamic limit $V \to \infty$ (while $N/V$ = constant) the surface-to-volume ratio approaches zero.

Equation (1.9) shows that the macroscopic energy can be obtained in terms of the molecular pair potential $u(r)$ and the radial distribution function (rdf) $g^{(2)}(r)$, both being microscopic two-body functions. The many-body problem discussed earlier is now reduced to a two-body problem. (Note, though, that the many-body correlations are sub-sumed in the definition for the pair correlation function (pcf).) It is easier to make approximations to two-body functions such as $g^{(2)}$ than to many-body functions. A number of ingenious integral equations have been proposed for the pcf but none exists for partition functions.

For systems with three-body forces, a similar derivation gives

$$\frac{U}{NkT} = \frac{3}{2} + \frac{\beta\rho}{2} \int_0^\infty dr_{12}\, 4\pi r_{12}^2 u(r_{12}) g(r_{12}) \tag{1.10}$$

$$+ \frac{\beta\rho}{6} \int dr_{13}\, dr_{23}\, u^{(3)}(r_{12}, r_{23}, r_{31}) g^{(3)}(r_{12}, r_{23}, r_{31})$$

where $g^{(3)}$ is the triplet correlation function and $u^{(3)}$ is the three-body potential with $r_{31} = |r_{12} - r_{23}|$.

## V.2.  The Virial Pressure

As before, the equilibrium pressure is given by

$$P = kT \frac{\partial \ln Z_N}{\partial V}\bigg|_{T,N} \tag{2.1}$$

Without loss of generality, we assume that the container is cubic and the volume $V = L^3$. We follow the method of Green [2] and scale the length in units of

$$r_i^* = \frac{r_i}{L} \tag{2.2}$$

The intermolecular potential $u(r_{ij})$ depends on

$$r_{ij} = |r_j - r_i| = \left[(r_{jx} - r_{ix})^2 + (r_{jy} - r_{iy})^2 + (r_{jz} - r_{iz})^2\right]^{1/2} \tag{2.3}$$

$$= \left\{ L^2 \left[(r_{jx}^* - r_{ix}^*)^2 + (r_{jy}^* - r_{iy}^*)^2 + (r_{jz}^* - r_{iz}^*)^2\right] \right\}^{1/2}$$

$$= L \ |r_j{}^* - r_i{}^*| = L \ r_{ij}{}^*$$

The partition function can be written as

$$Z_N = \frac{L^{3N}}{N!\Lambda^{3N}} \int_{V^*} d\mathbf{r}_1{}^* \cdots d\mathbf{r}_N{}^* \ \exp(-\beta V_N) \tag{2.4}$$

where $V^*$ is the unit volume. Since $L = V^{1/3}$,

$$\frac{\partial \ln Z_N}{\partial V} = \frac{1}{Z_N} \frac{\partial Z_N}{\partial V} \tag{2.5}$$

$$= \frac{1}{Z_N} \left[ \frac{NV^{N-1}}{N!\Lambda^{3N}} \int_{V^*} d\mathbf{r}^{*N} \ e^{-\beta V_N} + \frac{V^N}{N!\Lambda^{3N}} \int_{V^*} d\mathbf{r}^{*N} \ e^{-\beta V_N}(-\beta) \sum_{i<}\sum_j \frac{\partial u}{\partial V} \right]$$

$$= \frac{N}{V} - \frac{\beta V^N}{Q_N} \int d\mathbf{r}^{*N} \ e^{-\beta V_N} \left[ \sum_{i<}\sum_j \frac{\partial u}{\partial V} \right]$$

The derivative

$$\frac{\partial u(r_{ij})}{\partial V} = \frac{du(r_{ij})}{dr_{ij}} \frac{dr_{ij}}{dV} = u' \frac{d}{dV} (V^{1/3} r_{ij}{}^*) \tag{2.6}$$

$$= \frac{V^{-2/3}}{3} \ r_{ij}{}^* \ u' = \frac{r_{ij}}{3V} \ u'$$

where $u' = du(r)/dr$. Because the particles are identical, the summation involves $N(N-1)/2$ terms. Each term contributes the same value after integration. To see this, let any four particles be labeled $i,j,k,$ and $m$. When the distance $r_{ij}$ is equal to the distance $r_{km}$, $u(r_{ij}) = u(r_{km})$. The integration in (2.5) is over all possible configurations of the $N$ molecules (i.e., $\int d\mathbf{r}_1 \cdots d\mathbf{r}_N$). Each term in (2.5) then contributes the same value toward the pressure. There are totally $N(N-1)/2$ distinct binary pairs. Thus

$$\left. \frac{\partial \ln Z_N}{\partial V} \right|_{T,N} = \frac{N}{V} - \frac{\beta V^N N(N-1)}{2Q_N} \int d\mathbf{r}^{*N} \ e^{-\beta V_N} \frac{r_{12}}{3V} u'(r_{12}) \tag{2.7}$$

Upon restoring $r^* = r/V$, we get

$$\frac{P}{kT} = \frac{N}{V} - \frac{\beta}{6V} \int d\mathbf{r}_1 \ d\mathbf{r}_2 \ r_{12} \frac{du(r_{12})}{dr} \left[ \frac{N(N-1)}{Q_N} \int d\mathbf{r}_3 \cdots d\mathbf{r}_N e^{-\beta V_N} \right] \tag{2.8}$$

$$= \frac{N}{V} - \frac{\beta}{6V} \int d\mathbf{r}_1 \ d\mathbf{r}_2 \ r_{12} \frac{du(r_{12})}{dr} \ \rho^{(2)}(\mathbf{r}_1, \mathbf{r}_2)$$

where $\rho^{(2)}$ is the pair density. For an isotropic fluid, $\rho^{(2)}(r_1, r_2) = \rho^2 g(r_{12})$, and

$$\frac{PV}{NkT} = 1 - \frac{\rho}{6kT} \int d\mathbf{r}_{12} \, r_{12} \frac{du(r_{12})}{dr_{12}} \, g(r_{12}) \qquad (2.9)$$

or

$$\frac{\beta P}{\rho} = 1 - \frac{\beta \rho}{6} \int_0^\infty dr \, 4\pi r^3 \frac{du(r)}{dr} \, g(r) \qquad (2.10)$$

This equation is called the *virial (pressure) equation*. The pressure is given in terms of the rdf and the derivative of the pair potential. The quantity $\mathbf{r} \cdot \nabla u$ is called the *virial*. Since $g(r)$ gives the radial distribution of $N-1$ molecules around a central molecule, equation (2.10) gives the averaged virial ($\mathbf{r} \cdot \nabla u$ weighted by $g(r)$). The first term, 1, is the kinetic contribution to the pressure (i.e., the ideal-gas contribution); the second term is due to the interaction energies among the molecules. We call it the configurational contribution. When the potential energy $u(r)=0$, ($du/dr=0$), (2.10) reduces to $PV/NkT=1$, the ideal-gas equation. So the ideal gas is once more seen as noninteracting particles.

In the presence of three-body forces, pairwise additivity is no longer valid. Carrying out the same analysis gives

$$\frac{PV}{NkT} = 1 - \frac{\beta \rho}{6} \int_0^\infty d\mathbf{r}_{12} \, 4\pi r_{12}^3 \frac{du(r_{12})}{dr_{12}} g^{(2)}(r_{12}) \qquad (2.11)$$

$$- \frac{\beta \rho^2}{18} \int d\mathbf{r}_{12} \int d\mathbf{r}_{13} \left[ \left[ r_{12} \frac{du^{(3)}}{dr_{12}} + r_{23} \frac{du^{(3)}}{dr_{23}} + r_{31} \frac{du^{(3)}}{dr_{31}} \right] g^{(3)}(r_{12}, r_{23}, r_{31}) \right]$$

where $u^{(3)} \equiv u^{(3)}(r_{12}, r_{23}, r_{31})$ and $g^{(3)}$ is the triplet correlation function. It is also possible to derive higher-order correlation expressions for $\beta P/\rho$, where quadruplet interactions, etc., are present.

## V.3. *The Virial (Cluster) Coefficients*

We know from the equation of state studies of Kamerlingh Onnes [3] that the equilibrium pressure of a system can be expressed in a power series of density $\rho = N/V$:

$$\frac{P}{kT} = \rho + B_2 \rho^2 + B_3 \rho^3 + B_4 \rho^4 + B_5 \rho^5 + \cdots \qquad (3.1)$$

This is the so-called *virial (cluster) equation of state*. $B_2$ is called the second virial coefficient, $B_3$ the third virial coefficient, $B_4$ the fourth virial coefficient, etc. When $B_i = 0$ ($i=2,3,4,...$), we have $P = \rho kT$, the ideal-gas equation. Thus (3.1) corrects the ideal-gas behavior. Statistical mechanics expresses these virial coefficients in terms of intermolecular forces. Such study is the subject of the so-called imperfect-gas theories. Discussions could be found in standard texts (see, e.g., Mayer [4] and Hill [5]). We start with a grand canonical ensemble (see, e.g., Watts and McGee [6]). The partition function is

$$\Xi = \sum_{N=0}^{\infty} \frac{z_N}{N!} \int dr^N \exp[-\beta V_N(\mathbf{r}^N)]$$                                      (3.2)

where the activity $z$ is defined as $\beta\mu/\Lambda^3$. It has the units of density (or concentration). The partition function is related to the thermodynamic grand potential by

$$PV = kT \ln \Xi$$                                                                                                      (3.3)

Since

$$V_N(\mathbf{r}^N) = \sum_{i< j}\sum u(r_{ij}) + \sum_{i< j< k}\sum\sum u^{(3)}(r_i, r_j, r_k) + \cdots$$                     (3.4)

we could write out the first few terms of (3.2):

$$\Xi = 1 + z \int dr_1 + \frac{z^2}{2} \int dr_1\, dr_2\, e^{-\beta u(r_{12})}$$                                          (3.5)

$$+ \frac{z^3}{6} \int dr_1\, dr_2\, dr_3\, e^{-\beta[u(r_{12}) + u(r_{13}) + u(r_{23}) + u^{(3)}(\mathbf{r}_1, \mathbf{r}_2, \mathbf{r}_3)]} + \cdots$$

The Mayer functions are defined as

$$f_{12} \equiv \exp[-\beta u(r_{12})] - 1, \qquad f_{123} \equiv \exp[-\beta u^{(3)}(\mathbf{r}_1, \mathbf{r}_2, \mathbf{r}_3)] - 1$$       (3.6)

Equation (3.5) can be rewritten as

$$\Xi = 1 + zV + \frac{1}{2}z^2V^2(1 + I_2) + \frac{1}{6}z^3V^3(1 + 3I_2 + 3I_2^2 + I_3) + \cdots$$                          (3.7)

where the $I_i$'s are defined as

$$I_2 \equiv V^{-2} \int dr_1 \int dr_2 f_{12}$$                                                                           (3.8)

and

$$I_3 \equiv V^{-3} \int dr_1 \int dr_2 \int dr_3 \,[f_{12}f_{23}f_{31} + f_{123}(1 + f_{12})(1 + f_{23})(1 + f_{31})]$$       (3.9)

Therefore,

$$\frac{PV}{kT} = \ln\left[1 + zV + \frac{z^2V^2}{2}(1 + I_2) + \frac{1}{6}z^3V^3(1 + 3I_2 + 3I_2^2 + I_3) + \cdots\right]$$       (3.10)

Recall that for power series the logarithmic function can be expanded as

$$\ln (1+x) = x - \frac{x^2}{2} + \frac{x^3}{3} - \frac{x^4}{4} + - \cdots \qquad (3.11)$$

for $x<1$. Thus for $zV<<1$, (3.10) can be expanded as

$$\frac{PV}{kT} = \ln \Xi = zV + \frac{1}{2} z^2V^2(1 + I_2) + \frac{1}{6} z^3V^3(1 + 3I_2 + 3I_2^2 + I_3) \qquad (3.12)$$

$$- \frac{1}{2} z^2V^2 - \frac{1}{2} z^3V^3(1 + I_2) + \frac{1}{3} z^3V^3 + \cdots$$

The pressure is now expressed in terms of activity, $z$. However, we wish to express $PV/kT$ in terms of density. To find the relation between $z$ and $\rho$, we note that the singlet density $\rho^{(1)}$ in a GCE is defined as

$$\rho^{(1)}(\mathbf{r}_1) = \Xi^{-1} \sum_{N \geq 1} \frac{z^N}{(N-1)!} \int d\mathbf{r}_2 \cdots d\mathbf{r}_N \ \exp(-\beta V_N) \qquad (3.13)$$

$$= \Xi^{-1} \left[ \frac{z}{0!} + \frac{z^2}{1!} \int d\mathbf{r}_2 \ exp[-\beta u(r_{12})] \right.$$

$$\left. + \frac{z^3}{2!} \int d\mathbf{r}_2 \ d\mathbf{r}_3 \ exp[-\beta(u(r_{12}) + u(r_{23}) + u(r_{32}) + u^{(3)}(\mathbf{r}_1, \mathbf{r}_2, \mathbf{r}_3))] + \cdots \right]$$

The derivative of $\ln \Xi$ is

$$\left. \frac{\partial \ln \Xi}{\partial z} \right|_{T,V} = \Xi^{-1} \sum_{N \geq 1} \frac{z^{N-1}}{(N-1)!} \int d\mathbf{r}_1 \ d\mathbf{r}_2 \cdots d\mathbf{r}_N \exp(-\beta V_N) \qquad (3.14)$$

$$= \Xi^{-1} \left[ \frac{z^0}{0!} \int d\mathbf{r}_1(1) + \frac{z}{1!} \int d\mathbf{r}_1 \ d\mathbf{r}_2 \ exp[-\beta u(r_{12})] \right.$$

$$\left. + \frac{z^2}{2!} \int d\mathbf{r}_1 \ d\mathbf{r}_2 \ d\mathbf{r}_3 \ exp[-\beta(u(r_{12}) + u(r_{23}) + u(r_{31}) + u^{(3)}(\mathbf{r}_1, \mathbf{r}_2, \mathbf{r}_3))] + \cdots \right]$$

Comparison of (3.13) and (3.14) shows that

$$\int d\mathbf{r}_1 \ \rho^{(1)}(\mathbf{r}_1) = z \left[ \frac{\partial \ln \Xi}{\partial z} \right]_{T,V} \qquad (3.15)$$

In the absence of an external potential (i.e., in a homogeneous phase), $\int d\mathbf{r}_1 \rho^{(1)}(\mathbf{r}_1) = \rho V$. Thus

$$\rho = \frac{z}{V}\left[\frac{\partial \ln \Xi}{\partial z}\right]_{T,V}$$

(3.16)

We differentiate (3.13) to get

$$\frac{\partial \ln \Xi}{\partial z} = V + zV^2I_2 + \frac{1}{2}\, z^2V^3(3I_2^2 + I_3) + \cdots$$

(3.17)

Thus,

$$\rho = z + z^2VI_2 + \frac{1}{2}z^3V^2(3I_2^2 + I_3) + \cdots$$

(3.18)

$\rho$ is expressed in a series of $z$, while we wish to express $z$ in a series of $\rho$. The reversion formula in algebra says if

$$y = a_1x + a_2x^2 + a_3x^3 + a_4x^4 + \cdots$$

(3.19)

and conversely

$$x = b_1y + b_2y^2 + b_3y^3 + b_4y^4 + \cdots$$

(3.20)

then the coefficients are related by

$$b_1 = \frac{1}{a_1}, \qquad b_2 = -\frac{a_2}{a_1^3}, \qquad b_3 = \frac{2a_2^2 - a_1a_3}{a_1^5}, \quad \text{etc.}$$

(3.21)

Identifying $a_1 = 1$, $a_2 = VI_2$, $a_3 = V^2(3I_2^2 + I_3)/2$, etc., we have

$$z = \rho - VI_2\rho^2 + \frac{1}{2}\, V^2(I_2^2 - I_3)\rho^3 + \cdots$$

(3.22)

Substitution of (3.22) into (3.12) gives

$$\frac{PV}{kT} = \rho V - \frac{1}{2}\, I_2\rho^2V^2 - \frac{1}{3}\, I_3\rho^3V^3 + \cdots$$

(3.23)

Comparison with the definition of virial coefficients, (3.1), shows

$$B_2 = \frac{1}{2V}\int d\mathbf{r}_1\, d\mathbf{r}_2\, f_{12}$$

(3.24)

$$= -\frac{1}{2}\int_0^\infty dr_{12}\, 4\pi r_{12}^2\Big[\exp(-\beta u(r_{12})) - 1\Big]$$

and

$$B_3 = -\frac{1}{3} V \int dr_1\, dr_2\, dr_3\, [f_{12} f_{23} f_{31} + f_{123}(1 + f_{12})(1 + f_{23})(1 + f_{31})] \qquad (3.25)$$

Higher virial coefficients can be similarly obtained. The expressions are lengthy. For the sake of simplicity, we utilize a pictorial representation: *cluster diagrams*. The fundamentals of the cluster representation will be introduced below. We express the Mayer function $f_{12}$ by a bond connecting two labeled white circles:

$$f_{12} = \exp[-\beta u(r_{12})] - 1 \equiv \underset{1}{\text{o}} \!\!-\!\!\!-\!\! \underset{2}{\text{o}} \qquad (3.26)$$

The three-body Mayer function $f_{123}$ is represented by

$$f_{123} = \exp[-\beta u^{(3)}(\mathbf{r}_1, \mathbf{r}_2, \mathbf{r}_3)] - 1 \equiv \qquad \qquad (3.27)$$

A *diagram* is defined as a collection of circles (or vertices) connected by linear bonds (representing binary functions, e.g., the two-body Mayer functions), triangles (representing ternary functions), and rectangles (representing quaternary functions), etc. Each diagram corresponds to a mathematical integral. The rules of correspondence are as follows:

i.   The circles (or vertices) stand for the arguments (position vectors $\mathbf{r}_1, \mathbf{r}_2,...$) of the binary function.

ii.  As the position $\mathbf{r}$ is integrated, a black circle is put in its place. It is called a *field point*, or an unlabeled point. The integral is made dimensionless by multiplication by the number density $\rho$ or activity $z$ (called a $\rho$-vertex or $z$-vertex, respectively). We shall use the $\rho$-vertices:

$$\bullet = \rho \int d\mathbf{r} \qquad (3.28)$$

iii. When the argument is not integrated, we represent $\mathbf{r}$ by a white circle o called the *root point* or labeled point:

$$\underset{i}{\text{o}} = \mathbf{r}_i, \qquad i = 1,2,3,... \qquad (3.29)$$

iv.  A collection of vertices connected (or unconnected) by linear bonds, triangles, or rectangles is interpreted as a product. Thus

$$= 6(f_{14}f_{15}f_{23}f_{235}) \tag{3.30}$$

Whenever the position $\mathbf{r}_i$ is integrated, the circle is filled:

$$\rho \int dr_j \, f_{ij} \, f_{jk} = \; \circ\!\!-\!\!\bullet\!\!-\!\!\circ \tag{3.31}$$

Note that we do not use the permutation convention here (no division by factorials). In terms of cluster diagrams, $B_2$ as given by (3.24) can be written as

$$B_2 = -\frac{1}{2V\rho^2} \; \bullet\!\!-\!\!\bullet \tag{3.32}$$

Similarly,

$$B_3 = -\frac{1}{3V\rho^3}\left[\; \triangle + \triangle + 3\,\triangle + 3\,\triangle + \triangle \;\right] \tag{3.33}$$

We could also express $B_4$, $B_5$, etc., in terms of cluster diagrams. For simplicity, we assume pairwise additivity (therefore $u^{(3)} = 0$, $u^{(4)} = 0$,...). Then

$$B_4 = -\frac{1}{8V\rho^4}\left[\; 3\,\square + 6\,\square + \boxtimes \;\right] \tag{3.34}$$

$$B_5 = \frac{1}{30V\rho^5}\left[\; 12\; + 60\; + 10\; + 60\; + 30\; \right. \tag{3.35}$$

$$\left. + 10\; + 15\; + 30\; + 10\; + \; \right]$$

To evaluate virial coefficients higher than the third is a difficult numerical problem. Recently, numerical procedures (e.g., the Monte Carlo (MC) method) have been devised, and calculations have been carried out for simple pair potentials (e.g., for the hard spheres, up to $B_{10}$ [7]).

We note that in the derivation of the virial expansion (3.12), the activity $z$ must be small in order to guarantee convergence. Thus the virial cluster equation is valid only at relatively low densities. Lebowitz and Penrose [8] studied the convergence of the virial expansion. The radius of convergence in the infinite volume expansion is bounded from below by $0.28952/(x+1)B$, where

$$\ln x \equiv - \text{Min } L^{-1} \sum_{i<j\leq L} \frac{2u(r_{ij})}{kT} \tag{3.36}$$

and

$$B \equiv \int d\mathbf{r} \; [e^{-\beta u(r)} - 1] \tag{3.37}$$

The irreducible Mayer cluster integrals $\beta_k$ are also bounded from above by

$$\beta_k = \frac{1}{k} \left[ \frac{(x+1)B}{0.28952} \right]^k \tag{3.38}$$

where k = the rank of the cluster.

## V.4. The Isothermal Compressibility

In the discussion of the GCE, we derived the expression for the average number, $<N>$,

$$<N> = kT \frac{\partial \ln \Xi}{\partial \mu} \bigg|_{T,V} \tag{4.1}$$

The fluctuations (or variance) around the mean number are

$$<N^2> - <N>^2 = (kT)^2 \frac{\partial \ln \Xi}{\partial \mu^2} \bigg|_{V,T} = kT \frac{\partial <N>}{\partial \mu} \bigg|_{V,T} \tag{4.2}$$

From thermodynamics, we know that

$$d\mu = vdP \qquad \text{(at constant T)} \tag{4.3}$$

$$\frac{\partial \mu}{\partial v} \bigg|_{T} = v \frac{\partial P}{\partial v} \bigg|_{T} \tag{4.4}$$

Since the specific volume $v = V/N$, $dv = -(V/N^2) \, dN$,

$$-\frac{N^2}{V}\left[\frac{\partial\mu}{\partial N}\right]_{V,T} = v\left[\frac{\partial P}{\partial v}\right]_{N,T} \tag{4.5}$$

we have

$$\left.\frac{\partial N}{\partial\mu}\right|_{V,T} = -\frac{N^2}{V^2}\left.\frac{\partial V}{\partial P}\right|_{N,T} \tag{4.6}$$

If we define the isothermal compressibility $K_T$ to be

$$K_T = -\frac{1}{V}\left.\frac{\partial V}{\partial P}\right|_{N,T} \tag{4.7}$$

then

$$\left.\frac{\partial N}{\partial\mu}\right|_{V,T} = \frac{N^2}{V}K_T \tag{4.8}$$

or, from (4.2)

$$<N^2> - <N>^2 = <N>\rho kTK_T \tag{4.9}$$

On the other hand, the pair density $\rho^{(2)}$ in the GCE was defined as

$$\rho^{(2)}(\mathbf{r}_1, \mathbf{r}_2) = \Xi^{-1}\sum_{N\geq 2}\frac{z^N}{(N-2)!}\int dr_3\cdots dr_N\exp(-\beta V_N) \tag{4.10}$$

Integration of $\rho^{(2)}$ over $\mathbf{r}_1$ and $\mathbf{r}_2$ gives

$$\int dr_1\, dr_2\, \rho^{(2)}(\mathbf{r}_1, \mathbf{r}_2) = \Xi^{-1}\sum_{N\geq 2}N(N-1)\left[\frac{z^N Q_N}{N!}\right] \tag{4.11}$$

However, the factor $(\Xi)^{-1}(z^N Q_N/N!)$ is the probability $P_N$ of having $N$ particles in the open system. Thus

$$\int dr_1\, dr_2\, \rho^{(2)}(\mathbf{r}_1, \mathbf{r}_2) = <N^2 - N> = <N^2> - <N> \tag{4.12}$$

Comparison with (4.9) gives

$$<N^2> = <N> + \int dr_1\, dr_2\, \rho^{(2)}(\mathbf{r}_1, \mathbf{r}_2) = <N>^2 + <N>\rho kTK_T \tag{4.13}$$

Since $\rho^{(2)} = \rho^2 g^{(2)}$, we arrive at

$$\rho k T K_T - 1 = \frac{\rho}{V} \int d\mathbf{r}_1 \, d\mathbf{r}_2 \, g^{(2)}(r_1, r_2) - \langle N \rangle \qquad (4.14)$$

For an isotropic fluid,

$$\rho k T K_T - 1 = \rho \int_0^\infty dr \, 4\pi r^2 [g(r) - 1] = \rho \int_0^\infty dr \, 4\pi r^2 h(r) \qquad (4.15)$$

This is the *isothermal compressibility* equation. Knowledge of the pcf leads directly to the isothermal compressibility $K_T$. One can further integrate $K_T$ to obtain the pressure. Two methods are now at our disposal for calculating the pressure: the virial (pressure) equation (2.10) and the isothermal compressibility equation (4.15). For an exact rdf, the two pressures (either from (2.10), or from (4.15)) are the same, i.e., *internally consistent*. However, for approximate rdf, the two pressures differ. The magnitude of the difference can be used as a measure of errors in the proposed approximation. In section V.8, we shall discuss the consistency conditions of the pcfs.

## V.5. *The Inverse Isothermal Compressibility*

The inverse isothermal compressibility $\Pi_T$ is defined as

$$\Pi_T \equiv \beta \frac{\partial P}{\partial \rho}\bigg|_T \qquad (5.1)$$

It is related to the isothermal compressibility by

$$\Pi_T = (\rho k T \, K_T)^{-1} \qquad (5.2)$$

$\Pi_T$ can be calculated directly from the dcfs $C(r)$. Let us examine the OZ equation for $C(\mathbf{r}_i, \mathbf{r}_j)$:

$$h(\mathbf{r}_1, \mathbf{r}_2) = C(\mathbf{r}_1, \mathbf{r}_2) + \rho \int d\mathbf{r}_3 \, h(\mathbf{r}_1, \mathbf{r}_3) \, C(\mathbf{r}_3, \mathbf{r}_2) \qquad (5.3)$$

For isotropic functions, the three-dimensional Fourier transforms of $h$ and $C$ are given by

$$\tilde{h}(k) = \int d\mathbf{r} \, e^{i\mathbf{k}\cdot\mathbf{r}} h(\mathbf{r}) = \frac{4\pi}{k} \int_0^\infty dr \, r \sin(kr) h(r) \qquad (5.4)$$

where $\mathbf{r} = \mathbf{r}_2 - \mathbf{r}_1$ and $r = |\mathbf{r}|$. Also

$$\tilde{C}(k) = \int d\mathbf{r} \, e^{-i\mathbf{k}\cdot\mathbf{r}} C(\mathbf{r}) = \frac{4\pi}{k} \int_0^\infty dr \, r \sin(kr) C(r) \qquad (5.5)$$

Applying the Fourier transform to (5.3) gives

$$\tilde{h}(k) = \tilde{C}(k) + \rho\tilde{h}(k)\tilde{C}(k) \tag{5.6}$$

Rearrangement gives

$$\tilde{h}(k) = \frac{\tilde{C}(k)}{1 - \rho\tilde{C}(k)} \tag{5.7}$$

and

$$1 + \rho\tilde{h}(k) = \frac{1}{1 - \rho\tilde{C}(k)} \tag{5.8}$$

Now as the wave vector $k$ goes to zero

$$\lim_{k \to 0} \rho\tilde{h}(k) = \lim_{k \to 0} 4\pi\rho \int_0^\infty dr \; r^2 \frac{\sin(kr)}{kr} h(r) \tag{5.9}$$

$$= 4\pi\rho \int_0^\infty dr \; r^2 h(r) = \rho k T K_T - 1$$

Namely,

$$\rho k T K_T = 1 + \rho\tilde{h}(0) = \frac{1}{1 - \rho\tilde{C}(0)} \tag{5.10}$$

Thus,

$$1 - \rho\tilde{C}(0) = (\rho k T K_T)^{-1} = \Pi_T \tag{5.11}$$

or

$$\beta \frac{\partial P}{\partial \rho}\bigg|_T = 1 - \rho \int_0^\infty dr \; 4\pi r^2 C(r) \tag{5.12}$$

noting that $\tilde{C}(0) = 4\pi \int dr \; r^2 C(r)$. We remark that eqs. (4.15) and (5.12) are independent of the pairwise additivity assumption since, in the derivation, there was no need of such hypotheses. When triplet forces are present, (4.15) and (5.12) remain the same, in contrast to the three-body virial (pressure) equation (2.11), where additional terms appear.

We know from acoustics [9] that the velocity of sound, $c$, is given by

$$c^2 = \frac{C_p}{C_v} \frac{\partial P}{\partial \rho}\bigg|_T = \frac{C_p}{C_v} k T \Pi_T \tag{5.13}$$

Thus,

$$c^2 = \frac{C_p}{C_v} k T \left[ 1 - \rho \int_0^\infty dr \; 4\pi r^2 C(r) \right] \tag{5.14}$$

Knowing $C(r)$, we can calculate the velocity of sound.

## V.6. *Chemical Potential*

We denote the Helmholtz free energy of an $N$-body system by $A(N)$. In thermodynamics, the chemical potential $\mu$ is given by

$$\mu = \frac{\partial A(N)}{\partial N}\bigg|_{T,V} \tag{6.1}$$

For large $N$, we can approximate the derivative by a difference

$$\frac{\partial A(N)}{\partial N}\bigg|_{T,V} \approx A(N) - A(N-1) \tag{6.2}$$

Since

$$A(N) = -kT \ln\left[\frac{Q_N}{N!\Lambda^{3N}}\right] = -kT \ln Q_N + kT \ln N! + NkT \ln \Lambda^3 \tag{6.3}$$

and

$$A(N-1) = -kT \ln Q_{N-1} + kT \ln (N-1)! + (N-1)kT \ln \Lambda^3 \tag{6.4}$$

$$-\frac{\mu}{kT} = \ln\left[\frac{Q_N}{Q_{N-1}}\right] - \ln N - \ln \Lambda^3 \tag{6.5}$$

Let us single out an "outstanding" molecule with label "1" and rewrite the total potential as

$$V_N(\mathbf{r}_1, \ldots, \mathbf{r}_N) = \sum_{j=2}^{N} \xi u(\mathbf{r}_1, \mathbf{r}_j) + \sum_{2=k<m}^{N}\sum u(\mathbf{r}_k, \mathbf{r}_m) \tag{6.6}$$

introducing a coupling parameter $\xi$, $0 \leq \xi \leq 1$. When $\xi = 0$, molecule 1 is decoupled and the total potential is free of 1: $V_N = V_{N-1}$. As $\xi = 1$, we have fully restored the $N$-particle system. $\xi$ is called the charging parameter. It can also be shown that

$$Q_N(\xi = 0) = VQ_{N-1} \tag{6.7}$$

Thus,

$$\ln\left[\frac{Q_N(\xi=1)}{Q_{N-1}}\right] = \ln\left[\frac{Q_N(\xi=1)}{Q_N(\xi=0)}\right] + \ln V = \int_0^1 d\xi\, \frac{\partial \ln Q_N(\xi)}{\partial \xi} + \ln V \tag{6.8}$$

But

$$\frac{\partial \ln Q_N(\xi)}{\partial \xi} = -\frac{1}{NkT} \int d\mathbf{r}_1 \, d\mathbf{r}_2 \, u(r_{12})\rho^{(2)}(\mathbf{r}_1, \mathbf{r}_2; \xi) \tag{6.9}$$

Thus,

$$\frac{\mu}{kT} = \ln(\rho\Lambda^3) + \frac{1}{NkT} \int_0^1 d\xi \int_V d\mathbf{r}_1 \int_V d\mathbf{r}_2 \, u(\mathbf{r}_1, \mathbf{r}_2)\rho^{(2)}(\mathbf{r}_1, \mathbf{r}_2; \xi) \tag{6.10}$$

or, for isotropic fluids,

$$\frac{\mu}{kT} = \ln(\rho\Lambda^3) + \frac{\rho}{kT} \int_0^1 d\xi \int_0^\infty dr \, 4\pi r^2 u(r)g(r; \xi) \tag{6.11}$$

We note that in order to calculate $\mu$ from (6.11), we must evaluate the rdf $g(r; xi)$ at different values of $\xi$ $(0 \leq \xi \leq 1)$ according to (6.6). Only one particle has been singled out by the charging parameter. The first term on the RHS of (6.11) is the ideal-gas contribution (see, e.g., eq. (3.20)). The second term is the correction due to interaction, $u(r)$. From Chapter III, we know that for an ideal gas

$$\frac{\mu^*}{kT} = \ln P^* + \ln\frac{\Lambda^3}{kT} \qquad \text{(ideal gas)} \tag{6.12}$$

As in thermodynamics, we define, accordingly, the fugacity $f$ of a nonideal gas as

$$\frac{\mu}{kT} \equiv \ln f + \ln\frac{\Lambda^3}{kT} \tag{6.13}$$

Comparison of (6.13) with (6.11) gives

$$\ln f = \ln(\rho kT) + \beta\rho \int_0^1 d\xi \int_0^\infty dr \, 4\pi r^2 u(r)g(r; \xi) \tag{6.14}$$

In terms of the fugacity coefficient $\phi \equiv f/P$, we have

$$\ln \phi = \beta\rho \int_0^1 d\xi \int_0^\infty dr \, 4\pi r^2 u(r)g(r; \xi) - \ln\left[\frac{P}{\rho kT}\right] \tag{6.15}$$

Therefore if the interaction vanishes (i.e., $u=0$), $f = \rho kT = P^*$ and $\phi=1$. The corrections $f$ and $\phi$ are due to the interaction potential $u(r)$.

## V.7.  *The Potential Distribution Theorem*

Recent advances have prompted frequent applications of the potential distribution theorem (PDT), first proposed by Widom [10], in computer simulation of the chemical potential. This approach uses a *test particle*; i.e., a particle (at $\mathbf{r}_N$) is singled out

$$V_N(1,2,...,N-1,N) = \sum_{1=i<j}^{N} u(i,j) \tag{7.1}$$

$$= V_{N-1}(1,2,...,N-1) + \sum_{k=1}^{N-1} u(k,N)$$

We denote the potential field generated by particle $N$ (at $\mathbf{r}_N$) by

$$U(N) = \sum_{k=1}^{N-1} u(k,N) \tag{7.2}$$

Particle $N$ is also called the *source particle*. The configurational integral for the remaining $N-1$ particles is written as

$$Q_{N-1} = \int d1\ d2 \cdots d(N-1) e^{-\beta V_{N-1}} \tag{7.3}$$

Now if we treat the potential $U$ as an external potential (i.e., a potential imposed from outside), the change in the configurational integral is

$$Q_{N-1}[U] = \int d1\ d2 \cdots d(N-1)\ e^{-\beta V_{N-1}} e^{-\beta U} \tag{7.4}$$

$$\equiv Q_{N-1} <\exp[-\beta U]>_{N-1}$$

where we have defined an ensemble average $< \cdots >_{N-1}$ for the *external* potential $U$. On the other hand, the $N$-body configurational integral is

$$Q_N = \int d1\ d2 \cdots d(N-1)\ dN\ e^{-\beta V_N} \tag{7.5}$$

$$= \int d1 \cdots d(N-1)\ dN\ e^{-\beta V_{N-1}} e^{-\beta U}$$

$$= \int dN\ Q_{N-1}[U]$$

Noting (7.4), we have

$$Q_N = \int dN\ Q_{N-1} <\exp[-\beta U]>_{N-1} \tag{7.6}$$

$$= Q_{N-1} \frac{\int d1\ d2 \cdots d(N-1)\ dN\ \exp[-\beta V_{N-1}]\exp[-\beta U]}{Q_{N-1}}$$

$$= Q_{N-1} V \frac{\int d1\ d2 \cdots d(N-1)\ \exp[-\beta V_{N-1}]\exp[-\beta U]}{Q_{N-1}}$$

$$= Q_{N-1}V<\exp[-\beta U]>_{N-1}$$

In the last equality we have integrated over the coordinate $dr_N$, assuming a homogeneous fluid. (All coordinates were expressed relative to position $r_N$. Such translation is allowed since $N$ is a test particle.) On the other hand, the chemical potential is given by (6.2). Thus,

$$\exp[-\beta\mu] = \frac{Q_N}{NQ_{N-1}\Lambda^3} \tag{7.7}$$

Comparison with (7.6) gives

$$\beta\mu = \ln[\rho\Lambda^3] - \ln<\exp[-\beta U]>_{N-1} \tag{7.8}$$

This is the *potential distribution theorem*: the chemical potential is given by the ensemble average of a test particle over the $N$–1-body system. In practice, it is more convenient to work with the full $N$-body system. Shing and Gubbins [11] introduced the inverse PDT. They write

$$<\exp[-\beta U]>_{N-1} = \frac{\int d1\ d2 \cdots d(N-1)\exp[-\beta V_{N-1}]\exp[-\beta U]}{\int d1d2 \cdots d(N-1)\exp[-\beta V_{N-1}]} \tag{7.9}$$

$$= \frac{\int d1\ d2 \cdots d(N-1)\exp[-\beta V_N]}{\int d1\ d2 \cdots d(N-1)\exp[-\beta V_N]\exp[\beta U]}$$

Integrating with respect to $r_N$ while keeping the ratio constant, we obtain

$$<\exp[-\beta U]>_{N-1} = \frac{\int d1\ d2 \cdots d(N-1)\int dN\ \exp[-\beta V_N]}{\int d1\ d2 \cdots d(N-1)\int dN\ \exp[-\beta V_N]\exp[\beta U]} \tag{7.10}$$

$$= \frac{1}{<\exp[\beta U]>_N}$$

Thus the PDT (7.8) can be written in terms of $N$-body ensemble averages:

$$\beta\mu = \ln[\rho\Lambda^3] + \ln<\exp[\beta U]>_N \tag{7.11}$$

This formula has been used in the simulation of chemical potentials.

**POTENTIAL DISTRIBUTION UNDER THE EXTERNAL POTENTIAL w(r)**

Furthermore, for nonuniform systems with a bona fide external potential $w(r)$ (e.g. a *wall*, or an electric field), the $N$-body potential should be written as

$$V_N[w] \equiv \sum_{1=i<j}^{N} u(i,j) + \sum_{l=1}^{N} w(l) \tag{7.12}$$

$$= V_{N-1}[w] + \sum_{k=1}^{N-1} u(k,N) + w(N)$$

with the $N-1$-body nonuniform potential $V_{N-1}[w]$ defined the same way by replacing $N$ by $N-1$ in eq. (7.12). The configurational part of the canonical partition function $Q[w]$ is now defined as

$$Q_N[w] \equiv \int d1 \ d2 \cdots d(N-1) \ dN \ e^{-\beta V_N[w]} \tag{7.13}$$

$$= \int d1 \cdots d(N-1) \ dN \ e^{-\beta V_{N-1}} e^{-\beta U(N)} e^{-\beta w(N)} \cdot \exp[-\beta \sum_{l}^{N-1} w(l)]$$

Similarly,

$$Q_{N-1}[w] \equiv \int d1 \ d2 \cdots d(N-1) \ e^{-\beta V_{N-1}[w]} \tag{7.14}$$

$$= \int d1 \ d2 \cdots d(N-1) \ e^{-\beta V_{N-1}} \cdot \exp[-\beta \sum_{l}^{N-1} w(l)]$$

The ensemble average of any dynamic variable $B(r^N)$ in the $N$-body system is defined as usual

$$<B>_{N,w} \equiv \frac{1}{Q_N[w]} \int d1 \cdots d(N-1) \ dN \ B(\{N\}) \ e^{-\beta V_N[w]} \tag{7.15}$$

The subscript $w$ is to emphasize the nonuniformity. The definition for ensemble averages $<B>_{N-1,w}$ in the $N-1$-body system is the same upon changing $N$ to $N-1$. We shall derive the potential distribution theorem under the influence of a wall potential $w(\cdot)$. Let us examine the definition of the singlet density in an $N$-body system

$$\rho^{(1)}(N) \equiv \frac{N}{Q_N[w]} \int d1 \cdots d(N-1) \ e^{-\beta V_N[w])} \tag{7.16}$$

$$= \frac{N}{Q_N[w]} \int d1 \cdots d(N-1) \ e^{-\beta V_{N-1}[w]} e^{-\beta U(N)} e^{-\beta w(N)}$$

Applying the definition of the $N-1$-body configurational integral $Q_{N-1}[w]$, we have

$$\rho^{(1)}(N) = \frac{N}{Q_N[w]} e^{-\beta w(N)} Q_{N-1}[w] <e^{-\beta U(N)}>_{N-1,w} \tag{7.17}$$

Thus,

$$<e^{-\beta U(N)}>_{N-1,w} = \rho^{(1)}(N) \, e^{\beta w(N)} \frac{Q_N[w]}{N Q_{N-1}[w]} \tag{7.18}$$

Noting (7.7), we get by placing the $N$th particle at $\mathbf{r}$

$$<e^{-\beta U(\mathbf{r})}>_{N-1,w} = \rho^{(1)}(\mathbf{r})\Lambda^3 \, e^{\beta w(\mathbf{r}) - \beta\mu} \tag{7.19}$$

This is the *potential distribution theorem* for *inhomogeneous systems* (in this case, due to the wall potential $w(\mathbf{r})$). We have earlier introduced a one-body direct correlation function $C^{(1)}$ (eq. (IV.4.6)) as

$$C^{(1)}(\mathbf{r}) \equiv \ln [\rho^{(1)}(\mathbf{r})\Lambda^3] - \beta\mu + \beta w(\mathbf{r}) \tag{7.20}$$

Thus it is related to the potential distribution theorem by

$$C^{(1)}(\mathbf{r}) = \ln<\exp[-\beta U(\mathbf{r})]>_{N-1,w} \tag{7.21}$$

The singlet direct correlation function plays an important role in the study of the structure of uniform and nonuniform fluids. For example, the hypernetted chain integral equation is *generated* by $C^{(1)}$ as a functional of the singlet density $\rho^{(1)}$ (see next chapter). On the other hand, the Percus-Yevick equation is *generated* by $\exp(C^{(1)})$. The two-body direct correlation functions is the functional density derivative of $C^{(1)}$. The usefulness of $C^{(1)}$ could be gauged by its definition. It is the nonuniform counterpart of the thermodynamic chemical potential (the *nonuniform chemical potential*). Examination of the partition function in a grand canonical ensemble

$$\Xi[w] = \sum_{N\geq 0}^{\infty} \frac{1}{N!\Lambda^{3N}} \int d\{N\} \, \exp[-\beta V_N] \exp[-\beta\sum_{l=1}^{N} (w(l) - \mu)] \tag{7.22}$$

shows that the external potential $w(\cdot)$ counteracts the chemical potential in the exponent $(\mu - w(\cdot))/kT$. Thus the two potentials are *affine* (akin to each other). In the limit $w(r)=0$, $C^{(1)} = \ln (\rho\Lambda^3) - \beta\mu$. Therefore for homogeneous systems, $C^{(1)}$ reduces to the *nonideal-gas* part of the chemical potential. J.R. Henderson [12] showed that eq. (7.20) is equivalent to the Kirkwood formula (6.11) for isotropic fluids. In addition, it gives the mechanical stability (hydrostatic equilibrium) conditions for nonuniform fluids. We shall discuss these equivalences below.

## CHARGING THE CHEMICAL POTENTIAL

We shall derive (6.11) from the PDT discussed above. Let us consider another variation of the N-body potential

$$V_N(\xi) = V_{N-1} + \xi U(N), \qquad 0 \leq \xi \leq 1 \tag{7.23}$$

where $U(N)$ is defined as before (eq. (7.2)). Thus when $\xi = 0$, we have $V_{N-1}$; and when $\xi = 1$, we recover $V_N$. In this case,

$$C^{(1)}(\mathbf{r};\xi) = \ln <\exp(-\beta\xi U(\mathbf{r})>_{N-1} \tag{7.24}$$

Thus the nonideal part of the singlet dcf is

$$C^{(1)}(\mathbf{r})\Big|_{\xi=1} = \int_0^1 d\xi \, \frac{dC^{(1)}(\mathbf{r};\xi)}{d\xi} \tag{7.25}$$

Carrying out the differentiation with respect to $\xi$, we get

$$-kT \, C^{(1)}(\mathbf{r}) = \int_0^1 d\xi \, \frac{<U(\mathbf{r}) \, \exp[-\beta\xi U(\mathbf{r})]>_{N-1}}{\exp[-\beta\xi U(\mathbf{r})]>_{N-1}} \tag{7.26}$$

$$= \int_0^1 d\xi \, \frac{<\delta(\mathbf{r},N)U(N)>_N}{<\delta(\mathbf{r},N)>_N}$$

where we have upgraded to an $N$-body ensemble. For $U$ given by (7.2), eq. (7.26) is simply

$$-kT \, C^{(1)}(\mathbf{r}) = \int_0^1 d\xi \int d\mathbf{r}' \, u(\mathbf{r},\mathbf{r}') \, \rho^{(2)}(\mathbf{r},\mathbf{r}';\xi)/\rho^{(1)}(\mathbf{r}) \tag{7.27}$$

This is the Kirkwood formula when the definition of $C^{(1)}$ (7.20) is applied.

## MECHANICAL EQUILIBRIUM

Let $v = x$, y or z (one of the three Cartesian coordinates). The gradient of the singlet dcf in the $v$ direction with respect to the argument $\mathbf{r}$ of $C^{(1)}$ could be calculated as

$$-kT \, \nabla^v C^{(1)}(\mathbf{r}) = \frac{<\nabla^v U(\mathbf{r})\exp[-\beta U(\mathbf{r})]>_{N-1}}{<\exp[-\beta U(\mathbf{r})]>_{N-1}} \tag{7.28}$$

$$= \frac{<\delta(\mathbf{r},\mathbf{r}_N) \, \nabla^v_{\mathbf{r}_N} V_N(\mathbf{r}^N)>_N}{<\delta(\mathbf{r},\mathbf{r}_N)>_N}$$

$$= \frac{<\sum_i \delta(\mathbf{r},\mathbf{r}_i) \, \nabla^v_{\mathbf{r}_i} V_N(\mathbf{r}^N)>_N}{<\sum_i \delta(\mathbf{r},\mathbf{r}_i)>_N} = \frac{\nabla^\mu P_c^{\mu v}(\mathbf{r})}{\rho^{(1)}(\mathbf{r})}$$

where

$$\nabla^\mu P_c^{\mu\nu}(\mathbf{r}) \equiv \langle\sum_i \delta(\mathbf{r},\mathbf{r}_i)\, \nabla_{\mathbf{r}_i}^\nu V_N(\mathbf{r}^N)\rangle_N \tag{7.29}$$

is the gradient of the configurational part of the pressure tensor. The only other contribution to the pressure tensor is from the kinetic (ideal-gas) energy, $kT\rho^{(1)}\delta^{\mu\nu}$. Thus from the inhomogeneous $C^{(1)}$ (eq. (7.20), we obtain

$$\nabla^\mu P_c^{\mu\nu}(\mathbf{r}) = -\rho^{(1)}(\mathbf{r})\, \nabla^\nu w(\mathbf{r}) \tag{7.30}$$

This is an expression of the *mechanical equilibrium* (or hydrostatic stability) condition in an inhomogeneous fluid under the influence of $w(\mathbf{r})$. It is a consequence of the conservation of *linear* momentum. We note that in hydrostatics, the liquid pressure is given by

$$\nabla P = -\rho g \tag{7.31}$$

noting that the gradient of the potential of gravity is $\nabla^z(gz) = g$. Thus the PDT is an integrated form of the hydrostatic condition. It also says that the constancy of chemical potential implies mechanical equilibrium. These concepts are important in the studies of nonuniform fluids.

## V.8. *Helmholtz Free Energy*

Let us consider an $N$-body system with the total potential energy containing a charging parameter:

$$V_N(\mathbf{r}^N; \xi) = \sum_{i<j}^N \xi u(r_i, r_j) \tag{8.1}$$

In other words, when $\xi =0$, we have an ideal gas (involving no interaction), and when $\xi =1$, we have the full system. The partition function in the canonical ensemble is

$$Z_N(\xi) = \frac{1}{N!\Lambda^{3N}} \int dr^N \exp\left[-\beta\sum_{i<j}^N \xi u(r_i, r_j)\right] \tag{8.2}$$

and the Helmholtz free energy is related to $Z_N$ by

$$A(\xi) = -kT \ln Z_N(\xi) = -kT[\ln Q_N(\xi) - \ln(N!\Lambda^{3N})] \tag{8.3}$$

where $\ln[V^N/N!\Lambda^{3N}]$ is the ideal-gas contribution and $\ln Q_N(\xi)/V^N$ is the configurational correction. We write

$$A(\xi=1) = \int_0^1 d\xi\, \frac{\partial A(\xi)}{\partial \xi} + A(\xi=0) \tag{8.4}$$

Carrying out the differentiation gives

$$\frac{\partial A(\xi)}{\partial \xi} = Q_N^{-1} \int d\mathbf{r}_1 \, d\mathbf{r}_2 \, u(\mathbf{r}_1, \mathbf{r}_2) \frac{N(N-1)}{2} \int d\mathbf{r}^{N-2} \exp[-\beta \sum \sum \xi u(\mathbf{r}_i, \mathbf{r}_j)] \qquad (8.5)$$

$$= \frac{1}{2} \int d\mathbf{r}_1 \, d\mathbf{r}_2 \, u(\mathbf{r}_1, \mathbf{r}_2) \rho^{(2)}(\mathbf{r}_1, \mathbf{r}_2; \xi)$$

$$= \frac{\rho N}{2} \int_0^\infty dr \, 4\pi r^2 u(r) g(r; \xi)$$

Therefore,

$$A = A(\xi=1) = A(\xi=0) + \int_0^1 d\xi \, \frac{\partial A(\xi)}{\partial \xi} \qquad (8.6)$$

$$= NkT[\ln(\rho \Lambda^3) - 1] + \frac{\rho N}{2} \int_0^1 d\xi \int_0^\infty dr \, 4\pi r^2 u(r) g(r; \xi)$$

Knowing the rdf $g(r;\xi)$ at different values of $\xi$ $(0 \le \xi \le 1)$, we can use (8.6) to calculate the Helmholtz free energy. We note that the charging parameter $\xi$ is attached to all pairs $\xi u(r_{ij})$ not limited to one molecule. In practice, the use of $\xi$ can be incorporated into the temperature. Note that the pair potential $u(r)$ appears as a product $\beta u(r)$ in the partition function. With $\xi$, we simply include $\xi \beta u$. The changes in $\xi$ are effectively changes in temperature (via the product $\xi \beta$). The rdf at different temperatures are obtained from

$$g(r; \xi=x) = g(r) \rfloor_{\text{at } kT/x} \qquad (8.7)$$

For example, if $\xi=0.5$, we evaluate a $g(r)$ at a temperature twice the original temperature, i.e., $2kT$. Equation (8.6) is thermodynamically equivalent to the Gibbs-Helmholtz relation

$$\frac{A - A_0}{NkT} = \int d(\frac{1}{kT}) \frac{U}{N} \qquad (8.8)$$

## V.9. *The Hiroike Consistency*

The total differential of $A/T$ in thermodynamics is given by

$$d(\frac{A}{T}) = -\frac{P}{T} \, dV - \frac{U}{T^2} \, dT \qquad (9.1)$$

Since this is an exact differential, the partial derivatives must satisfy

$$\frac{\partial}{\partial T} \left[ \frac{P}{T} \right]_{V,N} = \frac{\partial}{\partial V} \left[ \frac{U}{T^2} \right]_{T,N} \qquad (9.2)$$

We can obtain the pressure from the virial (pressure) equation (2.10) and the energy from (1.9). We expect (9.2) to be satisfied by $P$ and $U$ thus obtained. However, this is the case only when the rdf used is *exact*. When $g(r)$ is obtained using some approximate equations, the equality (9.2) may not be satisfied. As a consequence, the Helmholtz free energy derived from integration of the pressure differs from that derived from the energy. An element of inconsistency is present.

Hiroike [13] examined the problem and derived the necessary and sufficient conditions of consistency. Note that from (8.6)

$$\frac{A(\xi)}{NkT} = \ln \rho \Lambda^3 - 1 + \frac{\beta \rho}{2} \int_0^\xi d\eta \int_0^\infty dr \, 4\pi r^2 u(r) g(r; \eta) \tag{9.3}$$

On the other hand, if we take the functional derivative of $A$, as given by (8.2) and (8.3), in terms of the partition function $Z_N$ with respect to the pair potential $u(s)$, we get

$$\frac{\delta A(\xi)}{\delta u(s)} = \frac{N\rho}{2} \xi g(s; \xi) \tag{9.4}$$

We can also differentiate (9.3) functionally with respect to $u(s)$:

$$\frac{\delta A(\xi)}{\delta u(s)} = \frac{N\rho}{2} \int_0^\xi d\eta \int dr \left[ \frac{\delta u(r)}{\delta u(s)} g(r; \eta) + u(r) \frac{\delta g(r; \eta)}{\delta u(s)} \right] \tag{9.5}$$

$$= \frac{N\rho}{2} \int_0^\xi d\eta \int dr \left[ \delta(r, s) g(r; \eta) + u(r) \frac{\delta g(r; \eta)}{\delta u(s)} \right]$$

$$= \frac{N\rho}{2} \int_0^\xi d\eta \, g(s; \eta) + \frac{N\rho}{2} \int_0^\xi d\eta \int dr \, u(r) \frac{\delta g(r; \eta)}{\delta u(s)}$$

Combining (9.4) and (9.5) gives

$$\xi g(s; \xi) = \int_0^\xi d\eta \, g(s; \eta) + \int_0^\xi d\eta \int dr \, u(r) \frac{\delta g(r; \eta)}{\delta u(s)} \tag{9.6}$$

Taking partial derivatives with respect to $\xi$ gives

$$g(s; \xi) + \xi \frac{\partial g(r; \xi)}{\partial \xi} = g(s; \xi) + \int dr \, u(r) \frac{\delta g(r; \xi)}{\delta u(s)} \tag{9.7}$$

or

$$\xi \frac{\partial g(s; \xi)}{\partial \xi} = \int dr \, u(r) \frac{\delta g(r; \xi)}{\delta u(s)} \tag{9.8}$$

Equation (9.8) is the *Hiroike consistency condition*, necessary and sufficient, for the equality (9.2). Not all approximate theories of the rdf satisfy the Hiroike consistency condition. For example, the Percus-Yevick (PY) equation does not satisfy (9.8), but the hypernetted chain (HNC) equation does. Also satisfaction of the Hiroike condition does not necessarily guarantee good performance of the theory: e.g., the PY equation performs better than the HNC equation for potentials with short-range repulsive forces. On the other hand, the HNC equation is more accurate for Coulomb forces. Thus examination of individual cases is indispensable in discriminating among the theories.

## V.10. *The Pressure Consistency Conditions*

The pressure consistency— that the pressure derived from the virial expression is equal to the one derived from the isothermal compressibility— is important for correlation functions derived from integral equations. These equations are approximate. Thus the correlation functions are not exact, and inconsistencies result. We give below the sufficient and necessary conditions for consistency. The method follows the formalism of Lebowitz and Percus [14] for inhomogeneous systems.

The virial expression when the triplet interaction is present assumes the form

$$\beta P^v = \rho - \frac{\beta}{6V} \int d1 \ d2 \ r_{12} u'(1,2)[F_2(1,2) + \rho^2] \tag{10.1}$$

$$- \frac{\beta}{18V} \int d1 \ d2 \ d3 \ K(1,2,3)\{\rho^3 + \rho[F_2(1,2) + F_2(2,3) + F_2(3,1)] + F_3(1,2,3)\}$$

where

$$K(1,2,3) = \sum_{1 \le i < j < 3} r_{ij} \frac{du^{(3)}(1,2,3)}{dr_{ij}} \tag{10.2}$$

whereas the total potential $V_N$ is given by

$$V_N(1,..., N) = \sum_{1 \le i < j < N} u^{(2)}(i,j) + \sum_{1 \le l < m < n < N} u^{(3)}(l,m,n) \tag{10.3}$$

$P^v$ is the virial pressure, $V$ is the volume, $\beta = (kT)^{-1}$, $k$ is the Boltzmann constant, $T$ is the absolute temperature, $\rho$ is the number density, $u^{(2)}$ is the pair potential with $u' = du^{(2)}/dr$, $u^{(3)}$ is the triplet potential, and $F_n$ ($n=2,3,4,...$) is the $n$-body Ursell function [15]; i.e., for $n=2$,

$$F_2(12) = \rho^{(2)}(12) - \rho^{(1)}(1)\rho^{(1)}(2) \tag{10.4}$$

and for $n=3$,

$$F_3(123) = \rho^{(3)}(123) \tag{10.5}$$

$$- \rho^{(2)}(12)\rho^{(1)}(3) - \rho^{(2)}(23)\rho^{(1)}(1) - \rho^{(2)}(31)\rho^{(1)}(2)$$

$$+ 2\rho^{(1)}(1)\rho^{(1)}(2)\rho^{(1)}(3)$$

The compressibility-fluctuation pressure $P^c$ obtained from (4.15) is independent of the assumption of the decomposition of the total potential. Heuristically, the requirement of pressure consistency is equivalent to requiring that the fluctuation effects on the virial pressure be equal to those on $P^c$. This is achieved by setting $\partial P^v/\partial \rho = \partial P^c/\partial \rho$ at constant $\beta$ and V for all $\rho$. The result is

$$\rho \int dr \, C(r) - \frac{\beta}{6} \int dr \, ru'(r) \left[ 2\rho + \rho^{-1} \left( 1 - \rho \int dx \, C(x) \right) \right. \tag{10.6}$$

$$\left. \cdot \left[ 2F_2(r) + \int ds \, F_3(r, s, |r-s|) \right] \right]$$

$$- \frac{\beta}{6} \int dr \, ds \, K(r, s, |r-s|) \left\{ \rho^3 + [1 + 2\left(1 - \rho \int dx \, C(x) \right)] \right.$$

$$\left. \cdot \frac{1}{3} \left[ F_2(r) + F_2(s) + F_2(|r-s|) \right] \right\}$$

$$- \frac{\beta}{18} \int dr \, ds \, K(r, s, |r-s|) \left( 1 - \rho \int dx \, C(x) \right)$$

$$\cdot \int dt \, [F_3(r, t, |r-t|) + F_3(t, s, |s-t|) + F_3(|r-t|,|t-s|,|r-s|)]$$

$$- \frac{\beta}{18\rho} \int dr \, ds \, K(r, s, |r-s|) \left( 1 - \rho \int dx \, C(x) \right)$$

$$\cdot \left[ 3F_3(r, s, |r-s|) + \int dt \, F_4(r, s, t, |r-t|) \right] = 0$$

The derivatives of the Ursell functions $\partial F_2/\partial \rho$ and $\partial F_3/\partial \rho$, necessary in arriving at eq. (10.6), can be found by the inhomogeneous methodology

$$\frac{\partial F_2(1,2)}{\partial \rho} = \left[ 1 - \rho \int dr \, C(r) \right] \rho^{-1} \left[ F_2(1,2) + \int d3 \, F_3(1, 2, 3) \right] \tag{10.7}$$

$$\frac{\partial F_3(1, 2, 3)}{\partial \rho} = \left[ 1 - \rho \int dr \, C(r) \right] \rho^{-1} \left[ 3F_3(1, 2, 3) + \int d4 \, F_4(1, 2, 3, 4) \right] \tag{10.8}$$

The consistency condition (10.6) requires knowledge of the four-body Ursell function $F_4$ through eq. (10.8); and if eq. (10.6) is truncated after the second term on the LHS, it is the consistency condition for the pair interaction case.

## V.11. *The Cluster Series of the RDF*

The cluster expansion of the rdf $g(r_{12})$ can be carried out within either the CE or the GCE. We shall expand the rdf in density series [16]

$$g^{(2)}(r_1, r_2) = \exp[-\beta u(r_1, r_2)] \left[ 1 + \rho g_1(r_1, r_2) + \rho^2 g_2(r_1, r_2) + \cdots \right] \tag{11.1}$$

$$= \exp[-\beta u(r_1, r_2)] \left[ 1 + \sum_{m=1}^{\infty} \rho^m g_m(r_1, r_2) \right]$$

where $g_m(\mathbf{r}_1, \mathbf{r}_2)$ are the expansion coefficients and are functions of the root points $\mathbf{r}_1$ and $\mathbf{r}_2$. To show this, we use the definition of the pcf in a CE. Assuming pairwise additivity, we obtain

$$g^{(2)}(\mathbf{r}_1, \mathbf{r}_2) = \frac{N(N-1)}{\rho^2 Q_N} \int d\mathbf{r}_3 \cdots d\mathbf{r}_N \, e^{-\beta u(\mathbf{r}_1, \mathbf{r}_2)} e^{-\beta u(\mathbf{r}_1, \mathbf{r}_3)} \cdots \tag{11.2}$$

$$= \frac{N(N-1)}{\rho^2 Q_N} e^{-\beta u(\mathbf{r}_1, \mathbf{r}_2)} \int d\mathbf{r}_3 \cdots d\mathbf{r}_N \, (1 + f_{13}) \cdots (1 + f_{N-1,N})$$

$f_{ij}$ being the Mayer factor. The first Boltzmann factor $\exp[-\beta u(\mathbf{r}_1, \mathbf{r}_2)]$ separates out because it is independent of the integrated variables $d\mathbf{r}_3 d\mathbf{r}_4 \cdots d\mathbf{r}_N$. When the products in $1 + f_{ij}$ are multiplied, the following cluster series ensues:

$$g^{(2)}(\mathbf{r}_1, \mathbf{r}_2) = \frac{N(N-1)}{\rho^2 Q_N} e^{-\beta u(12)} \tag{11.3}$$

$$\int d3 \cdots dN \left[ 1 + f_{13}f_{14} + f_{13}f_{15} + f_{13}f_{34} + f_{34}f_{35} + f_{34}f_{56} + \cdots \right.$$

$$+ f_{13}f_{14}f_{15} + f_{13}f_{34}f_{45} + f_{34}f_{45}f_{35}$$

$$\left. + f_{34}f_{56}f_{78} + \cdots + f_{13}f_{14}f_{15} \cdots f_{N-1,N} \right]$$

$Q_N$ should also be expanded. Collecting terms of the same power and comparing with (11.1) gives

$$g_1(12) = \int d3 \, f_{13}f_{23} = \rho^{-1} \; \text{o—o—o} \tag{11.4}$$

and

$$g_2(12) = \frac{1}{2}[g_1(12)]^2 + \int d3 \, d4 \, f_{13}f_{24}f_{34} + 2 \int d3 \, d4 \, f_{13}f_{23}f_{24}f_{34} \tag{11.5}$$

$$+ \frac{1}{2} \int d3 \, d4 \, f_{13}f_{23}f_{14}f_{24}f_{34}$$

which can be represented in diagrams as

$$\rho^2 g_2(12) = \quad + 2 \quad + \frac{1}{2} \quad + \frac{1}{2} \tag{11.6}$$

Higher-order $g_m$ were worked out by Ree, Keeler and McCarthy [17]. Table V.1 gives the cluster diagrams for $g_3$. For hard spheres, $g_2$ has been evaluated by Nijboer and van Hove [18]. They compared the exact $g_2$ (evaluated according to (11.5)) and the $g_2$ given by the Yvon-Born-Green approximation. The $g_2$ term in the density expansion contributes significantly at high densities. Comparison with the virial expansion of pressure shows that it is needed when the fourth virial coefficient $B_4$ is important. The correspondence between the virial coefficients and the $g_m$ terms is like this: retention of $g_1$ terms gives correct $B_3$; retention of $g_2$ gives correct $B_4$, etc. This can be demonstrated by direct integration of the rdf according to the virial (pressure) equation (2.10). We also notice that as the density approaches zero, eq. (11.1) gives

$$\lim_{\rho \to 0} g^{(2)}(12) = \exp[-\beta u(12)] \tag{11.7}$$

Thus we obtain the low density limit of rdf. If we take the logarithm of (11.7), we have

$$u(r_{12}) = \lim_{\rho \to 0} (-kT) \ln g(r_{12}) \tag{11.8}$$

*Table V.1.  The Cluster Series for the Radial Distribution Functions*

| Cluster Diagrams |
|---|

$\rho g_1 =$

$2\rho^2 g_2 =$

$6\rho^2 g_3 =$

That is, the logarithm of the rdf yields the pair potential $u(r)$ at low densities. For dense fluids, this concept can be extended, and we define a *potential of mean force* due to Kirkwood [19]:

$$W^{(2)}(r) \equiv -kT \ln g(r) \tag{11.9}$$

Heuristically, it says that the average interaction energy between pairs of molecules in the dense fluid is given by the logarithm of the rdf. Similarly, we define higher-order potentials of mean force as

$$W^{(3)}(123) \equiv -kT \ln g^{(3)}(123) \tag{11.10}$$

and

$$W^{(n)}(\{n\}) \equiv -kT \ln g^{(n)}(\{n\}) \tag{11.11}$$

In terms of $W^{(n)}$, the Kirkwood superposition is given by

$$W^{(3)}(123) = W^{(2)}(12) + W^{(2)}(23) + W^{(2)}(13) \tag{11.12}$$

Translated into triplet correlation functions, (11.12) becomes

$$g^{(3)}(123) = g^{(2)}(12)g^{(2)}(23)g^{(2)}(13) \tag{11.13}$$

Thus the potential of mean force of three particles, according to Kirkwood, is approximated by the sum of the potentials of mean force of pairs— namely there is no extra contribution beyond pairs. The validity of such an approximation will be discussed shortly.

The correlation functions $C(r)$ and $y(r)$ also possess cluster diagrams. From the OZ relation, the clusters for $C(r)$ can be deduced. Rushbrooke and Scoins [20] gave

$$C(12) = \qquad\qquad\qquad\qquad\qquad\qquad\qquad\qquad\qquad\qquad\qquad (11.14)$$

to order $O(\rho^2)$. Higher-order terms can be similarly derived. Since $y(r) = g(r)\exp(\beta u(r))$, $y(r)$ is simply given by

$$y(12) = 1 + \sum_{m=1}^{\infty} \rho^m g_m(12) \tag{11.15}$$

where the $g_m$ are given above. For the total correlation function $h(r)$ we write

$$h(12) = g(12) - 1 = (1 + f(12)) \left[ 1 + \sum_{m=1}^{\infty} \rho^m g_m(12) \right] - 1 \tag{11.16}$$

It is a simple exercise to find the clusters.

## V.12. *Thermodynamic Properties of Molecular Fluids*

Chapter IV discussed the correlation functions of polar and polyatomic fluids. Two representations were given: (i) the spherical harmonic expansions of the angular pcf, and (ii) the ss pcf. It is also possible to obtain the thermodynamic properties of the fluid from these representations.

### THERMODYNAMICS OF POLAR FLUIDS

The spherical harmonic representation is given in eq. (IV.7.18). To obtain the internal energy, one starts with formula (1.7) and generalizes to the anisotropic case:

$$\frac{U}{NkT} = \frac{3}{2} + \frac{\rho}{2kTV} \frac{1}{\Omega^2} \tag{12.1}$$

$$\cdot \int d\mathbf{r}_1 \, d\mathbf{r}_2 \, d\omega_1 \, d\omega_2 \, u(\mathbf{r}_1, \mathbf{r}_2, \omega_1, \omega_2) g^{(2)}(\mathbf{r}_1, \mathbf{r}_2, \omega_1, \omega_2)$$

where $\Omega = 4\pi$ for linear molecules and $8\pi^2$ for nonlinear molecules. We substitute the spherical harmonic expansions of $u$ and $g^{(2)}$ into (12.1) and use the orthogonality conditions of the spherical harmonics, i.e.,

$$\int_0^\pi d\theta \sin\theta \int_0^{2\pi} d\phi \, Y_{LM}{}^*(\theta, \phi) Y_{L'M'}(\theta, \phi) = \delta_{LL'}\delta_{MM'} \tag{12.2}$$

to get

$$\frac{U}{NkT} = \frac{3}{2} + \frac{\rho}{2kT} \sum_{(L)(L')(M)} \int_0^\infty dr \, 4\pi r^2 \, u_{LL'M}'(r) g_{LL'M}(r) \tag{12.3}$$

where $u_{LL'M}'(r)$ are the spherical harmonic coefficients of the angle-dependent pair potential $u(r_1, r_2, \omega_1, \omega_2)$, and $g_{LL'M}'(r)$ are the coefficients for $g^{(2)}$. The summation on $M$ is from $-\min(L,L')$ to $+\min(L,L')$. Similar derivations could be carried out for the virial pressure. The results are presented in Table V.2. We have also listed formulas for the isothermal compressibility $K_T$, mean squared torque $<\tau^2>$, and Kirkwood angular correlation parameters $G_L$, these being related to the isothermal compressibility ($L$=0), the dielectric constant ($L$=1), and the optical Kerr constant ($L$=2) of an anisotropic fluid.

## THERMODYNAMICS OF POLYATOMICS

The site-site correlation functions contain sufficient information for the calculation of the energy $U$ and the isothermal compressibility. However, additional information is needed for the virial pressure.

### Internal Energy

Energy is calculated from

$$\frac{U}{NkT} = \frac{3}{2} + \frac{\rho}{2kT} \sum_{\alpha}^{m}\sum_{\gamma}^{m} \int_0^\infty dr^{\alpha\gamma}\, 4\pi(r^{\alpha\gamma})^2 u_{\alpha\gamma}(r^{\alpha\gamma}) g_{\alpha\gamma}^{(2)}(r^{\alpha\gamma}) \tag{12.4}$$

where $u_{\alpha\gamma}$ is the site-site pair potential, $g_{\alpha\gamma}^{(2)}$ is ss pcf, and $r^{\alpha\gamma}$ is the distance between site $\alpha$ in one molecule and site $\gamma$ in a second molecule. There are $M$ sites in each molecule.

Table V.2.  *Thermodynamic Properties in Terms of Spherical Harmonic Coefficients*

**Configurational Internal Energy**

$$\frac{U^c}{N} = \frac{1}{2}\rho \sum_{LL'M} \int dr\, u_{LL'M}(r) g_{LL'M}(r)$$

**Pressure**

$$\frac{P}{\rho kT} = 1 - \frac{\rho}{6kT} \sum_{LL'M} \int dr\, r\, \frac{du_{LL'M}(r)}{dr}\, g_{LL'M}(r)$$

**Isothermal Compressibility**

$$K_T = \frac{1}{\rho kT} + \frac{1}{kT} \int dr[g_{000}(r)-1]$$

**Mean Squared Torque**

$$<\tau^2> = -\rho kT \sum_{LL'M} L(L+1) \int dr\, g_{LL'M}(r) u_{LL'M}(r)$$

**Kirkwood Factor,**  *(L=0,1,2)*

$$G_L^* = \frac{1}{2L+1}\rho \sum_M (-1)^M \int dr\, g_{LLM}(r)$$

*$G_0$, $G_1$, $G_2$ are related to the isothermal compressibility, dielectric constant, and optical Kerr constant, respectively.*

*Isothermal Compressibility*

$$\rho k T K_T - 1 = \rho \int_0^\infty dr_{12} \, 4\pi r_{12}^2 (\overline{g_{cc}}(r_{12}) - 1) \tag{12.5}$$

where $g_{cc}(\mathbf{r})$ is the center-of-mass to center-of-mass correlation function (with the center of mass considered as a special site in the polyatomic, even though there may not actually be an atom there). $\overline{g_{cc}}(r)$ is the orientationally averaged center-of-mass correlation function.

*Virial Pressure*

The virial pressure cannot be calculated from the ss pcf alone since the pressure is given in terms of the ensemble average of the virial:

$$P = \rho k T - \frac{N\rho}{6} < \sum_\alpha^m \sum_\gamma^m \mathbf{R}_{12} \cdot \left[ \frac{\partial u_{\alpha\gamma}(r_{12}^{\alpha\gamma})}{\partial \mathbf{R}_{12}} \right] >_\Omega \tag{12.6}$$

where $< \cdots >_\Omega$ is an ensemble angle average, $\mathbf{R}_{12}$ is the center-to-center vector and $r_{12}^{\alpha\gamma}$ is the site-to-site vector. Equation (12.6) can be written in terms of site-site distances:

$$P = \rho k T - \frac{N\rho}{6} < \sum_\alpha \sum_\gamma (\mathbf{r}_{ij}^{\alpha\gamma} - \mathbf{L}_j^\gamma + \mathbf{L}_i^\alpha) \cdot \frac{\mathbf{r}_{ij}^{\alpha\gamma}}{|\mathbf{r}_{ij}^{\alpha\gamma}|} \left[ \frac{\partial u_{\alpha\gamma}(r^{\alpha\gamma})}{\partial r^{\alpha\gamma}} \right] >_\Omega \tag{12.7}$$

Now this ensemble average involves angle-dependent site-site correlation functions which are not available from the site-site correlation function $g_{\alpha\gamma}^{(2)}$. Carrying out the ensemble averaging gives

$$P = \rho k T - \frac{\rho^2}{6} \sum_\alpha^m \sum_\gamma^m \int dr_{12}^{\alpha\gamma} \frac{\partial u_{\alpha\gamma}}{\partial r_{12}^{\alpha\gamma}} \, g_{\alpha\gamma}^{(2)}(r_{12}^{\alpha\gamma}) < (\mathbf{r}_{12}^{\alpha\gamma} - \mathbf{L}_2^\gamma + \mathbf{L}_1^\alpha) \cdot \frac{\mathbf{r}_{12}^{\alpha\gamma}}{|\mathbf{r}_{12}^{\alpha\gamma}|} >_\Omega \tag{12.8}$$

where

$$< \left[ \mathbf{r}_{12}^{\alpha\gamma} - \mathbf{L}_2^\gamma + \mathbf{L}_1^\alpha \right] \cdot \frac{\mathbf{r}_{12}^{\alpha\gamma}}{|\mathbf{r}_{12}^{\alpha\gamma}|} >_\Omega = \tag{12.9}$$

$$\frac{\int d\mathbf{L}_1^\alpha \, d\mathbf{L}_2^\gamma \left[ \mathbf{r}_{12}^{\alpha\gamma} - \mathbf{L}_2^\gamma + \mathbf{L}_1^\alpha \right] \frac{\mathbf{r}_{12}^{\alpha\gamma}}{|\mathbf{r}_{12}^{\alpha\gamma}|} g^{(2)}(\mathbf{r}_{12}^{\alpha\gamma} - \mathbf{L}_2^\gamma + \mathbf{L}_1^\alpha, \Omega_1, \Omega_2)}{\int d\mathbf{L}_1^\alpha \, d\mathbf{L}_2^\gamma \, g^{(2)}(\mathbf{r}_{12}^{\alpha\gamma} - \mathbf{L}_2^\gamma + \mathbf{L}_1^\alpha, \Omega_1, \Omega_2)}$$

and $g^{(2)}(\mathbf{r}_{12}^{\alpha\gamma} - \mathbf{L}_2^\gamma + \mathbf{L}_1^\alpha, \Omega_1, \Omega_2)$ is the angular pcf (IV.7.14). Nezbeda and Smith [21] have investigated the additional functions needed for calculating the virial pressure. They shifted the angles from the center-center frame to the site-site frame (See Fig.IV.9) via a rotation over the Euler angles, $\alpha$, $\beta$, and $\gamma$. The angular pcf $g(\mathbf{r}_1, \mathbf{r}_2, \omega_1, \omega_2)$ can now be expressed in terms of the new coordinates as $G(r_1^\alpha, r_2^\gamma, \omega_1', \omega_2')$, where $\omega_i$'s are the Euler angles referring to the site-site axis. In terms of the new angular pcf, the virial pressure can be calculated (for linear molecules) as

$$\frac{\beta P}{\rho} = 1 - \frac{\beta \rho}{6} \sum_{\alpha=1}^{m}\sum_{\gamma=1}^{m} \int dr_{12}^{\alpha\gamma}\, 4\pi (r_{12}^{\alpha\gamma})^2 \frac{\partial u_{\alpha\gamma}}{\partial r_{12}^{\alpha\gamma}} \qquad (12.10)$$

$$\left\{ r_{12}^{\alpha\gamma} g_{\alpha\gamma}(r_{12}^{\alpha\gamma}) - 3^{-1/2}[L_\alpha G_{100}^{\alpha\gamma}(r_{12}^{\alpha\gamma}) + L_\gamma G_{100}^{\gamma\alpha}(r_{12}^{\alpha\gamma})] \right\}$$

where

$$G_{100}^{\alpha\gamma}(r) = \frac{1}{4\pi} \int d\omega_1^\alpha\, d\omega_2^\gamma\, G(r_{12}^{\alpha\gamma}, \omega_1^\alpha, \omega_2^\gamma) Y_{10}(\omega_1^\alpha) Y_{00}(\omega_2^\gamma) \qquad (12.11)$$

is the coefficient of expansion of the rotated angular pcf. Formula (12.10) is valid for linear molecules. We note that, in general, $G_{100}^{\alpha\gamma}(r) \neq G_{100}^{\alpha\gamma}(r)$. $L_\alpha$ in (12.10) are the bond radii (from the center of mass to the atoms). This expression indicates clearly that the site-site correlation functions do not contain sufficient information on angular dependence. At least two more spherical harmonic coefficients, $G_{100}^{\alpha\gamma}$ and $G_{100}^{\gamma\alpha}$, are needed for calculating the virial pressure of linear molecules. Additional angular dependence is needed for nonlinear molecules.

## V.13. *Approximations for High-Order Correlation Functions*

Correlation functions of order greater than two contribute significantly to the properties of dense fluids and to a number of thermodynamic derivatives. They also appear in integral equations, such as the Born-Bogoliubov-Green-Kirkwood-Yvon equations. Direct evaluation from the definition (IV.2.9) is not feasible. The information we have is based on results from computer simulation. Alder [22] has produced the hard sphere triplet correlation function $g^{(3)}$ over several triangular geometries. Rahman [23], Krumhansl and Wang [24], Ravaché, Mountain, and Streett [25] have simulated $g^{(3)}$ for the LJ fluid. The pcf and its pressure or density derivatives obtained from scattering experiments also provide information on the global (but not detailed) behavior of $g^{(3)}$. Egelstaff et al. [26], and Ravaché and Mountain [27] [28] have proposed approximations to $g^{(3)}$ and compared the derivatives with experimental data. In this section, we shall discuss the cluster expansions of the triplet correlation function and a number of widely used approximations to $g^{(3)}$.

### CLUSTER EXPANSIONS

Salpeter [29] and later Henderson [30] gave the cluster expansion for $g^{(3)}$ as

$$g^{(3)}(1, 2, 3) = \exp[-\beta(u(12) + u(13) + u(23))][1 + \sum_{n=1}^{\infty} \rho^n \tau_n(1, 2, 3)] \qquad (13.1)$$

where $\tau_n$ are the coefficients of expansion. In the limit $\rho \to 0$, the triplet correlation function reduces to the product of the Boltzmann factors

$$\lim_{\rho \to 0} g^{(3)}(1, 2, 3) = \exp[-\beta(u_{12} + u_{13} + u_{23})] \qquad (13.2)$$

We write $u_{ij}$ for $u(ij)$. Henderson expressed $\tau_1$ in terms of the cluster integrals

$$\rho\tau_1(123) = \text{[diagram]} + \text{[diagram]} + \text{[diagram]} + \text{[diagram]} \tag{13.3}$$

We saw earlier ((11.4)) that

$$\rho g_1(12) = \text{[diagram: o——•——o]} \tag{13.4}$$

Thus

$$\tau_1(123) = g_1(12) + g_1(13) + g_1(23) + \rho^{-1} \text{[diagram]} \tag{13.5}$$

The next higher term, $\tau_2$, is

$$\tau_2(123) = g_2(12) + g_2(13) + g_2(23) \tag{13.6}$$

$$+ g_1(12)g_1(13) + g_1(12)g_1(23) + g_1(13)g_1(23)$$

$$+ \rho^{-1}\,[g_1(12) + g_1(13) + g_1(23)]\ \text{[diagram]}$$

$$+ \rho^{-2}\left[\ \text{[diagram]} + \text{[diagram]} + \text{[diagram]} + \text{[diagram]} + \text{[diagram]}\right.$$

$$\text{[diagram]} + \text{[diagram]} + \text{[diagram]} + \text{[diagram]} + \text{[diagram]}$$

$$\left. + \text{[diagram]} + \text{[diagram]} + \tfrac{1}{2}\Big[\ \text{[diagram]} + \text{[diagram]}\ \Big]\right]$$

Higher coefficients are lengthy expressions and will not be given here. In practice, the triplet correlation function has been approximated in a number of ways. Kirkwood proposed the superposition approximation

$$g_K^{(3)}(123) = g^{(2)}(12)g^{(2)}(13)g^{(2)}(23) \tag{13.7}$$

where the subscript K indicates the Kirkwood approximation. The cluster expansions of $\tau_1$ and $\tau_2$ under (13.7) are

$$\tau_1(123) \approx g_1(12) + g_1(13) + g_1(23) \tag{13.8}$$

$$\tau_2(123) \approx g_2(12) + g_1(12)g_1(13) + g_2(13) + g_1(12)g_1(23) \tag{13.9}$$

$$+ g_2(23) + g_1(13)g_1(23)$$

Comparison with the exact expressions (13.4) and (13.5) shows that the superposition approximation misses one term in $\tau_1$ and several terms in $\tau_2$. Henderson (*Ibid.*)

introduced an expansion in terms of $h$-bonds, $h(12)$ being the total correlation function

$$g^{(3)}(123) = g^{(2)}(12)g^{(2)}(13)g^{(2)}(23)[1 + \sum_{n=1}^{\infty} \rho^n \gamma_n(123; \rho)] \tag{13.10}$$

The two first expansion coefficients, $\gamma_1$ and $\gamma_2$, are given by

$$\rho\gamma_1(123) = \rho \int d4\ h(14)h(24)h(34) = \tag{13.11}$$

where the dashed bond o– – –o represents the $h(12)$ function. Also

$$\gamma_2 = \int d4\ d5\ [h_{14}h_{24}h_{15}(1 + h_{25})h_{35}h_{45} \tag{13.13}$$

$$+ \text{ terms obtained by cyclic permutations of 123}]$$

$$+ \frac{1}{2} \int d4d5\ h_{14}h_{24}h_{34}h_{15}h_{25}h_{35}(1 + h_{45})$$

Since $h(12)$ also possesses a cluster expansion in Mayer bonds, substitution of the clusters for $h(12)$ into (2.10 and 2.11) and matching terms shows that the expansion (13.9) is equivalent to (13.1). In addition, terms higher than second order in density also appear, since the cluster expansion for $h(12)$ contains all orders. Henderson has shown that expansion (13.9) gives an efficient resummation of $g^{(3)}$. Thus truncation beyond $\gamma_1$ or $\gamma_2$ will give rise to more cluster diagrams than, say, the Kirkwood superposition. This can be seen in (13.9); upon setting $\gamma_n = 0$, we recover the superposition equation (13.6).

Henderson evaluated the integrals, $\tau_1$ and $\tau_2$, for hard spheres. He compared the superposition results (13.7) and (13.8) and the $h$-bond expansions (13.10 and 11). The superposition approximation gives poor $\tau_1$ and $\tau_2$ at small $r$ (for equilateral triangular geometry, $r = r_{12} = r_{13} = r_{23}$), whereas the truncated $\gamma$-expansion gives almost exact agreement.

## References

[1]    J.G. Kirkwood, J. Chem. Phys. **3**, 300 (1935).

[2]    H.S. Green, *The Molecular Theory of Fluids* (North Holland, Amsterdam, 1952), pp.51-53.

[3]    H. Kamerlingh Onnes, Comm. Phys. Lab. Leiden, No.71 (1901).

[4]    J.E. Mayer, J. Chem. Phys. **5**, 67 (1937); also *Handbuch der Physik*, edited by S. Flügge, vol. 12 (Springer, Berlin, 1958), chapter 2.

[5]    T.L. Hill, *Statistical Mechanics* (McGraw-Hill, New York, 1956).

[6]   R.O. Watts and I.J. McGee, *Liquid State Chemical Physics* (John Wiley, New York, 1976).

[7]   K.W. Kratky, Physica, **78A**, 584 (1977).

[8]   J.L. Lebowitz and O. Penrose, J. Math. Phys. **5**, 841 (1964).

[9]   L.D. Landau and E.M. Lifshitz, *Fluid Mechanics* (Addison-Wesley, Reading, Massachusetts, 1959), pp. 246ff.

[10]  B. Widom, J. Chem. Phys. **39**, 2808 (1963).

[11]  K.S. Shing and K.E. Gubbins, Mol. Phys. **46**, 1109 (1982).

[12]  J.R. Henderson, Mol. Phys. **48**, 715 (1983).

[13]  K. Hiroike, J. Phys. Soc. Japan, **15**, 771 (1960).

[14]  J.L. Lebowitz and J.K. Percus, J. Math. Phys. **4**, 116, 248, and 1495 (1963).

[15]  H.D. Ursell, Proc. Cambridge Phil. Soc. **23**, 685 (1927).

[16]  J.E. Mayer and E. Montroll, J. Chem. Phys. **9**, 2 (1941).

[17]  F.H. Ree, N. Keeler, and S.L. McCarthy, J. Chem. Phys. **44**, 3407 (1966).

[18]  B.R.A. Nijboer and L. van Hove, Phys. Rev. **85**, 777 (1952).

[19]  J.G. Kirkwood, J. Chem. Phys. **3**, 300 (1935).

[20]  G.S. Rushbrooke and H.I. Scoins, Proc. Roy. Soc. **216A**, 203 (1953). Also, G.S. Rushbrooke and M. Silbert, Mol. Phys. **12**, 505 (1967).

[21]  I. Nezbeda and W.R. Smith, J. Chem. Phys. **75**, 4060 (1981).

[22]  B.J. Alder, Phys. Rev. Letters, **12**, 317 (1964).

[23]  A. Rahman, Phys. Rev. Letters, **12**, 575 (1964).

[24]  J.A. Krumhansl and S.S. Wang, J. Chem. Phys. **56**, 2034 (1972).

[25]  H.L. Raveché, R.D. Mountain, and W.B. Streett, J. Chem. Phys. **57**, 4999 (1972); Ibid. **61**, 1970 (1974).

[26]  P.A. Egelstaff, D.I. Page, and C.R.T. Heard, J. Phys. C **4**, 1453 (1971).

[27]  H.L. Raveché and R.D. Mountain, in *Progress in Liquid Physics*, edited by C.A. Croxton, (John Wiley, New York, 1978), p.469.

[28]  H.L. Raveché and R.D. Mountain, J. Chem. Phys. **53**, 3101 (1970).

[29]  E.E. Salpeter, Ann. Phys. **5**, 183 (1958).

[30]  D. Henderson, J. Chem. Phys. **46**, 4306 (1967).

[31]  G. Stell, Mol. Phys. **16**, 209 (1969).

## Exercises

1. Experimental argon $P$-$V$-$T$ behavior is given in NBS publications. Use the virial cluster equation of state with second virial term only to calculate the pressures assuming the Lennard-Jones potential (for argon, $\varepsilon=119.8$ K and $\rho = 3.405$A). Find the upper limit in density and lower limit in temperature such that the agreement is within 10%.

2. In problem 1, add the third virial coefficient, $B_3$. Find again the state conditions in order to achieve a 10% agreement.

3. There exist approximations to triplet correlation functions other than the Kirkwood formula. One of these was derived from one-dimensional fluids. Survey the literature and make a comparison of the different formulas.

4. Find the cluster series for $h(12)$ and $y(12)$. Compare the algebraic difference in clusters for $h(12) - C(12)$ with $y(12) - 1$. What diagrams are missing from $y(12) - 1$?

5. Find the hydrostatic condition (7.30) for an ideal gas in a gravitational field.

6. Show that the rdf obtained from the HNC integral equation satisfies the Hiroike consistency condition [31].

# CHAPTER VI

# INTEGRAL EQUATION THEORIES

The distribution functions discussed earlier can be obtained by several means. The integral equations are one of the fast methods for obtaining the distribution functions. Recall that the distribution functions are defined in terms of the interaction potential. Mathematically, the potential and the distribution functions are interrelated. It is not surprising that certain relations, in this case an integral equation, exists between these two quantities (see the following chart). With the advent of electronic computers in the 1950s, solution of the integral equations became feasible. The number crunching required by these nonlinear integral equations was met by the increased computing power of the computers. At the same time, computer simulation methods, solving coupled dynamic equations or generating ensemble configurations for hundreds and thousands of molecules, evolved rapidly. These approaches are now standard methods in molecular studies.

| *Knowledge of* *the pair potential,* *u(r)* | **through the** <br> integral equations | *Determination of* *the pair correlation function,* *g(r)* |
|---|---|---|

As mentioned earlier, there are at least three ways to obtain the structure of a liquid: (i) by integral equations; (ii) by computer simulation using Monte Carlo (MC) or molecular dynamics (MD) methods; and (iii) by scattering experiments. In practice, however, these methods are applied differently. Methods (i) and (ii) deal with hypothetical fluid models based on certain assumed potentials of interaction, usually expressed in the form of mathematical functions (e.g., the Lennard-Jones (LJ) potential). On the other hand, method (iii) is carried out directly on real liquids (e.g., bromine and butane) whose potentials of interaction are not well known. Recent developments indicate that these three methods are converging, as the mathematical representation of the interactions in real fluids has become more sophisticated and thus realistic. Table VI.1 gives a comparison of these methods and delineates their interrelationships.

*Table VI.1. Methods for Obtaining the Structure of Fluids*

| Method | | |
|---|---|---|
| 1 | 2 | 3 |
| *Scattering Experiments (X-ray or neutron diffraction)* | *Computer Simulations (Monte Carlo or Molecular Dynamics Methods)* | *Integral Equations (Percus-Yevick or Hypernetted Chain Equations)* |
| **Molecular Interaction Potentials** | | |
| Real fluid interaction potentials: usually complicated and involving many-body forces. | Idealized model potentials: e.g., hard sphere, Lennard-Jones, and dipole-dipole potentials. | Idealized potentials. |
| **Properties Generated** | | |
| Structure factors and scattering functions. | Time-dependent and independent properties (e.g., radial distribution functions, velocity autocorrelations, mean squared torque, virial pressure and energy). | Radial distribution functions. |
| **Quality of Results** | | |
| Results are exact for real fluids (e.g., $N_2$, and $CH_4$) except for experimental uncertainties. Detailed molecular motions are difficult to decipher. | Results are exact for the given potential model. Information is only as good as the potential model is realistic. | Results are only approximate for the given potential model. |
| **Utility of the Method** | | |
| Method giving the structure of real fluids. Detailed information not possible to obtain. Interplay of different physical factors difficult to differentiate. | Method giving exact and *detailed* information on the potential model (e.g., on vortex motions in particle diffusion, and capillary waves at interfaces). However, the results have no bearing on real fluids, if the potential model used is unrealistic. Computer time requirement is extremely large for good statistics. | Method valuable for theoretical understanding of properties of model fluids. Useful in formulating correlations of physical properties. Fast solutions could be obtained on computers. Results not exact for the models used, even less dependable for estimating real fluid behavior. |

In the integral representation of correlation functions, exact equations are not solvable, whereas solvable equations are not exact. As an example, the Born-Bogoliubov-Green-Kirkwood-Yvon (BBGKY) hierarchy is exact. However, the equation is not *closed* and cannot be solved without making assumptions on higher-order correlation functions. On the other hand, the Percus-Yevick (PY) and hypernetted chain (HNC) equations are solvable numerically for simple model potentials. Unfortunately, the correlation functions obtained therefrom are not *exact*. Despite the inexactitudes, these theories have provided valuable information, sometimes with surprising accuracy, on the structure of fluids ranging from hard spheres to electrolyte solutions. These integral equations originate either from resummation of cluster series or truncation of certain expansions or hierarchies. The degree of approximation could be gauged by comparing the particular cluster diagrams with the exact ones. Any missing graphs or redundant ones are indications of error. To acquire a feel for the integral equations we examine first the low density state.

At low densities, the pair correlation function (pcf) is given by the Boltzmann factor

$$\lim_{\rho \to 0} g(12) = \exp[-\beta(12)] \tag{1}$$

The pressure of such a system can be calculated from the virial theorem

$$\frac{\beta P}{\rho} = 1 - \frac{\beta \rho}{6} \int dr \, 4\pi r^3 \, \frac{du(r)}{dr} \, \exp[-\beta u(r)] \tag{2}$$

This formula is equivalent to the second virial equation

$$\beta P = \rho + B_2 \rho^2 \tag{3}$$

At higher densities, (1) is no longer sufficient. The pcf contains density terms higher than the Boltzmann factor. Additional cluster diagrams become important and new methods will have to be devised to include them. Prominent among these methods are the integral equations: PY, HNC, Yvon-Born-Green (YBG) and Kirkwood (KW) equations. These equations are formulated in such a way as to include certain classes but not all of the higher clusters. The ingenuity lies in the inclusion of the *important* ones. The methods of derivation are varied. A number of equations are based on the Ornstein-Zernike (OZ) equation. We call them OZ-based equations (e.g., the reference interaction site model). Others are relations derived from hierarchical relations on the probability functions (e.g., the YBG equation). Still others are due to resummations of the cluster series (e.g., the HNC equation). These equations are approximate formulations of the *N*-body problem. The accuracy of the integral equations depends on the state conditions and the types of potential models studied. For example, PY theory is fairly accurate for predicting hard-sphere properties but is less satisfactory for potentials with long-range forces. On the other hand, HNC theory performs well for ionic solutions but less so for hard-core fluids. These differences are subtle and not always easy to foretell. Actual tests should be carried out.

As shall be seen, the integral equations are not always an accurate method. However, the computer time required for their solution is quite small. For this and other reasons, these theories have been much used in the study of liquid structure. In the following, we shall introduce four such equations: PY, HNC, YBG, and KW. Their applications to model potentials will be explored subsequently. In addition, the mean spherical approximation (MSA) will be investigated. Chapter VII is devoted to the study of polar fluids. There the theories of linearized hypernetted chain (LHNC) and quadratic hypernetted chain (QHNC) equations will be introduced. All these equations could be extended to mixtures. Finally, a number of second-order theories will be examined.

## VI.1. *The Percus-Yevick Generating Functional*

Originally, the PY equation was derived from consideration of the generalized coordinates in crystalline structures [1]. Alternatively, Lebowitz and Percus [2] have shown that the integral equations could be derived from the expansions of the *generating functionals*. This is an attractive approach because it allows systematic derivation of different equations by a unified method and offers means of improving existing equations. It also opens the way for new integral equations, through proper choice of new expansion functionals. The PY equation is an example of the generating functional method.

The fluid that Lebowitz and Percus considered is an inhomogeneous one. The density is nonuniform due to the presence of an external field, $w(\cdot)$. The grand canonical ensemble (GCE) is used to represent the system. The partition function is

$$\Xi[w] = \sum_{N=0}^{\infty} \frac{z^N}{N!} \int d\{N\} \exp[-\beta V_N(\{N\})]\exp[-\beta \sum_{i=1}^{N} w(i)] \tag{1.1}$$

The argument $[w]$ denotes the inhomogeneity. The singlet density under the influence of $w$ is given by

$$\rho^{(1)}(1;w) = \frac{1}{\Xi} \sum_{N=1}^{\infty} \frac{z^N}{(N-1)!} \int d\{N-1\} \exp[-\beta V_N(\{N\})] \exp[-\beta \sum_{i=1}^{N} w(i)] \tag{1.2}$$

We multiply $\rho^{(1)}(1)$ by a factor $e^+(\cdot) \equiv \exp[+\beta w(\cdot)]$ to give

$$\rho^{(1)}(1; w)\exp[\beta w(1)] = \exp[C^{(1)}(1;w)] \exp(\beta\mu)/\Lambda^3 \tag{1.3}$$

$$= \frac{1}{\Xi} \sum_{N=1}^{\infty} \frac{z^N}{(N-1)!} \int d\{N-1\} \exp[-\beta V_N(\{N\})] \exp[-\beta \sum_{i=2}^{N} w(i)]$$

This is the *Percus-Yevick generating functional*. Recall that the singlet dcf stands for the (inhomogeneous) chemical potential. Thus the generating functional is the *nonuniform* equivalent to the activity $\exp(\beta\mu)/\Lambda^3$. Taylor's expansion for functionals says that given a functional $G$ with successive derivatives, the expansion with respect to a reference state $f_0$ is given by

$$G[f] = G[f_0] + \frac{1}{1!} \int d1 \frac{\delta G[f_0]}{\delta f(1)} [f(1) - f_0(1)] \tag{1.4}$$

$$+ \frac{1}{2!} \int d1 d2 \frac{\delta^2 G[f_0]}{\delta f(1)\delta f(2)} [f(1) - f_0(1)][f(2) - f_0(2)] + \cdots$$

where $f(\cdot)$ is the independent function and $\delta^n G/\Pi_{j=1}^{n}\delta f(j)$ is the $n$th functional derivative, $n=1,2,3,...$ By the same token, the PY generating functional can be expanded as

$$\exp[C^{(1)}] = \exp[C_0^{(1)}] + \int d2 \frac{\delta \exp[C_0^{(1)}]}{\delta\rho^{(1)}(2)} [\rho^{(1)}(2) - \rho_0^{(1)}(2)] \tag{1.5}$$

$$+ \frac{1}{2} \int d3 \, d2 \frac{\delta^2 \exp[C_0^{(1)}]}{\delta\rho^{(1)}(3)\delta\rho^{(1)}(2)} [\rho^{(1)}(3) - \rho_0^{(1)}(3)][\rho^{(1)}(2) - \rho_0^{(1)}(2)]+ \cdots$$

Note that the subscript 0 denotes a reference state, and all the derivatives are to be evaluated at this state. For us, the reference state is generated by a reference external potential $w$. To find the actual expressions for $\delta^n\exp[C^{(1)}]/\Pi_{j=1}^{n}\delta\rho^{(1)}(j)$, we need the following results on higher-order direct correlation functions $C^{(m)}(\{m\})$

$$C^{(m)}(\{m\}; w) = \frac{\delta^{m-1} C^{(1)}(1;w)}{\delta\rho^{(1)}(2; w) \cdots \delta\rho^{(1)}(m; w)}, \qquad m=2,3,\ldots \qquad (1.6)$$

Thus

$$\frac{\delta \exp[C^{(1)}(1;w)]}{\delta\rho^{(1)}(2)} = e^{C^{(1)}(1;w)} C^{(2)}(1,2;w) \qquad (1.7)$$

$$= \rho^{(1)}(1) \Lambda^3 e^{\beta(w-\mu)} C^{(2)}(1,2; w)$$

Similarly,

$$\frac{\delta^2 \exp[C^{(1)}]}{\delta\rho^{(1)}(3)\delta\rho^{(1)}(2)} = \exp[C^{(1)}]\left[\frac{\delta^2 C^{(1)}}{\delta\rho^{(1)}(3)\delta\rho^{(1)}(2)} + \frac{\delta C^{(1)}}{\delta\rho^{(1)}(3)}\frac{\delta C^{(1)}}{\delta\rho^{(1)}(2)}\right] \qquad (1.8)$$

$$= \rho^{(1)}(1) \Lambda^3 e^{\beta[w(1)-\mu]}\left[C^{(3)}(1,2,3; w) + C^{(2)}(1,3; w)\ C^{(2)}(1,2; w)\right]$$

Substitution into (1.5) gives

$$\exp[C^{(1)}(1)] = \exp[C_0^{(1)}(1)]\left\{1 + \int d2\ C^{(2)}(1, 2; w_0)[\rho^{(1)}(2) - \rho_0^{(1)}(2)]\right. \qquad (1.9)$$

$$+ \frac{1}{2}\int d3\ d2\ [C^{(3)}(1,2,3; w_0) + C^{(2)}(1,3; w_0)C^{(2)}(1,2; w_0)]$$

$$\left.\cdot[\rho^{(1)}(3) - \rho_0^{(1)}(3)][\rho^{(1)}(2) - \rho_0^{(1)}(2)] + \cdots\right\}$$

When a particle, called the *test* particle, is placed at the origin $\mathbf{r}_0$ of the coordinate system, it generates a potential field

$$w(\mathbf{r}_i) = u(\mathbf{r}_0, \mathbf{r}_i) \qquad (1.10)$$

$u(\mathbf{r}_0, \mathbf{r}_i)$ is a pair potential identical to those in $V_N(\{N\}) = \sum\sum u(ij)$. We distinguish two external potentials generated by two types of *test* particles: the reference $w_0(i)$ generated by a test particle of the reference type which interacts with the bath particles via the pair potential $u_0(0i)$ (i.e., $w_0(i) = u_0(0i)$), and the full $w(i)$ generated by a particle of the same type as those in the bath that interacts with the bath molecules by $u(0i)$. The situation is easily understood when put in terms of mixtures. Test particles of different species are placed at the origin. They interact with the bath particles via different potentials. It has been shown by Percus [3] that the pcf under the influence of test particles is related to the pcf of the homogeneous case by

$$\rho^{(1)}(1;\,w_{\text{test part.}}) = \frac{\rho^{(2)}(01)}{\rho^{(1)}(0)} = \rho^{(1)}(1)g(01) \tag{1.11}$$

Thus (1.9) becomes

$$g(01)e^{\beta u(01)} = y(01) = y_0(01)\left\{1 + \int d2\; C^{(2)}(12;\,w_0)\rho^{(1)}(2)[g(02) - g_0(02)]\right. \tag{1.12}$$

$$+ \frac{1}{2}\int d3\;d2\;[C^{(3)}(123;\,w_0) + C^{(2)}(13;\,w_0)C^{(2)}(12;\,w_0)]$$

$$\left. \cdot\rho^{(1)}(3)\rho^{(1)}(2)[g(03) - g_0(03)][g(02) - g_0(02)] + \cdots\right\}$$

Eq. (1.12) is the *second order Percus-Yevick equation* (PY2) based on a reference fluid with pair potential $u_0$. Since $\exp[C^{(1)}]$ is the nonuniform equivalent of the activity $\exp(-\beta\mu)$, the PY expansion is similar to expansions, in thermodynamics, of the activity in terms of density. The success (or failure) of such expansions depends on how the terms are resummed. To simplify (1.12), the interaction potential $u_0$ is set to zero. All $g_0(0i) = 1$, and all dcf reduce to the uniform ones. The singlet density $\rho^{(1)} = \rho$. To first order, we have

$$g(01)e^{\beta u(01)} = y(01) \approx 1 + \rho\int d2\; C^{(2)}(12)h(20) \qquad \text{(PY)} \tag{1.13}$$

Equation (1.13) is an integral equation of the convolution type. It is called the *Percus-Yevick* (PY) equation. Comparison with the OZ equation

$$g(01) - C^{(2)}(01) = 1 + \rho\int d2\; C^{(2)}(12)h(20) \tag{1.14}$$

shows that, for PY, the following relation holds (writing $C(01)$ for $C^{(2)}(01)$):

$$g(01)e^{\beta u(01)} \approx g(01) - C(01) \tag{1.15}$$

or alternatively,

$$C(01) \approx g(01) - y(01) = h(01) - y(01) + 1 = y(01)f(01) \tag{1.16}$$

Combination of (1.14) and (1.16) is sufficient for solution of $g(01)$. For example, eliminating the dcf after substitution gives

$$g(01)e^{\beta u(01)} = 1 + \rho\int d2\; g(12)e^{\beta u(12)}[e^{-\beta u(12)} - 1][g(20) - 1] \tag{1.17}$$

This equation relates the pcf $g(01)$ directly to the pair potential $u(01)$. Given the pair potential, we can, in principle, solve for $g$. Except for extremely simple potentials (e.g., the hard-sphere potential), the solution is carried out numerically. The numerical solution is based on a frame of reference called the *bipolar coordinates*.

## VI.2. *Bipolar Coordinates*

Kihara [3] introduced the bipolar coordinates in order to treat coupled three-body problems (for example, in evaluating the third virial coefficients). These coordinates are particularly useful in dealing with integral equations of the convolution type

$$y(\mathbf{r}_1, \mathbf{r}_2) = 1 + \rho \int d\mathbf{r}_3 \; C(\mathbf{r}_1, \mathbf{r}_3) h(\mathbf{r}_3, \mathbf{r}_2) \tag{2.1}$$

In Cartesian coordinates (Figure VI.1), we set $\mathbf{r}_3 = (x, y, z)$ and write eq. (2.1) as

$$y(r_{12}) = 1 + \rho \int_{-\infty}^{+\infty} dx \int_{-\infty}^{+\infty} dy \int_{-\infty}^{+\infty} dz \; C(r_{13}) h(r_{32}) \tag{2.2}$$

where $r_{12} = |\mathbf{r}_2 - \mathbf{r}_1|$, $r_{13} = |\mathbf{r}_3 - \mathbf{r}_1|$, and $r_{32} = |\mathbf{r}_2 - \mathbf{r}_3|$. Letting

$$r = r_{12}, \qquad s = r_{13}, \qquad t = r_{32} \tag{2.3}$$

we can rewrite eq. (2.2) as

$$y(r) = 1 + \rho \int_{-\infty}^{\infty} dx \int_{-\infty}^{\infty} dy \int_{-\infty}^{\infty} dz \; C(s) h(t) \tag{2.4}$$

Since the distances $s$ and $t$ refer to two centers (molecules 1 and 2), we designate the framework of reference *bipolar coordinates*. The third coordinate is the inclination angle $\psi$. The next step is to transform the variables from $x$-$y$-$z$ to $s$-$t$-$\psi$. The two sets are related via geometry:

*Radial distance* 1–3 $\qquad s^2 = x^2 + y^2 + z^2$ $\qquad\qquad$ (2.5)

*Radial distance* 3–2 $\qquad t^2 = x^2 + (r - y)^2 + z^2$

*Inclination angle* $\qquad \psi = \tan^{-1}\dfrac{x}{z}$

The derivation given below differs somewhat from that of Kihara: a rigorous mathematical proof is given. The differential volume $dx \, dy \, dz$ is transformed to $ds \, dt \, d\psi$ through the Jacobian

$$dx \, dy \, dz = J(xyz, st\psi) \; ds \, dt \, d\psi \tag{2.6}$$

where

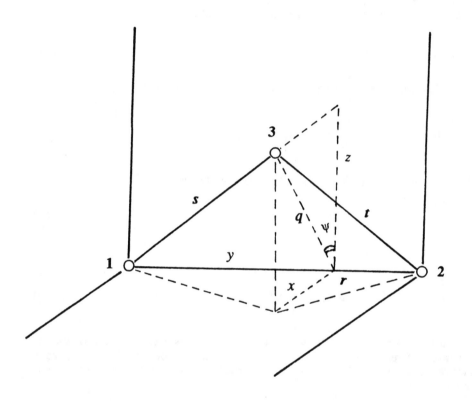

FIGURE VI.1.   The bipolar coordinates formed by three molecules 1, 2 and 3. The x, y, and z distances of molecule 3 are referred to molecule 1 as the origin. The radii s for $r_{13}$, t for $r_{23}$, and the inclination angle $\psi$ form the coordinates of the bopolar system. The distance $r = r_{12}$ is fixed.

$$J(xyz,st\psi) = \det \begin{bmatrix} \dfrac{\partial x}{\partial s} & \dfrac{\partial x}{\partial t} & \dfrac{\partial x}{\partial \psi} \\[2mm] \dfrac{\partial y}{\partial s} & \dfrac{\partial y}{\partial t} & \dfrac{\partial y}{\partial \psi} \\[2mm] \dfrac{\partial z}{\partial s} & \dfrac{\partial z}{\partial t} & \dfrac{\partial z}{\partial \psi} \end{bmatrix} \tag{2.7}$$

The partial derivatives can be evaluated from the defining relations (2.5):

$$\frac{\partial s}{\partial x} = \frac{x}{s}, \qquad \frac{\partial t}{\partial x} = \frac{x}{t}, \qquad \frac{\partial \psi}{\partial x} = \frac{z \sin^2 \psi}{x^2} \tag{2.8}$$

$$\frac{\partial s}{\partial y} = \frac{y}{s}, \qquad \frac{\partial t}{\partial y} = \frac{y-r}{t}, \qquad \frac{\partial \psi}{\partial y} = 0$$

$$\frac{\partial s}{\partial z} = \frac{z}{s}, \qquad \frac{\partial t}{\partial z} = \frac{z}{t}, \qquad \frac{\partial \psi}{\partial z} = \frac{-\sin^2 \psi}{x}$$

Since the determinant of the inverse matrix is the inverse of the determinant of the given matrix,

$$J(st\psi, xyz) = \det \begin{bmatrix} \dfrac{x}{s} & \dfrac{y}{s} & \dfrac{z}{s} \\[2mm] \dfrac{x}{t} & \dfrac{y-r}{t} & \dfrac{z}{t} \\[2mm] \dfrac{z \sin^2 \psi}{x^2} & 0 & \dfrac{-\sin^2 \psi}{x} \end{bmatrix} = \frac{r}{st} \tag{2.9}$$

we have

$$J(xyz, st\psi) = \text{inverse of (2.9)} = \frac{st}{r} \tag{2.10}$$

The convolution equation can now be written as

$$y(r) = 1 + \rho \int_0^\infty ds \int_{|r-s|}^{r+s} dt \int_0^{2\pi} d\psi \, \frac{st}{r} \, C(s)h(t) \tag{2.11}$$

$$= 1 + \frac{2\pi\rho}{r} \int_0^\infty ds \, sC(s) \int_{|r-s|}^{r+s} dt \, th(t)$$

where the integration with respect to $\psi$ has been carried out as long as the integrand is independent of $\psi$. The limits of integration are determined through the following considerations. Note that the differential volume can be written as

$$\frac{st}{r} \, ds \, dt \, d\psi = (ds)(q \, d\psi)(\frac{st}{rq} \, dt) \tag{2.12}$$

upon introducing the distance $q$ (*spikes* from the y-axis, see Figure VI.1). For fixed $s$ and $\psi$, the picture of the differential area $(ds)(st\ dt/rq)$ is given in Figure VI.2. We claim that the differential area $(ds)(s\ dA)$ is given by $(ds)(st\ dt/rq)$, where $A$ is the angle subtended by $t$. From cosine law, we know that

$$t^2 = r^2 + s^2 - 2rs \cos A \tag{2.13}$$

The total differential at constant $r$ and $s$ is

$$2t\ dt = 2\ rs \sin A\ dA \tag{2.14}$$

From geometry, $q = s \sin A$. Thus

$$s\ dA = \frac{t\ dt}{r \sin A} = \frac{st\ dt}{rs \sin A} = \frac{st}{rq}\ dt \tag{2.15}$$

Integrating over $t$ at constant $s$ and $\psi$ creates the situation of Figure VI.3. It is apparent that as $t$ goes from $|r-s|$ to $r+s$, molecule 3 traces out a semicircle. The second integration over $\psi$ (from 0 to $2\pi$) forms the surface of a sphere with radius $s$. Finally, the integration over $s$ (from 0 to $\infty$) covers the entire three-dimensional volume.

## VI.3. *Numerical Techniques*

In bipolar coordinates, the PY equation assumes the form

$$y(r) = 1 + \frac{2\pi\rho}{r} \int_0^\infty ds\ sC(s) \int_{|r-s|}^{r+s} dt\ th(t) \tag{3.1}$$

If the roles of $s$ and $t$ are interchanged, we have the equivalent formula

$$y(r) = 1 + \frac{2\pi\rho}{r} \int_0^\infty dt\ th(t) \int_{|r-s|}^{r+s} ds\ sC(s) \tag{3.2}$$

Either eq. (3.1 or 3.2) coupled with the PY assumption

$$C(r) = h(r) - y(r) + 1 \tag{3.3}$$

forms a complete set of equations sufficient for the determination of the radial distribution function (rdf). For simple potentials (e.g., the hard-sphere potential), one can solve (3.1) by Laplace transforms. An analytical solution is obtained. For more complicated potentials, numerical solutions must be sought. A number of schemes have been developed. We shall introduce three such methods below.

### PICARD'S METHOD

Picard's method refers to the solution by *iterations*. An initial guess of the

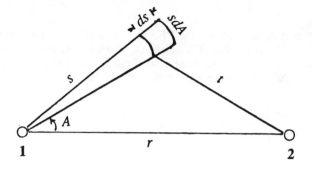

FIGURE VI.2. *The differential area formed by ds and s dA at fixed s and* $\psi$.

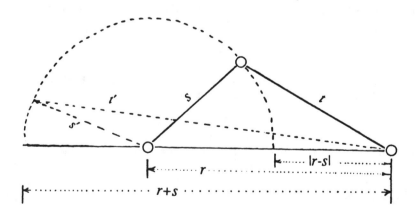

FIGURE VI.3. *Variations of t from |r−s| to r+s at constant r, s, and* $\psi$.

solution function is substituted into the RHS of equation. An output function is calculated for the LHS. Next this output is substituted into the RHS as a new input to yield a second output. The process is repeated until certain convergence criteria are satisfied. For PY, the convolution integral is evaluated in two parts:

$$y(r) = 1 + \frac{2\pi\rho}{r} \int_0^\infty ds\ s\ C(s)\ [E(r+s) - E(|r-s|)] \qquad (3.4)$$

and

$$E(x) \equiv \int_0^x dt\ t\ h(t) \qquad (3.5)$$

The direct and total correlation functions are related by

$$C(r) = h(r) - y(r) + 1 \qquad (3.6)$$

The pair potential is known in advance. It enters (3.6) via the definition $h(r) \equiv y(r)\exp[-\beta u(r)]-1$. The algorithm of solution commences with an initial guess $h^{[0]}(r)$, iterating through equations (3.4-6) until the so-called Neumann sequence $h^{[0]}, h^{[1]}, h^{[2]},..., h^{[n]},...$ converges to a *fixed point* within prescribed limits. The steps are:

1. Fix the pair potential $u(r)$, temperature T, and density $\rho$.

2. Choose initial guess $h^{[0]}(r)$. The initial choice is important in the present method for numerical convergence, especially at high densities. For dilute gas, a good guess is

$$g^{[0]}(r) = e^{-u(r)/kT} \qquad (3.7)$$

For dense fluids, there is no sure guidance. In practice, one builds up the densities from low to high, using the previous rdf as input for the next higher density.

3. Obtain $E(x)$ from (3.5). Here one can construct a table of $E$ values at different $x$ for a given $h(r)$, and store them in memory for later use. One may use any standard quadratures for numerical integration (or the usual trapezoidal and Simpson's methods).

$$E^{[0]}(x) = \int_0^x dt\ th^{[0]}(t) \qquad (3.8)$$

$C(r)$ is obtained from

$$C^{[0]}(r) = y^{[0]}(r)[e^{u(r)/kT} - 1] \qquad (3.9)$$

where $y^{[0]} = g^{[0]}\exp(u/kT)$

4. Obtain $y(r)$ from (3.4):

$$y^{[1]}(r) = 1 + \frac{2\pi\rho}{r} \int\limits_{o}^{b} ds \; sC^{[0]}(s)[E^{[0]}(r+s) - E^{[0]}(|r-s|)] \tag{3.10}$$

Since we cannot go to infinity numerically, the upper limit $b$ is chosen sufficiently large so that $C(b)$ is very small ($\cong 10^{-7}$). Normally, $b = 6\sigma\text{-}10\sigma$ is adequate for LJ molecules and $20\sigma$ for Coulomb electrostatic forces. ($\sigma$ corresponds to the molecular size).

5. To guarantee convergence, mix $y^{[1]}$ with $y^{[0]}$ before the the next iteration according to

$$y_{\text{mix}}^{[1]}(r) = \alpha y^{[0]}(r) + (1-\alpha)y^{[1]}(r) \tag{3.11}$$

where $0 < \alpha < 1$ is the mixing parameter. In numerical analysis, this is called *relaxation*. For low density states, $\alpha$ is small (say, 0.4); for high density states, $\alpha$ is large (e.g., $\alpha=0.9$ for $\rho^*=0.7$). It is used to insure numerical stability. The judicious choice could only be made from experience in working with the equations.

6. This $y_{\text{mix}}^{[1]}$ is used as input in (3.8) and (3.9) to get $E^{[1]}$ and $C^{[1]}$ (note that $g^{[1]}(r) = y_{\text{mix}}^{[1]}(r)\exp[-u(r)/kT]$). Steps (iv) to (vi) are repeated. A sequence $y^{[0]}$, $y^{[1]}$, $y^{[2]}$,..., $y^{[n]}$,... is generated. If the Neumann sequence is convergent, we shall have the solution $y(r)$. Numerically, we impose the Cauchy condition that if

$$|y^{[n+1]}(r) - y^{[n]}(r)| < \delta \qquad \text{for all } r \tag{3.12}$$

where $\delta$ is a small number (e.g., $\delta = 0.0001$), we consider the sequence convergent and the solution $y(r)$ is set to $y^{[n+1]}(r)$. Other correlation functions are obtained from

$$g(r) = y(r)e^{-u(r)/kT} \tag{3.13}$$

and

$$C(r) = g(r) - y(r) \tag{3.14}$$

If we had used (3.2), the procedure would have been the same. Only $C(t)$ will appear in (3.5) for $E(x)$, and $h(s)$ in (3.4). However, there is an advantage in our choice. The dcf, $C(r)$, is a short-ranged function in comparison to $h(r)$. The upper limit $b$ in the numerical integration can therefore be smaller, a time-saving choice.

## BROYLES' METHOD

Broyles [4] used a similar iterative procedure. He multiplied (3.1) by $r$

$$ry(r) = r + 2\pi\rho \int\limits_{0}^{\infty} ds \; sC(s) \int\limits_{|r-s|}^{r+s} dt \; th(t) \tag{3.15}$$

and differentiated it with respect to $r$ to get

$$\frac{dH(r)}{dr} = 1 + 2\pi\rho \int_0^\infty ds\, f(s)H(s)$$ (3.16)

$$\cdot\left[H(r+s)e(r+s) - H(|r-s|)e(|r-s|)\, sgn(r-s) - 2s\right]$$

where

$$H(r) \equiv ry(r), \qquad f(r) \equiv e^{-\beta u(r)} - 1, \qquad e(r) \equiv e^{-\beta u(r)}$$ (3.17)

and

$$sgn(x) = \frac{x}{|x|}$$

We used the Leibniz rule in the differentiation. $H(r)$ is obtained by integration:

$$H(r) = \int_0^r dz\, \frac{dH(z)}{dz}$$ (3.18)

The iterative procedure is as before: starting with an initial guess for $H^{[0]}$, substituting into the RHS of (3.16) to obtain $dH^{[1]}/dr$, obtaining $H^{[1]}$ from (3.18), then mixing according to

$$H^{[1]}_{mix}(r) = \alpha\, H^{[0]}(r) + (1 - \alpha)H^{[1]}(r)$$ (3.19)

where $0 \leq \alpha \leq 1$. The steps are repeated until the Cauchy condition is satisfied. The correlation functions are recovered from

$$g(r) = H(r)\frac{e^{-\beta u(r)}}{r}$$ (3.20)

$$C(r) = H(r)\frac{e^{-\beta u(r)} - 1}{r}$$ (3.21)

Other solution procedures exist, and each attesting to the authors' numerical ingenuity. Baxter [5] devised a Wiener-Hopf solution for the OZ equation. Gillan [6] proposed a Newton-Raphson method which is very efficient and is independent of the initial guesses. We shall discuss this method next.

## GILLAN'S METHOD

Gillan's method consists essentially in dividing the pair correlation function into two parts, a *coarse* part and a *fine* part, then applying the Newton-Raphson (NR) method to speed up convergence. It has recently been refined by Labik et al. [7] and applied to the reference interaction site model by Monson [8] and Enciso [9] for polyatomic molecules. A remarkable feature of Gillan's method is its insensitivity to initial guesses for the numerical solution, a step critical in earlier methods. In addition, for most state conditions, the convergence is fast. These advantages have made this method the prime choice in solution of integral equations.

Earlier, Watts [10] has used a Newton-Raphson procedure on discretized integral equations. Gillan divided the pcf in two parts: one slowly varying, the other small and rapidly oscillating. The outline of the method is given below. A computer program is also provided in the Appendices.

The iterative procedure is based on the indirect correlation function (icf) $\gamma(r)$ defined by

$$\gamma(r) \equiv h(r) - C(r) \tag{3.22}$$

(Note that the total correlation $h(r)$ is the sum, $h(r) = C(r) + \gamma(r)$, of the direct correlation $C(r)$ and the indirect correlation $\gamma(r)$). The OZ relation in the Fourier space could be written as

$$\tilde{\gamma}(k) = \frac{\rho \tilde{C}(k)^2}{1 - \rho \tilde{C}(k)} \tag{3.23}$$

where tilde indicates Fourier transforms. In the new method the icf is divided into

$$\gamma(r) = \gamma^c(r) + \Delta\gamma(r) \tag{3.24}$$

or in discretized form

$$\gamma_i = \gamma_i^c + \Delta\gamma_i \tag{3.25}$$

where $\gamma_i \equiv \gamma(r=i\delta r)$, $i=0,1,2,3,...,N$, and $\delta r$ is the discretized grid size. $\gamma^c$ is the coarse part of the icf to be defined below, and $\Delta\gamma_i$ is the fine part. The coarse part is expressed as an expansion (spectral decomposition) in terms of a set of orthogonal basis functions $P^\alpha$ $\alpha = 1,2,3,...$ with expansion coefficients $a_\alpha$

$$\gamma_i^c \equiv \sum_\alpha a_\alpha P_i^\alpha \tag{3.26}$$

Thus

$$\gamma_i = \sum_\alpha a_\alpha P_i^\alpha + \Delta\gamma_i \tag{3.27}$$

In addition

$$\sum_i P_i^\alpha \Delta\gamma_i = 0, \qquad \forall \alpha \tag{3.28}$$

In order to close the simultaneous set of equations, we need one additional relation between the icf and the dcf. This is furnished by the usual integral equations. For example in the PY closure

$$C(r) \approx [1 + \gamma(r)] \, f(r) \tag{3.29}$$

or in HNC

$$C(r) \approx e^{-\beta u(r)+\gamma(r)} - \gamma(r) - 1 \tag{3.30}$$

For arbitrarily chosen initial guess $\gamma_i$, eqs. (3.23 and 29) might not be satisfied simultaneously. Thus an iterative procedure commonly used in solving integral equations will be implemented. The procedure is repeated until certain convergence criteria are satisfied. The iterations are a combination of the Newton-Raphson steps (on the coarse part $\gamma_i{}^c$) and Picard steps (on the fine part $\Delta\gamma_i$). Each combination (Newton-Raphson + Picard steps) is called a *refinement cycle*. The final solution will be denoted by asterisks: its expansion coefficients are $\{a^*{}_1, a^*{}_2,...\}$, and the fine part is $\Delta\gamma^*{}_i$.

The coarse function subspace is spanned by a small number $\nu \approx 10$ of basis functions $P^\alpha$, (these functions will be specified later). Initially, we choose arbitrary but reasonable $\gamma(r)$ and $\{a_\alpha\}$. If the choice of $\{a_1, a_2,...\}$ is the proper set, we have the final solution. Substitution into (3.23) and (3.29), PY, for example, should give the same $\gamma(r)$. Any difference indicates improper answer. A new set $\{a_1, a_2,...\}$ is to be produced via a Newton-Raphson procedure. The process is repeated until convergence is achieved. Suppose that the output derived through (3.23 and 29) from the input of an initial $\gamma_i$ is

$$\gamma_i = \sum_\alpha a'_\alpha P_i^\alpha + \Delta\gamma_i \tag{3.31}$$

If $\gamma$ is the exact solution, we should have

$$d_\alpha \equiv a_\alpha - a'_\alpha = 0 \tag{3.32}$$

and

$$\Delta\gamma'_i = \Delta\gamma_i \tag{3.33}$$

In most cases, first few iterations will not satisfy the above two conditions. To proceed, we fix the $\Delta\gamma_i$ arbitrarily at the initial values and generate new expansion coefficients $a_\alpha$ until they satisfy (3.32). These new $a$'s are obtained by a Newton-Raphson formula

$$\overline{a}_\alpha = a_\alpha - \sum_\beta (J^{-1})_{\alpha\beta} \, d_\beta \tag{3.34}$$

where J is the Jacobian

$$J_{\alpha\beta} \equiv \frac{\partial d_\alpha}{\partial a_\beta} \tag{3.35}$$

evaluated at the old estimates $\{a_\alpha\}$. The derivatives are taken at constant $a_{\gamma(\neq\beta)}$ and constant $\Delta\gamma_i$. These corrective iterations on the coarse part are called NR cycles. The cycles are continued until the new $a$'s do not change from their previous values (Cauchy condition). Let the primed quantities represent outputs from an NR cycle, we have for the final $a^*$'s

$$\gamma^*_i = \sum_\alpha a^*_\alpha P^\alpha_i + \Delta\gamma'_i \qquad (3.36)$$

compared with the previous input

$$\gamma^*_i = \sum_\alpha a^*_\alpha P^\alpha_i + \Delta\gamma_i \qquad (3.37)$$

Note that the expansion coefficients stay the same after the NR cycle while the fine part $\Delta\gamma'$ changes. These $a^*$'s are consistent with the input $\Delta\gamma$. However, these $\Delta\gamma$ do not yet satisfy (3.33). To obtain the final solution we carry out Picard iterations on $\Delta\gamma$; i.e. the output $\Delta\gamma'_i$ is used as the new input for the next NR cycle on $a$'s. The entire procedure looks like this

*Series of refinement cycles = [fixed $\Delta\gamma$] $\rightarrow$ NR Cycles on $a$'s $\rightarrow$ [new $\Delta\gamma$] $\rightarrow$ NR Cycles on $a$'s $\rightarrow$ [new $\Delta\gamma'$] $\rightarrow$ ...*

The cycles continue until condition (3.33) is satisfied to with a small tolerance. After a number of cycles (in practice, 6 to 7 refinements), the final solution is obtained. There are at least two technical details worth discussing: (1) the spectral decomposition, and (2) evaluation of the Jacobian matrix.

### Basis Functions

The choices of basis functions are literally infinite. Almost any orthogonal set of functions such as Fourier series, Hermite polynomials, etc., will suffice. Labik et al. have selected the Fourier series. Gillan has chosen the roof functions. Out of $N$ grid points, we select a subset of $\nu$ points with labels $i_\alpha$ ($\alpha = 1,2,...,\nu$) in ascending order (with $i_1 > 1$ and $i_2 > i_1$, etc.) These points serve as nodes for the basis function $P^\alpha_i$, which are defined as

$$P^\alpha_i = 0 \qquad \text{for } 1 \leq i \leq i_{\alpha-2} \qquad (3.38)$$

$$= \frac{i - i_{\alpha-2}}{i_{\alpha-1} - i_{\alpha-2}} \qquad \text{for } i_{\alpha-2} \leq i \leq i_{\alpha-1}$$

$$= \frac{i_\alpha - i}{i_\alpha - i_{\alpha-1}} \qquad \text{for } i_{\alpha-1} \leq i \leq i_\alpha$$

$$= 0 \qquad \text{for } i_\alpha \leq i \leq n$$

for $\alpha \geq 2$ (with $i_0 = 1$); for the special case $\alpha = 1$, we define

$$P^1_i = \frac{i_1 - i}{i_1 - 1}, \qquad \text{for } 1 \leq i \leq i_1 \qquad (3.39)$$

$$= 0, \qquad\qquad \text{for } i_1 \leq i \leq n$$

These definitions may appear complicated. However, they define simple functions of tri-angular shape (the *roofs*) with apices centered at $i_{\alpha-1}$ and falling linearly to zero beyond the neighboring nodes ($i_{\alpha-2}$ and $i_\alpha$). (See Figure VI.4). To find the spectral coefficients $\{a_\alpha\}$ we construct the matrix $\mathbf{B}$

$$B_{\alpha\beta} \equiv \left[ \sum_j P_j^\alpha P_j^\beta \right]^{-1} \tag{3.40}$$

The conjugate basis functions $Q_i^\alpha$ are defined as

$$Q_i^\alpha \equiv \sum_\beta B_{\alpha\beta} P_i^\beta \tag{3.41}$$

Thus the spectral coefficients are obtained as

$$a_\alpha \equiv \sum_i Q_i^\alpha \gamma_i \tag{3.42}$$

It is easy to show that

$$\sum_i Q_i^\alpha \Delta\gamma_i = 0 \tag{3.43}$$

Figure VI.4 shows such a decomposition for an example $\gamma(r)$. The fine part is small in comparison with the coarse part, as it should be.

|Jacobian J|

The Jacobian is a matrix of derivatives of $d_\alpha$ with respect to $a_\beta$. Therefore the functional relationship between $\mathbf{d}$ and $\mathbf{a}$ must be established. We use PY as an illustra-tion. The sequence of functional relationships is

$$C_i = (1 + \gamma_i)[\exp(-\beta u_i) - 1] \tag{3.44}$$

$$\tilde{C}_j = \frac{4\pi\delta r}{k_j} \sum_{j=1}^{N/2-1} r_i \sin(k_j r_i) C_i \tag{3.45}$$

$$\tilde{\gamma}_j = \frac{\rho \tilde{C}_j^2}{1 - \rho \tilde{C}_j} \tag{3.46}$$

$$\gamma_m = \frac{\delta k}{2\pi^2 r_m} \sum_{j=1}^{N/2-1} k_j \sin(k_j r_m) \tilde{\gamma}_j \tag{3.47}$$

where

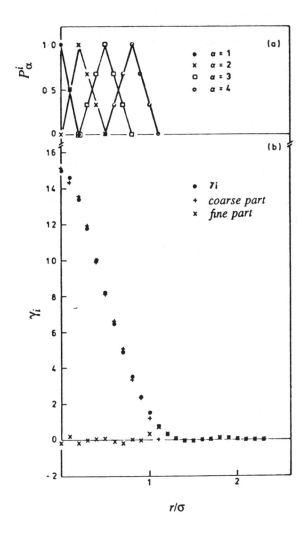

FIGURE VI.4. The roof functions for ν =4. Also shown are the coarse and fine parts of the indirect correlation function γ. (Gillan, Mol. Phys., 1979)

$$r_i \equiv i\delta r, \qquad k_j \equiv \frac{2\pi j}{N\delta r} \tag{3.48}$$

and $N$ is the number of grid points spanning the chosen range of r. These equations relate the initial guess $\gamma$ in (3.27) (thus the initial $a_\alpha$) to the output $\gamma'$ of (3.31) (i.e., $a'_\alpha$). Since $d_\alpha = a_\alpha - a'_\alpha$, one could evaluate the partial derivatives $\partial d_\alpha/\partial a_\beta$

$$J_{\alpha\beta} = \delta_{\alpha\beta} - \partial a'_\alpha/\partial a_\beta \tag{3.49}$$

$$= \delta_{\alpha\beta} - \sum_i \frac{\partial a'_\alpha}{\partial \gamma_i} \cdot \frac{\partial \gamma_i}{\partial a_\beta}$$

$$= \delta_{\alpha\beta} - \sum_{ij} Q_i^\alpha \frac{\partial \gamma_i}{\partial \gamma_j} P_j^\beta$$

The derivative $\partial \gamma/\partial \gamma$ could be obtained from the set of equations (3.44-47). The Fourier transforms are included in the differentiation

$$\frac{\partial \gamma_i}{\partial \gamma_j} = \sum_{m=0}^{N/2-1} \frac{\partial \gamma_i}{\partial \tilde{\gamma}_m} \frac{\partial \tilde{\gamma}_m}{\partial \tilde{C}_m} \frac{\partial \tilde{C}_m}{\partial C_j} \frac{\partial C_j}{\partial \gamma_j} \tag{3.50}$$

The results is

$$\frac{\partial \gamma_i}{\partial \gamma_j} = \frac{\delta r r_j}{\pi r_i} [\exp(-\beta u_j) - 1](D_{i-j} - D_{i+j}) \tag{3.51}$$

for $i \neq 0$, where

$$D_l = \delta k \sum_{m=0}^{N/2-1} \cos(k_m r_l) \left[ \frac{2\rho \tilde{C}_m}{1-\rho\tilde{C}_m} + \left[ \frac{\rho\tilde{C}_m}{1-\rho\tilde{C}_m} \right]^2 \right] \tag{3.52}$$

For the special case $i=0$

$$\frac{\partial \gamma_0}{\partial \gamma_j} = \frac{2\delta r r_j}{\pi} [\exp(-\beta u_j) - 1] E_j \tag{3.53}$$

where

$$E_j = \delta k \sum_{m=0}^{N/2-1} k_m \sin(k_m r_j) \left[ \frac{2\rho\tilde{C}_m}{1-\rho\tilde{C}_m} + \left[ \frac{\rho\tilde{C}_m}{1-\rho\tilde{C}_m} \right]^2 \right] \tag{3.54}$$

The above derivation was based on the PY assumption. To apply the HNC approximation, one replaces the factor $\exp(-\beta u_j) - 1$ by $\exp(-\beta u_j + \gamma_j)$ in eqs. (3.51 and 54). The expansion has been based on the *roof* functions. When other basis functions are chosen (e.g., the Fourier series of Labik et al.), a similar development could be made.

The evaluation of the Jacobian in the Newton-Raphson approach was straightforward for pure fluids. However, for mixtures, the task of finding the derivatives is increased manifold. Differentiation is still possible, but tedious. Recently Lee [11] has proposed a *regula falsi* method that avoids explicit evaluation of the Jacobians. The speed of convergence is comparable to the NR method. Thus for mixture systems, the latter method is more convenient to use.

*Examples*

Gillan has applied the above scheme to inverse sixth power and Lennard-Jones potentials using the PY and HNC equations. The fluids were in the dense liquid state (for LJ, near the triple point) and at temperatures from $kT/\varepsilon = 0.7$ to 2. A number of tests were carried out: (i) tests of the total number of NR cycles required (typically 10 cycles at low densities and 30 cycles at high densities); and (ii) tests of initial guesses for $\gamma_i$.

As mentioned earlier, the results were insensitive to initial guesses, provided some care was exercised. Gillan used the dcf $C(r)$ of hard spheres (from PY solution) as initial guesses. The equivalent diameters $d$ used for the inverse sixth and LJ potentials were $d/\sigma = 0.8$ and 0.9, respectively. Other diameters (0.2, 0.5 and 0.7) were chosen, the speed of convergence was not appreciably affected. This feature is most desirable for high density states, as otherwise the convergence could be painfully slow in the conventional Picard method (normally 200 to 300 iterations would be needed, that is, with a good initial guess).

Another useful technical information was that seven basis functions, in this case the roof functions, were used for the LJ potential, with nodes at $r/\sigma = 0.2, 0.4, 0.65, 0.95, 1.1, 1.2$ and 1.5. The number of grids was $N = 512$, and $\delta r/\sigma = 0.05$. Convergence was established when the root mean squared deviations of $|(\gamma' - \gamma)^2|$ was less than $10^{-5}$. For LJ, the initial values $\gamma$ near the potential minimum should be treated with care. In order to avoid divergence, $\gamma$ was set to $-1$ between $1.00 \le r \le 1.4$. Successive changes in $a_\alpha$ should also be limited in the NR cycles. The values $\bar{a}_\alpha$ given by eq. (3.34) was normalized by the root mean squares $\Delta^2 \equiv \sum_\alpha \Delta a_\alpha^2$ before input to the next NR cycle. The correction was triggered by the upper limit $\Delta^2 \le 5$.

Labik et al. [12] used sine functions instead of the roof functions. They obtained threefold to ninefold faster convergence than the Gillan scheme. This was made possible in part by avoiding evaluating the Fourier transforms in the NR cycles.

## VI.4. *The Hypernetted Chain Equation*

The HNC approximation corresponds to the following requirement

$$C(r) \approx h(r) - \ln y(r) \qquad \text{(HNC)} \qquad (4.1)$$

A connection between the PY and the HNC approximation can be made. For $y(r) \cong 1$ (e.g., at low densities), the logarithmic term can be expanded as

$$\ln y = \ln[1 + (y - 1)] = (y - 1) - \frac{1}{2}(y - 1)^2 + \frac{1}{3}(y - 1)^3 - + \cdots \qquad (4.2)$$

Thus,

$$C(r) = h(r) - [y(r) - 1] + \left[ \frac{(y - 1)^2}{2} - \frac{(y - 1)^3}{3} + - \cdots \right] \qquad (4.3)$$

If the terms in brackets are small, HNC reduces to PY. Equation (4.1) coupled with the OZ relation (1.16) gives a complete set of equations for the solution of $g(r)$.

## THE GENERATING FUNCTIONAL

The HNC equation could also be derived from the formalism of generating functionals [13]. The functional in this case is precisely the singlet direct correlation function

$$C^{(1)}(1) = \ln \rho^{(1)}(1)\Lambda^3 + \beta[w(1) - \mu] \tag{4.4}$$

defined earlier in (IV.4.6). Taylor's expansion gives

$$\ln \rho^{(1)}(1;w) + \beta w(1) = \ln \rho^{(1)}(1;w_0) \tag{4.5}$$

$$+ \beta w_0(1) + \int d2 \ C^{(2)}(1,2;w_0) \ [\rho^{(1)}(2;w) - \rho^{(1)}(2;w_0)]$$

$$+ \frac{1}{2} \int d3 \ d2 \ C^{(3)}(1,2,3;w_0) \ [\rho^{(1)}(3;w) - \rho^{(1)}(3;w_0)] \ [\rho^{(1)}(2;w) - \rho^{(1)}(2;w_0)] + \cdots$$

where we have used the relation (1.6). This expansion is the nonuniform equivalent to the density expansion of the uniform chemical potential in thermodynamics. Suppose, as before, that $w(i)$ is generated by a test particle at the origin 0, i.e., $w(i)=u(0i)$, where $u$ is the pair potential, then we have

$$\ln y(01) = \ln y_0(01) + \int d2 \ C^{(2)}(12;w_0)\rho^{(1)}(2)[g^{(2)}(20) - g_0^{(2)}(20)] \tag{4.6}$$

$$+ \frac{1}{2} \int d2 \ d3 \ C^{(3)}(123;w_0)\rho^{(1)}(2)\rho^{(1)}(3)[g^{(2)}(30) - g_0^{(2)}(30)][g^{(2)}(20) - g_0^{(2)}(20)] + \cdots$$

This is the *second-order hypernetted chain equation* (HNC2). If the reference potential $w_0 = 0$ (i.e., for uniform fluids), $u_0 = 0$ and $g_0^{(2)} = 1$. Thus to first order,

$$\ln y(01) \approx \rho \int d2 \ C^{(2)}(02)h(21) \qquad \text{(HNC)} \tag{4.7}$$

This is the *hypernetted chain* (HNC) equation. Comparison with the OZ relation shows that $\ln y(1,0) = h(1,0) - C(1,0)$, thus proving eq. (4.1).

## BIPOLAR COORDINATES

To carry out the numerical solution, we transform (4.7) into bipolar coordinates. The result is

$$\ln y(r) = \frac{2\pi\rho}{r} \int_0^\infty ds \ sC(s) \int_{|r-s|}^{r+s} dt \ th(t) \tag{4.8}$$

Furthermore $C(r)$ and $h(r)$ are related by (4.1). The method of solution is the same as the one described in section VI.3. It is also possible to derive the HNC equation from other methods, such as the resummation of the clusters [14] [15] [16] [17]. Morita [18] arrived at (4.1) by ignoring a certain class of bridge diagrams in the cluster expansion for $g(r)$. Both PY and HNC were derived from the singlet direct correlation function $C^{(1)}$ and are thus related to the (nonuniform) chemical potential. These two integral equations have proven to be successful in the theories for liquids. The YBG equation to be introduced below has a different origin. It is based on a force balance on the molecular level: the BBGKY hierarchy.

## VI.5  BBGKY Hierarchy and the YBG Equation

The Born-Bogoliubov-Green-Kirkwood-Yvon (BBGKY) hierarchy is an exact relation connecting a low order correlation function to a higher one. It is obtained by differentiating the n-tuplet density function $\rho^{(n)}$ with respect to the position, $r_1$, of the first particle:

$$\rho^{(n)}(r_1, r_2,..., r_n) \equiv \frac{1}{Q_N} \frac{N!}{(N-n)!} \int dr_{n+1} \cdots dr_N \, e^{-\beta V_N} \tag{5.1}$$

and

$$\frac{\partial \rho^{(n)}}{\partial r_1} = \frac{1}{Q_N} \frac{N!}{(N-n)!} \int dr_{n+1} \cdots dr_N \, e^{-\beta V_N} \left[ \sum_{j=2}^{N} \left( -\beta \frac{\partial u(r_{1j})}{\partial r_1} \right) \right] \tag{5.2}$$

The total potential $V_N$ is assumed to be pairwise additive. The summation over $j$ can be separated into two sets, $2 \leq j \leq n$ and $n < j \leq N$; i.e.,

$$\sum_{j=2}^{N} (\cdot) = \sum_{j=2}^{n} (\cdot) + \sum_{j=n+1}^{N} (\cdot) \tag{5.3}$$

The first set involves $u(r_{1j})$ independent of the integrated variables and can be taken out of the integral. The second set must stay within. Applying the definition of $\rho^{(n+1)}$, we have

$$\frac{\partial \rho^{(n)}(r^n)}{\partial r_1} = -\rho^{(n)}(r_n) \sum_{j=2}^{n} \beta \frac{\partial u(r_{1j})}{\partial r_1} - \beta \int dr_{n+1} \, \rho^{(n+1)}(r^{n+1}) \frac{\partial u(r_{1,n+1})}{\partial r_1} \tag{5.4}$$

For uniform fluids,

$$\frac{\partial g^{(n)}(r^n)}{\partial r_1} = -g^{(n)}(r^n) \sum_{j=2}^{n} \beta \frac{\partial u(r_{1j})}{\partial r_1} - \rho \int dr_{n+1} \, \beta \frac{\partial u(r_{1,n+1})}{\partial r_1} g^{(n+1)}(r^{n+1}) \tag{5.5}$$

Equations(5.4) and (5.5) are the *BBGKY hierarchy*. For $n=1$ and $n=2$,

$$\frac{\partial \rho^{(1)}(r_1)}{\partial r_1} = -\int dr_3 \, \beta \frac{\partial u(r_{13})}{\partial r_1} \rho^{(2)}(r_1, r_3) \tag{5.6}$$

and

$$\frac{\partial \rho^{(2)}(\mathbf{r}_1, \mathbf{r}_2)}{\partial \mathbf{r}_1} = -\rho^{(2)}(\mathbf{r}_1, \mathbf{r}_2)\, \beta\, \frac{\partial u(r_{12})}{\partial \mathbf{r}_1} \tag{5.7}$$

$$-\int d\mathbf{r}_3\, \beta\, \frac{\partial u(r_{13})}{\partial \mathbf{r}_1}\, \rho^{(3)}(\mathbf{r}_1, \mathbf{r}_2, \mathbf{r}_3)$$

For uniform fluids, the singlet density $\rho^{(1)}(\mathbf{r}_1)$ is independent of the vector $\mathbf{r}_1$. Thus $\partial \rho^{(1)}/\partial \mathbf{r}_1 = 0$, and

$$\rho \int d\mathbf{r}_3\, \beta\, \frac{\partial u(r_{13})}{\partial \mathbf{r}_1}\, g^{(2)}(\mathbf{r}_1, \mathbf{r}_3) = 0 \tag{5.8}$$

We multiply (5.8) by $g^{(2)}(\mathbf{r}_1, \mathbf{r}_2)$ and subtract the product from (5.7):

$$\frac{\partial g^{(2)}(\mathbf{r}_1, \mathbf{r}_2)}{\partial \mathbf{r}_1} = -g^{(2)}(\mathbf{r}_1, \mathbf{r}_2)\, \beta\, \frac{\partial u(r_{12})}{\partial \mathbf{r}_1} \tag{5.9}$$

$$-\rho \int d\mathbf{r}_3\, \beta\, \frac{\partial u(r_{13})}{\partial \mathbf{r}_1}\left[ g^{(3)}(\mathbf{r}_1, \mathbf{r}_2, \mathbf{r}_3) - g^{(2)}(\mathbf{r}_1, \mathbf{r}_3)g^{(2)}(\mathbf{r}_1, \mathbf{r}_2)\right]$$

Equation (5.9) is exact for uniform fluids. As the gradient $-\nabla u$ gives the pair forces, it expresses the balance of microscopic forces (*mechanical equilibrium*) in the fluid, To solve the equation, we need information of $g^{(3)}$. If we use the next higher hierarchy from (5.5), we will reintroduce one more unknown quantity, $g^{(4)}$. Therefore the sequence of equations is not *closed*. Kirkwood [19] proposed the superposition approximation for triplet correlation functions

$$g^{(3)}(\mathbf{r}_1, \mathbf{r}_3, \mathbf{r}_3) = g^{(2)}(\mathbf{r}_1, \mathbf{r}_2)g^{(2)}(\mathbf{r}_2, \mathbf{r}_3)g^{(2)}(\mathbf{r}_3, \mathbf{r}_1) \tag{5.10}$$

Substitution into (5.9) gives the approximate *Yvon-Born-Green* (YBG) equation

$$\frac{\partial \ln y(\mathbf{r}_1, \mathbf{r}_2)}{\partial \mathbf{r}_1} = -\rho \int d\mathbf{r}_3\, \beta\, \frac{\partial u(r_{13})}{\partial \mathbf{r}_1}\, g(\mathbf{r}_1, \mathbf{r}_3)h(\mathbf{r}_3, \mathbf{r}_2), \tag{5.11}$$

where $y$ is the background correlation function. Its validity depends critically on the superposition approximation.

## BIPOLAR COORDINATES

Equation (5.11) could be transformed into bipolar coordinates with the result

$$\ln y(r) = \frac{\pi \rho}{r} \int_0^\infty dt\, th(t)\left[ K(|r-t|) - K(r+t)\right] \tag{5.12}$$

where

$$K(x) \equiv \int_x^\infty ds \; \beta u'(s) g(s)(s^2 - x^2), \qquad x \geq 0 \qquad (5.13)$$

and $u'(r) = du(r)/dr$. We observe that eqs. (5.12 & 13) can be solved by the numerical procedure outlined in section VI.3 The mathematical steps leading up to eq.(5.12) are quite interesting. We shall pursue it in some detail below. The gradient operator $\partial/\partial \mathbf{r}_1$ in the Cartesian coordinates is

$$\frac{\partial}{\partial \mathbf{r}_1} = \left[ \frac{\partial}{\partial x_1}, \frac{\partial}{\partial y_1}, \frac{\partial}{\partial z_1} \right] \qquad (5.14)$$

The derivative for an isotropic pair potential can be obtained from the chain rule

$$\frac{\partial u(r_{12})}{\partial \mathbf{r}_1} = \frac{\partial u(r_{12})}{\partial r_{12}} \frac{\partial r_{12}}{\partial \mathbf{r}_1} \qquad (5.15)$$

Since

$$r_{12} = |\mathbf{r}_{12}| = |\mathbf{r}_2 - \mathbf{r}_1| = \sqrt{(x_2 - x_1)^2 + (y_2 - y_1)^2 + (z_2 - z_1)^2}$$

we have

$$\frac{\partial r_{12}}{\partial x_1} = \frac{1}{2 r_{12}} \, 2(x_2 - x_1)(-1) = -\frac{x_2 - x_1}{r_{12}} \qquad (5.16)$$

Similarly, for $\partial r_{12}/\partial y_1$ and $\partial r_{12}/\partial z_1$. We have

$$\frac{\partial r_{12}}{\partial \mathbf{r}_1} = \left[ -\frac{(x_2 - x_1)}{r_{12}}, -\frac{(y_2 - y_1)}{r_{12}}, -\frac{(z_2 - z_1)}{r_{12}} \right] = -\frac{\mathbf{r}_2 - \mathbf{r}_1}{r_{12}} = -\frac{\mathbf{r}_{12}}{r_{12}} \qquad (5.17)$$

Referring to Figure VI.1, we set $\mathbf{r}_{12} = \mathbf{r}$, $\mathbf{r}_{13} = \mathbf{s}$, and $\mathbf{r}_{23} = \mathbf{t}$; (5.11) is now

$$-\frac{\mathbf{r}}{r} \frac{d \ln y(r)}{dr} = \rho \int d\mathbf{r}_3 \, \beta \, \frac{\partial u(s)}{\partial s} \, \frac{\mathbf{s}}{s} g(s) h(t) \qquad (5.18)$$

Next we form the dot (inner) product with $\mathbf{r}$:

$$-r \frac{d \ln y(r)}{dr} = \rho \int d\mathbf{r}_3 \, \beta \, \frac{du(s)}{ds} \, \frac{\mathbf{r} \cdot \mathbf{s}}{s} \, g(s) h(t) \qquad (5.19)$$

From the definition of dot product, $\mathbf{r} \cdot \mathbf{s} = rs \cos A$, where $A$ is the angle between $\mathbf{r}$ and $\mathbf{s}$. Also from the cosine law, $\cos A = (r^2 + s^2 - t^2)/2rs$, so

$$-\frac{d \ln y(n)}{dr} = \rho \int d\mathbf{r}_3 \, \frac{r^2 + s^2 - t^2}{2rs} \, \beta u'(s) g(s) h(t) \qquad (5.20)$$

When we transform into bipolar coordinates, we obtain

$$-\frac{d\ln y(r)}{dr} = \frac{2\pi\rho}{r}\int_0^\infty ds\ s\beta u'(s)g(s)\int_{|r-s|}^{r+s} dt\ \frac{r^2+s^2-t^2}{2rs}th(t) \tag{5.21}$$

$$= \pi\rho\int_0^\infty ds\ \beta u'(s)g(s)\int_{|r-s|}^{r+s} dt\ \frac{r^2+s^2-t^2}{r^2} th(t)$$

Integration with respect to $r$ gives

$$-\int_a^b dr\ \frac{d\ln y(r)}{dr} = -\ln y(b) + \ln y(a) \tag{5.22}$$

$$= \pi\rho\int_0^\infty ds\ \beta u'(s)g(s)\int_a^b dr\int_{|r-s|}^{r+s} dt\ th(t)\left[1+\frac{s^2}{r^2}-\frac{t^2}{r^2}\right]$$

From the properties of $y(r)$ for isotropic fluids, we know that $y(r)=1$ as $r\to\infty$ . Thus ln $y(b)=0$ as $b\to\infty$ . The lower limit $\int_{r-s}$ can be relaxed to $r-s$ by redefining $h(-t)=h(t)$. Since the whole integrand is an odd function of $t$, the contribution from the negative part of $r-s$ will cancel that from the positive part, thus leaving the effective contribution to the integral come from $|r-s|$. Next we interchange the order of integration from $\int dr\int dt$ to $\int dt\int dr$. The region of integration can be separated into three subregions (see Figure VI.5), $A$, $B$ and $C$. Thus

$$\int_a^b dr\int_{r-s}^{r+s} dt = \int_{a-s}^{a+s} dt\int_a^{t+s} dr \qquad \text{(region } A) \tag{5.23}$$

$$+ \int_{a+s}^{b-s} dt\int_{t-s}^{t+s} dr \qquad \text{(region } B)$$

$$+ \int_{b-s}^{b+s} dt\int_{t-s}^{b} dr \qquad \text{(region } C)$$

Combining (5.23) with (5.22) and letting $b\to\infty$ gives the following integrals

**Region B**

$$\int_{a+s}^{b-s} dt\ th(t)\int_{t-s}^{t+s} dr\ (1+s^2r^{-2}-t^2r^{-2}) = \int_{a+s}^{b-s} dt\ th(t)\left[r-\frac{s^2}{r}+\frac{t^2}{r}\right]_{r=t-s}^{r=t+s} = 0 \tag{5.24}$$

since $[r-s^2r^{-1}+t^2r^{-1}]_{r=t-s}^{r=t+s} = 0$.

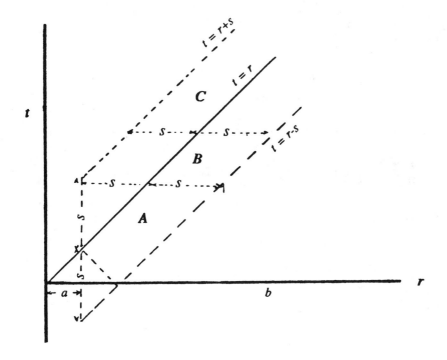

FIGURE VI.5. *The regions of integration, A, B and C, in the r-t plane. The distance s is fixed.*

*Region C*

$$\lim_{b \to \infty} \int_{b-s}^{b+s} dt \; th(t) \left[ r - \frac{s^2}{r} + \frac{t^2}{r} \right]_{r=t-s}^{r=b} = 0 \tag{5.25}$$

since for fixed $s$, $\int_{-\infty}^{\infty} dt(\cdot) = 0$. The only integral surviving the integration is from region A.

*Region A*

$$\int_{a-s}^{a+s} dt \; th(t) \left[ r - \frac{s^2}{r} + \frac{t^2}{r} \right]_{r=a}^{r=t+s} = \int_{a-s}^{a+s} dt \; th(t) \frac{s^2 - (a-t)^2}{a} \tag{5.26}$$

Replace $a$ by $r$ in eq. (5.22) to get

$$\ln y(r) = \pi \rho \int_0^{\infty} ds \; \beta u'(s) g(s) \int_{r-s}^{r+s} dt \; th(t) \frac{s^2 - (r-t)^2}{r} \tag{5.27}$$

We interchange the order of integration again from $\int ds \int dt$ to $\int dt \int ds$.

$$\int_0^{\infty} ds \int_{r-s}^{r+s} dt = \int_{-\infty}^{+\infty} dt \int_{|r-t|}^{+\infty} ds \tag{5.28}$$

$$= \int_0^{+\infty} dt \int_{|r-t|}^{+\infty} ds - \int_0^{-\infty} dt \int_{|r-t|}^{+\infty} ds$$

Now if we make the change of variables $t'=-t$ (thus $dt'=-dt$), (5.28) becomes

$$= \int_0^{+\infty} dt \int_{|r-t|}^{+\infty} ds + \int_0^{+\infty} dt' \int_{|r+t'|}^{+\infty} ds \tag{5.29}$$

And equation (5.27) is

$$\ln y(r) = \frac{\pi \rho}{r} \int_0^{\infty} dt \; th(t) \int_{|r-t|}^{\infty} ds \; \beta u'(s) g(s)[s^2 - (r-t)^2] \tag{5.30}$$

$$- \frac{\pi \rho}{r} \int_0^{\infty} dt' \; t'h(t') \int_{|r+t'|}^{\infty} ds \; \beta u'(s) g(s)[s^2 - (r+t')^2]$$

Define a kernel $K(x)$ as

$$K(x) \equiv \int\limits_{|x|}^{\infty} ds \ \beta u'(s)g(s)(s^2 - x^2) \tag{5.31}$$

Equation (5.30) can be written as

$$\ln y(r) = \frac{\pi \rho}{r} \int\limits_{0}^{\infty} dt \ th(t)[K(|r-t|) - K(r+t)] \tag{5.32}$$

which is precisely eq. (5.12).

## THE GENERATING FUNCTIONAL

It is also possible to derive the YBG equation by the method of functional expansion. The generating functional in this case is

$$G[\rho^{(1)}] = \rho^{(1)}(1; w) \frac{\partial}{\partial \mathbf{r}_1} \beta w(1) \tag{5.33}$$

The functional derivative relation needed is

$$\frac{\delta \rho^{(1)}(1; w) \nabla_1 \beta w(1)}{\delta[\ln \rho^{(1)}(2; w) + \beta w(2)]} = C^{-1}(12) \nabla_1 \beta w(1) - \rho^{(1)}(1) \nabla_1 \left[ \frac{1}{\rho^{(1)}(1)} \right] h^{-1}(12) \tag{5.34}$$

where the gradient operator is $\nabla_1 \equiv \partial/\partial \mathbf{r}_1$. $C^{-1}$ denotes the matrix inverse of the dcf $C(12)$. The Taylor expansion is

$$\rho^{(1)}(1; w) \nabla_1 \beta w(1) = \rho^{(1)}(1; w_0) \nabla_1 \beta w_0(1) \tag{5.35}$$

$$+ \int d2 \left\{ C^{-1}(12; w_0) \nabla_1 \beta w_0(1) - \rho^{(1)}(1; w_0) \nabla_1 \left[ \frac{h^{-1}(12; w_0)}{\rho^{(1)}(1; w_0)} \right] \right\}$$

$$\cdot \left\{ \ln \rho^{(1)}(2; w) + \beta w(2) - \ln \rho^{(1)}(2; w_0) - \beta w_0(2) + \cdots \right\}$$

For zero reference potential, $w_0 = 0$, $\nabla_1 w_0 = 0$. On the other hand, a test particle of the same species as that in the fluid produces a potential $w(i) = u(0i)$. Note that $\rho^{(1)}(1; w) = \rho g(01)$ and $\rho^{(1)}(1; w_0) = \rho$

$$\rho g(01) \nabla_1 \beta u(01) = -\rho \int d2 \ \ln y(20) \nabla_1 \left[ \frac{h^{-1}(12)}{\rho} \right] \tag{5.36}$$

Since $\nabla_1 h(r_{12}) = -\nabla_2 h(r_{12})$ for isotropic fluids

$$g(01)\nabla_1 \beta u(01) = \frac{1}{\rho} \int d2 \ \ln \ y(20)\nabla_2 h^{-1}(12) \tag{5.37}$$

Multiplying by $h(31)$ and integrating with respect to $\int d1$ (upon interchanging the gradient operation with integration), we obtain

$$\int d1 \ h(31)g(10)\nabla_1 \beta u(10) = \frac{1}{\rho} \int d1 \int d2 \ h(31)\left[\nabla_2 h^{-1}(12)\right] \ln \ y(20) \tag{5.38}$$

$$= \frac{1}{\rho} \int d2 \ \ln \ y(20) \ \nabla_2\left[\int d1 \ h(31)h^{-1}(12)\right]$$

$$= \frac{1}{\rho} \int d2 \ \ln \ y(20)\nabla_2\delta_D(32)$$

$$= -\frac{1}{\rho} \int d2 \ \delta_D(32)\nabla_2 \ \ln \ y(20)$$

$$= -\frac{1}{\rho} \ \nabla_3 \ \ln \ y(30)$$

which proves eq. (5.11). We have used the property of the Dirac delta as a *distribution* (according to L. Schwartz [20]), i.e.,

$$\int d2 \ f(12)\nabla_2\delta_D(23) = -\int d2 \ \delta_D(23)\nabla_2 \ f(12) = -\nabla_3 \ f(13) \tag{5.39}$$

for arbitrary $f(12)$ with a compact support (i.e., vanishing at large distances).

## VI.6. *The Kirkwood Equation*

Another approximate integral equation closely associated with the YBG equation is the Kirkwood equation (abbreviated KW). It could be derived the same way as YBG.

$$-\ln \ y(r_{12}) = \rho \int_0^1 d\xi \int dr_3 \ \beta u(r_{13})g(r_{13}; \ \xi)h(r_{32}) \tag{6.1}$$

where $0 \leq \xi \leq 1$ is a coupling parameter linking molecule 1 to all other molecules:

$$V_N(r_1, \ldots, r_N;\xi) = \sum_{i=2}^{N} \xi u(r_{1i}) + \sum_{2=i}^{N}\sum_{<j} u(r_{ij}) \tag{6.2}$$

and $g(r_{12}; \ \xi)$ is the rdf corresponding to the value of $\xi$ in (6.2). When $\xi = 0$, molecule 1 is removed from the system; while $\xi = 1$, molecule 1 is restored. Going over to bipolar coordinates, we have

$$\ln y(r) = \frac{\pi\rho}{r} \int\limits_{0}^{\infty} dt \ th(t)[K(|r-t|) - K(r+t)] \tag{6.3}$$

where

$$K(x) = -2 \int\limits_{0}^{1} d\xi \int\limits_{x}^{\infty} ds \ s\beta u(s)g(s;\xi) \tag{6.4}$$

There is certain similarity between the KW equation and the HNC equation, since the HNC could also be written in the form of (6.3) with $K_{HNC}(x)$

$$K_{HNC}(x) = 2 \int\limits_{x}^{\infty} ds \ s \ C(s) \tag{6.5}$$

If we rewrite (6.4) as

$$K_{KW}(x) = 2 \int\limits_{x}^{\infty} ds \ s[-\beta u(s) \int\limits_{0}^{1} d\xi \ g(s;\xi)] \tag{6.6}$$

comparison with HNC shows that KW makes the assumption

$$C(s) \approx -\beta u(s) \int\limits_{0}^{1} d\xi \ g(s;\xi) \tag{6.7}$$

Of course, the approximation

$$C(r) \approx -\beta u(r), \qquad r>d \tag{6.8}$$

is the basis of the MSA. KW proposes modification of the MSA (6.8) by a multiplicative factor $\int d\xi \ g(r;\xi)$.

## VI.7. *The Mean Spherical Approximation*

Lebowitz and Percus [21] proposed MSA based on the study of Ising models, For hard spheres, MSA coincides with the PY approximation. Given the pair potential,

$$u(r) = \infty, \qquad r \leq d \tag{7.1}$$

$$= u_{soft}(r), \qquad r>d$$

we know that the rdf, from its definition, must obey

$$g(r) = 0, \qquad r \leq d \qquad \text{(exact)} \tag{7.2}$$

For $r>d$, $g(r)$ is not known analytically. On the other hand, if we examine the cluster expansion of dcf at the zeroth order (for a rigorous demonstration, see, e.g., Stell [22]), we could write

$$C(r) = e^{-\beta u(r)} - 1 + \cdots \qquad (7.3)$$

At large $r$, $u(r) \to 0$; thus we can expand the exponential

$$\lim_{r \to \infty} C(r) = -\beta u(r) \qquad (7.4)$$

In MSA, the assumption is made such that for all $r>d$,

$$C(r) \approx -\beta u_{soft}(r), \qquad r>d \qquad \text{(approximate)} \qquad (7.5)$$

Physically, it says that the direct correlation between molecules is given by the interaction potential. Through the OZ relation

$$g(r) - C(r) - 1 = \rho \int ds\, C(|\mathbf{r} - \mathbf{s}|)h(s) \qquad (7.6)$$

which convolutes $h$ with the $C$ function, we have sufficient information to obtain $g(r)$ outside $d$ and $C(r)$ inside $d$. Assumption (7.5) is not exact. Furthermore, MSA is asymmetrical with respect to treatment of $h$ and $C$ (i.e., the convolutions $h*C \neq C*h$ due to the special form (7.5)). So far the theory has been restricted to molecules with a hard core for which condition (7.2) is valid. Recently, the hard core assumption has been relaxed by Rosenfeld and Ashcroft [23] and Rosenfeld [24] in a soft sphere MSA. MSA has been extensively studied for model potentials, ranging from dipole-dipole interaction, Coulomb potential to the reference interaction site approximation (RISA).

The advantages of MSA are as follows: (i) It makes maximum use of available information (i.e., $g(r)$ inside core and $C(r)$ outside core). (ii) It yields, for simple potentials, analytical formulas for correlation functions (e.g., for $C(r)$) and thermodynamic functions. For hard spheres, MSA is equivalent to the PY theory

$$C(r) = g(r) - y(r) = g(r)[1 - e^{\beta u(r)}] \qquad (7.7)$$

When $-\beta u_{HS}(r)$ is zero for $r>d$, $C(r)=0$.

## VI.8. *Numerical Results for Model Potentials*

Given a pair potential $u(r)$, we ought to be able to utilize the integral equations discussed so far to obtain the correlation functions $g(r)$ and $C(r)$. It is desirable to illustrate their applications at this point to some simple model potentials. We shall examine the hard-sphere (HS), square-well (SW), and Lennard-Jones (LJ) potentials. In recent years, simulation data have been generated for these potentials. It is possible to compare the integral equation results with these data.

Integral equations could be evaluated according to the following criteria. First, they should yield accurate structural information as compared to simulation data. Next, they should yield correct thermodynamic properties, such as virial coefficients, internal energy, and pressure. Third, they ought to be internally consistent, i.e., yielding consistent pressures and energies (see sections IV.8 and 9). Finally, they should be useful in predicting real fluid properties. The last requirement is the ultimate utility of integral equation theories, while the other criteria are tests of validity of the theory. All these aspects have been subjects of research in recent years. For example, the LJ has been used to represent thermodynamic and transport properties of argon. Except for extreme conditions, the agreement is generally satisfactory (see Chapter IX). The LJ potential with effective parameters has also been used in the PY theory for calculating the *P-V-T* properties of methane, ethane and propane. On the other hand, hard spheres, as a theoretical model, have no real counterpart in nature. However, they are used in perturbation theories as a reference fluid. The PY equation is able to give an accurate description of HS properties. The SW potential is an important theoretical tool since it possesses both repulsive and attractive forces. There have been recent attempts to develop equations of state for real fluids based on the SW model [25].

We shall use the notation *theory-model* to denote the combination of equations and models; thus PY-HS is used to denote the application of PY theory to hard spheres.

## HARD SPHERES

One of the simplest improvements over the ideal-gas behavior is the hard sphere model of molecules. The molecules are given an impenetrable core, i.e., an excluded volume. Interestingly enough, this feature alone is sufficient in many cases to represent the structure of liquids. The physical interpretation is that in dense gases and liquids, the molecules are closely packed and the small distances of separation imply that in molecular dynamics the repulsive part of the potential is sampled most frequently by neighboring pairs. As a consequence, the molecules *see* one another as being *harshly* repulsive. The steric effects dominate the structure in the liquid state (provided there are no other strong short-range forces, such as hydrogen bonds). Hard spheres play an important part in the modern theories of liquids. In Chapter VIII, we shall return to the topics of hard spheres and hard-core fluids.

PY equation has been solved (by Wertheim [26] and Thiele [27]) for the hard sphere model by analytical means. The procedure of solution itself was a remarkable exercise in mathematical analysis. Because the development was lengthy, we shall not repeat it here. We quote the results below and refer the reader to the original literature for details. As to the HNC-HS, the solution must be sought numerically since the HNC is highly nonlinear. The YBG equation has also been solved for HS. Among all three equations, the PY gives the best overall performance. Several authors have given a zonal calculation of the Wertheim solution for the pcfs at different densities. For hard spheres, the temperature has no influence on the properties because of the *infinite* repulsive forces. Figure VI.6 shows the pcf at the density $\rho^* = \rho d^3 = 0.8$ from PY and simulation. The YBG results are rather poor and are not shown here. At low densities, both PY and HNC yield reasonable structural information. At $\rho^*=0.9$, discrepancies show up. The contact value $g(d)$, which is important for the pressure values, is overestimated by HNC and underestimated by PY. The oscillations are also slightly out of phase compared with simulation data. These discrepancies will affect the thermodynamic properties. Verlet and Weis [28] have devised a method for constructing more accurate rdf for HS. The method will be discussed later.

The virial coefficients of the hard spheres are known up to $B_{10}$. Theories, PY, HNC, YBG, and KW all give correct second and third virial coefficients (see Table VI.2.) From $B_4$ onward, the errors in the approximations start to show up. The virial (*v*) pressure route gives results differing from those from the isothermal compressibility (*c*). The difference arises due to the inexactitudes in the integral equations. The PY values of 0.2500 (*v*) and 0.2969 (*c*) for $B_4$ are better than the HNC values of 0.4453 (*v*) and

*Table VI.2. Virial Coefficients for Hard Spheres*

| Theory | $C/B^2$ | $D/B^3$ | $E/B^4$ | $F/B^5$ | $G/B^5$ |
|--------|---------|---------|---------|---------|---------|
| **Exact** | **0.625** | **0.2896** | **0.1103** | **0.0386** | **0.0138** |
| SPT | 0.625 | 0.2969 | 0.1121 | 0.0449 | 0.0156 |
| CS | 0.625 | 0.2813 | 0.1094 | 0.0156 | 0.0132 |
| BG(V) | 0.625 | 0.2252 | 0.0475 | | |
| BG(C) | 0.625 | 0.3424 | 0.1335 | | |
| KW(V) | 0.625 | 0.1400 | | | |
| KW(C) | 0.625 | 0.4418 | | | |
| PY(V) | 0.625 | 0.2500 | 0.0859 | 0.0273 | 0.0083 |
| PY(C) | 0.625 | 0.2969 | 0.1211 | 0.0449 | 0.0156 |
| PY2(V) | 0.625 | 0.2869 | 0.1074 | | |
| PY2(C) | 0.625 | 0.2896 | 0.1240 | | |
| HNC(V) | 0.625 | 0.4453 | 0.1447 | 0.0382 | |
| HNC(C) | 0.625 | 0.2092 | 0.0493 | 0.0281 | |
| HNC2(C) | 0.625 | 0.2896 | 0.122 | | |

0.2092 (*c*), the simulation value being 0.2896. The YBG and KW results are poor. Similar observation holds for $B_5$, $B_6$, and $B_7$. On the other hand, the second-order PY equation (PY2) proposed by Verlet [29] gives drastic improvements over all theories. In fact, the fourth virial coefficient is given exactly by PY2.

The compressibility $Z = \beta P/\rho$ is shown in Figure VI.7. The virial pressure equation (IV.2.10) simplifies to

$$\frac{P}{\rho kT} = 1 + \frac{4\pi\rho d^3}{6} g(d^+) \tag{8.1}$$

due to the derivative $\beta u'(r) = - \delta_D(r,d)$, the Dirac delta function. $g(d^+)$ is the contact value of the rdf at the collision distance. Figure VI.7 shows that PY(*v*) and PY(*c*) values bracket the exact curve. Numbers from HNC do the same but with larger deviations. The YBG values underestimate the compressibility at high densities ($\rho d^3 \geq 0.5$). As far as pressure values are concerned, we can rank the theories as

PY > HNC > YBG

**THE SQUARE-WELL POTENTIAL**

A number of computer studies of the SW fluid have appeared since 1965. We now have fairly complete knowledge of its structure and thermodynamics. Three parameters characterize the SW potential: the hard-core diameter $d$, the well depth $\varepsilon$, and the distance of the attractive wall, expressed in units of $d$ as $\lambda d$. The most studied $\lambda$ value is $\lambda = 1.5$. On the other hand, Henderson et al. [30] [31] have also carried out simulations for $\lambda = 1.125$, 1.375, 1.50, 1.75, 1.85 and 2.0. The structure $g(r)$ is presented in Figure VI.8 for PY, HNC, and MSA at $\rho d^3 = 0.8$ and $\beta \varepsilon = 1.5$. It is seen that the MSA structure is quite satisfactory except for $r > \lambda d$. The HNC results are less accurate in the well region, $d < r < \lambda d$, but perform remarkably well at $r > \lambda d$. The PY-SW values are poor in $d < r < \lambda d$ but improve slightly beyond $r > \lambda d$.

Exact values of the thermodynamic functions are presented in Table VI.3 for $\lambda = 1.5$. Figure VI.9 shows the pressure as a function of temperature at $\rho^* = 0.80$ and

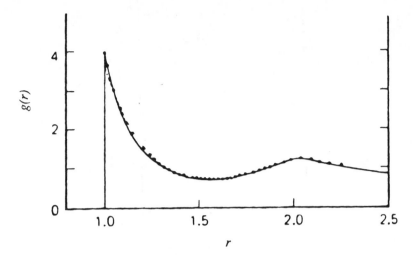

FIGURE VI.6. *Radial distribution functions g(r) for the hard sphere fluid at $\rho d^3 = 0.8$. •: simulation results. Full line: PY results. (Barker and Henderson, Rev. Mod. Phys., 1976)*

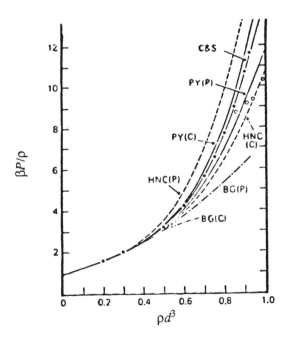

FIGURE VI.7. *Equation of state for hard spheres from different theories. The points • and O are machine simulation results. (Barker and Henderson, Rev. Mod. Phys., 1976)*

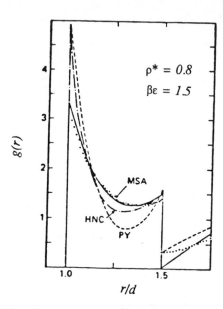

FIGURE VI.8. *Radial distribution function of the square-well fluid at* $\rho d^3 = 0.8$ *and* $\beta\varepsilon = 1.5$. *The well width is* $\lambda = 1.5$. $\cdots$ : *MC simulation data. Full line: MSA. Dashed line: PY. Dash-dot line: HNC. (Smith et al., J. Chem. Phys., 1977)*

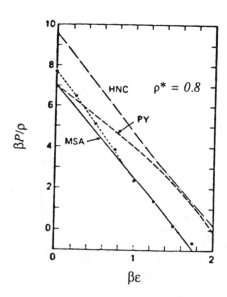

FIGURE VI.9. *Equation of state for square-well fluid with* $\lambda = 1.5$. *The points give the MC data. The curves give the theoretical results as marked. (Smith et al., J. Chem. Phys., 1977)*

λ=1.5. Results from PY, HNC, and MSA are compared. The MSA values are dependable at densities $\rho d^3 = 0.6$ and 0.8 The PY pressures, while worse than MSA, are slightly better than the HNC ones. For configurational energy, MSA again is a better theory. HNC is more accurate than PY. The ranking of performance in this case is

$$MSA > HNC > PY$$

Since the SW potential has an attractive well, there is a region where liquid and vapor phases coexist, and thus a critical point exists. Although the PY thermodynamics has been shown to be poor, the critical constants determined therefrom are reasonable. HNC values are unsatisfactory. PY2 theory gives the best agreement with MD data.

**THE LENNARD-JONES POTENTIAL**

The rdf generated from the integral equations have been compared with exact numerical simulation results as well as with experimental diffraction data. The general conclusions for the LJ fluid are:

1.  PY theory gives the best description for the LJ fluid, as compared to HNC and YBG equations, for most of the state conditions studied. Thus

$$PY > HNC > YBG$$

2.  PY theory fails in the liquid state with regard to the pressure and structure calculations. The PY energies are dependable up to $\rho^*=0.75$ ($T^*=1.07$) with an estimated error of +2%.

The above comments are to be taken with qualifications. Figure VI.10 shows the rdf calculated from PY at $T^* = 0.72$ and $\rho^*=0.85$. The first peak overshoots the MD value.

*Table VI.3. Monte Carlo Simulation of the Square-Well*
*Potential: Pressure and Energy Values ($\lambda = 1.5$)*

| $\rho\sigma^3$ | 0.5 | 0.6 | 0.7 | 0.8 |
|---|---|---|---|---|
| $\beta\varepsilon$ | | *PV/NkT* | | |
| 0.00 | 3.22 | 4.22 | 5.63 | 7.65 |
| 0.25 | 2.31 | 3.13 | 4.49 | 6.47 |
| 0.50 | 1.35 | 1.97 | 3.20 | 5.08 |
| 0.75 | 0.46 | 0.70 | 1.82 | 3.84 |
| 1.00 | -0.45 | -0.21 | 0.59 | 2.34 |
| 1.25 | | | -0.54 | 1.35 |
| 1.50 | | | | 0.18 |
| 1.75 | | | | -0.65 |
| $\beta\varepsilon$ | | $U^c/N\varepsilon$ | | |
| 0.00 | -3.342 | -4.135 | -4.907 | -5.616 |
| 0.25 | -3.482 | -4.267 | -5.028 | -5.729 |
| 0.50 | -3.636 | -4.415 | -5.174 | -5.861 |
| 0.75 | -3.892 | -4.586 | -5.352 | -5.991 |
| 1.00 | -4.069 | -4.774 | -5.487 | -6.188 |
| 1.25 | | | -5.610 | -6.302 |
| 1.50 | | | | -6.385 |
| 1.75 | | | | -6.459 |

Then the PY values oscillate out of phase (with a shorter period) with respect to the MD rdf. This phase *décalage* was already present in the HS case. This phenomenon also explains why the PY pressures are higher. Results from HNC and YBG are considerably poorer.

Table VI.4 compares the thermodynamic calculations of PY and HNC at two isotherms ($T^* = 2.74$, and $1.35$). It is seen that $U^c/N\varepsilon$ of PY is in good agreement with simulation. For a number of isotherms, the pressure values from the virial equation are better than those from the compressibility equation. The critical points predicted by various theories are compared in Table VI.5. We observe that PY($c$), PY2($v$), and HNC($c$) are good for $T_c$ predictions, PY2($c$) and PY($v$) are good for $\rho_c$ predictions, and PY($v$) and PY2($v$) are good for $Z_c$ predictions.

Since the PY energy is reliable for LJ fluid, Chen et al. [32] proposed a thermodynamic route based on the internal energy. For example, the free energy can be obtained from the Gibbs-Helmholtz relation

$$\frac{A^c}{T} = \int d(\frac{1}{T})U^c \tag{8.2}$$

and the pressure from

$$P = 1 - \left.\frac{\partial A^c}{\partial V}\right|_{T,N} \tag{8.3}$$

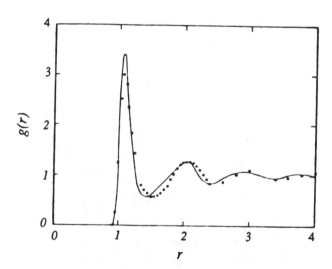

*FIGURE VI.10. Radial distribution function of the Lennard-Jones fluid at $kT/\varepsilon = 0.72$ and $\rho\sigma^3 = 0.85$. The dotted line is the simulation results; the full line is calculated from PY. (Barker and Henderson, Rev. Mod. Phys., 1976)*

Table VI.4. Thermodynamic Properties of the Lennard-Jones Fluid

| kT/ε | ρσ³ | MD | PY(E) | PY(V) | PY(C) | HNC(V) | HNC(C) |
|------|-----|-----|-------|-------|-------|--------|--------|
| | | | **Compressibility, $\beta P/\rho$** | | | | |
| 2.74 | 0.65 | 2.22 | 2.23 | - | - | - | - |
| | 0.70 | 2.64 | - | 2.645 | 2.44 | 3.16 | 2.21 |
| | 0.95 | 6.15 | 6.31 | - | - | - | - |
| | 1.00 | 7.39 | - | 6.67 | 6.81 | 9.11 | 5.10 |
| 1.35 | 0.30 | 0.35 | 0.36 | 0.40 | - | - | - |
| | 0.5 | 0.30 | 0.33 | 0.53 | - | 0.56 | - |
| | 0.65 | 0.80 | 0.85 | 1.26 | - | 1.49 | - |
| | 0.70 | 1.17 | - | 1.69 | - | 2.09 | - |
| | 0.85 | 3.37 | 3.24 | - | - | - | - |
| 1.00 | 0.65 | -0.25 | -0.22 | - | - | - | - |
| | 0.85 | 2.27 | 2.14 | - | - | - | - |
| 0.72 | 0.85 | 0.40 | 0.33 | - | - | - | - |
| | 0.90 | 1.60 | 1.59 | - | - | - | - |

| kT/ε | ρσ³ | MC | PY | HNC |
|------|-----|-----|-----|-----|
| | | **Configurational Energy, $U/N\varepsilon$** | | |
| 2.74 | 0.3 | -1.783 | -1.791 | -1.787 |
| | 0.55 | -3.207 | -3.238 | -3.127 |
| | 0.7 | -3.902 | -3.931 | -3.693 |
| | 0.80 | -4.281 | - | -3.852 |
| | 1.00 | -4.180 | -4.576 | -3.261 |
| 1.35 | 0.30 | -2.090 | -2.18 | - |
| | 0.50 | -3.372 | -3.38 | -3.37 |
| | 0.65 | -4.343 | -4.32 | -4.26 |
| | 0.70 | -4.684 | -4.65 | -4.52 |

where $c$ denotes configurational properties. The other thermodynamic properties can be obtained from the Maxwell relations. The pressure obtained from (6.3) (called PY($E$)) is quite accurate. Table VI.4 shows that $P$ is dependable in the liquid state (down to $T^*=1.00$, $\rho^*=0.90$, where −5% deviation is recorded). At $T^*=0.72$ (very close to the

Table VI.5. Critical Constants of the Lennard-Jones Fluid

| Theory | $kT_c/\varepsilon$ | $\rho_c\sigma^3$ | $Z_c$ |
|--------|------------|-------------|-------|
| YBG(C) | 1.58 ± 0.02 | 0.40 ± 0.03 | 0.48 ± 0.03 |
| YBG(V) | 1.45 ± 0.03 | 0.40 ± 0.05 | 0.44 ± 0.04 |
| PY(C) | 1.32 ± 0.02 | 0.28 ± 0.03 | 0.36 ± 0.02 |
| PY(V) | 1.25 ± 0.02 | 0.29 ± 0.03 | 0.36 ± 0.02 |
| HNC(C) | 1.39 ± 0.02 | 0.28 ± 0.03 | 0.30 ±0.04 |
| HNC(V) | 1.25 ± 0.02 | 0.26 ± 0.03 | 0.35 ±0.03 |
| PY2(C) | 1.33 ± 0.03 | 0.33 ± 0.04 | 0.34 ± 0.03 |
| PY2(V) | 1.36 ± 0.04 | 0.35 ± 0.03 | 0.31 ± 0.03 |
| **Exact** | 1.35 | 0.35 | 0.30 |
| Argon | 1.26 | 0.319 | 0.2912 |

triple point), $Z(MC) = 1.60$ at $\rho^* = 0.90$ while $Z(PY(E)) = 1.59$. (The phase diagram shows that this is in the pseudosolid, or subcooled liquid region.) On the fusion line at $T^* = 0.72$, $Z$ from $PY(E)$ is $-38\%$ off.

To improve the PY theory, Verlet [33] proposed a second-order theory, the PY2 equation. It performs better than PY in all respects (for hard spheres and LJ fluid in pressures or critical point calculations). However, PY2 remains inadequate as a quantitative theory for the liquid state. More accurate approaches are needed. This is furnished by the perturbation methods. The RHNC (reference HNC) equation of Lado [34] incorporates the bridge diagrams of hard spheres in the reference fluid. Improved results are obtained. It is also possible to formulate perturbations based on density expansions whereby the correlation function at one density is used as reference for correlations at other densities [35]. Better results could be obtained.

## VI.9. *Thermodynamic Relations from Integral Equations*

Due to the assumptions made in integral equations, the thermodynamic relations discussed in Chapter IV assume particularly simple forms. Among these are the chemical potential, the compressibility, and correlation functions.

### CHEMICAL POTENTIAL

Recall that the chemical potential $\mu$ can be calculated from eq. (IV.6.11), and the relation is exact. However, for PY it can be shown [36] that the following relation holds at low densities:

$$\frac{\mu}{kT} = \ln (\rho\Lambda^3) - \rho \int_0^\infty dr\, 4\pi r^2 \left[\frac{C(r)}{h(r) - C(r)}\right] \ln[g(r) - C(r)] \qquad \text{(PY)} \qquad (9.1)$$

For hard spheres, this formula is valid up to $\rho d^3 = 0.5$. In addition, Morita [37] has shown that in HNC the chemical potential is given by

$$\frac{\mu}{kT} = \ln(\rho\Lambda^3) - \rho \int_0^\infty dr\, 4\pi r^2 C(r) - \frac{1}{2}\, [h(0) - C(0)] \qquad (9.2)$$

$$+ \frac{1}{2}\, \rho \int_0^\infty dr\, 4\pi r^2 h(r)^2 \qquad \text{(HNC)}$$

It has been shown [38] that the HNC2 equation yields the formula

$$\beta\mu_{HNC2} = \ln \rho\Lambda^3 - \rho\tilde{C}(0) + \frac{1}{2}[h(0) - C(0)] - \frac{1}{2}\, \rho \int dx\, C(x)^2 \qquad (9.3)$$

$$- \frac{1}{3}\, \rho \int dx\, h(x)^3 - \frac{2}{3}\, \rho \int dx\, C(x)^3 + \rho \int dx\, C(x)^2 h(x)$$

$$- \rho^2 \int dx\, dy\, C(x)h(x)C(|x - y|) + \rho^2 \int dx\, dy\, C(x)^2 C(|x - y|)$$

$$+ \frac{1}{2} \rho^3 \int dx \, dy \, dz \, h^{(3)}(x', y', z') C(x) C(y)$$

$$+ \frac{1}{3} \rho^3 \int dx \, dy \, dz \, h^{(3)}(x', y', z') C(x) C(y) C(z)$$

$$- \frac{1}{2} \rho^4 \int dw \, dx \, dy \, dz \, h^{(3)}(x', y', z') C(x) C(y) C(|z - w|)$$

where $x' = |x - y|$, $y' = |y - z|$, and $z' = |z - x|$.

## PRESSURE

The compressibility equation (IV.4.15) can be integrated, according to Baxter [39], in the PY approximation to yield the pressure

$$\frac{P}{\rho kT} = 1 + \frac{1}{2} \rho \int_0^\infty dr \, 4\pi r^2 C(r) \left[ \frac{C(r)e(r)}{f(r)} - 2 \right] \tag{9.4}$$

$$+ \frac{1}{(2\pi)^3 \rho} \int_0^\infty dk \, 4\pi k^2 \left[ \rho \tilde{C}(k) + \ln(1 - \rho \tilde{C}(k)) \right] \qquad \text{(PY)}$$

where

$$e(r) = \exp[-\beta u(r)], \qquad f(r) = \exp[-\beta u(r)] - 1 \tag{9.5}$$

and

$$\tilde{C}(k) = \frac{4\pi}{k} \int_0^\infty dr \, rC(r) \sin(kr) \tag{9.6}$$

is the Fourier transform. A similar expression can be derived for mixtures [40]:

$$\frac{P}{\rho kT} = 1 + \frac{1}{2} \rho \sum_{i=1}^m \sum_{j=1}^m x_i x_j \int dr \, C_{ij}(r)[C_{ij}(r)e_{ij}(r)f_{ij}^{-1}(r) - 2] \tag{9.7}$$

$$+ \frac{1}{(2\pi)^3 \rho} \int dk \left[ \sum_{i=1}^m \rho_i \tilde{C}_{ii}(k) + \ln X(k) \right] \qquad \text{(PY)}$$

where $X(k)$ denotes the determinant

$$X(k) \equiv \det \left[ \delta_{ij} - \sqrt{\rho_i \rho_j} \tilde{C}_{ij}(k) \right] \tag{9.8}$$

$m$ is the number of components in the mixture, $x_j$ the mole fraction of component j, and $\delta_{ij}$ the Kronecker delta. For example, in a binary mixture of species A and B, $X(k)$ is the determinant of the $2 \times 2$ matrix

$$X(k) = \det \begin{bmatrix} 1 - \rho_A \tilde{C}_{AA}(k) & -\sqrt{\rho_A \rho_B} \tilde{C}_{AB}(k) \\ -\sqrt{\rho_B \rho_A} \tilde{C}_{BA}(k) & 1 - \rho_B \tilde{C}_{BB}(k) \end{bmatrix} \qquad (9.9)$$

The pressure obtained from (9.4) and (9.7) is equivalent to that from the isothermal compressibility equation (IV.4.15).

## CORRELATION FUNCTIONS

Under the PY approximation, certain equalities can be established for the correlation functions. Consider the expansion (1.14). If we set the second-order term to zero, then

$$\rho^2 \int d3 \, d2 \, [C^{(3)}(123) + C^{(2)}(13)C^{(2)}(12)]h(30)h(20) = 0 \qquad (9.10)$$

Rearrangement gives

$$\rho \int d3 \, d2 C^{(3)}(123)h(30)h(20) = -\rho^2 \int d3 \, d2 \, C^{(2)}(13)C^{(2)}(12)h(30)h(20) \qquad (9.11)$$

$$= -\rho^2 \int d3 \, C^{(2)}(13)h(30) \int d2 \, C^{(2)}(12)h(20)$$

$$= - [h(10) - C^{(2)}(10)]^2$$

Similarly, it can be shown that when higher-order terms are set to zero,

$$\rho^n \int d1 \, d2 \cdots dn \, C^{(n+1)}(123...n+1)h(10)h(20) \cdots h(n0) \qquad (9.12)$$

$$= (-1)^{n-1}(n-1)![h(n+1, 0) - C^{(2)}(n+1, 0)], \qquad n \geq 1 \qquad \text{(PY)}$$

When $n=1$, we have simply the OZ relation

$$\rho \int d1 \, C^{(2)}(12)h(10) = h(20) - C^{(2)}(20) \qquad (9.13)$$

Thus the PY assumption is consistent with OZ. For HNC, we could obtain parallel expressions

$$\rho^n \int d1 \cdots dn \, C^{(n+1)}(12 \cdots n+1)h(10) \cdots h(n0) = 0, \qquad n \geq 2 \qquad \text{(HNC)} \qquad (9.14)$$

For example, for $n=2$,

$$\rho^2 \int d1 \, d2 \, C^{(3)}(123)h(10)h(20) = 0 \qquad (9.15)$$

The case of $n=1$ should give back the OZ relation. However, instead we obtain

$$\rho \int d1 \ C^{(2)}(12)h(10) = 0 \qquad (9.16)$$

According to OZ, this quantity is not necessarily zero. Thus HNC does not satisfy the boundary condition OZ. (Note that we have excluded $n =1$ in eq. (9.14).) The convolution integral approaches zero only in the low density limit, where both $h(r)$ and $C(r)$ approach $f(r)$.

## VI.10. *Equations for Mixtures*

The integral equations discussed previously can be easily generalized to mixtures. A systematic way of deriving the mixture equations is again to use the method of functional differentiation. For example, in the HNC theory, the generating functional of species $\alpha$ $(1 \le \alpha \le m)$ can be expanded as

$$\ln \rho_\alpha^{(1)}(1; w) + \beta w_\alpha(1) = \ln \rho_\alpha^{(1)}(1; w^0) + \beta w_\alpha^0(1) \qquad (10.1)$$

$$+ \sum_{\gamma=1}^{m} \int d2 \ C_{\alpha\gamma}^{(1)}(1, 2; w^0)[\rho_\gamma^{(1)}(1; w) - \rho_\gamma^{(1)}(1; w^0)]+ \cdots$$

Simplification gives

$$\ln\left[g_{\alpha\beta}^{(2)}(1, 0)e^{\beta u_{\alpha\beta}(1, 0)}\right] = \sum_{\gamma=1}^{m} \rho_\gamma \int d2 \ C_{\alpha\gamma}(1, 2)h_{\gamma\beta}(2, 0) \qquad (10.2)$$

Going to bipolar coordinates, we have

$$\ln y_{\alpha\beta}(r) = \sum_{\gamma=1}^{m} \frac{2\pi\rho_\gamma}{r} \int_0^\infty ds \ s \ C_{\alpha\gamma}(s) \int_{|r-s|}^{r+s} dt \ t \ h_{\gamma\beta}(t) \qquad \text{(HNC)} \qquad (10.3)$$

Similar derivations can be made for

$$y_{\alpha\beta}(r) - 1 = \sum_{\gamma=1}^{m} \frac{2\pi\rho_\gamma}{r} \int_0^\infty ds \ s \ C_{\alpha\gamma}(s) \int_{|r-s|}^{r+s} dt \ t \ h_{\gamma\beta}(t) \qquad \text{(PY)} \qquad (10.4)$$

and

$$-\frac{d}{dr}\ln y_{\alpha\beta}(r) = \sum_{\gamma=1}^{m} \pi\rho_\gamma \int_0^\infty ds \ \frac{1}{kT} \frac{du_{\alpha\gamma}(s)}{ds} \ g_{\alpha\gamma}(s) \qquad (10.5)$$

$$\int_{|r-s|}^{r+s} dt \ th_{\gamma\beta}(t)\frac{r^2 +s^2 -t^2}{r^2} \qquad \text{(YBG)}$$

## MIXTURES OF KIHARA MOLECULES

As an example, we consider a mixture of argon and nitrogen whose molecular interaction is modeled by the Kihara (KH) potential

$$u_{\alpha\beta}(r) = 4\varepsilon_{\alpha\beta}\left[\left[\frac{\sigma_{\alpha\beta}-\delta_{\alpha\beta}}{r-\delta_{\alpha\beta}}\right]^{12} - \left[\frac{\sigma_{\alpha\beta}-\delta_{\alpha\beta}}{r-\delta_{\alpha\beta}}\right]^{6}\right] \tag{10.6}$$

where $\alpha,\beta = 1,2$, $\sigma_{\alpha\beta}$, $\delta_{\alpha\beta}$, and $\varepsilon_{\alpha\beta}$ are force constants. Let 1=Ar and 2=$N_2$, their values for the Ar-Ar, $N_2 - N_2$, and $Ar - N_2$ interactions are given in Table VI.10.

If we apply the PY theory, (10.4) becomes four equations

$$r(2\pi)^{-1}[y_{11}(r) - 1] = \rho_1 \int ds\, sC_{11} \int dt\, th_{11} + \rho_2 \int ds\, sC_{12} \int dt\, th_{21} \tag{10.7a}$$

$$r(2\pi)^{-1}[y_{12}(r) - 1] = \rho_1 \int ds\, sC_{11} \int dt\, th_{12} + \rho_2 \int ds\, sC_{12} \int dt\, th_{22} \tag{10.7b}$$

$$r(2\pi)^{-1}[y_{21}(r) - 1] = \rho_1 \int ds\, sC_{21} \int dt\, th_{11} + \rho_2 \int ds\, sC_{22} \int dt\, th_{21} \tag{10.7c}$$

$$r(2\pi)^{-1}[y_{22}(r) - 1] = \rho_1 \int ds\, sC_{21} \int dt\, th_{12} + \rho_2 \int ds\, sC_{22} \int dt\, th_{22} \tag{10.7d}$$

These relations coupled with the corresponding OZ relations

$$r(2\pi)^{-1}[h_{\alpha\beta}(r) - C_{\alpha\beta}(r)] = \sum_{\gamma=1}^{m} \rho_\gamma \int_0^\infty ds\, sC_{\alpha\gamma}(s) \int_{|r-s|}^{r+s} dt\, th_{\alpha\beta}(t) \tag{10.8}$$

form a complete set of equations for the solution of the unknown functions $y_{11}$, $y_{12}$, $y_{21}$, $y_{22}$, etc. Since $y_{12} = y_{21}$ by combinatorial symmetry, we actually need only three equations from (10.7a,b,c,d). For example (10.7a,b,d) will suffice. These sets of equations have been solved for the argon-nitrogen mixture [41]. Some of the results ($g_{\alpha\beta}(r)$ and $C_{\alpha\beta}(r)$) are presented in Figures VI.11 and 12. The rdfs correspond to pairs Ar-Ar, $N_2$-$N_2$, and Ar-$N_2$. The thermodynamics can also be calculated from the virial pressure and isothermal compressibility equation. Table VI.8 shows the results. National Bureau of Standards has reported a Strobridge equation for nitrogen [42], and it has been generalized to Ar-$N_2$ mixtures by Crain and Sonntag [43]. The equation empirically fits the experimental data. We see that except at very low temperatures, the agreement between the PY-KH results and the fitted experimental data is very close.

*Table VI.7. Force Constants for Kihara Potential*

| Parameter | Ar–Ar | $N_2 - N_2$ | $Ar - N_2$ |
|---|---|---|---|
| $\varepsilon/k$ (K) | 147.2 | 139.2 | 143.14 |
| $\sigma$ (A) | 3.314 | 3.526 | 3.420 |
| d (A) | 0.3682 | 0.705 | 0.5366 |

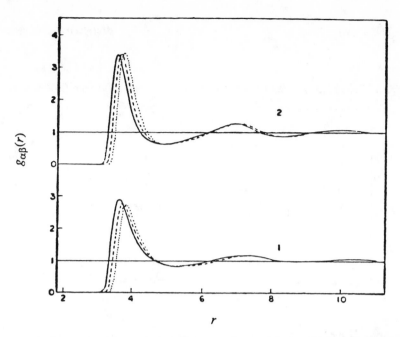

FIGURE VI.11. Radial distribution functions of mixtures of Kihara molecules: argon and nitrogen. Condition 1: $kT/\varepsilon = 0.798$, $\rho\sigma^3 = 0.538$, and $x_{Ar} = 0.306$. Condition 2: $kT/\varepsilon = 0.798$, $\rho\sigma^3 = 0.86$, and $x_{Ar} = 0.306$. Full line: $g_{Ar-Ar}$. Dashed line: $g_{Ar-N_2}$. Dotted line: $g_{N_2-N_2}$. (Lee and Hulburt, J. Chem. Phys., 1973)

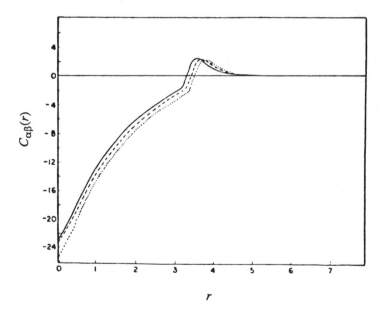

FIGURE VI.12. Direct correlation functions of mixtures of Kihara molecules- argon and nitrogen. $kT/\varepsilon = 0.798$, $\rho\sigma^3 = 0.86$, and $x_{Ar} = 0.306$. Full line: $C_{Ar-Ar}$. Dashed line: $C_{Ar-N_2}$. Dotted line: $C_{N_2-N_2}$. (Lee and Hulburt, J. Chem. Phys., 1973)

*Table VI.8. Pressures in the PY-KH Theory for Argon (1)- Nitrogen (2) Mixtures*

| $kT/\varepsilon$ | $x_1$ | $\rho\sigma^3$ | PY(V), atm | Strobridge, atm |
|---|---|---|---|---|
| 0.7982 | 0.20 | 0.0675 | 16.82 | 17.60 |
| | 0.206 | 0.5453 | 20.4 | 10.32 |
| | 0.406 | 0.5453 | 14.68 | -7.18 |
| | 0.79 | 0.55 | 0.99 | -68.6 |
| 1.074 | 0.10 | 0.0685 | 28.15 | 28.47 |
| 1.174 | 0.406 | 0.5453 | 266.84 | 262.85 |
| 1.38 | 0.799 | 0.10 | 53.55 | 54.0 |
| | | 0.20 | 93.3 | 96.0 |
| | | 0.40 | 172.7 | 185.4 |
| | | 0.60 | 381 | 408 |
| | | 0.70 | 631 | 689 |

## VI.11. *Second-Order Theories*

The knowledge that the integral equations of correlation functions can be derived from the derivatives of certain generating functionals enables us (i) to formulate perturbation theories for the correlation functions, and (ii) to extend existing integral equations to second order. We explore the second possibility here.

### THE PY2 EQUATION

Verlet [44] derived the PY2 equation essentially from (1.4). With $w_0 = 0$, the equation becomes

$$y(0, 1) = 1 + \rho \int d2 \, C(1, 2)h(2, 0) \tag{11.1}$$

$$+ \frac{1}{2} \rho^2 \int d3 \, d2 \, [C^{(3)}(1, 2, 3) + C(1, 3)C(1, 2)] \, h(3, 0)h(2, 0) + \cdots$$

To close this equation, we need a knowledge of $C^{(3)}(1, 2, 3)$. Using matrix inverses, Verlet transformed (11.1) to

$$y(0, 1) = 1 + \rho \int d2 \, C(0, 2)h(2, 1) \tag{11.2}$$

$$+ \frac{1}{2} \rho^2 \int d2 \, d3 \, d4 \, C(0, 2)C(0, 3)g(2, 3)\alpha(2, 3; 4)[\delta_D(1, 4) - \rho C(1, 4)]$$

where the three-body function $\alpha(2,3;4)$ is defined as

$$\alpha(2, 3; 4) \equiv \frac{g^{(3)}(2, 3, 4)}{g(2, 3)} - 1 - h(2, 4) - h(3, 4) \tag{11.3}$$

and $\delta_D$ is the Dirac delta function. The three-body function $\alpha(2,3;4)$ is unknown. A supplementary equation is obtained by expanding

$$\frac{\rho^{(2)}(1, 2; w)}{\rho^{(1)}(1)\rho^{(1)}(2)} \tag{11.4}$$

Thus

$$\alpha(2, 3; 4) = h(2, 4)h(3, 4) + g(2, 4)g(3, 4)\rho \int d5 \, C(4, 5)\alpha(2, 3; 5) \tag{11.5}$$

The $\alpha$-function is not symmetric in all its arguments (2,3,4) owing to the truncation (11.2). Equations (11.2 and 4) and the OZ relation form a closed set of equations, called the PY2 system, and have been solved for HS and LJ molecules. Definite improvements over the first-order theories were obtained. For example, the fifth virial coefficient given by PY2 is superior to the PY value (see Table VI.2). For the critical constants of the Lennard-Jones fluid, PY2 also gives better prediction (Table VI.5). The radial distribution functions given by PY2 are better than PY. However, the improvement is not substantial.

## THE HNC2 EQUATION

By the same method, the HNC2 theory is derived as

$$\ln y(0, 1) = h(0, 1) - C(0, 1) - \frac{1}{2}[h(0, 1) - C(0, 1)]^2 \tag{11.6}$$

$$+ \frac{1}{2} \rho^2 \int d2 \, d3 \, d4 \, C(0, 2)C(0, 3)g(2, 3)\alpha(2, 3; 4)[\delta_D(1, 4) - \rho C(1, 4)]$$

where the $\alpha$-function is obtained from the expansion to first order of

$$\ln [\rho^{(2)}(1, 2; w)] + \beta w(1) + \beta w(2) \tag{11.7}$$

The result is

$$\alpha(1, 2; 0) = 1 - g(0, 1) - g(0, 2) \tag{11.8}$$

$$+ g(0, 1)g(0, 2)\exp [\rho \int d3 \, C(0, 3)\alpha(1, 2; 3)]$$

Equation (11.8) is highly nonlinear. No attempts have been made to solve it numerically. For hard spheres, Verlet [45] has calculated the fifth virial coefficient, and the value is given in Table VI.2.

## THE YBG2 EQUATION

Lee, Ree, and Ree [46] obtained the YBG2 equation from the second and third hierarchies of the BBGKY relations:

$$-\nabla_1 g(r_{12}) = g(r_{12})\beta\nabla_1 u(r_{12}) + \rho \int dr_3 \, \beta\nabla_1 u(r_{13})g^{(3)}(r_1 r_2 r_3) \tag{11.9}$$

and

$$-\nabla_1 g^{(3)}(r_1 r_2 r_3) = g^{(3)}(r_1 r_2 r_3)[\nabla_1 u(r_{12}) + \nabla_1 u(r_{13})] \tag{11.10}$$

$$+ \rho \int dr_4 \; \beta \nabla_1 u(r_{14}) g^{(4)}(r_1 r_2 r_3 r_4)$$

The quadruplet correlation function $g^{(4)}$ is approximated by the form suggested by Fisher and Kopeliovich [47]

$$g^{(4)}(r_1 r_2 r_3 r_4) = \frac{g^{(3)}(123) g^{(3)}(124) g^{(3)}(134) g^{(3)}(234)}{g(12) g(13) g(14) g(23) g(24) g(34)} \tag{11.11}$$

where $g^{(3)}(ijk) = g^{(3)}(r_i r_j r_k)$, etc. Equations (11.9 and 10) become

$$\frac{d}{dr_{12}} \ln y(r_{12}) = - \rho \int dr_3 \, \cos(\theta_{213}) \left[ \frac{du(r_{13})}{dr_{13}} \right] g(r_{13})[X_{1,23} - 1] \tag{11.12}$$

and

$$\frac{\partial}{\partial \xi} \ln \left[ X_{2,13} g(r_{12}) e^{\beta u(r_{12}) + \beta u(r_{13})} \right] = - \rho \int dr_4 \, \cos(\theta_{214}) \left[ \frac{du(r_{14})}{dr_{14}} \right] g(r_{14}) \tag{11.13}$$

$$\cdot \left[ \frac{X_{1,24} X_{1,34} X_{3,24}}{g(r_{24})} - 1 \right]$$

where

$$X_{i,jk} \equiv \frac{g^{(3)}(ijk)}{g(ij) g(ik)} \tag{11.14}$$

and

$$\cos(\theta_{ijk}) \equiv \frac{r_{ij}^2 + r_{jk}^2 - r_{ik}^2}{2 r_{ij} r_{jk}} \tag{11.15}$$

The Fréchet derivative $\partial/\partial\xi$ is taken as the gradient $\nabla_1$ dotted along the $r_1 - r_2$ direction with fixed $\theta_{123}$ and $r_{23}$. The integral is dependent upon the paths taken in the integration. This causes some problems and ambiguities. Uehara et al. [48] solved YBG2 for two-dimensional hard disks and three-dimensional hard spheres. Close agreement with simulation data was obtained for moderately dense fluids.

In closing, we note one additional important development in integral equations, i.e., equations for interaction site models (ISM) of polyatomic fluids. These equations are based on an OZ relation generalized to polyatomics. They have been applied, with remarkable success, to interpretation of the structure of real fluids (such as methane, ethane, and bromine) obtained from neutron scattering experiments. Detailed discussions will be presented in Chapter XIV.

## References

[1]  J.K. Percus and G.J. Yevick, see IV. Ref. [5].

[2]  J.L. Lebowitz and J.K. Percus, see IV. Ref. [8].

[3]  T. Kihara, Rev. Mod. Phys. **25**, 831 (1953).

[4]  A.A. Broyles, J. Chem. Phys. **33**, 456 (1960).

[5]  R.J. Baxter, Phys. Rev. **154**, 170 (1967).

[6]  M.J. Gillan, Mol. Phys. **38**, 1781 (1979).

[7]  S. Labik, A. Malijevsky, and P. Vonka, Mol. Phys. **56**, 709 (1985).

[8]  P.A. Monson, Mol. Phys. **47**, 435 (1982).

[9]  E. Enciso, Mol. Phys. **56**, 129 (1985).

[10]  R.O. Watts, J. Chem. Phys. **48**, 50 (1968).

[11]  L.L. Lee, Mol. Phys. to appear (1987).

[12]   S. Labik, A. Malijevsky, and P. Vonka, *Ibid.* (1985).

[13]  J.K. Percus, in *The Equilibrium Theory of Classical Fluids*, edited by H.L. Frisch and J.L. Lebowitz (Benjamin, New York, 1964).

[14]  J.M.J. van Leeuwen, J. Groeneveld, and J. de Boer, Physica **25**, 792 (1959)

[15]  E. Meeron, J. Math. Phys. **1**, 192 (1960).

[16]  G.S. Rushbrooke, Physica **26**, 259 (1960).

[17]  L. Verlet, Nuovo Cimento **18**, 77 (1960).

[18]  T. Morita, Progr. Theor. Phys. **20**, 920 (1958); also, T. Morita and K. Hiroike, see IV. Ref. [6].

[19]  J.G. Kirkwood, see V. Ref. [1].

[20]  L. Schwartz, *Théories des Distributions*, Tomes 1 and 2 (Hermann et Cie., Paris, 1957).

[21]  J.L. Lebowitz and J.K. Percus, Phys. Rev. **144**, 251 (1966).

[22]  G. Stell, in *Statistical Mechanics: Part A. Equilibrium Techniques*, edited by B. J. Berne (Plenum, New York, 1977).

[23]  Y. Rosenfeld and N.W. Ashcroft, Phys. Rev. A20, 1208 (1979).

[24]  Y. Rosenfeld, J. Phys. C **13**, 3227 (1980).

[25]  S.B. Kanchanakpan, L.L. Lee, and C.H. Twu, in *Equations of State: Theories and Applications*, edited by K.C. Chao and R.L Robinson, Jr. (American Chemical Society Symposium Series 300, Washington D.C., 1986), pp.227-249.

[26]  M.S. Wertheim, Phys. Rev. Letters, **8**, 321 (1963).

[27]  E. Thiele, J. Chem. Phys. **38**, 1959 (1963).

[28]  L. Verlet and J.-J. Weis, Phys. Rev. **A5**, 939 (1972).

[29]  L. Verlet, Physica, **30**, 95 (1964); and *Ibid.*, **31**, 959 (1965).

[30]  D. Henderson, W.G. Madden, and D.D. Fitts, J. Chem. Phys. **64**, 5026 (1976).

[31]  D. Henderson, O.H. Scalise, and W.R. Smith, J. Chem. Phys. **72**, 2431 (1980).

[32]  M. Chen, D. Henderson, and J.A. Barker, Can. J. Phys. **47**, 2009 (1969).

[33]  L. Verlet, *Ibid.* (1965).

[34]  F. Lado, Phys. Rev. **A8**, 2548 (1973).

[35]  L.L. Lee, J. Chem. Phys. **62**, 4436 (1975).

[36]  L.L. Lee, J. Chem. Phys. **60**, 1197 (1974).

[37]  T. Morita, Progr. Theor. Phys. **23**, 829 (1960).

[38]  L.L. Lee, *Ibid.* (1974).

[39]  R.J. Baxter, J. Chem. Phys. **7**, 4855 (1967).

[40]  K. Hiroike and T. Morita, J. Chem. Phys. **52**, 5489 (1970).

[41]  L.L. Lee, J. Chem. Phys. **58**, 44 and 61 (1973).

[42]  T.R. Strobridge, National Bureau of Standards Technical Notes **129**, (1962).

[43]  R.W. Crain, Jr. and R.E. Sonntag, Adv. Cryogenic Eng. **11**, 379 (1966).

[44]  L. Verlet, *Ibid.* (1965).

[45]  L. Verlet, *Ibid.* (1965).

[46]  Y. Lee, F.H. Ree, and T. Ree, J. Chem. Phys. **48**, 3506 (1968); *Ibid.* **55**, 234 (1971).

[47]  I.Z. Fisher and B.L. Kopeliovich, Dokl. Akad. Nauk. USSR **133**, 81 (1960).

[48]  Y. Uehara, T. Ree, and F.H. Ree, J. Chem. Phys. **70**, 1876 (1979).

## Exercises

1.  Evaluate the pair correlation functions $g(r)$ for LJ fluids at the following conditions: $kT/\varepsilon = 2.0$, 1.5, 0.72, and $\rho\sigma^3 = 0.85$ using the Picard and Gillan methods. Apply both PY and HNC approximations. Compare the speeds of convergence of the two methods. In carrying out Picard's iterations, one must have *good* initial guesses. These could be obtained from literature sources or from Gillan's method first.

2.  From the rdfs obtained above, calculate the energy and pressure using the energy and virial pressure equations. Compare the pressure values with argon experimental data. (Note that for argon, $\varepsilon/k = 119.8\ K$, and $\sigma = 3.405$ A.) Which theory, PY or HNC, is better in reproducing the experimental values?

3.    Use the MSA to reproduce the data of Table VI.3 for the SW potential. Is the MSA method dependable for energy here?

4.    Use Baxter's formula (9.4) to evaluate the compressibility pressure for the LJ fluids listed in problem 1. Compare the results with the virial pressures.

5.    Reproduce selective points listed in Table VI.8 for mixtures of Kihara molecules using the PY integral equations.

6.    Consulting the article by Uehara et al. [48], apply the YBG2 equation to two-dimensional hard disks. Compare the results with simulation data.

7.    Use the formulas of PY and HNC (eqs. (1.16) and (4.1)) to find the cluster diagrams of the bcf $y(12)$ to second order in density. You can do this by rearranging (1.16) and (4.1), then substituting the exact clusters of $C(12)$ and $g(12)$ into the RHS to obtain $y(12)$ on the LHS. Which diagrams are missing from $y(12)$ in the approximate theories? Evaluate PY versus HNC based on your findings here.

# CHAPTER VII

# THEORIES FOR POLAR FLUIDS

$F$or molecules with spherical core and embedded with point dipoles and/or point quadrupoles, a number of integral equation theories have been developed. Some of these are quite successful. We shall discuss the mean spherical approximation (MSA), the linearized hypernetted chain (LHNC, or the generalized mean field GMF) theory, and the quadratic hypernetted chain (QHNC) equation here. For nonspherical and nonideal polar molecules, it is necessary to use other methods, such as the interaction site models with distributed charges or convex bodies with acentric charges. These topics are beyond the scope of this book. Current literature will be a good source of reference.

## VII.1. *The Integral Equations for Polar Fluids:*
## *MSA for Dipolar Spheres*

Wertheim [1] has solved the MSA for hard spheres with embedded ideal dipoles. The method of solution is very elegant. However, the results are not reliable as compared to simulation data. A discussion of the solution is nonetheless illuminating for pedagogical purposes. The pair potential in this case is

$$u(r_{12}) = \infty, \qquad\qquad\qquad\qquad r_{12} < d \qquad\qquad (1.1)$$

$$= -\frac{\mu^2}{r_{12}^3}\left[\frac{3(\hat{L}_1 \cdot r_{12})(\hat{L}_2 \cdot r_{12})}{r_{12}^2} - \hat{L}_1 \cdot \hat{L}_2\right], \qquad r_{12} > d$$

where $\mu$ is the dipole moment and $\hat{L}_i$ is the unit vector along the direction of the dipole moment in molecule $i$, $i = 1,2$, $r_{12} = r_2 - r_1$. We shall also denote the bracketed quantity on the RHS of (1.1) as $D(12)$:

$$D(12) \equiv \frac{3(\hat{L} \cdot r_{12})(\hat{L}_2 \cdot r_{12})}{r_{12}^2} - \hat{L}_1 \cdot \hat{L}_2 \qquad\qquad (1.2)$$

Furthermore, we define the function $\Delta(12)$ as

$$\Delta(12) \equiv \hat{\mathbf{L}}_1 \cdot \hat{\mathbf{L}}_2 \tag{1.3}$$

In two-body relative coordinates,

$$D(12) = 2c_1c_2 - s_1s_2c, \qquad\qquad \Delta(12) = c_1c_2 + s_1s_2c \tag{1.4}$$

where $c_i = \cos\theta_i$, $s_i = \sin\theta_i$, $i = 1,2$, and $c = \cos(\phi_2 - \phi_1)$; $\theta_i$ is the polar angle of dipole $i$ formed with the intermolecular axis, and $\phi_i$ is the azimuthal angle thereof. Wertheim made the assumption that the set of functions

$$\{1, \ D(12), \ \Delta(12)\} \tag{1.5}$$

forms a basis set that spans the function space for the correlation functions $h(12)$ and $C(12)$. This basis set is sufficient within MSA for dipolar interactions, but is not sufficient (i.e., Parseval conditions) for exact correlation functions (see below). In MSA, we expand $h(12)$ as

$$h(12) = h_s(r_{12}) + h_D(r_{12})D(12) + h_\Delta(r_{12})\Delta(12) \tag{1.6}$$

Likewise

$$C(12) = C_s(r_{12}) + C_D(r_{12})D(12) + C_\Delta(r_{12})\Delta(12) \tag{1.7}$$

Note that the set of functions (1.5) are orthogonal to one another; i.e.,

$$\int d\omega_1 \int d\omega_2 \, D(12) = \int_0^\pi d\theta_1 \sin\theta_1 \int_0^\pi d\theta_2 \sin\theta_2 \int_0^{2\pi} d\phi_2 \, D(12) = 0 \tag{1.8}$$

$$\int d\omega_1 \int d\omega_2 \, \Delta(12) = 0$$

and

$$\int d\omega_1 \int d\omega_2 \, D(12)\Delta(12) = 0$$

The normalization factors are

$$\int d\omega_1 \int d\omega_2 \, (1) = (4\pi)^2 \tag{1.9}$$

$$\int d\omega_1 \int d\omega_2 \, D(12)^2 = \frac{2}{3}(4\pi)^2$$

$$\int d\omega_1 \int d\omega_2 \, \Delta(12)^2 = \frac{1}{3}(4\pi)^2$$

Now as the expansions (1.6 and 7) are substituted into the Ornstein-Zernike (OZ) relation

$$h(12) - C(12) = \frac{\rho}{4\pi} \int d3 \, h(13)C(32) \tag{1.10}$$

where $d3 = dr_3 \, d\omega_3$, with $C(12)$ given by the MSA assumption

$$C(12) \approx -\beta u(12) = \frac{\beta\mu^2}{r_{12}^3} D(12), \qquad\qquad r_{12} > d \tag{1.11}$$

or in terms of projections on to the basis functions

$$h_s(r_{12}) = -1, \qquad h_D(r_{12}) = 0, \qquad h_\Delta(r_{12}) = 0, \qquad\qquad r < d \tag{1.12}$$

$$C_s(r_{12}) = 0, \qquad C_D(r_{12}) = \beta\mu^2 r_{12}^{-3}, \qquad C_\Delta(r_{12}) = 0, \qquad\qquad r > d$$

One arrives at the solution

$$h_s(r_{12}) = h^{HS}(r_{12}), \qquad\qquad C_s(r_{12}) = C^{HS}(r_{12}) \tag{1.13}$$

where the superscript HS indicates hard-sphere quantities. Equations (1.13) say that the symmetrical parts of the correlation functions are simply the HS ones. For $D$- and $\Delta$-components, Wertheim gave

$$h_D(r) = \hat{h}_D(r) - \frac{3}{r^3} \int_0^r dx \, x^2 \hat{h}_D(x) \tag{1.14}$$

with

$$\hat{h}_D(r) \equiv K\left[2h_s(r; 2K\rho) + h_s(r; -K\rho)\right] \tag{1.15}$$

The function $h_s(r; 2K\rho)$ is the symmetrical part of $h(12)$, i.e., $h^{HS}(r)$ at a density value of $2K\rho$. $h_s(r; -K\rho)$ is $h^{HS}(r)$ at the hypothetical negative density of $-K\rho$. (Note that the analytical solution for an HS correlation function is given in Chapter VIII.) The factor $K$ is defined as

$$K \equiv \int_d^\infty dr \, \frac{h_D(r)}{r} \tag{1.16}$$

which is also given by the solution to

$$\frac{4\pi\beta\mu^2\rho}{9} = \frac{1}{3}[q(2Ky) - q(-Ky)] \tag{1.17}$$

with the $q(x)$ function defined by

$$q(x) \equiv \frac{(1+2x)^2}{(1-x)^4} \tag{1.18}$$

The quantity on the LHS of (1.17) is also given the notation $\gamma \equiv (4\pi\beta\mu^2\rho)/9$, $y \equiv \pi\rho d^3/6$ being the packing fraction. The $\Delta$-component is

$$h_\Delta(r) = 2K\left[h_s(r;\ 2K\rho) - h_s(r;\ -K\rho)\right] \tag{1.19}$$

A completely similar set of equations holds for $C(12)$:

$$C_D(r) = \hat{C}_D(r) - \frac{3}{r^3}\int_0^r dx\ x^2 \hat{C}_D(x) \tag{1.20}$$

$$\hat{C}_D(r) = K\left[2C_s(r;\ 2K\rho) + C_s(r;\ -K\rho)\right]$$

$$C_\Delta(r) = 2K\left[C_s(r;\ 2K\rho) - (C_s(r;\ -K\rho)\right]$$

The thermodynamic properties can be obtained from the usual equations. The isothermal compressibility is

$$kT\frac{\partial\rho}{\partial P}\bigg|_T = 1 + \rho\int dr\ h_s(r) \tag{1.21}$$

$h_D(r)$ determines the pressure via

$$\frac{\beta P}{\rho} = 1 + 4\eta g_s(d^+) - \frac{4\pi}{3}\beta\mu^2\rho\int_d^\infty dr\ \frac{h_D(r)}{r} \tag{1.22}$$

The configurational energy (i.e., internal energy in excess of the ideal-gas contribution) is just the last term in the pressure expression (1.22):

$$\frac{U^c}{NkT} = -\frac{4\pi}{3}\beta\mu^2\rho\int_d^\infty dr\ \frac{h_D(r)}{r} \tag{1.23}$$

A second pressure expression could be obtained from integrating the energy. The result is

$$\frac{\beta P}{\rho} = \frac{\beta P^{HS}}{\rho} + \frac{3}{y}[J(Ky) - Ky\gamma] \qquad \text{(energy route)} \tag{1.24}$$

where

$$J(Ky) \equiv \frac{8}{3}(Ky)^2 \left[ \frac{(1+Ky)^2}{(1-2Ky)^4} + \frac{(2-Ky)^2}{8(1+Ky)^4} \right] \tag{1.25}$$

Using (1.16), we can also express the energy simply by

$$\frac{U^c}{NkT} = -3K\gamma \tag{1.26}$$

The Helmholtz free energy is calculated by

$$\frac{A}{NkT} = \frac{A^{HS}}{NkT} - \frac{3}{y}J(Ky) \tag{1.27}$$

and the entropy is

$$\frac{S}{Nk} - \frac{S^{HS}}{Nk} = \frac{3}{y}[J(Ky) - Ky\gamma] \tag{1.28}$$

Patey and Valleau [2] [3] as well as Verlet and Weis [4] have compared their simulation results with thermodynamic properties determined by MSA (see Table VII.1). They found MSA to be unsatisfactory. This is due partly to the deficiencies in the structures $h_D$ and $h_s$. Figure VII.1 shows that at $\rho^* = 0.9$ the PY solution for $h_s$ is not accurate at core; $h_D(d^+)$ and $h_\Delta(d^+)$ are also too small at contact. As a theory for polar fluids, MSA does not yield quantitative results.

For dipolar hard spheres, the dielectric constant $\varepsilon$ can be determined from $h_\Delta(r)$. Let us define the Kirkwood factor $g_K$ as

$$g_K \equiv 1 + \frac{\rho}{3} \int_0^\infty dr \, 4\pi r^2 h_\Delta(r) \tag{1.29}$$

The dielectric constant in MSA is given in terms of $g_K$ as

$$\frac{(\varepsilon-1)(2\varepsilon+1)}{9\varepsilon} = \gamma g_K \tag{1.30}$$

An equivalent expression for $\varepsilon$ is

$$\varepsilon = \frac{(1+4Ky)^2(1+Ky)^4}{(1-2Ky)^6} \tag{1.31}$$

The MSA formula can be compared with other existing theories for the dielectric constant. The Clausius-Mossotti formula is

$$\frac{\varepsilon-1}{\varepsilon+2} = \gamma \tag{1.32}$$

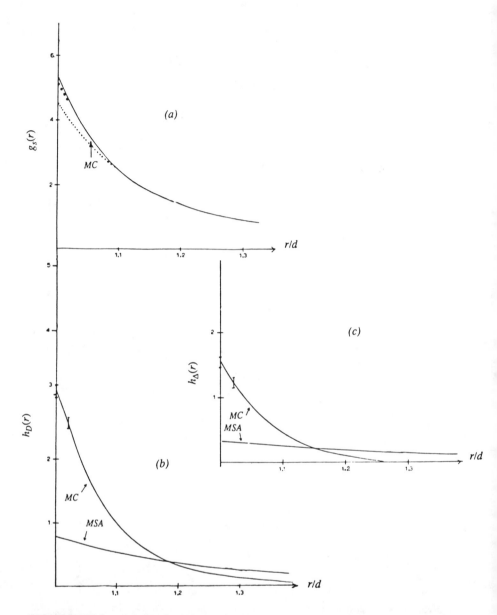

FIGURE VII.1. (a) Angle-averaged component $g_s$ of the angular pair correlation functions of dipolar hard spheres at $\rho^* = 0.9$ and $\mu^{*2} = 1.0$. Solid line = MC results. Dotted line = MSA results. Black circles = hard sphere $g(r)$ from MD. (b) Component $h_D$. Solid line = MC results. Broken line = MSA results. (c) Component $h_\Delta$. Legend is same as in (b). (Verlet and Weis, Mol. Phys., 1974)

Table VII.1.  Thermodynamic Properties of Dipolar Hard Spheres

| $kTd^3/\mu^2$ | Configurational Energy $-U/NkT$ at $\rho/\rho_{c.p.} = 0.59$ | | | |
|---|---|---|---|---|
| | $\mu$, Debye | MC | MSA | Padé* |
| 17.781 | 0.25 | 0.0064 ± 0.0004 | 0.0036 | 0.006 |
| 1.976 | 0.75 | 0.348 ± 0.006 | 0.226 | 0.368 |
| 0.711 | 1.25 | 1.79 ± 0.01 | 1.20 | 1.87 |
| 0.363 | 1.75 | 4.80 ± 0.04 | 3.21 | 4.64 |

| $kTd^3/\mu^2$ | Excess Compressibility $(\beta P - \beta P^{HS})/\rho$ at $\rho/\rho_{c.p.} = 0.59$ | | |
|---|---|---|---|
| | MC | MSA | Padé |
| 17.781 | 0.03 | -0.0036 | -0.0045 |
| 1.976 | -0.58 | -0.226 | -0.265 |
| 0.711 | -1.35 | -1.20 | -1.26 |
| 0.363 | -3.08 | -3.21 | -2.92 |

| $kTd^3/\mu^2$ | Dielectric Ratio $(\varepsilon -1)/(\varepsilon +2)$, and Mean Square Moment $<M^2>$** | | | |
|---|---|---|---|---|
| | MC | MSA | Onsager** | $<M^2>/N\mu^2$ |
| 17.781 | 0.065 | 0.065 | 0.065 | 0.975 |
| 1.976 | 0.467 | 0.482 | 0.435 | 0.792 |
| 0.711 | 0.819 | 0.799 | 0.698 | 0.500 |
| 0.363 | 0.993 | 0.917 | 0.823 | 0.309 |

*Padé is result obtained from Padé resummation on free energy according to Stell et al. [5]
**Onsager formula is as in Nienhuis and Deutch. [6]
***Mean square moment is related to the dielectric constant by $(\varepsilon -1)/(\varepsilon +2) = (4\pi<M^2>)/(9VkT)$ following Rahman and Stillinger [7] for spherical cutoff.

The Onsager formula is

$$\frac{(\varepsilon-1)(2\varepsilon+1)}{9\varepsilon} = \gamma \tag{1.33}$$

We note that the Clausius-Mossotti formula has no solution for $\gamma =1$, an incorrect behavior.  Onsager's equation corresponds to (1.30) when the Kirkwood factor $g_K =1$ (or $h_\Delta(r) = 0$), a simplification valid only at small $\gamma$.

Due to the unsatisfactory performance for thermodynamic properties, MSA is not used in calculations of polar fluid properties.  Other theories have been proposed: e.g., the perturbation theory is used for thermodynamic properties and the LHNC and QHNC theories are used for the structure of polar fluids.

## VII.2.  The LHNC and QHNC Equations

The LHNC theory proposed by Patey [8] is equivalent to the single superchain (SSC) theory of Wertheim [9] or the generalized mean field (GMF) theory of Henderson and Gray [10].  The QHNC equation [11] is, as the name implies, second order in the expansion of the HNC equation.  Both LHNC and QHNC have been applied to potentials with a spherical core plus polar and/or multipolar interactions

$$u(12) = u_0(r_{12}) + u_a(12) \tag{2.1}$$

where $12 = \mathbf{r}_1\omega_1\mathbf{r}_2\omega_2$, the positions and Euler angles of orientation of molecules 1 and 2, respectively; $r_{12} = |\mathbf{r}_2 - \mathbf{r}_1|$. The subscript 0 denotes properties belonging to the isotropic potential. The original HNC for such a fluid is

$$C(12) = h(12) - \ln y(12) \tag{2.2}$$

The term $y(12)$ can be written equivalently as

$$y(12) = g(12)e^{\beta u(12)} = g^{000}(r_{12})e^{\beta u(12)}\left[1 + \frac{g(12) - g^{000}(r_{12})}{g^{000}(r_{12})}\right] \tag{2.3}$$

where $g^{000}(r)$ is the spherically averaged part of $g(12)$:

$$g^{000}(r_{12}) \equiv (4\pi)^{-2} \int d\omega_1 \int d\omega_2 \, g(12) \tag{2.4}$$

Since $\ln(1+x) = x - x^2/2 + \cdots$, we could expand $\ln y$ as

$$\ln y(12) = \ln\left[g^{000}(r_{12})e^{\beta u(12)}\right] + \ln\left[1 + \frac{g(12) - g^{000}(r_{12})}{g^{000}(r_{12})}\right] \tag{2.5}$$

$$= \ln\left[g^{000}(r_{12})e^{\beta u(12)}\right] + \left[\frac{g - g^{000}}{g^{000}}\right] - \frac{1}{2}\left[\frac{g - g^{000}}{g^{000}}\right]^2 + \cdots$$

Thus HNC becomes

$$C(12) = h(12) - \ln\left[g^{000}(r_{12})e^{\beta u(12)}\right] \tag{2.6}$$

$$- \left[\frac{g(12) - g^{000}(r_{12})}{g^{000}(r_{12})}\right] + \frac{1}{2}\left[\frac{g(12) - g^{000}(r_{12})}{g^{000}(r_{12})}\right]^2 + \cdots$$

Truncation at the first order gives the LHNC equation, whereas truncation at the second order gives the QHNC equation. Equation (2.6) coupled with the OZ relation forms a closure and can be solved for $h(12)$ and $C(12)$.

## VII.3. *Applications of the LHNC and the QHNC to Hard Spheres with Embedded Dipoles and Quadrupoles*

The pair potential for hard spheres with dipolar and quadrupolar interactions is

$$u(12) = u^{HS}(r) - \frac{\mu^2}{r^3}\phi^{112}(12) + \frac{\mu Q}{2r^4}\left[\phi^{123}(12) - \phi^{213}(12)\right] + \frac{3Q^2}{4r^5}\phi^{224}(12) \tag{3.1}$$

where, in the convention of Blum [12] and Blum and Torruella [13], $\phi^{112}$ is the angular function for dipole-dipole interaction, $[\phi^{123} - \phi^{213}]$ that for dipole-quadrupole interaction, and $\phi^{224}$ for quadrupole-quadrupole interaction (see Table VII.2).

Table VII.2.  Rotation Matrix Representations of Dipole-dipole,
Dipole-quadrupole and Quadrupole-quadrupole Interactions

**Representations of the $\mu\mu$ potential:**

(i)    *In terms of two-body relative angles*

$$u_{\mu\mu}(12) = -\frac{\mu^2}{r^3}\,[s_1 s_2 c - 2c_1 c_2]$$

(ii)   *In terms of dipole and center-of-mass vectors*

$$u_{\mu\mu}(12) = -\frac{\mu^2}{r^3}[3(\hat{\mathbf{L}}_1 \cdot \mathbf{r}_{12})(\hat{\mathbf{L}}_2 \cdot \mathbf{r}_{12})/r_{12}^2 - \hat{\mathbf{L}}_1 \cdot \hat{\mathbf{L}}_2]$$

*Note:*

$$D(12) = 3(\hat{\mathbf{L}}_1 \cdot \mathbf{r}_{12})(\hat{\mathbf{L}}_2 \cdot \mathbf{r}_{12})/r_{12}^2 - \hat{\mathbf{L}}_1 \cdot \hat{\mathbf{L}}_2 = 2c_1 c_2 - s_1 s_2 c$$

*and*

$$\Delta(12) = \hat{\mathbf{L}}_1 \cdot \hat{\mathbf{L}}_2 = c_1 c_2 + s_1 s_2 c$$

(iii)  *In terms of spherical harmonics (see Rose [14])*

$$u_{\mu\mu}(12) = 4\pi \sum_{|m|\leq 1} (-1)^m E_{11m}(r) Y_{1m}(\theta_1 \phi_1) Y_{1\underline{m}}(\theta_2 \phi_2)$$

where

$$E_{11,-1} = \frac{1}{3}\frac{\mu^2}{r^3}$$

$$E_{110} = \frac{2}{3}\frac{\mu^2}{r^3}$$

$$E_{111} = \frac{1}{3}\frac{\mu^2}{r^3}$$

(iv)   *In terms of rotation matrices (see Blum [Ibid.])*

$$u_{\mu\mu}(12) = -\frac{\mu^2}{r^3}\Phi^{112}(12)$$

## Representations of the μQ Potential

(i) *In terms of two-body relative angles*

$$u_{\mu Q}(12) = \frac{3\mu Q}{2r^4}[c_1(3c_2^2 - 1) - c_2(3c_1^2 - 1) - 2(c_1 - c_2)s_1 s_2 c]$$

(ii) *In terms of spherical harmonics (Rose [Ibid.])*

$$u_{\mu Q}(12) = 4\pi \sum_{|m| \leq 1} E_{12m}(r)Y_{1m}(\theta_1\phi_1)Y_{2\underline{m}}(\theta_2\phi_2) + 4\pi \sum_{|m| \leq 1} E_{21m}(r)Y_{2m}(\theta_1\phi_1)Y_{1\underline{m}}(\theta_2\phi_2)$$

where

$$E_{120}(r) = \sqrt{\frac{3}{5}}\frac{\mu Q}{r^4} = E_{210}$$

$$E_{121}(r) = \frac{1}{\sqrt{5}}\frac{\mu Q}{r^4} = E_{211}$$

$$E_{ll'\underline{m}}(r) = E_{ll'm}, \qquad l,l' = 1,2; \; m = -1,0,+1$$

(iii) *In terms of rotation matrices (Blum)*

$$u_{\mu Q}(12) = \frac{\mu Q}{2r^4}[\Phi^{123} - \Phi^{213}(12)]$$

## Representations of the QQ Potential

(i) *In terms of two-body relative angles*

$$u_{QQ}(12) = \frac{3Q^2}{4r^5}[1 - 5c_1^2 - 5c_2^2 - 15c_1^2c_2^2 + 2(s_1 s_2 c - 4c_1 c_2)^2]$$

(ii) *In terms of spherical harmonics (Rose)*

$$u_{QQ}(12) = 4\pi \sum_{|m| \leq 2} E_{22m}(r)Y_{2m}(\theta_1\phi_1)Y_{2\underline{m}}(\theta_2\phi_2)$$

where

$$E_{220}(r) = \frac{6}{5}\frac{Q^2}{r^5}$$

$$E_{221}(r) = \frac{4}{5}\frac{Q^2}{r^5}$$

$$E_{222}(r) = \frac{1}{5}\frac{Q^2}{r^5}$$

$$E_{22\underline{m}}(r) = E_{22m}, \qquad m = 1,2$$

(iii)   *In terms of rotation matrices (Blum)*

$$u_{QQ}(12) = \frac{3Q^2}{4r^5} \Phi^{224}(12)$$

---

In LHNC (SSC, or GMF), (2.6) is truncated at first order; i.e.,

$$C(12) = h(12) - \ln\left[g^{000}(r_{12})e^{\beta u(12)}\right] - \left[\frac{g(12) - g^{000}(r_{12})}{g^{000}(r_{12})}\right] \tag{3.2}$$

and for QHNC, the second-order term is retained. A second relation is OZ:

$$h(12) - C(12) = \frac{\rho}{4\pi} \int d3\, h(13)C(32) \tag{3.3}$$

Before substituting (3.1) into (3.2), it is more convenient to work with the rotational invariants $\phi^{MNL}(\omega_1\omega_2\Omega_{12})$:

$$\phi^{MNL}(\omega_1\omega_2\Omega_{12}) = \sum_{\mu\nu\lambda} f^{MNL}\binom{MNL}{\mu\nu\lambda}\, D^M_{0\mu}(\omega_1)D^N_{0\nu}(\omega_2)D^L_{0\lambda}(\Omega_{12}) \tag{3.4}$$

where $\omega_1$ and $\omega_2$ are the Euler angles of dipoles (or multipoles) 1 and 2, and $\Omega_{12}$ is the Euler angle of the vector connecting the centers of the two dipolar molecules; $f^{MNL}$ is a constant, $\binom{MNL}{\mu\nu\lambda}$ is the so-called 3$j$ symbols (related in a simple way to the Clebsch-Gordan coefficients introduced earlier), and $D^M_{0\mu}$ is the rotation matrix. In terms of these $\phi^{MNL}$, the total correlation function $h(12)$ is expanded for the general case of dipole-dipole, dipole-quadrupole, and quadrupole-quadrupole interactions as

$$h(12) = h^{000}(r) + h^{110}(r)\phi^{110}(12) + h^{112}(r)\phi^{112}(12) + h^{121}(r)\phi^{121}(12) \tag{3.5}$$

$$+ h^{123}(r)\phi^{123}(12) + h^{211}(r)\phi^{211}(12) + h^{213}(r)\phi^{213}(12)$$

$$+ h^{220}(r)\phi^{220}(12) + h^{222}(r)\phi^{222}(12) + h^{224}(r)\phi^{224}(12)$$

The functions $h^{MNL}(r)$ are the coefficients of expansion. For dipole-dipole ($\mu\mu$) interaction, only the first three terms (i.e., $h^{000}$, $h^{110}$, and $h^{112}$) are needed. For quadrupole-quadrupole interaction (QQ), the terms $h^{000}$, $h^{220}$, $h^{222}$, and $h^{224}$ are needed. The three terms $h^{123}$, $h^{211}$, and $h^{213}$ are for dipole-quadrupole interactions. Note that for exact $h(12)$ the expansion (3.5) is incomplete. There are higher-order terms (e.g., $h^{440}$, $h^{448}$, $h^{620}$). However, in MSA, due to the assumption (1.11), only $h^{224}$ appears as the highest-order term in QQ interaction, or $h^{112}$ as the highest-order term in $\mu\mu$ interaction. In LHNC and QHNC, the same terms are retained.

A note of comparison: expansion (3.5) is analogous to the spherical harmonic expansion in space-fixed coordinates (see (IV.7.15)):

$$h(12) = \sum_M \sum_N \sum_L \sum_\mu \sum_\nu \sum_\lambda h(MNL; r)C(MNL, \mu\nu\lambda)Y_\mu^M(\omega_1)Y_\mu^N(\omega_2)Y_\lambda^{L*}(\Omega_{12}) \tag{3.6}$$

It can be shown that the coefficients $h^{MNL}(r)$ of Blum are related to $h(MNL;r)$ of Rose [15] in (3.6) by

$$h(MNL; r) = \frac{(4\pi)^{3/2}L!}{[(2L+1)(2M+1)(2N+1)]^{1/2}} \frac{1}{C(MNL, 000)} h^{MNL}(r) \tag{3.7}$$

except for the (220) and (224) terms, which should read

$$h(220; r) = -(\frac{2}{5^{1/2}})(4\pi)^{3/2}h^{220}(r) \tag{2.14}$$

$$h(224; r) = \frac{8}{45}(\frac{35}{2})^{1/2}(4\pi)^{3/2}h^{224}(r) \tag{2.15}$$

For dipolar spheres, $h^{000}(r)$, $h^{110}(r)$, and $h^{112}(r)$ are also related to the $h_s(r)$, $h_\Delta(r)$, and $h_D(r)$ used in the MSA. The coefficient $f^{MNL}$ have the values

$$f^{MNL} = \frac{L!}{\begin{bmatrix} MNL \\ 000 \end{bmatrix}} \tag{3.10}$$

except for

$$f^{220} = -2\sqrt{5} \tag{3.11}$$

$$f^{224} = 8\sqrt{\frac{35}{2}} \tag{2.18}$$

A similar expansion can be made for the dcf $C(12)$:

$$C(12) = C^{000}(r) + C^{110}(r)\phi^{110}(12) + C^{112}(r)\phi^{112}(12) + C^{121}(r)\phi^{121}(12) \tag{3.13}$$

$$+ C^{123}(r)\phi^{123}(12) + C^{211}(r)\phi^{211}(12) + C^{213}(r)\phi^{213}(12)$$

$$+ C^{220}(r)\phi^{220}(12) + C^{222}(r)\phi^{222}(12) + C^{224}(r)\phi^{224}(12)$$

When (3.5) and (3.13) are substituted into the QHNC equation (2.6), a set of algebraically lengthy, but manageable, equations results. For example, the dcfs are given by

$$C^{000}(r) = 0 + [h^{000}(r) - \ln g^{000}(r) - u_{HS}(r)/kT] + B^{000}(r)/g^{000(r)} \tag{3.14}$$

$$C^{112}(r) = \mu^2/kTr^3 + h^{000}(r)[\gamma^{112}(r) + \mu^2/kTr^3] + B^{112}(r)/g^{000}(r)$$

$$C^{123}(r) = -\mu Q/2kTr^4 + h^{000}(r)[\gamma^{123}(r) - \mu Q/2kTr^4] + B^{123}(r)/g^{000}(r)$$

$$C^{213}(r) = \mu Q/2kTr^4 + h^{000}(r)[\gamma^{213}(r) + \mu Q/2kTr^4] + B^{213}(r)/g^{000}(r)$$

$$C^{224}(r) = -3Q^2/4kTr^5 + h^{000}(r)[\gamma^{224}(r) - 3Q^2/4kT^5] + B^{224}(r)/g^{000}(r)$$

and all other projections satisfy

$$C^{MNL}(r) = 0 + h^{000}(r)\gamma^{MNL}(r) + B^{MNL}(r)/g^{000}(r) \qquad (3.15)$$

The $B^{MNL}(r)$ functions are related to the total correlation functions and their algebraic forms could be obtained from the original literature [16]. $\gamma^{MNL}$ are coefficients of the indirect correlation function $\gamma(r) \equiv h(r) - C(r)$. It is interesting to note that equation (3.14) gives a summary of the theories discussed so far: MSA is obtained by keeping only the first terms on the RHS of the equations (3.14), i.e., 0, $\mu^2/kTr^3$, $-\mu Q/2kTr^4$, $\mu Q/2kTr^4$, and $-3Q^2/4kTr^5$. LHNC results from retaining the sums of the first two terms on the RHS; and QHNC retains all three terms on the RHS.

## REMARKS ON THE QHNC APPROXIMATION

For fluids with dipolar and quadrupolar interactions, the MSA and LHNC theories are self-consistent in the sense that as the quadrupole moment $Q \rightarrow 0$, we have a purely dipolar fluid. The only nonzero terms are $h^{000}$, $h^{110}$, and $h^{112}$. Other coefficients disappear. However, in applying QHNC to dipolar fluids, terms such as $h^{220}$, $h^{222}$, and $h^{224}$ are nonzero when they are calculated from eq. (3.14) because these higher-order coefficients derive their values from $B^{224}$, etc. These $B$ coefficients according to QHNC are nonzero since they are based on the nonzero terms $h^{110}$ and $h^{112}$. On the other hand, for dipolar fluids we should have terms only up to $h^{112}$ according to (3.5). In other words, terms $h^{220}$, $h^{222}$, and $h^{224}$ must be zero for dipolar fluids in all three theories. The fact that $h^{220}$, $h^{222}$ and $h^{224}$ remain in QHNC for purely dipolar fluids poses an *inconsistency*. One of the consequences of this inconsistency is that for strongly dipolar fluids, i.e., in the $Q \rightarrow 0$ limit, the QHNC equation has no convergent solution. However, QHNC causes no such problem in the $\mu \rightarrow 0$ limit (i.e., in the limit of purely quadrupolar interactions). In this case, $h^{110}$ and $h^{112}$ are strictly zero. The inconsistency revolves around the recognition that, although the results of the QHNC formulated for mixed multipolar interactions (eq. (3.1)) are identical to those for purely quadrupolar fluids as $\mu \rightarrow 0$, no such identification is evident when the other limit $Q \rightarrow 0$ is taken. Thus one could not recover the purely dipolar results by using (3.14). In practice, however, this does not cause any problems. To obtain answers for purely dipolar fluids, one simply begin QHNC with strictly dipole-dipole interaction.

Applications of LHNC and QHNC to dipolar and/or quadrupolar fluids have been carried out and proved to be "quantitative" in many instances. Thermodynamic behavior was discussed by Levesque et al. [17] and Patey et al. [18] and will be presented below.

## VII.4. *Structure and Thermodynamics of Polar Fluids*

### DIPOLAR HARD SPHERES

Extensive comparison of MC results with the LHNC theory was made by Levesque et al. (*Ibid.*) Figures VII.2 and 3 show the projections $h_\Delta(r)$ and $h_D(r)$ at $\rho^* = 0.80$ and $\mu^{*2} = 2.75$. The LHNC results are seen to follow the MC data closely for the dipole-dipole potentials truncated at $r_c = 3.4d$ and $5.1d$. As to QHNC, the angle-averaged $g^{000}$ term is shown in Figure VII.4 for $\rho^* = 0.15$ and close agreement is

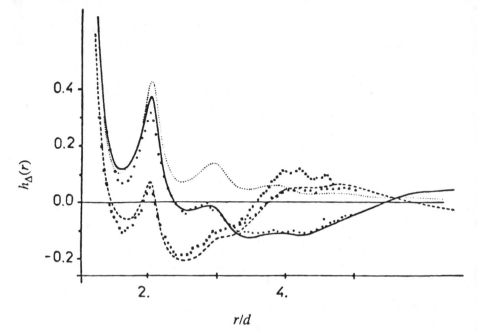

FIGURE VII.2. Comparison of MC and LHNC results for $h_\Delta$ at $\rho^* = 0.8$ and $\mu^{*2} = 2.75$. The symbols represent the following: (- - - -) LHNC for an infinite system with a potential cutoff at $r_c/d = 3.4$; (xxx) MC result for $h_\Delta^{sc}$, N=256, $r_c/d = 3.4$; ($\Delta\Delta\Delta$) MC result for $h_\Delta^{sc}$, N= 864, $r_c/d = 3.4$. Up to $r=r_c$ both sets of MC results coincide within statistical error; (———) LHNC for an infinite system with a potential outoff at $r_c/d = 5.1$; (•••) MC results for $h_\Delta^{sc}$, N=864, $r_c/d = 5.1$; (····) LHNC for an infinite system with an untruncated potential. (Levesque et al., Mol. Phys., 1977)

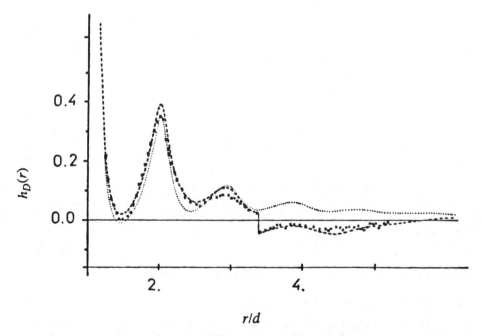

FIGURE VII.3. Comparison of MC and LHNC results for $h_D$ at $\rho^*=0.8$ and $\mu^{*2}=2.75$. The symbols have the same meaning as in FIGURE VII.2. (Levesque et al., Mol. Phys., 1977)

obtained. The nonpolar hard-sphere $g^{HS}$ is very different from $g^{000}$. But as the density is increased to 0.8 (at $\mu*^2 = 2.75$), $g^{000}$ approaches the $g(r)$ of hard spheres. This comparison drives home the point that, as the density is increased, the structure of the fluid is essentially determined by the repulsive (hard-sphere, in this case) part of the potential.

Internal energy of dipolar spheres is presented in Table VII.3. The LHNC theory is quite adequate at the given density.

Table VII.3. *The LHNC Theory and the Internal Energy of Dipolar Hard Spheres at* $\rho* = 0.8$: *(a) N=256, $r_c/d$=3.4. (b) N=864, $r_c/d$=5.1.*

| $\mu*^2$ | $-U/NkT$ (LHNC) | | $-U/NkT$ (MC) |
|---|---|---|---|
| | Untruncated potential | Truncated potential | |
| 0.50 | 0.329 | | $0.327 \pm 0.003$ (a) |
| 1.00 | 1.001 | | $1.000 \pm 0.003$ (a) |
| 2.75 | 4.180 | 4.330 | $4.300 \pm 0.02$ (b) |

## QUADRUPOLAR HARD SPHERES

For hard spheres plus point quadrupole-quadrupole interaction, the QHNC results are shown in Figure VII.5 for $\rho*=0.8344$ and $Q*^2=0.4$. Three components, $h^{220}$, $-h^{222}$, and $-h^{224}$, are displayed. The QHNC values of $h^{224}$ match fairly well the MC data. Figure VII.6 compares $h^{224}$ calculated from QHNC as well as LHNC, with QHNC performing slightly better. The MSA theory is seen to be inadequate. The MC peak at $r=1.5$ is not reproduced by any of the theories. We note that in QHNC, all $h^{220}$, $h^{222}$, and $h^{224}$ terms contribute to $C^{000}$. This is not the case for LHNC, where the isotropic part derives no contributions from the anisotropic components. Strictly speaking, this independence is incorrect. Also note that at large $Q*$, $g^{000}(r)$ differs considerably from the underlying $g^{HS}$. Figure VII.7 compares the projection $g^{000}$ with MC at $Q*^2=1.666$. $g^{000}$ from MC differs appreciably from $g^{HS}$. We make the following observation: although increase in density forces the spherically symmetrical $g^{000}$ to approach that of the hard spheres, the increase in $Q*$ tends to pull $g^{HS}$ and $g^{000}$ apart. In other words, the directional forces of quadrupole moment tend to orient the molecules away from spherical symmetry.

The internal energy calculated from all three theories, MSA, LHNC, and QHNC, were compared in Table VII.4 for quadrupolar spheres with $Q*^2$ ranging from 0.4 to 1.666. QHNC gives the best overall agreement with MC, outperforming MSA, and LHNC.

Table VII.4. *Internal Energy of Quadrupolar Hard Spheres at* $\rho* = 0.8344$

| | Configurational Energy $-U/NkT$ | | | | |
|---|---|---|---|---|---|
| $Q*^2$ | MSA | Padé | LHNC | QHNC | MC |
| 0.4 | 0.320 | 0.673 | 0.697 | 0.663 | $0.655 \pm 0.003$ |
| 1.0 | 1.811 | 3.378 | 3.396 | 3.228 | $3.182 \pm 0.014$ |
| 1.25 | 2.703 | 4.868 | 4.830 | 4.629 | $4.574 \pm 0.017$ |
| 1.6666 | 4.450 | 7.645 | 7.464 | 7.272 | $6.987 \pm 0.027$ |

## HARD SPHERES WITH EMBEDDED DIPOLES AND QUADRUPOLES

For hard spheres embedded with both point dipoles and linear quadrupoles, the interaction is given by eq. (3.1). LHNC and QHNC have been solved at three different combinations of polar moments: equal magnitudes $\mu*^2 = \mu^2/kTd^3 = 1.0 = Q*^2 = Q^2/kTd^5$; high $\mu* =1.5$ and low $Q*=0.5$; and high $\mu* =1.5$ and high $Q*=1.0$. The results of both QHNC and LHNC are, in general, less satisfactory than the purely dipolar and purely

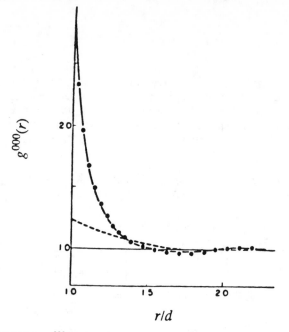

FIGURE VII.4. $g^{000}(r)$ for dipolar hard spheres at $\rho^*=0.15$ and $\mu^{*2}=2.0$. Solid line = QHNC approximation. Dashed line = pure hard-sphere function. Dots = MC results. (Patey et al., Mol. Phys., 1979)

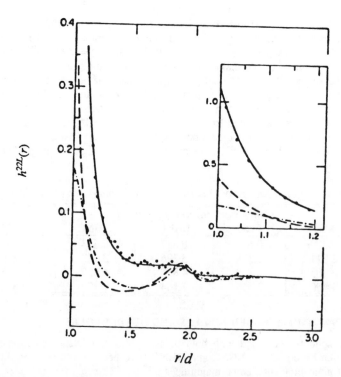

FIGURE VII.5. $h^{220}(r)$, $-h^{222}(r)$, and $-h^{224}(r)$ at $Q^2=0.4$ and $\rho^*=0.8344$. $\cdot -\cdot\cdot -\cdot\cdot -=$ $h^{220}(r)$. ---- $= -h^{222}(r)$. ----- $= -h^{224}(r)$. The dots are MC results for $-h^{224}(r)$. (Patey, Mol. Phys., 1978)

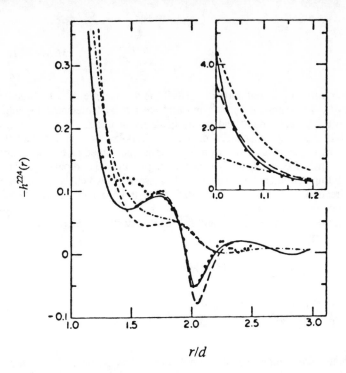

FIGURE VII.6. Comparison of $h^{224}(r)$ with MC calculations at $Q^{*2}=1.666$ and $\rho^*=0.8344$. -----: QHNC. — — —: LHNC. — · — · —: MSA. The dots are MC results. (Patey, Mol. Phys., 1978)

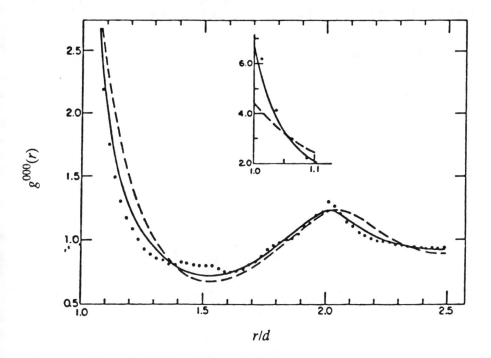

FIGURE VII.7. $\rho^*=0.8344$, $Q^{*2}=1.666$. Solid line = QHNC approximation. Dots = MC results. Broken line = $g^{HS}(r)$. (Patey, Mol. Phys., 1978)

quadrupolar cases discussed earlier. QHNC yields accurate $g^{000}$, $h^{123}$, and $h^{224}$, and so does LHNC. The components $h^{110}$ and $h^{112}$ are not well predicted by either theory, in contrast to the purely dipolar and purely quadrupolar cases. The $h^{110}$ term as given by MC exhibits a strong dependence on the quadrupolar strength $Q^*$, and its magnitude diminishes as $Q^*$ is increased. Physically this is understood as follows: strong $Q^*$ tends to increase the $T$-orientation. At this configuration, the dipolar product $\mu_1 \cdot \mu_2 = 0$. Hence quadrupolar interaction tends to decrease $h^{110}$, which is the component in the dipolar direction. At $\mu^*=1.5$ and $Q^*=0.5$, there is no solution from QHNC due to the inconsistency discussed earlier.

The components $h^{112}$, $h^{123}$, and $h^{224}$ contribute to the internal energy via

$$\frac{U^c}{NkT} = \frac{4\pi}{3}\rho^*\mu^{*2}\int_1^\infty dr\ \frac{h^{112}(r)}{r} + 8\pi\rho^*\mu^*Q^* \int_1^\infty dr\ \frac{h^{123}(r)}{r^2} \qquad (4.1)$$

$$+ \frac{112\pi}{15}\rho^*Q^{*2} \int_1^\infty dr\ \frac{h^{224}(r)}{r^3}$$

These three terms represent the dipole-dipole, dipole-quadrupole, and quadrupole-quadrupole contributions, respectively. The results from MSA, LHNC, and QHNC are compared with the MC values in Table VII.5. The Padé results are also shown. None of the theories MSA, LHNC, or QHNC gives adequate energy prediction. This is attributed to the poor results for $h^{112}$ and $h^{224}$.

The dielectric constant $\varepsilon$ for the potential (3.1) can be obtained from the component $h^{110}$ through (1.30), where the Kirkwood $g$-factor is given by

$$g_K = 1 + \frac{\rho}{3} \int dr\ 4\pi r^2 h^{110}(r) \qquad (4.2)$$

or from $h^{112}$ as

$$h^{112}(r) \to \frac{(\varepsilon - 1)}{4\pi\varepsilon\gamma r^3} \qquad \text{as } r \to \infty \qquad (4.3)$$

In computer simulation, one calculates the $g$-factor via the mean square moment

*Table VII.5. The Internal Energy for Hard Spheres with
Embedded Dipoles and Quadrupoles at $\rho^* = 0.8$*

| $\mu^*$ | $Q^*$ | \multicolumn{5}{c}{Configurational Energy $-U/NkT$} |
| | | Padé | MSA | LHNC | QHNC | MC (±0.02) |
|---|---|---|---|---|---|---|
| 1.0 | 1.0 | 4.40 | 3.44 | 5.35 | 4.97 | 4.64 |
| 1.5 | 0.5 | 3.93 | 3.11 | 4.27 | | 3.72 |
| 1.5 | 1.0 | 6.27 | 5.86 | 8.31 | 7.58 | 6.99 |

$$g_K = \frac{<M^2>}{N\mu^2} = 1 + \frac{N-1}{\mu^2}<\mu_1 \cdot \mu_2> \qquad (4.4)$$

Table VII.6. *The Mean Square Moment and Dielectric Constant for Hard Spheres with Embedded Dipoles and Quadrupoles ($\rho^* = 0.8$).*

| | | $<M^2>_{cube}/N\mu^2$ | Dielectric Constant $\varepsilon$ | | | |
|---|---|---|---|---|---|---|
| $\mu^*$ | $Q^*$ | MC($\pm 0.02$) | MSA | LHNC | QHNC | MC |
| 1.0 | 1.0 | 0.59 | 6.27 | 5.67 | 4.61 | 6.8 |
| 1.5 | 0.5 | 0.33 | 20.7 | 29.9 | | 15.6 |
| 1.5 | 1.0 | 0.30 | 16.29 | 12.68 | 10.33 | 10.2 |

where the angular brackets denote the canonical average. $M$ is the total moment of the system. The MC results for mixed dipoles and quadrupoles are listed in Table VII.6. We observe that MSA and LHNC seriously overestimate the dielectric constant. For purely dipolar spheres, the overestimation becomes even more drastic. In this case, QHNC is no better. Thus, as a theory for dielectric constants, none of the theories discussed gives a quantitative answer. Lately, Fries and Patey [19] solved the complete (reference) HNC equation and obtained a better prediction of $\varepsilon$ for purely dipolar spheres (see Table VII.7). This improvement is a step in the right direction. On the other hand, Patey et al. [20] relied on the LHNC theory to determine qualitatively the behavior of $\varepsilon$ with respect to different quadrupole moments. Figure VII.8 shows the dramatic reduction of $\varepsilon$ as $Q^*$ is increased. Their findings indicate two interesting points: (i) The simple dipolar model of dielectric constant found in literature is not adequate for most real fluids, since the quadrupolar moments have profound influence on $\varepsilon$. (ii) From moderate to high quadrupole moments, the simple theory of Onsager, eq. (1.33), is likely to give satisfactory answers, this being due to the disappearance of the integral $\int dr\ h^{110}(r)$ as $h^{110}$ gets smaller with increasing $Q^*$.

Table VII.7. *Comparison of the RHNC Dielectric Constant at $\rho^* = 0.8$ with Computer Simulation Results and Selected Theories.*

| $\mu^{*2}$ | MSA | RLHNC | RQHNC | RHNC | MC |
|---|---|---|---|---|---|
| 1.0 | 7.80 | 9.62 | 9.62 | 8.82 | $9.0\pm0.5^a$ |
| 2.0 | 20.0 | 50.0 | 51.7 | 31.7 | $31.6\pm4^b$ |
| 2.75 | 31.9 | 250.0 | 444.9 | 93.0 | $65\pm3^c$ |

*a*: Levesque, Patey, and Weis [21]
*b*: Patey, Levesque, and Weis [22]
*c*: This is from the Neumann-Steinhauser [23] reanalysis of Adams' [24] zero field EK calculations.

APPLICATION OF QHNC TO THE STOCKMAYER POTENTIAL

The Stockmayer potential is based on the LJ potential

$$u(12) = u_{LJ}(r_{12}) - \frac{\mu^2}{r^3}\Phi^{112}(12) \qquad (4.5)$$

MC data have been generated at $\rho^*=0.80$, $T^*=1.35$, $\mu^*=2.75$ and $\mu^*=1.0$. Figures VII.9,

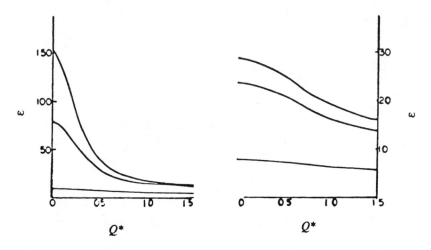

FIGURE VII.8. Variations of the dielectric coefficient ε with Q* at ρ* =0.8. (a) LHNC results. (b) MSA results. From top to bottom: curve 1 at μ* =1.6, curve 2 at μ* = 1.5, and curve 3 at μ* = 1.0. (Patey et al., Mol. Phys., 1979)

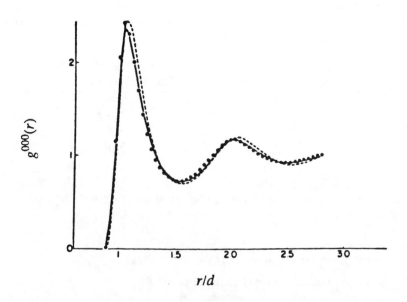

FIGURE VII.9. $g^{000}(r)$ for Stockmayer particles at ρ*=0.8, T*=1.35 and μ*²=2.75. Solid line = QHNC approximation. Dashed line = pure LJ fluid result. Dots = MC results. (Patey et al., Mol. Phys., 1979)

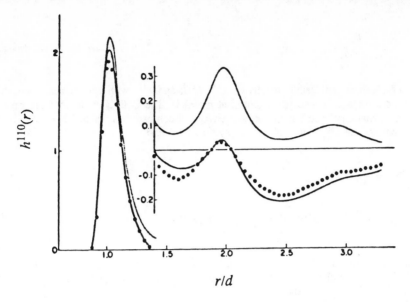

FIGURE VII.10. $h^{110}(r)$ for Stockmayer particles at $\rho^*=0.8$, $T^*=1.35$, and $\mu^{*2}=2.75$. Dots = MC results. Dotted line = QHNC result for an infinite system with an untruncated potential. (Patey et al., Mol. Phys., 1979)

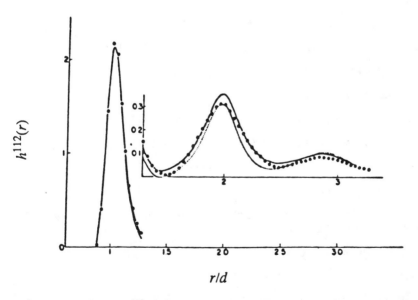

FIGURE VII.11. $h^{112}(r)$ for Stockmayer particles at $\rho^*=0.8$, $T^*=1.35$, and $\mu^{*2}=2.75$. The solid Solid line = QHNC approximation for a spherically truncated potential. Dots = MC results. Dotted line = QHNC results for an infinite system with an untruncated potential. (Patey et al., Mol. Phys., 1979)

10 & 11 compare the QHNC results of $g^{000}$, $h^{110}$, and $h^{112}$ with MC values. (The MC results were for a truncated $\mu\mu$ potential at $r/\sigma=3.3$). The QHNC predictions are seen to be quite satisfactory. The configurational energy is compared in Table VII.8. $U^c_{DD}$ is the dipolar part, while $U^c_{SS}$ is the spherically symmetrical part,

$$\frac{U^c}{NkT} = \frac{U^c_{SS}}{NkT} + \frac{U^c_{DD}}{NkT} \tag{4.6}$$

$$= \frac{\rho}{2kT} \int dr\, u^{000}(r) g^{000}(r) - \frac{\mu^2\rho}{3kT} \int dr\, \frac{h^{112}(r)}{r^3}$$

The LHNC results are also listed (for LHNC, $U^c_{SS}$ is simply the pure LJ internal energy). In QHNC, the dipolar interaction contributes to $U^c_{SS}$. The QHNC energy is quite dependable at the states studied.

*Table VII.8. Stockmayer Fluid at $\rho^* =0.8$ and $T^* =1.35$, $R_c =3.3\sigma$*
*(LHNC result for $-<U_{ss}>/NkT$ is 5.25 at all values of $\mu^*$).*

| | $-<U_{DD}>/NkT$ | | | $-<U_{ss}>/NkT$ | |
|---|---|---|---|---|---|
| $\mu^{*2}$ | LHNC | QHNC | MC | QHNC | MC |
| 1.0 | 1.01 | 1.03 | 1.02 | 5.14 | 5.14 |
| 2.75 | 4.28 | 4.57 | 4.50 | 4.77 | 4.76 |

## References

[1]    M.S. Wertheim, J. Chem. Phys. **55**, 4291 (1971).

[2]    G.N. Patey and J.P. Valleau, Chem. Phys, Letters **21**, 297 (1973).

[3]    G.N. Patey and J.P. Valleau, J. Chem. Phys. **61**, 534 (1974).

[4]    L. Verlet and J.-J. Weis, Mol. Phys. **28**, 665 (1974).

[5]    G. Stell, J.C. Rasaiah, and H. Narang, Mol. Phys. **23**, 393 (1972).

[6]    G. Nienhuis and J.M. Deutch, J. Chem. Phys. **55**, 4213 (1971).

[7]    A. Rahman and F.H. Stillinger, J. Chem. Phys. **55**, 3336 (1971).

[8]    G.N. Patey, Mol. Phys. **34**, 427 (1977).

[9]    M.S. Wertheim, Mol. Phys. **25**, 211 (1973); *Ibid.* **26**, 1425 (1973); *Ibid.* **33**, 95 (1977); and *Ibid.* **34**, 1109 (1977).

[10]   R.L. Henderson and C.G. Gray, Can. J. Phys. **56**, 571 (1978).

[11]   G.N. Patey, Mol. Phys. **35**, 1413 (1978).

[12]   L. Blum, J. Chem. Phys. **57**, 1862 (1972); and *Ibid.* **58**, 3295 (1973).

[13]   L. Blum and A.J. Torruella, J. Chem. Phys. **56**, 303 (1972).

[14]   M.E. Rose, see IV. Ref. [10]

[15]   M.E. Rose, see IV. Ref. [10].

[16] G.N. Patey, D. Levesque, and J.-J. Weis, Mol. Phys. **38**, 1635 (1979).

[17] D. Levesque, G.N. Patey, and J.-J. Weis, Mol. Phys. **34**, 1077 (1977).

[18] G.N. Patey, D. Levesque, and J.-J. Weis, Mol. Phys. **38**, 219 (1979).

[19] P.H. Fries and G.N. Patey, J. Chem. Phys. **82**, 429 (1985).

[20] G.N. Patey, D. Levesque, and J.-J. Weis, Mol. Phys. **38**, 1625 (1979).

[21] D. Levesque, G.N. Patey, and J.-J. Weis, Mol. Phys. **34**, 1077 (1977).

[22] G.N. Patey, D. Levesque, and J.-J. Weis, Mol. Phys. **45**, 733 (1982).

[23] N. Neumann and O. Steinhauser, Chem. Phys. Letters **95**, 417 (1983).

[24] D.J. Adams, Mol. Phys. **40**, 1261 (1980).

## Exercises

1. Prove the equivalence of the dipole-dipole expressions (i), (ii), (iii), and (iv) in Table VII.2.

2. Use LHNC and QHNC theories to calculate for the internal energies listed in Table VII.4 for quadrupolar hard spheres.

3. Consult reference Fries and Patey [19] and apply RHNC to dipolar spheres. Evaluate the dielectric constants for at least one condition listed in Table VII.7.

4. Find the numerical factors connecting the spherical harmonic coefficients to the coefficients of Blum as given in eq. (3.5). Prove the identities (2.14) and (2.15).

5. Show the orthogonality conditions (1.8) and the normalization conditions (1.9).

# CHAPTER VIII

# HARD SPHERES AND HARD-CORE FLUIDS

## VIII.1. *The Hard-Sphere Potential*

The study of fluids composed of hard spheres represents a first step in correcting the ideal-gas behavior. As discussed in Chapter III, the ideal gas is characterized as molecules without interaction energy. As a consequence, ideal-gas molecules have no *excluded volume* and no cohesive forces. This picture contradicts the behavior of real gases. As an improvement, one surrounds each molecules with an excluded volume, i.e., a hard core. The simplest of such molecules are hard spheres. The potential energy of interaction, $u_{HS}$, could be described by

$$u_{HS}(r) = +\infty, \qquad r \leq d \qquad\qquad (1.1)$$

$$= 0, \qquad r > d$$

where $r$ is the distance between the centers of two hard spheres. To see the consequence of (1.1) in terms of interaction forces, we use the formula

$$\mathbf{f}_{ij} = -\nabla u_{HS}(r_{ij}), \qquad r_{ij} = |\mathbf{r}_j - \mathbf{r}_i| \qquad\qquad (1.2)$$

where $\mathbf{f}_{ij}$ is the force on molecule $i$ by molecule $j$, and $\mathbf{r}_j$ is the position vector of molecule $j$. The gradient is zero everywhere except at $r = d$, where $\nabla u_{HS}(d) = -\infty$, because of the steep descent of the energy barriers. Therefore, the force $\mathbf{f}_{ij} = +\infty$ is an infinite repulsion. The picture is two colliding hard spheres interacting with an infinite force of repulsion at contact. For a collection of hard spheres, the total potential energy $V_N$ is

$$V_N(\mathbf{r}_1,..., \mathbf{r}_N) = \sum_{i<j} u_{HS}(r_{ij}), \qquad r_{ij} = |\mathbf{r}_j - \mathbf{r}_i| \qquad\qquad (1.3)$$

The total energy, or the Hamiltonian $H_N$, is the sum of the kinetic energy and the potential energy:

$$H_N(\mathbf{r}_1,...\mathbf{r}_N) = \frac{1}{2}\sum_j \frac{p_j^2}{m} + \sum_{i<j} u_{HS}(r_{ij}) \tag{1.4}$$

For a canonical ensemble, we can readily write down the partition function

$$Z_N = \frac{1}{N!\Lambda^{3N}} \int d\mathbf{r}^N \exp\left[-\beta\sum_{i<j}^N u_{HS}(r_{ij})\right] \tag{1.5}$$

At first glance, the HS potential might appear so simple that an analytical solution to the N-body problem is possible. Closer examination of (1.5) shows that even for $u_{HS}$, the configurational integral is insoluble analytically in three dimensions. Efforts have been made since the 1960s to characterize $Z_N$ and related quantities for hard spheres. Now the behavior of hard spheres is well understood. Modern liquid state theory started with the elucidation of hard-sphere properties. Earlier, liquid had always been thought of as a dense gas (the imperfect gas theory) or as a disordered solids (the cell model of liquids). As repulsive forces are dominant in dense liquids, hard core molecules give a realistic description of the liquid structure. In the following, we shall begin by discussing one-dimensional hard rods whose exact solution is known, proceeding to hard disks in two dimensions, then to hard spheres in three dimensions. For nonspherical molecules, we shall introduce several general classes of hard core fluids: dumbbells, spherocylinders, and polyatomics.

## VIII.2. *The Hard Rods in One Dimension*

The interaction energy $u_1(r)$ for hard rods is the same as (1.1) except that $r$ stays in one dimension.

$$u_1(r) = +\infty, \qquad\qquad r\le d \tag{2.1}$$

$$= 0, \qquad\qquad r>d$$

Salsburg, Zwanzig, and Kirkwood [1] as well as Tonks [2] solved this problem using Laplace transforms. We summarize their results in the following.

**EQUATION OF STATE: HARD RODS**

$$\frac{P}{\rho kT} = \frac{PL}{NkT} = \frac{1}{1 - \rho d} \qquad\qquad \text{(exact)} \tag{2.2}$$

where P is the pressure (in units of force), $L$ is the linear length of the system, $N$ is the number of hard rods in the system, $k$ is the Boltzmann constant, $T$ is the absolute temperature, and $\rho = N/L$ is the number density. [3] We remark that (2.2) is reminiscent of the van der Waals equation of state

$$\frac{P^{vdw}}{\rho kT} = \frac{1}{1-b\rho} - \frac{a\rho}{RT} \qquad (2.3)$$

The repulsion term in (2.3) is identical to (2.2) if we identify the covolume $b$ with the hard rod size $d$. Equation (2.2) partly explains the reason for success of the van der Waals equation. We shall have more to say on this.

## CHEMICAL POTENTIAL

$$\frac{\mu}{kT} = \ln\frac{\rho\Lambda}{1-\rho d} + \frac{\rho d}{1-\rho d} \qquad (2.4)$$

where $\Lambda$ is the de Broglie thermal wavelength.

## HELMHOLTZ FREE ENERGY

$$\frac{A}{NkT} = -\frac{\ln Z_N}{N} = \ln\frac{\rho\Lambda}{1-\rho d} - 1 \qquad (2.5)$$

## GIBBS FREE ENERGY

$$\frac{G}{NkT} = \ln\frac{\rho\Lambda}{1-\rho d} + \frac{\rho d}{1-\rho d} \qquad (2.6)$$

## ENTROPY

$$S = \frac{3}{2} Nk - Nk \ln\frac{\rho\Lambda}{1-\rho d} \qquad (2.7)$$

## INTERNAL ENERGY

$$U = \frac{1}{2} NkT \qquad (2.8)$$

## ENTHALPY

$$H = \frac{1}{2} NkT + \frac{NkT}{1-\rho d} \qquad (2.9)$$

## HEAT CAPACITIES

$$C_v = \frac{1}{2} k, \qquad\qquad C_p = \frac{3}{2} k \qquad (2.10)$$

We also know the radial distribution function of hard rods exactly.

## THE RADIAL DISTRIBUTION FUNCTION

$$g(r) = \frac{1}{\rho d} \sum_{k=1}^{\infty} H(r-k) \left[\frac{\rho d}{1-\rho d}\right]^{k} \frac{(r-k)^{k-1}}{(k-1)!} \exp\left[\frac{-\rho d}{1-\rho d}(r-k)\right] \qquad (2.11)$$

where $H(r-k)$ is the Heaviside function, which is zero for $r<k$ and unity for $r>k$. Note that $d$ has been taken as the unit of length in (2.11). Figure VIII.1 compares the molecular dynamics result with the exact expression (2.11). They are in total agreement. Although hard rod systems are of limited practical interests, they are of theoretical value in testing theories, such as the superposition approximation and virial coefficients.

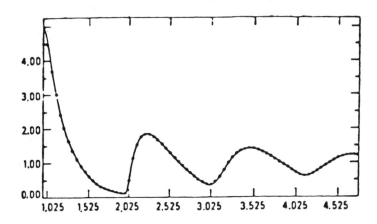

*FIGURE VIII.1. Radial distribution function of one-dimensional hard rods at $\rho^* = 0.80$. The line is results from equation (2.11). The dots are from molecular dynamics. (Haus and Raveché, J. Chem. Phys., 1978)*

## VIII.3. *The Hard Disks in Two Dimensions*

The analogue in two dimensions to the three dimensional hard spheres is circular disks with impenetrable boundaries (i.e., hard disks). The interaction potential $u_{HD}(r)$ is of the same form as (1.1) with **r** a 2-vector. The solution for the partition function (1.5) in two dimensions is not known analytically, in contrast to eq. (2.5) for one dimension. Information has to be obtained through computer simulations. Wood [4] compared the MC results with analytical approximations and gave the following formulas for the pressure of hard disks.

### EQUATION OF STATE

#### *The Fluid Branch*

$$\frac{PA}{NkT} = 1 + 1.81380\rho' \frac{1 - 0.356780\rho' + 0.021447\rho'^2}{1 - 1.775171\rho' + 0.787808\rho'^2} \qquad (3.1)$$

The above expression is restricted to $\rho' = \rho/\rho_0 \leq 1/1.4$, where $\rho_0 d^2 = 2/\sqrt{3}$ is the close packing value. Equation (3.1) is in the form of a Padé approximant. This equation was originally given by Ree and Hoover [5]. Its accuracy is within 0.1% as compared to the sixth virial polynomial at $\rho/\rho_0 = 0.25$. At high densities, the hard disks form a regular triangular lattice. We have then

**The Solid Branch (The Free Volume Expression)**

$$\frac{P}{\rho kT} = \frac{1}{1 - \sqrt{\rho'}}$$

(3.2)

for $\rho/\rho_0 > 0.798$. Note that in (3.1) and (3.2), $\rho = N/A$ is the number density, $N$ is the number of hard disks in the system, and $A$ is the *area* of the system.

That there is a fluid-solid phase transition for two-dimensional hard disks is now well established. The values of $\rho/\rho_0$ where the first order phase transition occurs are given in Table VIII.1.

Table VIII.1.  Fluid-Solid Phase Transition for Hard Disks

| $\rho/\rho_0$ (Fluid) | $\rho/\rho_0$ (Solid) | $P/\rho_0 kT$ |
|---|---|---|
| 0.761 | 0.798 | 8.08 |

Once the equation of state of hard disks is known, the other thermodynamic properties can be easily derived. We leave the derivation as an exercise. The radial distribution functions for hard disks have been simulated by Chae et al. [6] Physical systems that are two dimensional include monolayer adsorptives and interfaces between two phases.

## INTERNAL ENERGY

As in the one-dimensional case, hard core systems do not contribute to the internal energy except from kinetic motion. From the equipartition theorem, two degrees of freedom contribute twice $kT/2$

$$U = NkT$$

(3.3)

## CLOSE PACKING

Close packing of hard cores is a state of maximum density where all particles are closely packed. It is a consequence of purely geometrical arrangements. For hard disks at the highest density, a triangular lattice is formed. The density is

$$\rho_0 d^2 = \frac{2}{\sqrt{3}}$$

(3.4)

## VIRIAL COEFFICIENTS

The virial coefficients have been evaluated from the defining cluster integrals either analytically or by Monte Carlo integration. Table VIII.2 gives the known results (where $B, C, D,...$ etc. stand for the second, third, fourth,... virial coefficients).

An alternative, simplified, and therefore approximate, equation of state for hard disks was given by Henderson [7]

$$\frac{PA}{NkT} = \frac{1 + y^2/8}{(1-y)^2} \tag{3.5}$$

where $y = \pi \rho d^2/4$ is the packing fraction in two dimensions.

## VIII.4. *Hard Spheres: The PY Results*

The interaction potential for hard spheres is given by (1.1) with **r** a 3-vector. Since the model of hard spheres is idealized and does not exist in nature, we cannot obtain their properties form experiments. There are two other ways of obtaining this information: by computer simulation and/or by integral equations. The PY integral equation yields reasonable results for hard spheres. It is discussed first.

Wertheim [8] and Thiele [9] solved the PY equation analytically. The direct correlation function (dcf) and the equation of state have been determined. As there are two formulas for calculating the pressure: the virial equation and the compressibility equation, two pressure values are obtained from a single PY distribution function. This is due to the inexactitudes inherent in the PY approximation. For exact rdf (such as those obtained from simulations), both formulas give the same answer. The degree of discrepancy is a measure of the inaccuracies in the theory. For PY, the compressibility pressure is closer to machine results.

EQUATION OF STATE (PY)

*From Compressibility Equation*

$$\frac{P^c}{\rho kT} = \frac{1 + y + y^2}{(1-y)^3} \tag{4.1}$$

*From Virial Equation*

$$\frac{P^v}{\rho kT} = \frac{1 + 2y + 3y^2}{(1-y)^2} \tag{4.2}$$

where the superscripts $c$ and $v$ denote the compressibility and the virial methods, respectively, $\rho = N/V$ is the number density, and $y$, the packing fraction, is defined as

| Table VIII.2.  Virial Coefficients for Hard Disks | | |
|---|---|---|
| $B$ | $\pi d^2/2$ | |
| $C/B^2$ | $4/3 - \sqrt{3}/\pi$ | $= 0.7820044$ |
| $D/B^3$ | $2 - 4.5(\sqrt{3}/\pi) + 10/\pi^2$ | $= 0.5322318$ |
| $E/B^4$ | | $= 0.3335561$ |
| $F/B^5$ | | $= 0.19893$ |
| $G/B^6$ | | $= 0.1148$ |

$$y \equiv \frac{\pi}{6}\rho d^3 \qquad (4.3)$$

It gives the fraction of volume occupied by the hard spheres. Figure VI.7 gives a comparison of the two equations together with the exact results. We note that for hard spheres, temperature has no effect on the equation of state. This is due to the steepness of the HS potential. The factor $\beta u_{HS}(r_{ij})$ in the partition function is independent of $T$. We also note that in (4.1-2) the compressibility $Z = \beta P/\rho$ goes to infinity as $y \rightarrow 1$. However, simulation has shown that a phase transition occurs before this limit. Next, we examine the structure of the HS fluid.

## DIRECT CORRELATION FUNCTION (PY)

Wertheim and Thiele (IIbid.) gave the following solution for the dcf:

$$C(r) = -\lambda_1 - 6\lambda_2 yr^* - \frac{1}{2}\lambda_1 yr^{*3}, \qquad r \leq d \qquad (4.4)$$

$$= 0, \qquad r > d$$

where $r^* = r/d$ and

$$\lambda_1 = \frac{(1+2y)^2}{(1-y)^4} \qquad (4.5)$$

$$\lambda_2 = -\frac{(1+y/2)^2}{(1-y)^4}$$

$C(r)$ is negative for $r<d$ and zero for $r>d$. The rdf can be obtained from (4.4) via the OZ relation or, equivalently, from the Laplace or Fourier transform of $C(r)$. Wertheim (*Ibid.*) gave the expression for the rdf in terms of its Laplace transform

## RADIAL DISTRIBUTION FUNCTION (PY)

The Laplace transform of $rg(r)$ is given by

$$G(s) = d^{-2}\int_d^\infty dr \; e^{-rs}rg(r) \qquad (4.6)$$

Then

$$G(s) = \frac{sL(s)}{12y \; [L(s) + M(s)e^s]} \qquad (4.7)$$

where

$$L(s) = 12y[(1 + \frac{y}{2})s + (1+2y)] \qquad (4.8)$$

$$M(s) = (1-y)^2 s^3 + 6y(1-y)s^2 + 18y^2 s - 12y(1+2y) \tag{4.9}$$

Using the inverse Laplace transform formula, we get

$$rg(r) = \frac{1}{2\pi i} \int_{a-i\infty}^{a+i\infty} ds \, \frac{sL(s)e^{sr}}{12yL(s)+M(s)e^s} \tag{4.10}$$

Throop and Bearman [10] carried out a zoned inverse up to $r = 4d$. Smith and Henderson [11] gave tabulations up to $r = 5d$.

## CONTACT VALUES OF RDF (PY)

### *From Compressibility Equation (or Scaled Particle Theory)*

$$g(d) = \frac{1+y/2}{(1-y)^2} \tag{4.11}$$

### *From Virial Equation*

$$g(d) = \frac{4-2y+y^2}{(1-y)^2} \tag{4.12}$$

## DERIVATIVES OF RDF AT CONTACT

The background distribution function $y(r)$ for hard spheres is continuous to the second derivative. The derivatives of $ry(r)$ at $r = d$ have been calculated by Kim [12]:

$$\left[r^* y(r^*)\right]_{r^*=1} = \frac{1+y/2}{(1-y)^2} \tag{4.13}$$

$$\frac{d}{dr^*}\left[r^* y(r^*)\right]_{r^*=1} = \frac{1-5y-5y^2}{(1-y)^3} \tag{4.14}$$

$$\frac{d^2}{dr^{*2}}\left[r^* y(r^*)\right]_{r^*=1} = \frac{-3y(2-4y-7y^2)}{(1-y)^4} \tag{4.15}$$

$$\frac{d^3}{dr^{*3}}\left[r^* y(r^*)\right]_{r^*=1} = \frac{12y(1+3y-4y^3)}{(1-y)^5} \tag{4.16}$$

where $r^* = r/d$.

## INTERNAL ENERGY

The internal energy arises due to kinetic contributions

$$U = \frac{3}{2} NkT \tag{4.17}$$

Equations (4.1 and 2) give the third virial coefficient $B_3$ correctly. At high densities, $P^c$ is more accurate than $P^v$.

## VIII.5. *Simulation Results for Hard Spheres*

Both MC and MD simulations have been carried out for hard spheres. Our knowledge on the equilibrium properties of HS is now fairly extensive. In the following, we shall present the exact results from numerical calculation and from simulation.

### VIRIAL COEFFICIENTS

The virial coefficients for hard spheres are known up to $B_{10}$ [13]. Table VIII.3 presents the data. The second virial coefficient $B_2$ is expressed in units of volume per *molecule*. We use $B, C, D, E, F,...$ to denote second, third, fourth, fifth,... virial coefficients.

*Table VIII.3. The Virial Coefficients of Hard Spheres [14]*

| $B$ | $4\pi d^3/6$ | | $E/B^4$ | 0.11025 | (0.1094) | $H/B^7$ | 0.00445 |
|---|---|---|---|---|---|---|---|
| $C/B^2$ | 0.625 | | $F/B^5$ | 0.0389 | (0.0156) | $I/B^8$ | 0.00150 |
| $D/B^3$ | 0.28695 | (0.2813)* | $G/B^6$ | 0.0137 | (0.0132) | $J/B^9$ | 0.00051 |

*\*Numbers in parentheses are virial coefficients from the Carnahan-Starling equation of state.*

### EQUATION OF STATE

Based on the virial coefficients, a Padé approximant was developed by Carnahan and Starling [15] to represent the simulation data of hard spheres.

$$\frac{P^{CS}}{\rho kT} = \frac{1 + y + y^2 - y^3}{(1 - y)^3} \tag{5.1}$$

where the superscript CS indicates the Carnahan-Starling equation. The formula is highly accurate. It is obtained by taking a (1/3:2/3) average of the virial pressure (4.2) and the compressibility pressure (4.1):

$$\frac{P^{CS}}{\rho kT} = \frac{1}{3}\frac{P^v}{\rho kT} + \frac{2}{3}\frac{P^c}{\rho kT} \tag{5.2}$$

The CS equation reproduces exact third virial coefficients. For higher virial coefficients, it also gives accurate values (see Table VIII.3). We exhibit below the thermodynamic properties of hard spheres based on CS.

### CONTACT VALUES OF THE RDF (CS)

Since the contact value of the rdf $g(d)$ is related to the virial pressure by

$$\frac{P}{\rho kT} = 1 + \frac{4\pi\rho d^3}{6} g(d) \tag{5.3}$$

we obtain

$$g^{CS}(d) = \frac{4 - 2y}{4(1-y)^3} \tag{5.4}$$

**HELMHOLTZ FREE ENERGY (CS)**

Integration of the pressure equation (5.1) yields

$$\frac{A^{CS}}{NkT} = \ln(\rho\Lambda^3) - 1 + \frac{4y - 3y^2}{(1-y)^2} \tag{5.5}$$

**RADIAL DISTRIBUTION FUNCTIONS**

The rdf of hard spheres have been obtained by both MD [16] and MC methods. Table VIII.4 gives the MC results of Barker and Henderson [17].

**CLOSE PACKING**

Geometrically, spheres could be packed closely to a density of

$$\rho_0 d^3 = \sqrt{2} \qquad \text{or} \qquad y_0 = \frac{\pi}{6} \, \rho_0 d^3 = 0.74074 \tag{5.6}$$

This is the highest density under which hard spheres could be packed. However, even before reaching this value, hard spheres undergo a phase transition from the fluid state to the *solid* state.

**PHASE TRANSITION**

Hard spheres exhibit a first-order phase transition before attaining close packing. This behavior has been established by machine simulations [18] [19]. Table VIII.5 gives the densities at phase transition.

*Table VIII.5. Fluid-Solid Phase Transition for Hard Spheres*

| $\rho/\rho_0$ (Fluid) | $\rho/\rho_0$ (Solid) | $P/\rho_0 kT$ |
|---|---|---|
| $0.667 \pm 0.003$ | $0.736 \pm 0.003$ | $8.27 \pm 0.13$ |

# VIII.6. *Hard Sphere Mixtures*

Hard sphere mixtures are characterized by spheres of different diameters $d_i$, $i = 1, 2, \ldots, c$. The PY equation was solved by Lebowitz [20] for mixtures. We first present the analytical PY results. Simulation results will be given later.

**LEBOWITZ SOLUTION OF PY FOR BINARY MIXTURES**

The solution was carried out in Laplace space. The mathematics is interesting but lengthy. We shall present the results below.

Table VIII.4. Radial Distribution Functions for Hard Spheres [17]

| r/d \ $\rho d^3$ | 0·20 | 0·30 | 0·40 | 0·50 | 0·60 | 0·70 | 0·80 | 0·90 |
|---|---|---|---|---|---|---|---|---|
| 1·0173 | 1·3165 | 1·5062 | 1·7619 | 2·0615 | 2·4530 | 2·9510 | 3·6206 | 4·5179 |
| 1·0512 | 1·2797 | 1·4432 | 1·6732 | 1·9392 | 2·2415 | 2·5863 | 3·0149 | 3·4892 |
| 1·0840 | 1·2532 | 1·4074 | 1·5962 | 1·7970 | 2·0350 | 2·2855 | 2·5377 | 2·7722 |
| 1·1158 | 1·2275 | 1·3708 | 1·5188 | 1·6903 | 1·8713 | 2·0364 | 2·1669 | 2·2471 |
| 1·1467 | 1·2051 | 1·3335 | 1·4546 | 1·6008 | 1·7216 | 1·8191 | 1·8826 | 1·8767 |
| 1·1769 | 1·1914 | 1·2992 | 1·4075 | 1·5082 | 1·5986 | 1·6456 | 1·6540 | 1·5869 |
| 1·2062 | 1·1720 | 1·2673 | 1·3504 | 1·4301 | 1·4861 | 1·5008 | 1·4698 | 1·3566 |
| 1·2349 | 1·1629 | 1·2278 | 1·3054 | 1·3606 | 1·3891 | 1·3856 | 1·3200 | 1·1837 |
| 1·2629 | 1·1456 | 1·2110 | 1·2697 | 1·2967 | 1·3112 | 1·2767 | 1·2017 | 1·0512 |
| 1·2903 | 1·1279 | 1·1972 | 1·2258 | 1·2510 | 1·2395 | 1·1951 | 1·0980 | 0·9459 |
| 1·3172 | 1·1204 | 1·1673 | 1·1974 | 1·2087 | 1·1791 | 1·1210 | 1·0096 | 0·8638 |
| 1·3435 | 1·0968 | 1·1364 | 1·1651 | 1·1601 | 1·1245 | 1·0514 | 0·9491 | 0·7915 |
| 1·3693 | 1·0917 | 1·1235 | 1·1333 | 1·1203 | 1·0718 | 1·0056 | 0·8795 | 0·7472 |
| 1·3946 | 1·0668 | 1·1125 | 1·1171 | 1·0920 | 1·0416 | 0·9547 | 0·8483 | 0·7038 |
| 1·4195 | 1·0682 | 1·0950 | 1·0897 | 1·0617 | 0·9999 | 0·9214 | 0·8091 | 0·6756 |
| 1·4440 | 1·0642 | 1·0749 | 1·0704 | 1·0305 | 0·9767 | 0·8898 | 0·7722 | 0·6521 |
| 1·4680 | 1·0500 | 1·0659 | 1·0591 | 1·0097 | 0·9453 | 0·8653 | 0·7492 | 0·6336 |
| 1·4916 | 1·0453 | 1·0443 | 1·0297 | 0·9888 | 0·9219 | 0·8376 | 0·7377 | 0·6237 |
| 1·5149 | 1·0464 | 1·0350 | 1·0208 | 0·9685 | 0·9010 | 0·8202 | 0·7255 | 0·6186 |
| 1·5379 | 1·0247 | 1·0273 | 1·0010 | 0·9547 | 0·8808 | 0·8077 | 0·7138 | 0·6217 |
| 1·5604 | 1·0214 | 1·0288 | 0·9890 | 0·9314 | 0·8747 | 0·7938 | 0·7091 | 0·6259 |
| 1·5827 | 1·0117 | 1·0105 | 0·9745 | 0·9302 | 0·8567 | 0·7956 | 0·7113 | 0·6339 |
| 1·6047 | 1·0148 | 1·0076 | 0·9756 | 0·9180 | 0·8560 | 0·7878 | 0·7066 | 0·6489 |
| 1·6263 | 1·0173 | 0·9957 | 0·9549 | 0·9077 | 0·8479 | 0·7851 | 0·7152 | 0·6670 |
| 1·6477 | 1·0032 | 0·9846 | 0·9514 | 0·9028 | 0·8461 | 0·7834 | 0·7283 | 0·6893 |
| 1·6688 | 1·0106 | 0·9747 | 0·9507 | 0·9010 | 0·8384 | 0·7851 | 0·7381 | 0·7223 |
| 1·6897 | 1·0043 | 0·9804 | 0·9438 | 0·9069 | 0·8430 | 0·7892 | 0·7576 | 0·7478 |
| 1·7103 | 0·9975 | 0·9712 | 0·9337 | 0·8948 | 0·8430 | 0·7991 | 0·7723 | 0·7823 |
| 1·7306 | 0·9931 | 0·9733 | 0·9360 | 0·8925 | 0·8472 | 0·8065 | 0·7928 | 0·8195 |
| 1·7507 | 0·9816 | 0·9664 | 0·9350 | 0·8955 | 0·8527 | 0·8146 | 0·8176 | 0·8655 |
| 1·7706 | 0·9890 | 0·9652 | 0·9317 | 0·8889 | 0·8619 | 0·8283 | 0·8438 | 0·9081 |
| 1·7903 | 0·9928 | 0·9581 | 0·9354 | 0·8947 | 0·8705 | 0·8480 | 0·8678 | 0·9516 |
| 1·8097 | 0·9760 | 0·9592 | 0·9292 | 0·8893 | 0·8749 | 0·8615 | 0·8934 | 0·9942 |
| 1·8289 | 0·9852 | 0·9603 | 0·9292 | 0·9033 | 0·8867 | 0·8796 | 0·9307 | 1·0305 |
| 1·8480 | 0·9784 | 0·9602 | 0·9357 | 0·9132 | 0·9029 | 0·9034 | 0·9562 | 1·0660 |
| 1·8668 | 0·9760 | 0·9612 | 0·9350 | 0·9210 | 0·9100 | 0·9238 | 0·9810 | 1·0992 |
| 1·8855 | 0·9787 | 0·9701 | 0·9367 | 0·9241 | 0·9250 | 0·9411 | 1·0170 | 1·1258 |
| 1·9039 | 0·9840 | 0·9619 | 0·9384 | 0·9303 | 0·9375 | 0·9638 | 1·0444 | 1·1585 |
| 1·9222 | 0·9882 | 0·9621 | 0·9464 | 0·9370 | 0·9528 | 0·9908 | 1·0682 | 1·1822 |
| 1·9404 | 0·9869 | 0·9701 | 0·9452 | 0·9518 | 0·9676 | 1·0094 | 1·0869 | 1·2060 |
| 1·9583 | 0·9805 | 0·9640 | 0·9547 | 0·9613 | 0·9854 | 1·0299 | 1·1186 | 1·2293 |
| 1·9761 | 0·9863 | 0·9700 | 0·9654 | 0·9722 | 0·9982 | 1·0534 | 1·1423 | 1·2507 |
| 1·9937 | 0·9868 | 0·9714 | 0·9720 | 0·9823 | 1·0147 | 1·0769 | 1·1660 | 1·2721 |
| 2·0112 | 0·9878 | 0·9770 | 0·9760 | 0·9912 | 1·0250 | 1·1011 | 1·1923 | 1·2979 |
| 2·0285 | 0·9875 | 0·9866 | 0·9833 | 1·0001 | 1·0471 | 1·1131 | 1·2102 | 1·3019 |
| 2·0457 | 0·9891 | 0·9831 | 0·9883 | 1·0138 | 1·0602 | 1·1188 | 1·2131 | 1·3020 |
| 2·0628 | 0·9895 | 0·9816 | 0·9918 | 1·0178 | 1·0649 | 1·1297 | 1·2146 | 1·2878 |
| 2·0797 | 0·9951 | 0·9834 | 0·9978 | 1·0250 | 1·0667 | 1·1382 | 1·2085 | 1·2686 |
| 2·0964 | 0·9910 | 0·9933 | 0·9946 | 1·0278 | 1·0731 | 1·1419 | 1·2056 | 1·2467 |
| 2·1131 | 0·9904 | 0·9898 | 1·0036 | 1·0265 | 1·0738 | 1·1340 | 1·1900 | 1·2217 |
| 2·1296 | 0·9826 | 0·9882 | 1·0093 | 1·0312 | 1·0712 | 1·1341 | 1·1760 | 1·1924 |
| 2·1459 | 0·9902 | 0·9933 | 1·0069 | 1·0304 | 1·0726 | 1·1329 | 1·1644 | 1·1674 |
| 2·1622 | 0·9957 | 0·9969 | 0·9999 | 1·0387 | 1·0771 | 1·1188 | 1·1506 | 1·1422 |
| 2·1783 | 0·9921 | 0·9964 | 1·0112 | 1·0362 | 1·0735 | 1·1140 | 1·1320 | 1·1129 |
| 2·1943 | 0·9979 | 1·0030 | 1·0108 | 1·0396 | 1·0731 | 1·1091 | 1·1193 | 1·0787 |
| 2·2102 | 1·0013 | 0·9984 | 1·0113 | 1·0431 | 1·0715 | 1·0982 | 1·1011 | 1·0637 |
| 2·2260 | 0·9958 | 1·0021 | 1·0146 | 1·0369 | 1·0656 | 1·0911 | 1·0898 | 1·0300 |
| 2·2417 | 1·0047 | 0·9971 | 1·0155 | 1·0402 | 1·0603 | 1·0870 | 1·0735 | 1·0111 |
| 2·2572 | 0·9928 | 1·0106 | 1·0151 | 1·0401 | 1·0593 | 1·0745 | 1·0600 | 0·9932 |
| 2·2727 | 1·0004 | 0·9987 | 1·0171 | 1·0431 | 1·0559 | 1·0608 | 1·0450 | 0·9729 |

*Equation of State (PY)*

From Compressibility Equation

$$\beta P^c = \rho \frac{1 +y +y^2}{(1 -y)^3} - \frac{18}{\pi} \frac{y_1 y_2}{(1 -y)^3} \frac{(d_2 -d_1)^2}{d_1^3 d_2^3} \left[ d_1 +d_2 + d_1 d_2 \left( \frac{y_1}{d_1} + \frac{y_2}{d_2} \right) \right] \tag{6.1}$$

where the superscript $c$ denotes the compressibility equation; $\rho =\rho_1 +\rho_2$; $\rho_i =N_i/V$ is the number density of species $i$; $y_i=(\pi/6)\rho_i d_i^3$ is the partial packing fraction of species $i$. $y = y_1 + y_2$. We choose the diameter of sphere 2 to be greater than the diameter of sphere 1, $d_2 > d_1$.

From Virial Equation

$$\beta P^v = \beta P^c - \frac{18}{\pi} \frac{y}{(1 -y)^3} \left[ \frac{y_1}{d_1} + \frac{y_2}{d_2} \right]^3 \tag{6.2}$$

where the superscript $v$ denotes the virial equation.

*Contact Values of the RDF (PY)*

For a binary mixture, we have three contact values for the rdf:

$$g_{11}^v(d_1) = (1 -y)^{-2} \left[ 1 +\frac{y}{2} + \frac{3}{2} \frac{y_2}{d_2}(d_1 -d_2) \right] \tag{6.3}$$

$$g_{22}^v(d_2) = (1 -y)^{-2} \left[ 1 +\frac{y}{2} + \frac{3}{2} \frac{y_1}{d_1}(d_2 -d_1) \right]$$

$$g_{12}^v(d_{12}) = \frac{d_2 g_{11}(d_1) + d_1 g_{22}(d_2)}{d_1 +d_2}$$

where $d_{12} =(d_1 +d_2)/2$. The superscript $v$ is to indicate that these contact values yield the virial pressure $P^v$.

*Direct Correlation Function (PY)*

$$C_{11}(r) = -a_1 -b_1 r -cr^3, \qquad r<d_1 \tag{6.4}$$

$$= 0, \qquad r>d_1$$

$$C_{22}(r) = -a_2 -b_2 r -cr^3, \qquad r<d_2$$

$$= 0, \qquad r>d_2$$

$$C_{12}(r) = -a_1, \qquad r< \lambda=(d_2 -d_1)/2$$

$$= -a_1 - [bx^2 + 4\lambda c x^3 + c x^4]/r, \qquad \lambda \le r \le d_{12}$$

$$= 0, \qquad\qquad r > d_{12}$$

where $x = r - \lambda$. The constants are

$$a_i = \frac{\partial}{\partial \rho_i} [\beta P^c(\rho_1, \rho_2)] \qquad\qquad (6.5)$$

$$b_1 = -6 \left[ \frac{y_1}{d_1} g_{11}(d_1)^2 + \frac{y_2}{d_2^3} g_{12}(d_{12})^2 d_{12}^2 \right]$$

$$b_2 = -6 \left[ \frac{y_2}{d_2} g_{22}(d_2)^2 + \frac{y_1}{d_1^3} g_{12}(d_{12})^2 d_{12}^2 \right]$$

$$b = -6 \left[ \frac{y_1}{d_1^2} g_{11}(d_1) + \frac{y_2}{d_2^2} g_{22}(d_2) \right] d_{12} g_{12}(d_{12})$$

$$c = \frac{1}{2} \left[ a_1 \frac{y_1}{d_1^3} + a_2 \frac{y_2}{d_2^3} \right]$$

Using OZ, one can obtain the corresponding $g_{ij}(r)$. Throop and Bearman [21] have numerically inverted the Laplace transform for the rdf. Ashcroft and Langreth [22] gave the Laplace transforms of the dcf, which are useful in numerical calculations.

## MONTE CARLO SIMULATIONS

Monte Carlo simulations for the rdf of hard sphere mixtures have been carried out by Smith and Lea [23], Lee and Levesque [24], Rotenberg [25], Fries and Hansen [26], and Alder [27]. The results are presented in Table VIII.6. The MC pressure can be approximated by the CS rule:

$$\beta P^{CS} = \frac{1}{3} \beta P^v + \frac{2}{3} \beta P^c \qquad\qquad (6.8)$$

### Equation of State (CS)

$$\frac{\beta P^{CS}}{\rho kT} = \frac{1 + y + y^2}{(1-y)^3} - \frac{3y(x_1 + x_2 y) - x_3 y^3}{(1-y)^3} \qquad\qquad (6.9)$$

where

$$y = \sum_{i=1}^{c} \frac{\pi}{6} \rho_i d_i^3, \qquad\qquad \rho = \sum_{i=1}^{n} \rho_i \qquad\qquad (6.10)$$

$$x_1 = \sum_{i<j}^{c} \frac{s_{ij}(d_i + d_j)}{\sqrt{d_i d_j}}, \qquad\qquad x_2 = \sum_{i<j}^{c} s_{ij} \sum_{k=1}^{n} \frac{y_k}{y} \frac{\sqrt{d_i d_j}}{d_k}$$

Table VIII.6. Radial Distribution Functions for Hard Sphere Mixtures [24]

| $y = 0.490$ | | $x_1 = 0.50$ | | $d_{22}/d_{11} = 0.90$ | |
|---|---|---|---|---|---|
| $r$ | $g_{11}(r)$ | $s$ | $g_{12}(s)$ | $t$ | $g_{22}(t)$ |
| 0·910 | 0·000000 | 0·920 | 0·000000 | 0·910 | 4·950301 |
| 0·930 | 0·000000 | 0·940 | 0·000000 | 0·930 | 4·153722 |
| 0·950 | 0·000000 | 0·960 | 5·297763 | 0·950 | 3·506222 |
| 0·970 | 0·000000 | 0·980 | 4·405256 | 0·970 | 2·969437 |
| 0·990 | 0·000000 | 1·000 | 3·691420 | 0·990 | 2·562835 |
| 1·010 | 5·467156 | 1·020 | 3·110376 | 1·010 | 2·205385 |
| 1·030 | 4·525668 | 1·040 | 2·644135 | 1·030 | 1·910863 |
| 1·050 | 3·733440 | 1·060 | 2·265082 | 1·050 | 1·684950 |
| 1·070 | 3·130888 | 1·080 | 1·950900 | 1·070 | 1·479275 |
| 1·090 | 2·650246 | 1·100 | 1·714355 | 1·090 | 1·319804 |
| 1·110 | 2·246668 | 1·120 | 1·501557 | 1·110 | 1·183028 |
| 1·130 | 1·955819 | 1·140 | 1·330548 | 1·130 | 1·070782 |
| 1·150 | 1·672192 | 1·160 | 1·186620 | 1·150 | 0·983578 |
| 1·170 | 1·467301 | 1·180 | 1·062714 | 1·170 | 0·907777 |
| 1·190 | 1·298689 | 1·200 | 0·966696 | 1·190 | 0·845996 |
| 1·210 | 1·158013 | 1·220 | 0·887866 | 1·210 | 0·792112 |
| 1·230 | 1·047607 | 1·240 | 0·825163 | 1·230 | 0·746009 |
| 1·250 | 0·953954 | 1·260 | 0·775258 | 1·250 | 0·710943 |
| 1·270 | 0·874784 | 1·280 | 0·731637 | 1·270 | 0·678409 |
| 1·290 | 0·799467 | 1·300 | 0·689433 | 1·290 | 0·654284 |
| 1·310 | 0·753123 | 1·320 | 0·663160 | 1·310 | 0·634558 |
| 1·330 | 0·693494 | 1·340 | 0·639584 | 1·330 | 0·619434 |
| 1·350 | 0·655205 | 1·360 | 0·624304 | 1·350 | 0·608897 |
| 1·370 | 0·625933 | 1·380 | 0·615554 | 1·370 | 0·606643 |
| 1·410 | 0·610193 | 1·400 | 0·607526 | 1·390 | 0·608198 |
| 1·390 | 0·592129 | 1·420 | 0·608560 | 1·410 | 0·619754 |
| 1·430 | 0·578785 | 1·440 | 0·607915 | 1·430 | 0·629550 |
| 1·450 | 0·570310 | 1·460 | 0·614554 | 1·450 | 0·603628 |
| 1·470 | 0·570258 | 1·480 | 0·614926 | 1·470 | 0·648943 |
| 1·490 | 0·569295 | 1·500 | 0·627931 | 1·490 | 0·671143 |
| 1·510 | 0·573956 | 1·520 | 0·648699 | 1·510 | 0·696406 |
| 1·530 | 0·579615 | 1·540 | 0·670674 | 1·530 | 0·727273 |
| 1·550 | 0·597017 | 1·560 | 0·696514 | 1·550 | 0·760868 |
| 1·570 | 0·622170 | 1·580 | 0·721721 | 1·570 | 0·796138 |
| 1·590 | 0·648730 | 1·600 | 0·759013 | 1·590 | 0·840433 |
| 1·610 | 0·678706 | 1·620 | 0·802139 | 1·610 | 0·888517 |
| 1·630 | 0·717053 | 1·640 | 0·853351 | 1·630 | 0·940703 |
| 1·650 | 0·756306 | 1·660 | 0·901108 | 1·650 | 0·981239 |
| 1·670 | 0·796444 | 1·680 | 0·950356 | 1·670 | 1·031500 |
| 1·690 | 0·845083 | 1·700 | 1·000735 | 1·690 | 1·073678 |
| 1·710 | 0·893900 | 1·720 | 1·049477 | 1·710 | 1·111178 |
| 1·730 | 0·949496 | 1·740 | 1·095199 | 1·730 | 1·143979 |
| 1·750 | 1·015735 | 1·760 | 1·134290 | 1·750 | 1·163624 |
| 1·770 | 1·064278 | 1·780 | 1·168159 | 1·770 | 1·180207 |
| 1·790 | 1·107069 | 1·800 | 1·192587 | 1·790 | 1·205820 |
| 1·810 | 1·150530 | 1·820 | 1·212631 | 1·810 | 1·230695 |
| 1·830 | 1·187415 | 1·840 | 1·235085 | 1·830 | 1·243334 |
| 1·850 | 1·215814 | 1·860 | 1·257873 | 1·850 | 1·254651 |
| 1·870 | 1·243949 | 1·880 | 1·267429 | 1·870 | 1·254281 |
| 1·890 | 1·269845 | 1·900 | 1·270869 | 1·890 | 1·268779 |
| 1·910 | 1·284912 | 1·920 | 1·269237 | 1·910 | 1·230966 |
| 1·930 | 1·297389 | 1·940 | 1·269143 | 1·930 | 1·256002 |

$$x_3 = \left[ \sum_{i<j} \left[ \frac{y_i}{y} \right]^{2/3} \left[ \frac{\rho_i}{\rho} \right]^{1/3} \right]^3, \qquad S_{ij} = \frac{\sqrt{y_i y_j}}{y} \frac{(d_i - d_j)^2}{d_i d_j} \frac{\sqrt{\rho_i \rho_j}}{\rho}$$

### Contact Values of the RDF (CS)

The same CS rule can be applied to the contact values; i.e.,

$$g_{ij}^{CS}(d_{ij}) = \frac{1}{3} g_{ij}^v(d_{ij}) + \frac{2}{3} g_{ij}^c(d_{ij}) \tag{6.11}$$

where the $g_{ij}^v(d_{ij})$ values are the PY results of eqs. (6.3). The compressibility values $g_{ij}^c(d_{ij})$ are obtained from the scaled particle theory (SPT) of Lebowitz et al. [28]

$$g_{ij}^c(d_{ij}) = \frac{1}{1-y} + \frac{6z}{(1-y)^2} \frac{d_i d_j}{2(d_i + d_j)} + 12 \frac{z^2}{(1-y)^3} \left[ \frac{d_i d_j}{2(d_i + d_j)} \right]^2 \tag{6.12}$$

where, $z = \sum_{i=1}^{c} \rho_i d_i^2$. The contact values thus obtained have been tested against the MC results by Lee et al. [29] for different diameter ratios. Close agreement (within 4%) was obtained (see Table VIII.7).

*Table VIII.7. Hard Sphere Contact Values from the Carnahan-Starling Equation*

| | Conditions | | CS Rule | | | Monte Carlo Results | | |
|------|------------|-------|--------------|--------------|--------------|--------------|--------------|--------------|
| $y$ | $d_{22}/d_{11}$ | $x_1$ | $g_{11}(d_{11})$ | $g_{12}(d_{12})$ | $g_{22}(d_{22})$ | $g_{11}(d_{11})$ | $g_{12}(d_{12})$ | $g_{22}(d_{22})$ |
| 0.49 | 0.9 | 0.5 | 5.91 | 5.65 | 5.42 | 5.91 | 5.84 | 5.47 |
| 0.49 | 0.3 | 0.5 | 5.97 | 3.56 | 2.95 | 6.18 | 3.50 | 3.06 |
| 0.47 | 0.9 | 0.25 | 5.44 | 5.21 | 5.01 | 5.28 | 5.29 | 4.94 |

### Helmholtz Free Energy (CS)

Integration of the equation of state (6.9) yields

$$\frac{A^{CS} - A^*}{NkT} = \frac{3 - 3x_1 - 3x_2 - x_3}{2(1-y)^2} + \frac{3x_2 + 2x_3}{1-y} - (1-x_3)\ln(1-y) \tag{6.13}$$

$$- \frac{3}{2}(1 - x_1 + x_2 + x_3)$$

where A* is the ideal-gas value. The constants are the same as in (6.10).

## VIII.7. *Analytical Construction of the RDF for Hard Spheres*

### PURE HARD SPHERES

Since the amount of simulation results for the rdf of hard spheres is limited, it is desirable to have some analytical means of generating an accurate rdf at any conditions. The PY results are approximate and are not satisfactory at high densities. Verlet and

Weis (VW) [30] have proposed an analytical construction that is accurate at high densities. Their method is summarized in the following.

They compared the PY rdf with the simulation rdf of Alder and Hecht (Ibid.) and found the following deficiencies:

1.    At large $r$, the PY $g(r)$ oscillates out of phase with respect to the exact rdf.

2.    At contact ($r=d$), the PY $g(d)$ are uniformly too low.

To correct the first error, they adjusted the equivalent packing fraction $y_w$ in the Wertheim solution in such a way as to compensate for the difference in phase.

$$y_w = y - \frac{y^2}{16}$$                                                              (7.1)

where $y$ is the actual packing fraction of the system, $y=\pi\rho d^3/6$, and $y_w$ is a hypothetical packing fraction to be used in conjunction with the PY solution (4.4). The diameter in the PY $g(r/d_w)$ is related to the exact $d$ by

$$\frac{d_w^3}{y_w} = \frac{d^3}{y}$$                                                       (7.2)

For $r>1.6d$, the newly constructed PY $g_w(r/d_w)$ oscillates in phase with exact rdf. However, (7.1) lowers the effective density, $y_w<y$, and this makes the second discrepancy worse. VW added a correction term $\Delta g(r)$ to $g_w(r/d_w)$:

$$g(\frac{r}{d};y) = g_w(\frac{r}{d_w};y_w) + \Delta g(r)$$                                 (7.3)

where $\Delta g(r)$ has the parametric form

$$\Delta g(r) = \frac{A}{r}\exp[-\mu(r-d)]\cos\mu(r-d)$$                                    (7.4)

and $A$ and $\mu$ are constants to be determined. They are correlated with the packing fraction as

$$A = \frac{3}{4}\frac{y_w^2(1-0.7117y_w-0.114y_w^2)}{(1-y_w)^4}$$                          (7.5)

$$\mu d = \frac{24A/d}{y_w g_w(1;y_w)}$$                                                    (7.6)

where $g_w$ is evaluated at the contact, i.e. $r/d_w=1$. This construction is found to be very successful in reproducing the exact rdf to within 1% in error. The HS rdf is needed in a number of successful perturbation theories of liquids. This analytical method is very useful.

## HARD SPHERE MIXTURES

Lee and Levesque [31] and Henderson and Grundke [32] have generalized the above method to mixtures. The analytical solution for HS mixtures in the PY theory was given by Lebowitz (*Ibid.*). A procedure similar to VW can be formulated.

1. To obtain agreement at large $r$, the equivalent packing fraction $y_e$ is given by

$$y_e = y - \frac{y^2}{16} \tag{7.7}$$

This $y_e$ is to be used in the Lebowitz solution. $y$ is the actual packing fraction, $y = \sum \pi \rho_i d_i^3 / 6$. The hard sphere of diameter $d_1$ is taken as reference. The equivalent diameter $d_{1e}$ is calculated from

$$\frac{d_{1e}^3}{y_e} = \frac{d_1^3}{y} \tag{7.8}$$

For the other diameters, $j \neq 1$,

$$\frac{d_{je}}{d_{1e}} = \frac{d_j}{d_1}, \qquad j=2,3,...,c \tag{7.9}$$

As to the rule (7.7), there is no reason to expect that the same $y_e$ would give the best fit to all rdf, $g_{11}$, $g_{22}$, and $g_{12}$, etc. Lee and Levesque (*Ibid.*) tested (7.7) with respect to MC data. The choice of $y_e = y - y^2/16$ did not minimize the root mean squared differences between all $g_{11}$, $g_{12}$, and $g_{22}$ and their MC counterparts. Three $y_e$'s would have been required. However, (7.7) did not do injustice to $g_{12}$ and $g_{22}$. To simplify matters, a single $y_e$ value, consistent with the pure case, was chosen.

2. The correction of the core values was achieved by attaching a $\Delta g_{ij}(r)$ term:

$$g_{ij}(\frac{r}{d_1};y) = g_{ij}^{PY}(\frac{r}{d_{1e}};y_e) + \Delta g_{ij}(r) \tag{7.10}$$

where $i,j=1,2,...,c$, and

$$\Delta g_{ij}(r) = \frac{A_{ij}}{r} \exp[-\mu_{ij}(r - d_{ij})] \cos \mu_{ij}(r - d_{ij}) \tag{7.11}$$

with $d_{ij} = (d_i + d_j)/2$. To obtain a good fit, it was found that

$$\frac{A_{ij}}{d_{ij}} = g_{ij}^{CS}(d_{ij}) - g_{ij}^{PY}(\frac{d_{ij}}{d_{1e}};y_e) \tag{7.12}$$

and

$$\mu_{ij}d_{ij} = \frac{24A_{ij}/d_{ij}}{y_e g_{ij}^{PY}(d_{ije}/d_{1e};y_e)} \tag{7.13}$$

yield an accurate rdf at contact and for all $r$. $g_{ij}^{CS}$ is obtained via (6.11) at $r = d_{ij}$ and the actual packing fraction $y$. $g_{ij}^{PY}$ is the contact value at $r = d_{ije}$ from the Lebowitz solution (6.3) at $y_e$. Table VIII.7 shows the agreement obtained by the above method with the MC results. Thus for both pure and mixture HS, we are in a position to construct accurate rdfs at any state conditions.

## VIII.8. *Hard Convex Bodies: The Scaled Particle Theory*

The *cores* of some molecular fluids could be represented by simple geometrical shapes. For example, benzene molecules would have a hexagonal core, carbon tetrachloride a tetrahedral core, and ethylene is approximately dumbbell-like. However, there are molecules that do not have a simple representation, such as n-decane. For hard bodies of nonspherical shapes, such as ellipsoids, spherocylinders, tetrahedra, the scaled particle theory (SPT) developed by Reiss, Frisch, and Lebowitz [33] could be applied. This theory is based on an interpretation of the chemical potential as the reversible work required for inserting a solute molecule of given type into a bath of solvent molecules. Gibbons [34] has applied SPT to convex bodies. We summarize some of the results below.

A convex body is a geometrical volume whereby any line connecting two points of the body lies entirely in the same body. For example, a spherocylinder is convex, whereas a dumbbell is not. Gibbons and Boublik [35] derived the equations of state for hard convex bodies.

### EQUATION OF STATE

$$\frac{P}{\rho kT} = \frac{1}{1-\rho b} + \frac{\bar{r}s\rho}{(1-\rho b)^2} + \frac{1}{3}\frac{cs^2\rho^2}{(1-\rho b)^3} \tag{8.1}$$

where $\rho$ is the number density, $\bar{r}$ is the mean radius of curvature of the convex core, $s$ is the total surface area of a particle, $b$ is the volume of the convex core, and $c = \bar{r}^2$. Equation (8.1) is an approximate expression for hard cores. It gives good agreement at low densities with MC results. It gives exact second virial coefficients. For mixtures, the four geometric factors are given by

$$b = \sum x_i b_i, \qquad \bar{r} = \sum x_i \bar{r}_i, \qquad s = \sum x_i s_i, \qquad c = \sum x_i \bar{r}_i^2 \tag{8.2}$$

where $b_i$, $\bar{r}_i$, $s_i$ are the volume, mean radius of curvature, and surface area of the hard core of type $i$, respectively. For different hard-core shapes, their values are given in Table VIII.8. Thus for spheres, the volume $b_i = \pi d^3/6$, the surface area $s_i = \pi d^2$, and the mean radius of curvature is simply the radius of the sphere $\bar{r} = d/2$, $d$ being the diameter of the sphere. For hard spheres, eq. (8.1) reduces to the compressibility equation (4.1). Therefore the SPT theory is consistent with the compressibility approach.

Recently it has been found that for spherocylinders the SPT results overestimate higher virial coefficients. Improved formulae were proposed. We shall discuss the exact machine results in the next section. Here we present the thermodynamic properties base on eq. (8.1).

Table VIII.8. Geometric Factors for Convex Bodies

| Shape | Volume (b) | Surface Area (s) | Mean Radius of Curvature $(\bar{r})$ |
|---|---|---|---|
| **Sphere** (radius $=r$) | $\dfrac{4\pi r^3}{3}$ | $4\pi r^2$ | $r$ |
| **Parallelopiped** (Rectangular, with sides $x,y,z$) | $xyz$ | $2(xy+yz+zx)$ | $(x+y+z)/4$ |
| **Tetrahedron** (Regular; with side $= x$) | $\dfrac{x^3}{6\sqrt{2}}$ | $\sqrt{3}x^2$ | $3x\,\tan^{-1}(\sqrt{2})/(2\pi)$ |
| **Octahedron** (Regular; with side $= x$) | $\dfrac{x^3}{3\sqrt{2}}$ | $2\sqrt{3}x^2$ | $3x\,\cot^{-1}(\sqrt{2})/\pi$ |
| **Cylinder** (radius $= r$, side $= x$) | $\pi r^2 x$ | $2\pi r(r+x)$ | $(\pi r +x)/4$ |
| **Spherocylinder** (with radius $=r$ side $=x$) | $\dfrac{4}{3}\pi r^3 +\pi r^2 x$ | $4\pi r^2 +2\pi rx$ | $r +x/4$ |

## HELMHOLTZ FREE ENERGY

$$\frac{A^{\text{HC}} - A^*}{NkT} = - \ln(1 -\rho b) + \frac{\bar{r}s}{\bar{v}} \frac{1}{1 -\rho b} \tag{8.3}$$

$$+ \frac{cs^2}{3b^3}\left[\frac{1/2}{(1 -\rho b)^2} - \frac{1}{1 -\rho b}\right] - \frac{\bar{r}s}{b} + \frac{cs^2}{6b^2}$$

where $A^{\text{HC}}$ is the hard-core Helmholtz free energy and $A^*$ is the ideal-gas value.

## INTERNAL ENERGY

$$U = \frac{3}{2} NkT \tag{8.4}$$

## ENTHALPY

$$\frac{H}{NkT} = \frac{3}{2} + \frac{1}{1 -\rho b} + \frac{\bar{r}s\rho}{(1 -\rho b)^2} + \frac{1}{3} \frac{cs^2\rho^2}{(1 -\rho b)^3} \tag{8.5}$$

**ENTROPY**

$$\frac{S}{Nk} = \frac{5}{2} - \ln\frac{\rho\Lambda^3}{1-\rho b} - \frac{\bar{r}s}{b}\frac{1}{1-\rho b} \qquad (8.6)$$

$$-\frac{cs^2}{3b^2}\left[\frac{1/2}{(1-\rho b)^2} - \frac{1}{1-\rho b}\right] + \frac{\bar{r}s}{b} - \frac{cs^2}{6b^2}$$

**CHEMICAL POTENTIAL**

$$\beta\mu_i = \ln(\rho_i b_i) - \ln(1-\rho b) + \frac{\rho}{1-\rho b}(\bar{r}_i s + s_i \bar{r} + b_i) \qquad (8.7)$$

$$+ \frac{\rho^2}{2(1-\rho b)^2}[\bar{r}_i^2 s^2 + 2b_i \bar{r}s] + \frac{\rho^3 b_i cs^2}{3(1-\rho b)^3}$$

## VIII.9. *Hard Convex Bodies: Simulation Results*

A number of simulation results have been published on the virial coefficients of nonspherical molecules. MC calculations have been reported on convex bodies by Monson and Rigby [36], Nezbeda [37], Wojcik and Gubbins [38], and others (see review by Boublik and Nezbeda [39]). Molecular dynamics was performed by Robertus and Sando [40]. Due to the discontinuous nature of hard body potentials, the MD method is quite complex. Recently, it has been combined with MC in a hybrid method by Erpenbeck and Wood [41] for hard spheres for efficient sampling. Most of the results reported below are on hard spherocylinders. However, Nezbeda and Boublik [42] have investigated the equivalences among general convex bodies. They found that the HCBs form classes, such as linear molecules, disk-like molecules, and cubes. A principle could be valid for all HCBs within a class of shapes, if not across classes. We present some of the results below for the class of spherocylinders.

The dimensions of a spherocylinder (i.e., a cylinder with two hemispheres capped on both ends) are characterized by the diameter $d$ of the hemispheres and the height $L$ (from the center of one hemisphere to the center of the other). In comparison with the MC data on virial coefficients, the equation of Gibbons (8.1) is inaccurate for higher virial coefficients. Boublik [43] modified the SPT expression so that an equation of state similar to the CS equation for hard spheres was obtained.

**EQUATION OF STATE (Boublik)**

$$\frac{P}{\rho kT} = \frac{1}{1-y} + \frac{3\alpha y}{(1-y)^2} + \frac{3\alpha^2 y^2}{(1-y)^3} - \frac{\alpha^2 y^3}{(1-y)^3} \qquad (9.1)$$

where the geometric factor $\alpha$ is defined as

$$\alpha \equiv \frac{\bar{r}s}{3b} \qquad (9.2)$$

$$y = \rho b \qquad (9.3)$$

and $\overline{r}$, s, and b are as defined before. $y$ has the same physical meaning of packing fraction: the fraction of the volume occupied by the *bodies* of the molecules. The same equation (9.1) could be applied to mixtures when (8.2) is used for the geometric factors $\overline{r}$, s, c, and b. In eq. (9.1), the first three terms are Gibbons equation (8.1). The last term is a correction term. Therefore (9.1) will reduce to the CS equation (5.1) for hard spheres with $\alpha = 1$. The validity of (9.1) is demonstrated in Table VIII.9.

Equation (9.1) is an approximate equation. Another version of the equation of state for hard spherocylinders was proposed by Nezbeda [44].

**EQUATION OF STATE (Nezbeda)**

$$\frac{P}{\rho kT} = \frac{1}{1-y} + \frac{3\alpha y}{(1-y)^2} + \frac{(\alpha^2 + 4\alpha - 2)y^2}{(1-y)^3} - \frac{(5\alpha^2 - 4\alpha)y^3}{(1-y)^3} \tag{9.4}$$

It gives accurate pressures for $L/d$ up to 2 (see Table VIII.9).

Monson and Rigby (*Ibid.*) have simulated the spherocylinders to obtain the radial distribution functions in terms of their spherical harmonics coefficients $g_{LL'}{}^M$. Vieillard-Baron [45] using MC examined the anisotropy-isotropy phase transition for elongated spherocylinders, similar to the nematic phase transition in liquid crystals. The above results for spherocylinders are applicable to the class of linear molecules.

## VIII.10. *The Interaction Site Model for Fused Hard Spheres*

As remarked in Chapter IV, the interaction between two structured polyatomic molecules can be considered as the sum of interactions among the constituent atoms in the two molecules. This point of view is called the interaction site model (ISM) of molecular fluids. For a polyatomic molecule composed of $m$ hard spheres that have fused together in part or in whole, the interaction energy between molecules 1 and 2 can be written as the sum of the interactions of all sites (in this case, the centers of hard spheres) in molecule 1 with all sites (also the centers of hard spheres) in molecule 2. With the notation of Chapter IV, the pair potential $u(12)$ between the polyatomic 1 and the polyatomic 2 is given by

*Table VIII.9. Pressure of Prolate Hard Spherocylinders*

| $L^*$ | $y$ | Compressibility, $\beta P/\rho$ | | | |
| | | MC | Nezbeda | Boublik | Gibbons |
|---|---|---|---|---|---|
| 1.0 | 0.2 | 2.65±0.02 | 2.67 | 2.69 | 2.71 |
| | 0.2454 | 3.37±0.04 | 3.39 | 3.43 | 3.48 |
| | 0.3 | 4.48±0.07 | 4.56 | 4.65 | 4.77 |
| | 0.4 | 8.2±0.1 | 8.1 | 8.44 | 8.87 |
| | 0.5 | 15.2±0.2 | 15.3 | 16.4 | 17.84 |
| 2.0 | 0.2 | 3.07±0.03 | 3.06 | 3.15 | 3.18 |
| | 0.3 | 5.4±0.1 | 5.4 | 5.77 | 5.95 |
| | 0.35 | 7.17 | 7.24 | 7.93 | 8.28 |
| | 0.4 | 9.6 | 9.74 | 11.0 | 11.67 |
| | 0.45 | 13.0 | 13.25 | 15.5 | 16.73 |
| | 0.5 | 18.0 | 18.25 | 22.25 | 24.50 |

$$u(12) = \sum_{\alpha=1}^{m}\sum_{\gamma=1}^{m} u_{\alpha\gamma}(|\mathbf{r}_1^{\alpha} - \mathbf{r}_2^{\gamma}|) \tag{10.1}$$

For fused hard spheres, the site-site interaction is impulsive:

$$u_{\alpha\gamma}(r) = +\infty, \qquad\qquad r \le d_{cl}^{\alpha\gamma} \tag{10.2}$$

$$= 0, \qquad\qquad r > d_{cl}^{\alpha\gamma}$$

where $d_{cl}^{\alpha\gamma}$ is the distance of closest approach between site $\alpha$ of molecule 1 and site $\gamma$ of molecule 2. It depends, in general, on the overall geometry of the pair of molecules (e.g., orientations and structure of the polyatomics), and not just on the diameters of the respective sites ($d_{\alpha}$ and $d_{\gamma}$). Consider for example a polyatomic with the geometry of a neopentane molecule. The closest approach between the central carbon atom in one molecule and the central carbon atom in another is not the same as those for other carbon atoms. The obstruction of the methyl groups must be taken into account.

As before, the site-site pair correlation function $g_{\alpha\gamma}(\mathbf{r}, \mathbf{r'})$ (ss pcf) is defined as

$$g_{\alpha\gamma}(\mathbf{r}, \mathbf{r'}) = \rho^{-2}\left[\sum_{i<}^{N}\sum_{j} \delta(\mathbf{r}, \mathbf{r}_i^{\alpha})\delta(\mathbf{r'}, \mathbf{r}_j^{\gamma})\right] \tag{10.3}$$

In addition, there is an *intra*-molecular structure factor $\omega_{\alpha\gamma}$ whose Fourier transform is given by (see eq. (IV.8.14))

$$\tilde{\omega}_{\alpha\gamma}(k) = \frac{1}{4\pi} \int d\omega\, e^{ik(L_1^{\alpha}-L_1^{\gamma})} = \frac{\sin(kL_{\alpha\gamma})}{kL_{\alpha\gamma}} \tag{10.4}$$

In real space,

$$\omega_{\alpha\gamma}(r) = \delta_{\alpha\gamma}\,\delta_D(r,0) + (1-\delta_{\alpha\gamma})s_{\alpha\gamma}(r) \tag{10.5}$$

and

$$s_{\alpha\gamma}(r) = \frac{1}{4\pi L_{\alpha\gamma}^2}\, \delta_D(r, L_{\alpha\gamma}) \tag{10.6}$$

where $L_{\alpha\gamma}$ is the distance between site $\alpha$ and site $\gamma$ in the same molecule, $L_{\alpha\gamma} = |L_1^{\alpha} - L_1^{\gamma}|$, $\delta_{\alpha\gamma}$ is the Kronecker delta, and $\delta_D(r,0)$ is the Dirac delta. Equation (10.4) is valid for rigidly bonded molecules where bond stretching and bond angle rotations are forbidden. $\omega$ is the Euler angle spanned by the bond $L_1^{\alpha} - L_1^{\gamma}$. For flexible molecules, the intramolecular structure factor should be appropriately modified.

A third function to be defined in the site-site interaction approach is the site-site direct correlation function (ss dcf) through an OZ-like equation

$$h_{\alpha\delta}(\mathbf{r}_1^{\alpha}, \mathbf{r}_4^{\delta}) = \sum_{\beta=1}^{m}\sum_{\gamma=1}^{m}\int d\mathbf{r}_2^{\beta}\, d\mathbf{r}_3^{\gamma}\, \omega_{\alpha\beta}(\mathbf{r}_1^{\alpha}, \mathbf{r}_2^{\beta})C_{\beta\gamma}(\mathbf{r}_2^{\beta}, \mathbf{r}_3^{\gamma})\omega_{\gamma\delta}(\mathbf{r}_3^{\gamma}, \mathbf{r}_4^{\delta}) \tag{10.7}$$

$$+ \sum_{\beta=1}^{m}\sum_{\gamma=1}^{m}\int dr_2^\beta \, dr_3^\gamma \, \omega_{\alpha\beta}(r_1^\alpha, \, r_2^\beta)C_{\beta\gamma}(r_2^\beta, \, r_3^\gamma)h_{\gamma\delta}(r_3^\gamma, \, r_4^\delta)$$

where $h_{\alpha\delta}(r) = g_{\alpha\delta}(r) -1$. This equation is the OZ relation according to RISM (reference interaction site model) of Chandler and Andersen [46], and is one of many possible alternative formulations of OZ (see Chapter XIV).

Given a particular polyatomic, the solution for $h_{\alpha\gamma}$ could be obtained after imposing a closure condition on $C_{\alpha\gamma}$ and $h_{\alpha\gamma}$. Chandler et al. proposed the so-called reference interaction site approximation (RISA)

$$h_{\alpha\gamma}(r) = -1, \qquad\qquad r<d_{cl}^{\alpha\gamma} \qquad\qquad (10.8)$$

and

$$C_{\alpha\gamma}(r) \approx 0, \qquad\qquad r>d_{cl}^{\alpha\gamma} \qquad\qquad (10.9)$$

Relation (10.8) is exact from the definitions of $g_{\alpha\gamma}$. The second closure is similar to the MSA. As a further simplification, we assume that the closest approach among a pair of sites is $d_{cl}^{\alpha\gamma} = (d_\alpha + d_\gamma)/2$. Equations (10.7, 8 and 9) form a complete set of equations for the determination of $h_{\alpha\gamma}$ and $C_{\alpha\gamma}$. There are a number of ways of solving the equations: e.g., the method of functional variations and the Wiener-Hopf method. We discuss the variational method below.

## VARIATIONAL SOLUTION OF RISM

In this method, a functional of the ss dcf is constructed, and the solution $C_{\alpha\gamma}$ is the one that makes the functional stationary. Chandler et al. considered the functional

$$F_{RISM} \equiv \rho^2 \sum_{\alpha=1}^{m}\sum_{\gamma=1}^{m} \tilde{C}_{\alpha\gamma}(0) \qquad\qquad (10.10)$$

$$- \frac{1}{(2\pi)^3}\int d\mathbf{k}\left\{\mathrm{Tr}(\rho\tilde{\omega}(k)\tilde{C}(k)) + \ln \det[I - \rho\tilde{\omega}(k)\tilde{C}(k)]\right\}$$

This functional was constructed in such a way that its derivative with respect to the ss dcf gives the RISM pcf

$$\frac{\delta F_{RISM}}{\delta C_{\alpha\gamma}(\mathbf{r})} = \rho^2 g_{\alpha\gamma}(\mathbf{r}) \qquad\qquad (10.11)$$

When condition (10.8) is applied, we obtain

$$\frac{\delta F_{RISM}}{\delta C_{\alpha\gamma}(r)} = 0, \qquad\qquad r\leq d_{\alpha\gamma} \qquad\qquad (10.12)$$

The functional derivative $\delta F_{RISM}/\delta C\alpha\gamma$ is mathematically difficult to evaluate. In practice, it is replaced by partial derivatives with respect to the spectral coefficients of the projections of the ss dcf onto a set of basis functions. We shall consider the simplest case of a polyatomic, the diatomics, in the following section.

## VIII.11. *Hard Dumbbells*

We distinguish two types of hard dumbbells: the homonuclear hard dumbbells (i.e., two fused hard spheres of equal diameter) and heteronuclear hard dumbbells (two fused hard spheres of unequal diameters). The study of diatomic molecules, especially hard dumbbells, has advanced in recent years due to application of the ISM approach. Wertheim [47] has proposed a decomposition of the Mayer factor $f(12)$ in terms of its site-site components $f_{\alpha\gamma}$. It was possible to evaluate analytically all the graphs by making simplifying assumptions. The average Boltzmann factor $<e(12)>$ has been obtained for diatomics and special linear triatomics. On the other hand, the RISA equations (10.8 and 9) could be directly applied to diatomics. We discuss the homonuclear dumbbells first.

### HOMONUCLEAR HARD DUMBBELLS

To solve the RISA equation for symmetric diatomics, we need only to deal with one unique dcf, i.e., $C_{\alpha\alpha} = C_{\alpha\gamma} = C_{\gamma\gamma}$, since both sites are equivalent. First we represent the dcf in terms of basis functions $\phi_{ij}(r)$:

$$C_{\alpha\alpha}(r) = \sum_{i=1}^{n} \sum_{j=1}^{3} a_{ij}\phi_{ij}(r), \qquad\qquad r \leq d \qquad\qquad (11.1)$$

where

$$\phi_{ij}(r) \equiv H(x_j - r)\left[\frac{r - x_j}{d}\right]^{i-1} \qquad\qquad (11.2)$$

and $H(y)$ is the Heaviside function (i.e., $H(y) = 0$ for $y<0$, and $H(y) = 1$ for $y>0$). The parameters are $x_1 = d$, the diameter of a hard sphere (in the dumbbell), $x_2 = \min(d, L)$, $L$ being the bond length connecting the two sites, and $x_3 = \min(d, |d-L|)$. This choice of the $\phi_{ij}$ functions introduces naturally discontinuities of $C_{\alpha\alpha}$ at $r=d$, $r=L$, and $r=|d-L|$. $n$ assumes any value required for the solution of the RISA equations. In practice, $n=4$ is found to be sufficient. This gives totally twelve terms in the expansion (11.1). The variational principle (10.12) is equivalent to the requirements

$$\frac{\partial F_{RISM}}{\partial a_{ij}} = 0, \qquad\qquad 1\leq i\leq n, \qquad 1\leq j\leq 3 \qquad\qquad (11.3)$$

The coefficients $a_{ij}$ are determined from (11.3). The behavior of $C_{\alpha\alpha}(r)$ is depicted in Figure VIII.2. It is observed that $C_{\alpha\alpha}(r)$ has two cusps at $r=x_2=L$ ($L=0.3d$), and $r=x_3=d-L$. For $r>d$, $C_{\alpha\alpha}=0$, as required by the RISA conditions. The $C_{\alpha\alpha}(r)$ obtained is used to calculate the ss pcf $g_{\alpha\alpha}(r)$ via the OZ relation (10.7) or in Fourier space,

$$\tilde{h}_{\alpha\alpha}(k) = \frac{[1 + \tilde{\omega}(k)]^2 \tilde{C}_{\alpha\alpha}(k)}{1 - 2\rho[1 + \tilde{\omega}(k)]\tilde{C}_{\alpha\alpha}(k)} \qquad\qquad (11.4)$$

where

$$\tilde{C}_{\alpha\alpha}(r) = \frac{4\pi}{k} \int\limits_{-\infty}^{+\infty} dr\ rC_{\alpha\alpha}(r)\sin(kr) \tag{11.5}$$

is the Fourier transform of $C_{\alpha\alpha}(r)$, and $\tilde{\omega}(k)$ is given by (10.4). $h_{\alpha\alpha}(r)$ is obtained from the inverse transform

$$h_{\alpha\alpha}(r) = \frac{1}{2\pi^2} \int\limits_{-\infty}^{+\infty} dk\ k\tilde{h}_{\alpha\alpha}(k)\sin(kr) \tag{11.6}$$

The results for $g_{\alpha\alpha}$ are given in Figures VIII.3 and 4 for low and high densities and for $L/d = 0.3$. We note that the RISM equations reduce to the PY theory for hard spheres at $L = 0$. At intermediate L, $(0<L<\infty)$, it will be seen that the RISM is not accurate at low densities. It is also less accurate at very high densities. Thus RISM is best suited for medium density states. Lowden and Chandler [48] gave the exact low density form of $g_{\alpha\gamma}$ as

$$\lim_{\rho\to 0} g_{\alpha\gamma}(r) = \int d1\ d2\ \delta_D(\mathbf{r}_1^\alpha, 0)\delta_D(\mathbf{r}_2^\gamma, \mathbf{r})\exp[-\beta\sum_{\mu,\nu} u_{\mu\nu}(|\mathbf{r}_1^\mu - \mathbf{r}_2^\nu|)] \tag{11.7}$$

where $\int d1$ is the integral over the position and averaged angles; i.e., $\int d1 = (4\pi)^{-1}\int d\mathbf{r}_1\ d\omega_1$, $\omega_1$ being the Euler angle. Figure VIII.3 shows that at $\rho d^3 = 0$, the RISM overestimates the exact structure (11.7). The curves at higher densities resemble those for atomic fluids except for a cusp at $r=d+L$. This is due to the excluded volume effect of the other site (which is at a distance $d+L$ from the given site). The contact values increase with density while the heights of the cusp increase from low densities to medium densities but decrease from $\rho d^3 = 0.4$ onward for $L/d=0.6$.

Subsequently, Chandler, Hsu and Streett [49] carried out MC simulation for hard symmetric dumbbells at $\rho d^3 = 0.9$ and $L/d = 0.6$. They compared the RISM results with the MC results. At this high density, the RISM curve underestimates the exact results, especially at contact, and at the cusp. Otherwise, RISM is qualitatively correct. Streett and Tildesley [50] have made MC simulation of symmetric and asymmetric hard dumbbells (Figure VIII.5). The condition is $L/d = 0.4$, and $\rho b = 0.78$ ($b$ is the volume of one hard dumbbell). $C_{\alpha\alpha}(r)$ in this case shows one submolecular cusp at $0.6d$ and one cusp at supermolecular distance $r=d+L=1.4d$. There is also a discontinuity at $r=d$, as expected. Two discrepancies of RISM are exposed: $C_{\alpha\alpha}(r)$ is nonzero beyond $r=d$; and there is no cusp at $r = L = 0.4d$. The former indicates that the RISM assumption, $C_{\alpha\alpha}(r) = 0$ for $r>d$, is not valid.

### Properties of Homonuclear Hard Dumbbells

We have discussed the thermodynamics of polyatomics in section V.12. In terms of the ss pcf $g_{\alpha\gamma}(r)$, the energy is given by (V.12.4). However, the virial pressure cannot be obtained from $g_{\alpha\gamma}(r)$ alone. Other angle dependent quantities are required (see eq. (V.12.8)). Recently, Nezbeda and Smith [51] have investigated the additional orientational correlation functions needed in the virial pressure equation. For linear molecules, the result is given in eq. (V.12.10). With symmetric hard dumbbells, it simplifies to

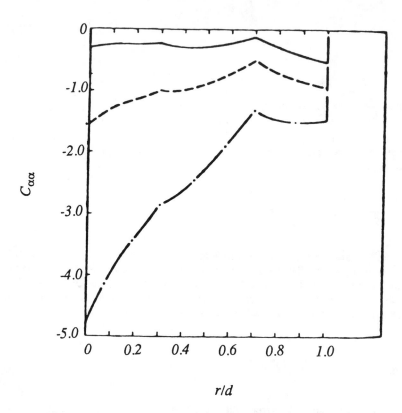

FIGURE VIII.2. Site-site direct correlation function $C_{\alpha\alpha}$ of symmetrical hard dumbbells as obtained from the RISA equation. The bond length is $L = 0.3d$. ———: zero density solution. - - - -: $\rho d^3 = 0.3$. — · — · —: $\rho d^3 = 0.5$. (Lowden and Chandler, J. Chem. Phys., 1973)

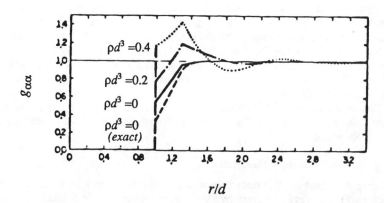

FIGURE VIII.3. Site-site pair correlation functions of symmetrical hard dumbbells at different densities from the RISA equation. The bond length is $L = 0.3d$. The dashed line is the exact low density solution. RISA does not give correct ss pcf at zero density. (Lowden and Chandler, J. Chem. Phys., 1973)

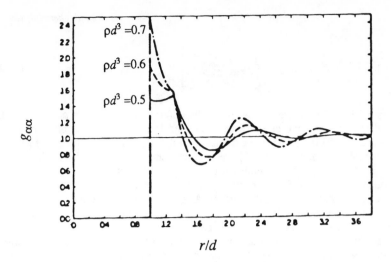

FIGURE VIII.4. Site-site pair correlation functions of hard dumbbells at moderate to high densities from the RISA equations. The bond length L = 0.3d. (Lowden and Chandler, J. Chem. Phys., 1973)

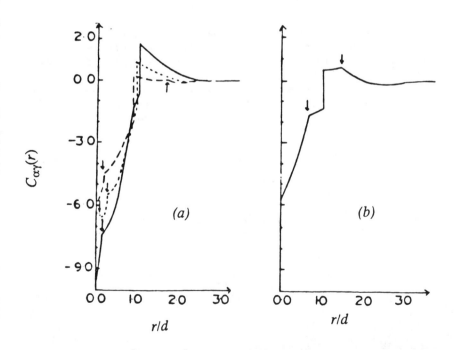

FIGURE VIII.5. Site-site direct correlation functions $C_{\alpha\gamma}$ for heteronuclear dumbbells (a); and homonuclear dumbbells (b). The arrows mark the positions of the cusps at the characteristic distances. ————: $C_{AA}$. -------: $C_{BB}$. · · · · · · : $C_{AB}$. (Streett and Tildesley, J. Chem. Phys., 1978)

$$\frac{\beta P}{\rho} = 1 + \frac{8\pi\rho}{3} d^2 \left\{ dg_{\alpha\alpha}(d) - \frac{1}{\sqrt{3}} [L_{\alpha\alpha} G^{\alpha\alpha}_{100}(d)] \right\} \qquad (11.8)$$

where $L_{\alpha\alpha}$ is the bond length (from one site to the other), and $d$ is the diameter of an HS atom. Recall that the pressure can also be calculated from the compressibility equation (V.12.5). Since the RISM gives approximate $g_{\alpha\gamma}(r)$, the pressures calculated from the compressibility (V.12.5) and from (11.8) are not necessary the same. Figure VIII.6 gives a comparison of the pressures obtained from (V.12.5) (denoted by $C$) and from an equivalent equation to (11.8) (denoted by $V$). The difference is appreciable at high densities, indicating the approximate nature of RISA for pressure calculation. We have also included the MC data of Freasier [52] at $L=0.6d$. We see that the $V$ (virial) curve is closer to the MC value than the $C$ (compressibility) curve.

A number of computer simulations have been carried out for both the thermodynamics and correlation functions of the homonuclear hard dumbbells. To summarize the simulation results, several equations of state have been proposed. Nezbeda [53] in 1977 recommended the equation

$$\frac{P}{\rho kT} = \frac{1}{1-y} + \frac{y}{(1-y)^2} \frac{9 + 27L^*/2 + 6L^{*2} + L^{*4}/2}{3(1 + 3L^*/2 - L^{*3}/2)} \qquad (11.9)$$

$$+ \frac{y^2}{3(1-y)^3} \left[ \frac{3 + 9L^*/2 + 2L^{*2} - L^{*4}/2}{1 + 3L^*/2 - L^{*3}/2} \right]^2$$

where $L^* = L/d$ is the reduced bond length. Prediction of the compressibility factor is compared with MC data in Table VIII.10. It is adequate at low densities (e.g., $y<0.2$) but deteriorates at higher densities. Boublik and Nezbeda [54] examined the relation between the equations for HCBs and those for dumbbells (which are nonconvex bodies). Equation (9.1) could be used for hard dumbbells if an *equivalent* spherocylinder could be found. The equivalence is achieved by requiring the geometric factor $\alpha$ to be derived from the actual volume $b$ and the actual surface area $s$ of the hard dumbbell:

$$b = \frac{\pi}{6} [1 + \frac{3}{2}L^* - \frac{1}{2}L^{*3}]d^3, \qquad s = \pi(1 + L^*)d^2 \qquad (11.10)$$

*Table VIII.10. Hard Dumbbells: Compressibility from MC Simulation*

| $L/d$ = $\rho d_e^3$ | 0.2 | 0.4 | Compressibility, $\beta P/\rho$ 0.6 | 0.8 | 1.0 |
|---|---|---|---|---|---|
| 0.2 | 1.56 | 1.59 | 1.63 (1.65)* | 1.70 | 1.83 (1.85) |
| 0.4 | 2.59 | 2.64 | 2.78 (2.84) | 3.01 | 3.36 (3.48) |
| 0.6 | 4.45 | 4.59 | 4.95 (5.14) | 5.48 | 6.40 (6.74) |
| 0.8 | 8.02 | 8.42 | 9.23 (9.92) | 10.54 | 12.64 (12.73) |
| 0.9 | 11.17 | 11.67 | 12.87 (–) | 14.88 | 18.06 (20.14) |

*The numbers in parentheses are from the equation of Nezbeda [53], eq. (11.9).

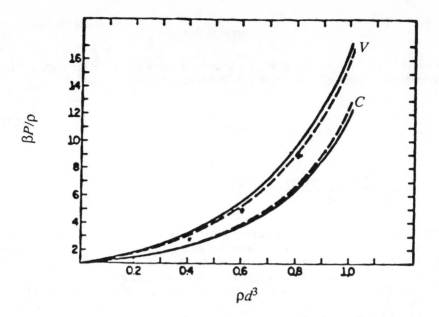

FIGURE VIII.6. *Compressibility factors* $\beta P/\rho$ *of hard dumbbells obtained from solutions of RISA.* $C$ = *pressure obtained from the compressibility equation;* $V$ = *pressure obtained from the virial equation.* *Solid line: bond length* $L = 0.3d$*; dashed line: bond length* $L = 0.6d$. •*: MC data.*

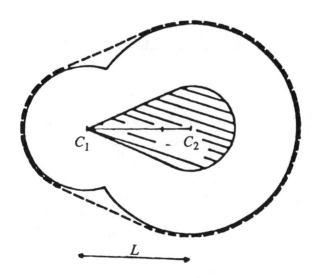

FIGURE VIII.7. *A circumscribed asymmetrical dumbbell. Dashed line depicts the circumscribing convex body. The shadowed area is the convex core. (Nezbeda and Boublik, Czech. J. Phys., 1977)*

where $L^*=L/d$. The mean radius of curvature is calculated from the corresponding "filled-in" hard dumbbell (see Figure VIII.7) treated as a spherocylinder

$$\bar{r} = \frac{d}{2}(1 + \frac{L^*}{2})$$                                                                 (11.11)

Thus

$$\alpha = \frac{\bar{r}s}{3b} = \frac{(1 + L^*)(2 + L^*)}{2 + 3L^* - L^{*3}}$$                                              (11.12)

The results are quite accurate, indicating great similarities between the spherocylinders and dumbbells (see Table VIII.11). From the general case of second virial coefficients, Boublik [55] has shown that a dumbbell could be approximated by a spherocylinder with the same surface areas $s$ and equal cylinder height and bond length $L$. The volume of the equivalent spherocylinder is increased by a factor of $\pi dL^3/12$. Then the radius of curvature $\bar{r}$ to use is the one from the spherocylinder. Since the second virial coefficient for hard bodies is half the average excluded volume $V_{excl.}$

$$2B_2 = V_{excl.} = 2(b + \bar{r}s) - \Delta b$$                                                          (11.13)

where $\Delta b$ is the volume correction noted above. It assumes different values for different polyatomics. This correction is usually very small in comparison with $b$ or $\bar{r}s$. Neglecting $\Delta b$ does not cause much loss in accuracy for hard dumbbells. In addition, Tildesley and Streett [56] also presented a semiempirical equation to summarize their simulation results on homonuclear hard dumbbells

$$\frac{P}{\rho kT} = \frac{1 + (1 + UL^* + VL^{*3})y + (1 + WL^* + XL^{*3})y^2 - (1 + YL^* + ZL^{*3})y^3}{(1 - y)^3}$$          (11.14)

Table VIII.11. *Hard Dumbbells and Hard Convex Bodies: Comparison of Pressures*

| $y$ | \multicolumn{5}{c}{Compressibility, $\beta P/\rho$} |||||
|---|---|---|---|---|---|
|  | MC | HCB, eq.(9.1) | Eq.(11.9) | RISM($v$) | RISM($c$) |
| $L^* =0.6$ |  |  |  |  |  |
| 0.1047 | 1.64 | 1.63 | 1.65 | 1.8 | 1.4 |
| 0.2094 | 2.84 | 2.77 | 2.84 | 3.2 | 2.2 |
| 0.3142 | 5.02 | 4.89 | 5.14 | 5.5 | 3.7 |
| 0.4189 | 9.24 | 9.15 | 9.92 | 9.8 | 6.4 |
| $L^* =1.0$ |  |  |  |  |  |
| 0.1047 | 1.83 | 1.80 | 1.85 |  |  |
| 0.2094 | 3.38 | 3.33 | 3.48 |  |  |
| 0.3142 | 6.37 | 6.31 | 6.74 |  |  |
| 0.4189 | 12.4 | 12.50 | 13.37 |  |  |
| 0.4712 | 17.5 | 18.02 | 20.14 |  |  |

where

$$U = 0.37836, \quad V = 1.07860, \quad W = 1.30376 \tag{11.15}$$

$$X = 1.80010, \quad Y = 2.39803, \quad Z = 0.35700$$

It fits the MC data on the compressibility to 0.4% on the average (with maximum error of 1.1%).

## HETERONUCLEAR HARD DUMBBELLS

The solution of the RISM for heteronuclear hard dumbbells has been carried out by Freasier, Jolly, and Bearman [57]. They also used the variations of the functional $F_{RISM}$ in (1.10). In this case, there are three distinct ss pcf $g_{\alpha\alpha}$, $g_{\alpha\gamma}$, and $g_{\gamma\gamma}$. The $C_{\alpha\gamma}(r)$s are expanded as

$$C_{\alpha\gamma}(r) = \sum_{i=1}^{4} \sum_{j=1}^{n^i_{\alpha\gamma}} H(x_{ij} - r) \left[ \left(\frac{r}{x_{ij}}\right) - 1 \right]^{i-1} a^{ij}_{\alpha\gamma} \tag{11.16}$$

where $a^{ij}_{\alpha\gamma}$ are the expansion coefficients, and $n^i_{\alpha\gamma}$ is the number of the $x_{ij}$ values for a particular $i$. This amounts to a power series expansion in $r$ for $C_{\alpha\gamma}$ between 0 and $d_{\alpha\gamma}$, modified to account for the discontinuities in the derivatives of $C_{\alpha\gamma}$ which are located at the points $x_{ij}$. Theoretically, the discontinuities are to be found from eq. (10.7). In practice, this is a difficult problem, even in the homonuclear case. Jolly et al. [58] found these $x_{ij}$ values by empirically inspecting the singularities in $h_{\alpha\gamma}$ and adding the locations of singularities successively to (11.16). Eventually 34 terms were used for (11.16). The variational equation (10.12) is replaced by

$$\frac{\partial F_{RISM}}{\partial a^{ij}_{\alpha\gamma}} = 0, \quad r < d_{\alpha\gamma} = (d_\alpha + d_\gamma)/2 \tag{11.17}$$

For $d_\gamma/d_\alpha = 1.5$ and $\rho d_e^3 = 0.7$ ($d_e$ is the equivalent diameter of a hard sphere with the volume of the dumbbell), the pcfs are qualitatively similar to the symmetrical case. Again the RISM results overestimate the exact structure at zero density. In comparison with MC results at higher densities, i.e., at $\rho d_e^3 = 0.9$ and $d_\alpha/d_\gamma = 0.790$ ($L/d_\gamma = 0.490$) and $d_\alpha/d_\gamma = 0.675$ ($L/d_\gamma = 0.346$), the RISM pcfs are quite reasonable.

The RISM method has been further applied to molecular fluids with structures simulating carbon tetrachloride, carbon disulfide, and benzene (see Lowden and Chandler *Ibid.*). Hard core RISM has been used recently as the reference system for structured molecules with ss LJ interactions [59]. Detailed discussions of the RISM approach are given in Chapter XIV.

### Properties of the Heteronuclear Hard Dumbbells

The equation of state for heteronuclear hard dumbbells is given by eq. (V.12.10). Jolly et al. [60] have made MC simulation of five models of the heteronuclear dumbbell with different diameter ratios. They compared the pressure obtained from the RISM compressibility equation (V.12.5) with the MC data. RISM consistently underestimates the MC data. The deviations reach −30% at $\rho d_e^3 = 0.7$.

Nezbeda and Boublik [61] proposed the Boublik eq. (9.1) to describe the pressure of hard dumbbells. They changed the hard heteronuclear dumbbell into a hard convex core by circumscribing a cone over the neck part of the dumbbell (see Figure VIII.7). The volume $b$ is the actual dumbbell volume:

$$b = \frac{\pi d_1^3}{12}[1 + \gamma^3 + 3(\gamma^2 a_+ + a_-) - 4(a_+^3 - a_-^3)] \tag{11.18}$$

The actual surface area is

$$s = \pi \gamma d_1^2(\frac{\gamma}{2} + a_+) + \pi d^2(\frac{1}{2} + a_-) \tag{11.19}$$

where

$$a_\pm = \frac{L^*}{2} \pm \frac{(\gamma^2 - 1)d_1}{8L}, \qquad \gamma = d_2/d_1 > 1, \qquad L^* = L/d_1 \tag{11.20}$$

The mean radius of curvature $\bar{r}$ is calculated as

$$\bar{r} = \frac{d_1}{4}\left[\gamma - 1 + L^* + \frac{(\gamma - 1)^2 d_1}{4L}\right] + \frac{d_1}{2} \tag{11.21}$$

$$\alpha = \frac{\bar{r}s}{3b} \tag{11.22}$$

When this $\alpha$ and $y = \rho V_m$ are substituted into eq. (9.1), good agreement with the MC data of Jolly et al. (*Ibid.*) is obtained (see Table VIII.12).

Table VIII.12. *Heteronuclear Hard Dumbbells: Pressures from MC*

| Model* | $d_2/d_1$ | $L/d_1$ | $\rho d_e^3$ | Compressibility, $\beta P/\rho$ MC | Eq.(9.1) filled dumbbells |
|--------|-----------|---------|--------------|------|---------------------------|
| I   | 1.5 | 0.75 | 0.4 | 2.75 | 2.69 |
| I   |     |      | 0.7 | 6.45 | 6.34 |
| II  | 1.5 | 0.5  | 0.4 | 2.66 | 2.57 |
| II  |     |      | 0.7 | 6.04 | 5.90 |
| III | 1.5 | 1.0  | 0.4 | 2.93 | 2.88 |
| III |     |      | 0.7 | 7.13 | 7.06 |
| IV  | 1.8 | 0.9  | 0.4 | 2.68 | 2.66 |
| IV  |     |      | 0.7 | 6.34 | 6.24 |
| V   | 1.2 | 0.6  | 0.4 | 2.78 | 2.70 |
| V   |     |      | 0.7 | 6.51 | 6.4  |

*Model numbers refer to the classifications of Jolly et al. [60].

# References

[1]  Z.W. Salsburg, R.W. Zwanzig, and J.G. Kirkwood, J. Chem. Phys. **21**, 1098 (1953).

[2]  L. Tonks, Phys. Rev. **50**, 955 (1963).

[3]  Lord Rayleigh, Nature, Lond., **45**, 80 (1891).

[4]  W.W. Wood, in *Physics of Simple Liquids*, edited by H.N.V. Temperley, J.S. Rowlinson, and G.S. Rushbrooke (North Holland, Amsterdam, 1968), Chapter 5.

[5]  F.H. Ree and W.G. Hoover, J. Chem. Phys. **40**, 939 (1964); *Ibid.* **46**, 4181 (1967).

[6]  D.G. Chae, F.H. Ree, and T. Ree, J. Chem. Phys. **50**, 1581 (1969).

[7]  D. Henderson, Mol. Phys. **30**, 971 (1975).

[8]  M.S. Wertheim, J. Math. Phys. **5**, 643 (1964).

[9]  E. Thiele, J. Chem. Phys. **39**, 474 (1963).

[10]  G.J. Throop and R.J. Bearman, J. Chem. Phys. **42**, 2408 (1965).

[11]  W.R. Smith and D. Henderson, Mol. Phys. **19**, 411 (1970).

[12]  S. Kim, Phys. Fluids **12**, 2046 (1969).

[13]  K.W. Kratky, Physica **87A**, 584 (1977); also J. Chem. Phys. **69**, 2251 (1978).

[14]  T. Boublik and I. Nezbeda, Coll. Czech. Chem. Commun. **51**, 2301 (1986).

[15]  N.F. Carnahan and K.E. Starling, J. Chem. Phys. **51**, 635 (1969).

[16]  B.J. Alder and C.E. Hecht, J. Chem. Phys. **50**, 2032 (1969).

[17]  J.A. Barker and D. Henderson, Mol. Phys. **21**, 187 (1971).

[18]  B.J. Alder and T.E. Wainwright, J. Chem. Phys. **27**, 1208 (1957).

[19]  W.W. Wood and J.D. Jacobson, J. Chem. Phys. **27**, 1207 (1957).

[20]  J.L. Lebowitz, Phys. Rev. **133**, A895 (1964).

[21]  G.J. Throop and R.J. Bearman, *Ibid.* (1965).

[22]  N.W. Ashcroft and D.C. Langreth, Phys. Rev. **156**, 685 (1967).

[23]  E.B. Smith and K.R. Lea, J. Chem. Soc., Faraday Trans. II **59**, 1535 (1963).

[24]  L.L. Lee and D. Levesque, Mol. Phys. **26**, 1351 (1973).

[25]  A. Rotenberg, J. Chem. Phys. **43**, 4377 (1965).

[26]  P.H. Fries and J.-P. Hansen, Mol. Phys. **48**, 891 (1983).

[27]  B.J. Alder, J. Chem. Phys. **40**, 2724 (1964).

[28]  J.L. Lebowitz, E. Helfand, and E. Praestgaard, J. Chem. Phys. **43**, 774 (1965).

[29]  L.L. Lee and D. Levesque, *Ibid.* (1973).

[30]  L. Verlet and J.-J. Weis, Phys. Rev. A5, 939 (1972).

[31]  L.L. Lee and D. Levesque, *Ibid.* (1973).

[32]  D. Henderson and E.W. Grundke, J. Chem. Phys. 63, 601 (1975).

[33]  H. Reiss, H.L. Frisch, and J.L. Lebowitz, J. Chem. Phys. 31, 369 (1959).

[34]  R.M. Gibbons, Mol. Phys. 17, 81 (1969); *Ibid.* 18, 809 (1970).

[35]  T. Boublik, Mol. Phys. 27, 1415 (1974); also J. Chem. Phys. 63, 4084 (1975).

[36]  P.A. Monson and M. Rigby, Mol. Phys. 35, 1337 (1978); also Chem. Phys. Letters 58, 122 (1978).

[37]  I. Nezbeda, Czech. J. Phys. B30, 601 (1980).

[38]  M. Wojcik and K.E. Gubbins, Mol. Phys. 53, 397 (1984).

[39]  T. Boublik and I. Nezbeda, *Ibid.* (1986).

[40]  D.W. Robertus and K.M. Sando, J. Chem. Phys. 67, 2585 (1977).

[41]  J.J. Erpenbeck and W.W. Wood, J. Stat. Phys. 35, 321 (1984).

[42]  I. Nezbeda and T. Boublik, Mol. Phys. 51, 1443 (1984)

[43]  T. Boublik, Mol. Phys. 27, 1415 (1974); also J. Chem. Phys. 63, 4084 (1975).

[44]  I. Nezbeda, Chem. Phys. Letters 41, 55 (1976).

[45]  J. Vieillard-Baron, Mol. Phys. 28, 809 (1974).

[46]  D. Chandler and H.C. Andersen, J. Chem. Phys. 57, 1930 (1972).

[47]  M.S. Wertheim, J. Chem. Phys. 78, 4619 (1983); *Ibid.* 78, 4625 (1983).

[48]  L.J. Lowden and D. Chandler, J. Chem. Phys. 59, 6587 (1973).

[49]  D. Chandler, C.S. Hsu, and W.B. Streett, J. Chem. Phys. 66, 5231 (1977).

[50]  W.B. Streett and D.J. Tildesley, J. Chem. Phys. 68, 1275 (1978).

[51]  I. Nezbeda and W.B. Smith, see V. Ref. [21].

[52]  B.C. Freasier, Chem. Phys. Letters 35, 280 (1975); also B.C. Freasier, D. Jolly, and R.J. Bearman, Mol. Phys. 31, 255 (1976).

[53]  I. Nezbeda, Mol. Phys. 33, 1287 (1977).

[54]  T. Boublik and I. Nezbeda, Chem. Phys. Letters 46, 315 (1977).

[55]  T. Boublik, Mol. Phys. 44, 1369 (1981).

[56]  D.J. Tildesley and W.B. Streett, Mol. Phys. 41, 85 (1980).

[57]  B.C. Freasier, D. Jolly, and R.J. Bearman, Mol. Phys. 32, 1463 (1976).

[58]  D. Jolly, B.C. Freasier, and R.J. Bearman, Chem. Phys. Letters 46, 75 (1977).

[59]  D.J. Tildesley, Mol. Phys. 41, 341 (1980).

[60]  D. Jolly, B.C. Freasier, and R.J. Bearman, Chem. Phys. Letters **46**, 75 (1977).

[61]  I. Nezbeda and T. Boublik, Czech. J. Phys. B**27**, 1071 (1977).

## *Exercises*

1.  Given the equation of state (3.5) for the hard-disk fluid, find the Helmholtz free energy, Gibbs free energy G, chemical potential $\mu$, enthalpy H, and entropy S.

2.  The Carnahan-Starling equation of state for hard spheres is very accurate. Derive the thermodynamic functions, $A$, $G$, $H$, and $\mu$, based on equation (5.1) for pure hard spheres and (6.9) for hard sphere mixtures.

3.  Use the Verlet-Weis construction for hard spheres to calculate the rdf at $y=\pi\rho d^3/6=0.8$. Compare your results with simulation data.

4.  The Nezbeda equation (9.4) for hard spherocylinders is very accurate. Derive the properties $A$, $S$, $G$, and $\mu$.

5.  Show that the volume of a dumbbell is given by (11.10).

6.  The Boublik equation (9.1) could be applied to mixtures upon interpreting the geometric factors $s$, $b$, and $\bar{r}$ as in (8.2). Derive the chemical potentials for individual components.

# CHAPTER IX

# THE LENNARD-JONES FLUID

In previous chapters, we have looked at some idealized models of real molecules. For hard spheres, we gave the molecule a volume (represented by a hard sphere diameter $d$). In this chapter, we shall add an attractive force. Real molecules possess repulsive forces (usually short-ranged for overlap forces and long-ranged for charge repulsion) as well as attractive forces (short-ranged for, e.g., hydrogen bonds and longer-ranged for dispersion and polar forces). A simple pair potential that possesses both the repulsive and the inverse sixth power London forces is the Lennard-Jones (LJ) potential [1].

$$u(r) = 4\varepsilon \left[ \left[ \frac{\sigma}{r} \right]^{12} - \left[ \frac{\sigma}{r} \right]^{6} \right] \tag{1}$$

where $r$ is the intermolecular separation and $\sigma$ and $\varepsilon$ are force constants characteristic of the molecular species. This potential has been extensively studied due to its simplicity combined with realism. Lennard-Jones fluid exhibits two first-order (liquid-vapor, and fluid-solid) phase transitions and one second-order phase transition (a critical point). At moderate densities, it simulates the noble gas behavior quite well. Characterization of the LJ fluid represents a major advance in statistical mechanics in understanding real fluid behavior.

## IX.1. *The Lennard-Jones Potential*

As given by eq. (1), the LJ potential is characterized by two force constants $\varepsilon$ and $\sigma$. The meaning of these constants is examined below. The LJ potential is plotted in Figure IX.1.

1.  The energy parameter $\varepsilon$ has the unit of energy and is the well depth of the LJ $u(r)$; i.e.,

$$\min u(r) = -\varepsilon \tag{1.1}$$

2.    The size parameter $\sigma$ has the unit of length and is the location where $u(r) = 0$; i.e.,

$$u(\sigma) = 0 \tag{1.2}$$

3.    The distance $r^*$ for the minimum of $u(r)$ $(= -\varepsilon)$ can be obtained by setting $du/dr = 0$ and is related to $\sigma$ by

$$r^* = \sqrt[6]{2}\ \sigma = 1.12246\sigma \tag{1.3}$$

Ordinarily, $\sigma$ is expressed in angstroms (Å) and $\varepsilon/k$ ($k$ being the Boltzmann constant) in Kelvin. They could be determined from the second virial coefficient and viscosity data of the substance. The values for some inert gases are presented in Table IX.1. For example, $\varepsilon/k$ for argon is 119.8 K and $\sigma = 3.405$ Å. $\sigma$ corresponds to the size of an argon molecule.

*Table IX.1.  Lennard-Jones Constants for Selected Compounds*

| Gas | $\sigma$ (A) | $\varepsilon/k$ (K) | Gas | $\sigma$ (A) | $\varepsilon/k$ (K) |
|---|---|---|---|---|---|
| Neon | 2.86 | 34.2 | Argon | 3.405 | 119.8 |
| Krypton | 3.67 | 167.0 | Nitrogen | 3.612 | 101.3 |
| Oxygen | 3.36 | 119.8 | Methane | 3.74 | 152.0 |
| Carbon monoxide | 3.62 | 104.2 | | | |

## IX.2.  *Thermodynamic Properties*

First we note that the properties of the LJ fluid are dependent on the temperature and the density. If we express the temperature in reduced units

$$T^* = \frac{kT}{\varepsilon} \tag{2.1}$$

and density

$$\rho^* = \rho\sigma^3 \tag{2.2}$$

an examination of the pair potential shows that all LJ fluids have the same properties at the same reduced $T^*$ and $\rho^*$. This result is the basis of the *corresponding states theory*.

**THE CRITICAL POINT**

Verlet [2] estimated the critical temperature and density by the molecular dynamics method:

$$T_c^* = \frac{kT_c}{\varepsilon} = 1.35 \qquad (cf.\ argon\ 1.26) \tag{2.3}$$

$$\rho_c^* = \rho_c\sigma^3 = 0.35 \qquad (cf.\ argon\ 0.319)$$

$$Z_c = \frac{P_c}{\rho_c k T_c} = 0.30 \qquad (cf. \text{ argon } 0.291)$$

Numbers in parentheses are experimental argon values. We see that the critical temperature of the LJ fluid is about +7% above experimental value, and $\rho_c$ is +9.7% above. On the other hand, Adams [3] has made MC calculations in the grand canonical ensemble, He estimated the critical point from the vapor pressure line and scaling laws with the results: $T_c^* = 1.30 \pm 0.02$, $\rho_c^* = 0.33 \pm 0.03$, and $P_c^* = P_c \sigma^3/\varepsilon = 0.13 \pm 0.02$. These estimates put the LJ critical values closer to those of argon.

## PHASE DIAGRAM

Hansen and Verlet [4] investigated the fluid-solid phase boundaries for the LJ fluid. Nicolas et al. [5] have compiled and correlated existing computer simulation data. The phase boundaries are displayed in a *P–T* diagram, Figure IX.2. Both fusion and condensation lines are exhibited.

## THE TRIPLE POINT

Hansen and Verlet [6] have determined by the MC method the triple point of an LJ fluid:

$$T_{tp}^* = 0.68 \pm 0.02, \qquad \rho_{tp}^* = 0.85 \pm 0.01 \qquad (2.4)$$

For comparison, the triple point of argon (NBS 1969 [7]) is

$$(2.5)$$

| | |
|---|---|
| $T_{tp} = 83.8$ K | or $T^* = 0.6995$ |
| $\rho_{tp \ (liq)} = 35.4126$ gmole/L | or $\rho^* = 0.8420$ |
| $\rho_{tp \ (vap)} = 0.01354$ gmole/L | or $\rho^* = 0.00241$ |
| $P_{tp} = 0.67979$ atm | or $Z = \dfrac{P}{\rho_L k T} = 0.00279$ |

The agreement with experiment is remarkable.

## EQUATION OF STATE

Nicolas et al. [8] have also fitted the simulation data of pressure, energy, and second virial coefficient to an empirical formula: the 33-parameter modified Benedict-Webb-Rubin equation

$$P^* = \rho^* T^* + \rho^{*2}(x_1 T^* + x_2 T^{*1/2} + x_3 + x_4 T^{*-1} + x_5 T^{*-2}) \qquad (2.6)$$

$$+ \rho^{*3}(x_6 T^* + x_7 + x_8 T^{*-1} + x_9 T^{*-2}) + \rho^{*4}(x_{10} T^* + x_{11} + x_{12} T^{*-1})$$

$$+ \rho^{*5}(x_{13}) + \rho^{*6}(x_{14} T^{*-1} + x_{15} T^{*-2})$$

$$+ \rho^{*7}(x_{16} T^{*-1}) + \rho^{*8}(x_{17} T^{*-1} + x_{18} T^{*-2}) + \rho^{*9}(x_{19} T^{*-2})$$

$$+ \rho^{*3}(x_{20} T^{*-2} + x_{21} T^{*-3})\exp(-\gamma \rho^{*2}) + \rho^{*5}(x_{22} T^{*-2} + x_{23} T^{*-4})\exp(-\gamma \rho^{*2})$$

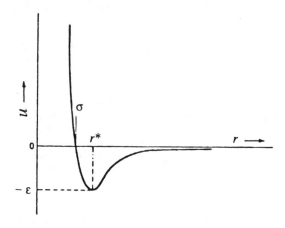

FIGURE IX.1. Lennard-Jones potential. Potential minimum –ε occurs at r*.

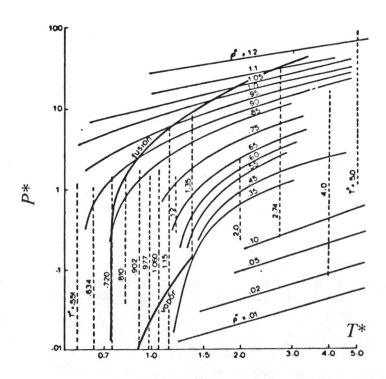

FIGURE IX.2. Pressure-temperature diagram for Lennard-Jones fluid. The phase boundaries are shown in dark lines; the thin lines are iso-chores. Results are from simulations. $P^* = P\sigma^3/\varepsilon$, $T^* = kT/\varepsilon$. (Nicolas et al., Mol. Phys., 1979)

$$+\rho^{*7}(x_{24}T^{*-2}+x_{25}T^{*-3})\exp(-\gamma\rho^{*2}) +\rho^{*9}(x_{26}T^{*-2}+x_{27}T^{*-4})\exp(-\gamma\rho^{*2})$$

$$+\rho^{*11}(x_{28}T^{*-2}+x_{29}T^{*-3})\exp(-\gamma\rho^{*2}) +\rho^{*13}(x_{30}T^{*-2}+x_{31}T^{*-3}+x_{32}T^{*-4})\exp(-\gamma\rho^{*2})$$

where $T^* = kT/\varepsilon$, $\rho^* = \rho\sigma^3$, $P^* = P\sigma^3/\varepsilon$. $x_i$ are constants. Their values are listed in Table IX.2. (Note that $\gamma$ has been set to 3.0.)

Table IX.2.  *Parameters in the Equation of State for LJ Fluids*

| $i$ | $x_i$ | $i$ | $x_i$ |
|---|---|---|---|
| 1 | −0.44480725E-01 | 17 | 0.10591298E+02 |
| 2 | 0.72738221E+01 | 18 | 0.49770046E+03 |
| 3 | −0.14343368E+02 | 19 | −0.35338542E+03 |
| 4 | 0.38397096E+01 | 20 | 0.45036093E+04 |
| 5 | −0.20057745E+01 | 21 | 0.77805296E+01 |
| 6 | 0.19084472E+01 | 22 | 0.13567114E+05 |
| 7 | −0.57441787E+01 | 23 | −0.85818023E+01 |
| 8 | 0.25110073E+02 | 24 | 0.16646578E+05 |
| 9 | −0.45232787E+04 | 25 | −0.14092234E+02 |
| 10 | 0.89327162E-02 | 26 | 0.19386911E+05 |
| 11 | 0.98163358E+01 | 27 | 0.38585868E+02 |
| 12 | −0.61434572E+02 | 28 | 0.33800371E+04 |
| 13 | 0.14161454E+02 | 29 | −0.18567754E+03 |
| 14 | 0.43353841E+02 | 30 | 0.84874693E+04 |
| 15 | 0.11078327E+04 | 31 | 0.97508689E+02 |
| 16 | −0.35429519E+02 | 32 | −0.14483060E+02 |

At medium densities ($\rho^* = 0.35 \sim 0.45$), the values predicted by (2.6) are within 0.04 (in $P^*$) and 0.03 (in $U^*$) in error. At higher densities ($\rho^* \approx 1.0$), the deviations are 0.08 in $P^*$ and 0.05 in $U^*$. The equation is to be used for interpolative calculations and is not recommended for extrapolations beyond the ranges $0<\rho^*<1.2$ and $0.5<T^*<6$. Equation (2.6) could also be used to establish the vapor-liquid coexistence line (through equality of the chemical potentials). Note that the internal energy is derived from the pressure by

$$U^* = \int_0^{\rho^*} d\rho^* \; \rho^{*-2}\left[P^*-T^*\left(\frac{\partial P^*}{\partial T^*}\right)_{\rho^*}\right] \tag{2.7}$$

and the reduced second virial coefficient B* is

$$B^* = \frac{3}{2\pi}\frac{\partial Z}{\partial \rho^*}\bigg|_{\rho^*=0} \tag{2.8}$$

$$= \frac{3}{2\pi}(x_1+x_2T^{*-1/2}+x_3T^{*-1}+x_4T^{*-2}+x_5T^{*-3})$$

The other thermodynamic properties can be derived from (2.6) via thermodynamic relations.

## IX.3. *Distribution Functions*

We have discussed the integral equation methods for calculating the correlation functions of the LJ model in Chapter VI. Most of the commonly used equations, PY, HNC, or YBG, are inadequate in the liquid state. On the other hand, the rdfs of LJ fluid have been determined by MC and MD simulations. Some of the rdfs determined by Verlet [9] are given in Table IX.3. As $T^*$ is lowered and $\rho^*$ is increased, oscillations in rdf intensify, e.g., the first $g_{max} = 3.279$, while the first minimum = 0.518 at $T^* = 0.591$ and $\rho^* = 0.880$, exhibiting considerable variations from unity.

At lower densities and higher temperatures, the PY equation gives reasonable rdfs. However, at high $\rho^*$ and low $T^*$, the PY results are not reliable. For example, as one study [10] showed, the PY rdf is reasonable up to $T^* = 1.827$ and $\rho^* = 0.65$. At a lower temperature, $T^* = 1.036$, ($\rho^* = 0.65$), PY gives a first peak of 2.45, too high as compared to the MD $g_{max}=2.26$. As the density is increased to $\rho^* = 0.85$ (with $T^* = 0.72$), the PY result becomes inadequate (see Figure IV.10). Table IX.4 presents the equilibrium properties. At $T^* = 1.827$ and $\rho^* = 0.65$, the MD energy $\beta U^c/N = -2.26\pm0.02$. PY gives $-2.22$, a reasonable prediction. At $T^* =1.036$ and $\rho^* = 0.65$, MD $\beta U^c/N =-4.36\pm0.02$. Again PY gives $\beta U^c/N = -4.30$. However, the pressures given by PY have deteriorated (to 0.56; cf. MD $-0.11$). Despite the disagreement, PY is reliable for $T^*>2$, and $\rho^*< 0.65$, especially in energy calculations. The usefulness of the energy route to thermodynamic properties has been confirmed by the calculations of Chen et al. [11] Henderson et al. [12] and Barker and Henderson [13].

Two methods are now available for calculating accurate correlation functions: (a) integral equations such as the reference hypernetted chain equation (RHNC) of Lado or the HMSA equation of Zera and Hansen [14], and (b) the functional density expansions. The first method utilizes a judiciously chosen closure relation in addition to the OZ relation for correlation functions. The pcf of LJ is obtained through a universality condition on the bridge function $B(r)$, valid for all simple potentials (see, e.g., RHNC), or via a self-consistent interpolation between the HNC and MSA closures (see, e.g., HMSA). The second method is based on a functional expansion, with density as the expansion parameter, around *any* suitable reference fluid. We shall discuss RHNC first.

### THE REFERENCE HYPERNETTED CHAIN EQUATION

Lado (*Ibid.*) presented an RHNC equation based on the universality of the bridge functions (see Rosenfeld and Ashcroft [15]). The idea is that the bridge function $B(r)$ is short-ranged and is almost the same for different pair potentials. Figure IX.3 shows the bridge functions for hard spheres [16] and for one-component plasma [17] based on computer simulation. Within short ranges, $B(r)$ exhibits the same universal behavior. For longer ranges, there exist nonuniversal but small corrections. The tail part could be rectified by, say, the MSA method [18]. From cluster analysis, the pcf is given exactly by

$$g(r) = \exp[-\beta u(r) + h(r) - C(r) + B(r)] \tag{3.1}$$

The same equation could be written for the hard spheres:

$$g^{HS}(r) = \exp[-\beta u^{HS}(r) + h^{HS}(r) - C^{HS}(r) + B^{HS}(r)] \tag{3.2}$$

Now if we assume the equality of $B(r)$ and $B^{HS}(r)$, eq. (3.1) becomes

Table IX.3. Pair Correlation Functions of Pure Lennard-Jones Fluid from Molecular Dynamics Simulation: $T^* = kT/\varepsilon$, $\rho^* = \rho\sigma^3$. (Verlet, Phys. Rev., 1968) [9]

| $r/\sigma$ \ $\rho^*$ $T^*$ | 0.880 1.095 | 0.880 0.936 | 0.880 0.591 | 0.850 2.888 | 0.850 2.202 | 0.850 1.273 | 0.850 1.127 | 0.850 0.880 |
|---|---|---|---|---|---|---|---|---|
| 0.84 | 0.000 | 0.000 | 0.000 | 0.007 | 0.001 | 0.000 | 0.000 | 0.000 |
| 0.88 | 0.001 | 0.000 | 0.000 | 0.145 | 0.060 | 0.004 | 0.002 | 0.003 |
| 0.92 | 0.086 | 0.048 | 0.003 | 0.716 | 0.490 | 0.135 | 0.085 | 0.030 |
| 0.96 | 0.688 | 0.520 | 0.169 | 1.545 | 1.310 | 0.799 | 0.610 | 0.412 |
| 1.00 | 1.871 | 1.691 | 1.128 | 2.093 | 2.076 | 1.846 | 1.750 | 1.511 |
| 1.04 | 2.701 | 2.682 | 2.592 | 2.210 | 2.350 | 2.508 | 2.560 | 2.546 |
| 1.08 | 2.785 | 2.899 | 3.279 | 2.069 | 2.228 | 2.587 | 2.710 | 2.871 |
| 1.12 | 2.453 | 2.584 | 3.032 | 1.834 | 1.973 | 2.309 | 2.413 | 2.594 |
| 1.16 | 2.003 | 2.111 | 2.440 | 1.586 | 1.699 | 1.944 | 2.009 | 2.151 |
| 1.20 | 1.596 | 1.682 | 1.864 | 1.372 | 1.438 | 1.597 | 1.643 | 1.744 |
| 1.24 | 1.286 | 1.337 | 1.401 | 1.196 | 1.223 | 1.323 | 1.349 | 1.386 |
| 1.28 | 1.058 | 1.093 | 1.081 | 1.058 | 1.068 | 1.107 | 1.122 | 1.121 |
| 1.32 | 0.891 | 0.905 | 0.853 | 0.949 | 0.945 | 0.947 | 0.955 | 0.946 |
| 1.36 | 0.781 | 0.770 | 0.713 | 0.872 | 0.859 | 0.840 | 0.824 | 0.817 |
| 1.40 | 0.699 | 0.677 | 0.617 | 0.818 | 0.798 | 0.758 | 0.739 | 0.742 |
| 1.44 | 0.650 | 0.621 | 0.563 | 0.778 | 0.754 | 0.699 | 0.682 | 0.664 |
| 1.48 | 0.616 | 0.595 | 0.532 | 0.751 | 0.724 | 0.672 | 0.640 | 0.622 |
| 1.52 | 0.606 | 0.574 | 0.520 | 0.742 | 0.719 | 0.661 | 0.627 | 0.597 |
| 1.56 | 0.612 | 0.583 | 0.518 | 0.741 | 0.720 | 0.652 | 0.625 | 0.599 |
| 1.60 | 0.631 | 0.601 | 0.548 | 0.759 | 0.732 | 0.667 | 0.644 | 0.626 |
| 1.64 | 0.663 | 0.645 | 0.590 | 0.782 | 0.757 | 0.696 | 0.677 | 0.656 |
| 1.68 | 0.720 | 0.702 | 0.651 | 0.817 | 0.792 | 0.738 | 0.725 | 0.701 |
| 1.72 | 0.790 | 0.771 | 0.733 | 0.863 | 0.839 | 0.794 | 0.781 | 0.757 |
| 1.76 | 0.873 | 0.857 | 0.830 | 0.913 | 0.891 | 0.854 | 0.854 | 0.831 |
| 1.80 | 0.960 | 0.946 | 0.951 | 0.967 | 0.954 | 0.920 | 0.930 | 0.907 |
| 1.84 | 1.040 | 1.039 | 1.058 | 1.019 | 1.016 | 0.992 | 1.009 | 0.986 |
| 1.88 | 1.113 | 1.114 | 1.153 | 1.075 | 1.071 | 1.062 | 1.079 | 1.072 |
| 1.92 | 1.180 | 1.178 | 1.217 | 1.118 | 1.174 | 1.127 | 1.137 | 1.142 |
| 1.96 | 1.221 | 1.236 | 1.261 | 1.143 | 1.153 | 1.188 | 1.191 | 1.198 |
| 2.00 | 1.250 | 1.268 | 1.286 | 1.156 | 1.172 | 1.220 | 1.231 | 1.240 |
| 2.04 | 1.268 | 1.284 | 1.306 | 1.151 | 1.180 | 1.232 | 1.248 | 1.267 |
| 2.08 | 1.257 | 1.277 | 1.305 | 1.135 | 1.165 | 1.225 | 1.241 | 1.269 |
| 2.12 | 1.223 | 1.248 | 1.288 | 1.108 | 1.131 | 1.203 | 1.214 | 1.246 |
| 2.16 | 1.168 | 1.194 | 1.244 | 1.084 | 1.097 | 1.160 | 1.172 | 1.207 |
| 2.20 | 1.103 | 1.132 | 1.179 | 1.046 | 1.064 | 1.112 | 1.122 | 1.151 |
| 2.24 | 1.044 | 1.066 | 1.100 | 1.016 | 1.029 | 1.056 | 1.068 | 1.090 |
| 2.28 | 0.981 | 0.994 | 1.015 | 0.992 | 0.998 | 1.007 | 1.011 | 1.021 |
| 2.32 | 0.936 | 0.931 | 0.945 | 0.968 | 0.969 | 0.967 | 0.960 | 0.967 |
| 2.36 | 0.896 | 0.890 | 0.876 | 0.952 | 0.946 | 0.934 | 0.924 | 0.923 |
| 2.40 | 0.869 | 0.851 | 0.825 | 0.939 | 0.930 | 0.906 | 0.892 | 0.881 |
| 2.60 | 0.894 | 0.878 | 0.842 | 0.950 | 0.937 | 0.933 | 0.894 | 0.876 |
| 2.80 | 1.044 | 1.046 | 1.059 | 1.016 | 1.016 | 1.017 | 1.026 | 1.023 |
| 3.00 | 1.088 | 1.101 | 1.129 | 1.035 | 1.045 | 1.071 | 1.080 | 1.095 |
| 3.20 | 1.002 | 1.006 | 1.010 | 1.002 | 1.005 | 1.012 | 1.014 | 1.021 |
| 3.40 | 0.941 | 0.930 | 0.911 | 0.980 | 0.975 | 0.959 | 0.951 | 0.942 |
| 3.60 | 0.978 | 0.972 | 0.960 | 0.992 | 0.988 | 0.976 | 0.974 | 0.966 |
| 3.80 | 1.032 | 1.037 | 1.051 | 1.009 | 1.011 | 1.017 | 1.023 | 1.026 |
| 4.00 | 1.026 | 1.033 | 1.047 | 1.008 | 1.011 | 1.023 | 1.026 | 1.034 |
| 4.20 | 0.988 | 0.986 | 0.981 | 0.998 | 0.998 | 0.997 | 0.994 | 0.995 |
| 4.40 | 0.978 | 0.972 | 0.959 | 0.994 | 0.992 | 0.984 | 0.980 | 0.974 |
| 4.60 | 1.000 | 0.999 | 0.992 | 0.999 | 0.998 | 0.996 | 0.996 | 0.994 |
| 4.80 | 1.016 | 1.020 | 1.029 | 1.003 | 1.005 | 1.009 | 1.012 | 1.015 |
| 5.00 | 1.005 | 1.008 | 1.012 | 1.001 | 1.003 | 1.006 | 1.007 | 1.010 |

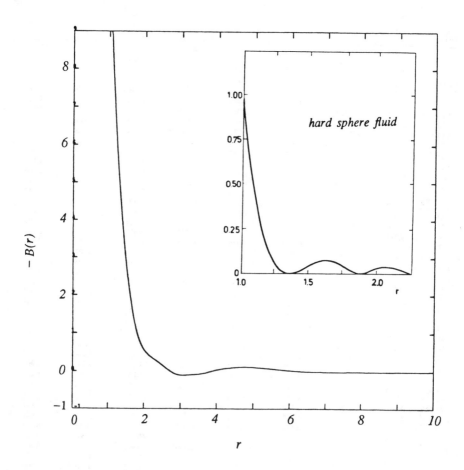

FIGURE IX.3. The bridge function B(r) for hard sphere fluid (inset) and one-component plasma. Results are from computer simulation. (Labik, Mol. Phys., 1987; and DeWitt et al. 1987)

$$g(r) = \exp[-\beta u(r) + h(r) - C(r) + B^{HS}(r)] \tag{3.3}$$

This is the *RHNC equation*. It is assumed that one has means of obtaining the bridge function for hard spheres. This is the case when simulation data are available. Otherwise, one could use the Verlet-Weis construction. To match the HS fluid to the LJ fluid, one has the option of selecting a hard sphere diameter. To achieve thermodynamic consistency (see Hiroike's condition discussed earler), the following minimum free energy condition is proposed

$$\rho d \int d\mathbf{r} \ [g(r) - g^{HS}(r)] \ \frac{\partial B^{HS}(r)}{\partial d} = 0 \tag{3.4}$$

where $d$, the diameter of the hard sphere, is determined by annulling the integral. This equation has been applied to the LJ fluid [19] with the reference HS pcf calculated from the Verlet-Weis construction. The results at $\rho\sigma^3 = 0.85$ and $T^* = 0.719$ are presented in Figure IX.4. Excellent agreement is obtained.

*Table IX.4. Thermodynamic Properties of LJ Fluids*

| $T^* = 1.827$ | | $\rho^* = 0.65$ | |
|---|---|---|---|
| | MD | PY Pert. | PY |
| $\beta U/N$ | −2.26 | −2.22 | −2.22 |
| $\beta P/\rho$ | 1.58 | 1.78 | 1.78 |
| $\rho kTK_T - 1$ | −0.96 | −0.84 | −0.83 |
| $T^* = 1.036$ | | $\rho^* = 0.65$ | |
| | MD | PY Pert. | PY |
| $\beta U/N$ | −4.36 | −4.38 | −4.30 |
| $\beta P/\rho$ | −0.11 | 0.20 | 0.56 |
| $\rho kTK_T - 1$ | −1.03 | −0.85 | −0.58 |
| $T^* = 1.070$ | | $\rho^* = 0.75$ | |
| | MD | PY Pert. | PY |
| $\beta U/N$ | −4.83 | −4.77 | −4.72 |
| $\beta P/\rho$ | 0.90 | 1.03 | 1.81 |
| $\rho kTK_T - 1$ | −0.99 | −1.10 | −0.76 |

## THE DENSITY EXPANSION EQUATION

The functional expansion methods introduced in Chapter VI could be used to formulate a perturbation type expansion of the PY and/or the HNC generating functionals. Let us choose the PY generating functional first. Expansion around a reference state gives

$$\rho_i^{(1)}(1;w) = \rho_i^{(1)}(1;w_0)\left\{1 + \int d2 \ C_{ii}(12;w_0)[\rho_i^{(1)}(2;w) - \rho_i^{(1)}(2;w_0)]\right. \tag{3.5}$$

$$\left. + \int d2 \ C_{ij}(12;w_0) \ [\rho_j^{(1)}(2;w) - \rho_j^{(1)}(2;w_0)] + \ \cdots \right\}$$

where the subscripts $i$ and $j$ represent here different chemical species. This equation contains dcfs that are under the influence of a test particle of the reference type $j$. The

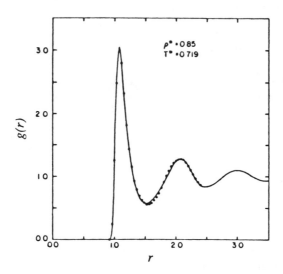

FIGURE IX.4. Reference hypernetted chain (RHNC) results for the pair correlation functions of Lennard-Jones fluid. $T^* = 0.719$, $\rho^* = 0.85$. Points: MD results. Full line: RHNC calculations (Lado et al., Phys. Rev., 1983) [19]

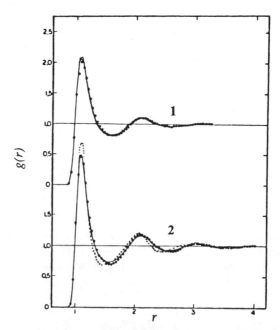

FIGURE IX.5. Pair correlation functions for Lennard-Jones fluid obtained from PY and density expansion methods. Curve 1: $T^* = 1.827$, $\rho^* = 0.65$. Curve 2: $T^* = 1.07$, $\rho^* = 0.75$. Circles: MD results [9]. Full lines: PYP (density expansion) Dotted lines: PY calculations. (Lee, J. Chem. Phys., 1975) [10]

external potential corresponding to the *j*-particle is denoted by $w_0$. The inhomogeneous dcfs could further be expanded functionally and resummed afterwards (see Lee [20] for details). The result is

$$\frac{y(12)}{y_0(12)} \approx \frac{1 + h(12) - C(12)}{1 + h_0(12) - C_0(12)} \tag{3.6}$$

Thus if one knows the reference fluid correlation functions (subscripted 0), the full fluid correlations could be obtained via eq. (3.6) plus the OZ relation. In the derivation, the reference fluid was implicit. In other words, we could choose any convenient reference in order to effect the expansion. For example, one could choose instead the same fluid at a different state as reference. The scaling properties of the LJ fluid were examined [21] by taking, e.g., the correlation functions of LJ at $\rho^* = 0.65$ to generate the pcf at $\rho^* = 0.75$ (both at the same temperature, $T^* = 1.0$). The results are shown in Figure IX.5. The agreement, compared with MD data, is remarkable. The straight PY results are also shown. The peak obtained from PY is too high. The oscillations are also out of phase. These deficiencies are corrected in the scaling calculations. Of course, when the two chosen states are identical, the scaling results are exact. It is also possible to carry out temperature scaling [22] or use other reference fluids.

## IX.4. *Mixtures of LJ Molecules*

LJ mixtures are much used in the studies of the effects of size and energy differences of the constituent molecules on the mixture properties, namely the *mixing rules*. There exist in literature many simplified mixing rules, such as the conformal solutions, van der Waals one-fluid theory, or local composition models. It is not clear, to begin with, which mixing rules are effective in reproducing the bulk properties of mixtures containing molecules of different sizes and shapes interacting with polar and nonpolar forces. Molecules interacting with the LJ potential are the simpliest models for testing the predicted size and energy variations of these mixture theories. This will be the major topic of the next chapter on solution thermodynamics. Here we examine some simulation results.

Singer and McDonald [23] have made MC calculations for mixtures of LJ molecules. They simulated Ar-Kr mixtures at 115.8 K, Ar-$CH_4$ at 91.0 K, CO-$CH_4$ at 91.0 K, Ar-$H_2$ at 83.8 K, Ar-CO at 83.8 K, and $O_2$-$N_2$ at 83.8 K. The force constants used for each component have been given in Table IX.1. Recently, Hoheisel and Lucas [24], and Shukla et al. [25] have made extensive calculations for LJ mixtures to determine their thermodynamic properties and test the mixing rules. Haile [26] has calculated the free energies using the Kirkwood charging procedure. Shing et al. [27] investigated the infinite dilution chemical potentials by application of the umbrella sampling techniques.

An additional information needed in a mixture is the cross-interaction parameters $\varepsilon_{12}$ and $\sigma_{12}$. For example, in the Ar-Kr mixture, though the Ar-Ar and Kr-Kr interactions are the same as in the pure fluid cases, the cross interaction Ar-Kr must be specified independently. The usual combining rules are of the Berthelot-Lorentz (BL) type:

$$\sigma_{12} = \frac{1}{2}(\sigma_{11} + \sigma_{22}), \qquad \varepsilon_{12} = \xi\sqrt{\varepsilon_{11}\varepsilon_{22}} \tag{4.1}$$

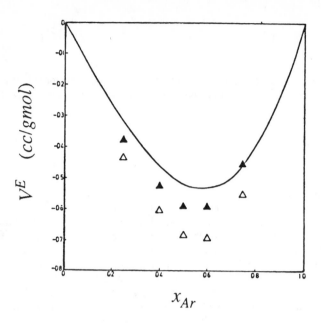

FIGURE IX.6. *Excess volume of LJ mixtures (Ar+Kr). Open circles =
MC simulations [23] with the binary parameter* $\xi = 1.0$. *Filled triangles:
MC results for* $\xi = 0.989$. *Curve: experimental data for Ar-Kr mixtures.
(McDonald, Mol. Phys., 1972)*

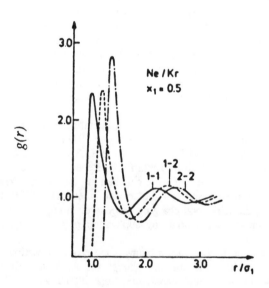

FIGURE IX.7. *Pair correlation functions from molecular dynamics simulation of an equimolar Ne-Kr mixture. 1 = Neon, 2 = Krypton. (Hoheisel
and Deiters, Mol. Phys., 1979) [29]*

where $\xi$ is chosen to be 1 here. We remark that these rules have no rigorous physical basis, though they are commonly used in mixture theories. To see this, we must realize that the LJ potential is only approximate in its mathematical representation of real interactions (in particular, the repulsive twelfth power is arbitrary). The constants $\varepsilon$ and $\sigma$ are empirically fitted to the second virial coefficient data, viscosity data, or other experimental information. As to the cross interaction, first there is no reason that the 1-2 interaction should obey the same LJ force law; second, if it did, it is not necessary that $\sigma_{12}$ and $\varepsilon_{12}$ obey (4.1). For rare gases, the BL rules are found to be insufficient (this could be seen in the excess properties). For example, for the Ar-Kr mixture, the excess volume data are displayed in Figure IX.6. Using strictly BL rules, one gets the points denotes by open triangles. They fail to follow the experimental curve. By adjusting $\xi$, the binary interaction parameter, one could achieve a better agreement with experimental values. This led McDonald [28] to consider the source of error being in the BL rules. We shall also point out that the excess properties are very sensitive quantities. For the LJ mixture, the excess $H^E$ is about 0.5% to 1% of the bulk values. This requires an accuracy of at least 0.1% of any theory in order to predict the excess quantities correctly.

Hoheisel and Deiters [29] simulated the neon-krypton mixture with the MD method. The following potential parameters were used (Table IX.5). Note that their choice of the cross-interaction parameters (Ne-Kr) was not the usual BL rules. Rather they corrected the BL values by an empirical equation of state that describes the mixture behavior correctly. The correction to $\varepsilon_{12}$(BL) is about 20%, to $\sigma_{12}$(BL) is 0.1%. The Ne-Kr mixture is of interest because it shows the changes in structure due to the disparate size ratio $\sigma_{Kr}/\sigma_{Ne}$ =1.28. The rdfs of one state are shown in Figure IX.7. The peaks are situated in the order of the collision distances (i.e., $(\sigma_i +\sigma_j)/2$). The Kr-Kr rdf has the highest peak.

*Table IX.5. LJ Parameters for Neon-Krypton Mixtures*

|  | Ne-Ne | Ne-Kr | Kr-Kr |
|---|---|---|---|
| $\varepsilon/k$ (K) | 34.2 | 64.6 | 167.0 |
| $\sigma$ (A) | 2.86 | 3.26 | 3.67 |

It is useful to have simulation data on mixture pcfs in order to verify theoretical calculations. Haile and Gupta [30] have used the $N$-$P$-$T$ ensemble to simulate LJ mixtures. Some of their rdfs are presented in Table IX.6.

## IX.5. *The Significance of the LJ Potential for Real Gases*

The LJ potential is the simplest potential that exhibits both the repulsive and attractive features of a real potential. For this reason, we can use the LJ results to

1.  *Predict the thermodynamic properties, especially quantities derived from the internal energy, of rare gases.* Table IX.7 gives a comparison of the thermodynamic properties of the LJ fluid with those from experiments on argon. The agreement is very close. The internal energy values of LJ are reliable for most of the conditions tabulated. The fluid-solid transition is compared in Table IX.8. The fusion densities are well predicated (about 2%). Also the heat of fusion is of the correct order (about +15% in maximum deviation). The fusion pressures are predicted to within 14% (maximum deviation). The vapor-liquid transition is compared in Table IX.9 at $T^* = 0.75$, and $T^* = 1.15$. Again reasonable results are obtained.

2.  *Estimate thermodynamic properties of weakly anisotropic molecules (such as methane, nitrogen, and oxygen).* For these molecules, the LJ potential is by no means exact. We can consider the LJ as an "effective" potential [31]. Under

*Table IX.6. Radial Distribution Functions for Lennard-Jones Mixtures from the Isothermal-isobaric Molecular Dynamics Simulations.* $\sigma_{22}/\sigma_{11}=1.25$, $\sigma_{12} = 0.5(\sigma_{11} +\sigma_{22})$, $\varepsilon_{11} = \varepsilon_{12} = \varepsilon_{22}$. $x_1 = 0.75$, $P\sigma_{11}^3/\varepsilon_{11} =0.5$, $kT/\varepsilon_{11} = 1.0$, $\rho\sigma_{11}^3 = 0.6256$. *(Gupta, Thesis, 1984)*

| r | $g_{11}(r)$ | $g_{12}(r)$ | $g_{22}(r)$ |
|---|---|---|---|
| 0.8750 | 0.00002 | 0.0 | 0.0 |
| 0.9000 | 0.00428 | 0.0 | 0.0 |
| 0.9250 | 0.04917 | 0.0 | 0.0 |
| 0.9500 | 0.24728 | 0.0 | 0.0 |
| 0.9750 | 0.72187 | 0.00002 | 0.0 |
| 1.0000 | 1.38065 | 0.00080 | 0.0 |
| 1.0250 | 2.02237 | 0.01428 | 0.0 |
| 1.0500 | 2.44724 | 0.09398 | 0.0 |
| 1.0750 | 2.64032 | 0.33116 | 0.0 |
| 1.1000 | 2.61940 | 0.79230 | 0.00044 |
| 1.1250 | 2.48290 | 1.36958 | 0.00360 |
| 1.1500 | 2.28142 | 1.91810 | 0.03312 |
| 1.1750 | 2.06792 | 2.30898 | 0.14958 |
| 1.2000 | 1.85013 | 2.49392 | 0.42690 |
| 1.2250 | 1.65420 | 2.53290 | 0.85680 |
| 1.2500 | 1.48313 | 2.42915 | 1.37119 |
| 1.2750 | 1.33453 | 2.26776 | 1.82290 |
| 1.3000 | 1.20639 | 2.07561 | 2.17499 |
| 1.3250 | 1.10822 | 1.88010 | 2.32583 |
| 1.3500 | 1.01523 | 1.69347 | 2.37416 |
| 1.3750 | 0.94403 | 1.53631 | 2.34224 |
| 1.4000 | 0.87969 | 1.38483 | 2.19210 |
| 1.4250 | 0.83167 | 1.26547 | 2.00439 |
| 1.4500 | 0.79007 | 1.15719 | 1.83013 |
| 1.4750 | 0.75958 | 1.07009 | 1.68379 |
| 1.5000 | 0.73393 | 0.98569 | 1.51444 |
| 1.5250 | 0.71213 | 0.91752 | 1.39131 |
| 1.5500 | 0.69649 | 0.87033 | 1.26244 |
| 1.5750 | 0.68627 | 0.82281 | 1.14488 |
| 1.6000 | 0.68883 | 0.79099 | 1.07287 |
| 1.6250 | 0.69208 | 0.75596 | 0.99675 |
| 1.6500 | 0.69649 | 0.73235 | 0.91369 |
| 1.6750 | 0.71109 | 0.72122 | 0.85330 |
| 1.7000 | 0.72160 | 0.70491 | 0.81991 |
| 1.7250 | 0.73447 | 0.70276 | 0.77174 |
| 1.7500 | 0.76201 | 0.70065 | 0.73476 |
| 1.7750 | 0.78630 | 0.69940 | 0.71102 |
| 1.8000 | 0.81544 | 0.70977 | 0.68132 |
| 1.8250 | 0.85059 | 0.71989 | 0.67826 |
| 1.8500 | 0.88555 | 0.73198 | 0.66655 |
| 1.8750 | 0.91502 | 0.75796 | 0.67412 |
| 1.9000 | 0.95136 | 0.77881 | 0.67059 |
| 1.9250 | 0.98277 | 0.81175 | 0.67544 |
| 1.9500 | 1.01704 | 0.83658 | 0.70378 |
| 1.9750 | 1.05464 | 0.87788 | 0.70468 |

moderate conditions, LJ results give reasonable estimates of their properties. In fact, this is the basis of the two-parameter corresponding states theory for many substances.

Table IX.7. Comparison of LJ and Argon Fluid Properties

| T | V | P (atm) | | −U (cal/mol) | |
|---|---|---|---|---|---|
| (K) | (cc/mol) | Argon | MC | Argon | MC |
| 97.0 | 28.48 | 214 | 200±14 | 1386 | 1413 |
| 108.0 | 28.48 | 451 | 443±12 | 1360 | 1387 |
| 117.0 | 28.48 | 619 | 605±16 | 1334 | 1372 |
| 127 | 30.92 | 367 | 331±13 | 1219 | 1261 |
| 136.0 | 30.30 | 585 | 607±14 | 1214 | 1264 |
| T | V | Argon | MD | Argon | MD |
| 86.1 | 27.98 | 95 | 91 | - | 1457 |
| 105.4 | 27.98 | 513 | 507 | 1389 | 1414 |
| 128.3 | 31.71 | 295 | 295 | 1190 | 1230 |

That the LJ fluid predicts the behavior of argon well is remarkable in light of the fact that properties could be generated numerically on a computer. On the other hand, we caution that LJ is a simple model for molecules. Even for argon, there is room for improvement. Figure IX.8 presents the $P–V–T$ isotherms of an LJ fluid versus those for argon. The supercritical pressures are well predicted by the LJ potential. However, for liquid pressures, the discrepancies are large. Barker, Fisher, and Watts [32] proposed an argon potential that includes the three-body forces. They obtained overall better predictions of argon properties (see Table IX.10).

Table IX.8. Fluid-Solid Transition in Argon and LJ Fluid

| | $T^*$ | $P^*$ | $\rho^*_{fluid}$ | $\rho^*_{solid}$ | $\Delta V^*$ | $\lambda^*$ |
|---|---|---|---|---|---|---|
| LJ | 2.74 | 32.2 | 1.113 | 1.179 | 0.050 | 2.69 |
| Ar | 2.74 | 37.4 | ... | ... | ... | 2.34 |
| LJ | 1.35 | 9.00 | 0.964 | 1.053 | 0.087 | 1.88 |
| Ar | 1.35 | 9.27 | 0.982 | 1.056 | 0.072 | 1.63 |
| LJ | 1.15 | 5.68 | 0.936 | 1.024 | 0.091 | 1.46 |
| Ar | 1.15 | 6.09 | 0.947 | 1.028 | 0.082 | 1.44 |
| LJ | 0.75 | 0.67 | 0.875 | 0.973 | 0.135 | 1.31 |
| Ar | 0.75 | 0.59 | 0.856 | 0.967 | 0.133 | 1.23 |

*$\lambda^*$ is the reduced heat of fusion.

Table IX.9. Liquid-Vapor Transition for Argon and LJ Fluid

| | $T^*$ | $P^*$ | $\rho^*_{gas}$ | $\rho^*_{liquid}$ | $\Delta H^{**}$ |
|---|---|---|---|---|---|
| LJ | 0.75 | 0.0025 | 0.0035 | 0.825 | 6.62 |
| Ar | 0.75 | 0.0031 | 0.0047 | 0.818 | 6.50 |
| LJ | 1.15 | 0.0597 | 0.073 | 0.606 | 4.34 |
| Ar | 1.15 | 0.0664 | 0.093 | 0.579 | 3.73 |

*All quantities are reduced by combinations of $\varepsilon$ and $\sigma$.
**$\Delta H^*$ is reduced heat of vaporization.

Table IX.10. BFW Potential for Argon: Pressures and Energies

| V (cc/mol) | T (K) | U(MC) (cal/mol) | U(Ar) (cal/mol) | P(MC) (bar) | P(Ar) (bar) |
|---|---|---|---|---|---|
| Solid on the melting line | | | | | |
| 23.75 | 63.10 | −1786 | ... | 1 | 0 |
| 24.03 | 108.12 | −1664 | ... | 1028 | 1051 |
| 22.09 | 180.15 | −1525 | ... | 4964 | 4999 |
| 21.47 | 197.78 | −1459 | ... | 6593 | 6140 |
| 19.41 | 323.14 | −941 | ... | 15988 | 15354 |
| Fluid on the melting line | | | | | |
| 23.66 | 180.15 | −1297 | ... | 4907 | 4999 |
| 23.10 | 201.32 | −1236 | ... | 6143 | 6335 |
| 21.31 | 273.11 | −940 | ... | 11645 | 11380 |
| 20.46 | 323.14 | −664 | ... | 15513 | 15354 |
| Fluid State | | | | | |
| 27.04 | 100.00 | −1423 | −1432 | 655 | 661 |
| 30.65 | 140.00 | −1213 | −1209 | 588 | 591 |
| 48.39 | 150.87 | −679 | −689 | 48 | 51 |
| 91.94 | 150.87 | −462 | −481 | 50 | 50 |

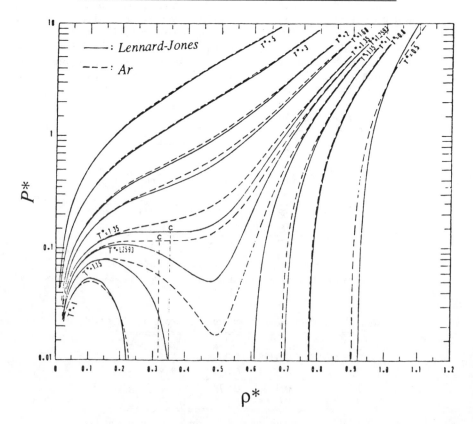

FIGURE IX.8. Pressure isotherms of Lennard-Jones and real fluid argon. Solid lines: LJ isotherms (eq. (2.6)). Dashed lines: experimental argon isotherms. C = critical point.

## References

[1]  F. London, Trans. Faraday Soc. **33**, 8 (1937); and Z. Phys. Chem. B11, 222 (1930).

[2]  L. Verlet, Phys. Rev. **159**, 98 (1967).

[3]  D.J. Adams, Mol. Phys. **37**, 211 (1979).

[4]  J.-P. Hansen and L. Verlet, Phys. Rev. **184**, 151 (1969).

[5]  J.J. Nicolas, K.E. Gubbins, W.B. Streett, and D.J. Tildesley, Mol. Phys. **37**, 1429 (1979).

[6]  J.-P. Hansen and L. Verlet, *Ibid.* (1969).

[7]  A.L. Gosman, R.D. McCarty, and J.G. Hust, NBS 27, NSRDS-NBS 27 (U.S. Department of Commerce, Washington, 1969).

[8]  J.J. Nicolas et al. *Ibid.* (1979).

[9]  L. Verlet, Phys. Rev. **165**, 201 (1968).

[10]  L.L. Lee, J. Chem. Phys. **62**, 4436 (1975).

[11]  M. Chen, D. Henderson, and J.A. Barker, Can. J. Phys. **47**, 2009 (1969).

[12]  D. Henderson, J.A. Barker, and R.O. Watts, IBM J. Res. Dev. **14**, 668 (1970).

[13]  J. A. Barker and D. Henderson, Rev. Mod. Phys. **48**, 587 (1976).

[14]  F. Lado, Phys. Rev. A8, 2548 (1973); and G. Zerah and J.-P. Hansen, J. Chem. Phys. **84**, 1336 (1986).

[15]  Y. Rosenfeld and N.W. Ashcroft, Phys. Rev. A20, 1208 (1979).

[16]  A. Malijevsky and S. Labik, Mol. Phys. **60**, 663 (1987); and F.H. Ree, R.N. Keeler, and S.L. McCarty, J. Chem. Phys. **44**, 3407 (1966).

[17]  P.D. Poll, N.W. Ashcroft, and H.E. DeWitt, "Theory of the one-component plasma bridge function", presented at the West Coast Statistical Mechanical Conference, June 9 (1987).

[18]  S.M. Foiles, N.W. Ashcroft, and L. Reatto, J. Chem. Phys. **80, 4441 (1984)**.

[19]  F. Lado, S.M. Foiles, and N.W. Ashcroft, Phys. Rev. A28, 2374 (1983).

[20]  L.L. Lee, J. Chem. Phys. **62**, 4436 (1975).

[21]  L.L. Lee, *Ibid.* (1975).

[22]  L.L. Lee, *Ibid.* (1975).

[23]  I.R. McDonald, Mol. Phys. **23**, 41 (1972); J.V.L. Singer and K. Singer, Mol. Phys. **19**, 279 (1970); K. Singer, Chem. Phys. Lett. **3**, 164 (1969); and I.R. McDonald, Chem. Phys. Lett. **3**, 241 (1969).

[24]  C. Hoheisel and K. Lucas, Mol. Phys. **53**, 51 (1984).

[25]  K.P. Shukla, M. Luckas, H. Marquardt, and K. Lucas, Fluid Phase Equil. **26**, 129 (1986).

[26]  J.M. Haile, Fluid Phase Equil. **26**, 103 (1986).

[27]  K.S. Shing and K.E. Gubbins, Mol. Phys. **43**, 717 (1981); and *Ibid.* **46**, 1109 (1982).

[28]  I.R. McDonald, *Ibid.* (1972).

[29]  C. Hoheisel and U. Deiters, Mol. Phys. **37**, 95 (1979).

[30]  J.M. Haile and S. Gupta, J. Chem. Phys. **79**, 3067 (1983).

[31]  F.H. Stillinger, J. Chem. Phys. **57**, 1780 (1972); and J. Phys. Chem. **74**, 3677 (1970).

[32]  J.A. Barker, R.A. Fisher, and R.O. Watts, Mol. Phys. **21**, 657 (1971).

## Exercises

1.  Plot the LJ potentials for neon, argon, and krypton, and compare their behavior.

2.  Use the RHNC and density expansion theories to calculate the pcfs at (i) $T^*$ =1.827 and $\rho^*$ =0.65; and (ii) $T^*$ = 1.07 and $\rho^*$ =0.75. Compare with the MD results.

3.  Use the PY equation to determine the vapor-liquid phase boundaries (spinodal lines) of the LJ fluid simulating argon. Compare with Figure IX.2.

4.  Reproduce the rdfs in Table IX.6 for the LJ mixture, using the RHNC integral equations. The reference hard sphere rdfs could be obtained from the Lee-Levesque construction outlined in Chapter VIII.

# CHAPTER X

# SOLUTION THERMODYNAMICS

$\mathbf{F}$or calculation of the properties of mixtures in the liquid state, two methods are currently being utilized: (i) the equations of state and (ii) activity coefficient models. Recently, equations of state have been increasingly applied to liquid mixtures. One of the major problems to be resolved in this approach is the proper use of *mixing rules* to account for the composition dependence. For example, the van der Waals equation

$$\frac{P}{RT} = \frac{1}{v-b} - \frac{a}{RTv^2} \tag{1}$$

can be applied to mixtures by making parameters $a$ and $b$ depend on composition:

$$b = \sum_i x_i b_i^0 \tag{2}$$

and

$$a = \sum_i \sum_j x_i x_j \, a_{ij} \tag{3}$$

where

$$a_{ij} = \sqrt{a_i^0 a_j^0} \tag{4}$$

The superscript 0 indicates pure fluid properties. Relations (2) and (3) are called *mixing rules*, and relation (4) is called the *combining rule*. It is legitimate to ask why (2) and (3) were chosen as the mixing rules. Are there justifications, theoretical or otherwise, for doing so? Or more importantly, are there better choices? These and other related questions are examined in this chapter. For example, we may write the second virial coefficient of (1) as

$$B_2 = b - \frac{a}{RT} = \sum_i x_i b_i^0 - \frac{1}{RT} \sum_i \sum_j x_i x_j a_{ij} \tag{5}$$

As we know from molecular theory, $B_2$ for mixtures should be quadratic in composition:

$$B_2 = \sum_i \sum_j x_i x_j B_{ij} \tag{6}$$

Rule (2) does not satisfy this condition. A better choice would have been

$$b = \sum_i \sum_j x_i x_j b_{ij} \tag{7}$$

Equation (7) combined with eq. (3) would satisfy eq. (6). (N.B. There is also interrelation between combining rules and mixing rules. For additive covolumes $b_{ij}$, (2) and (7) are actually equivalent.)

   As the above example has just shown, progress could be made in the van der Waals equation by introducing molecular arguments. Other examples of interest are (i) the conformal solution model (CSM) and (ii) the local composition model (LCM). We shall discuss these models below.

   At the outset, we remark that most of the mixing rules in use are not *exact*, because they were developed based on assumptions, such as ideal or regular solutions. Upon comparison with rigorous molecular theories, these rules emerge as approximations. For example, in order for the ideal-solution assumptions to be valid, the exchange energy of Guggenheim [1] must be zero. This is, in general, not the case for real solutions. Some rules are more *realistic* than others, and thus more useful as tools for calculation. Comparison of different sets of mixing rules is an important part of mixture study. Without statistical formulas as a guide, comparisons tend to be inconclusive, since the alternative, i.e., direct comparison with real mixtures, camouflages the inaccuracies of the mixing rules as compensation of errors is built into other "adjustable" parameters (e.g., the binary interaction parameters).

   If molecular theories provide the exact rules for mixtures, why do we not adopt the molecular formulas and dispense with the empirical ones? This is not entirely the case because mixing effects are contained implicitly in the pair correlation functions (pcfs), the mathematical solution for which is not available for general potentials. On the other hand, computer simulations provide exact information, but they do not provide the rules so much as yielding numerical "data" for testing such rules. By the 1970s, simulation information had become more abundant and available for more systems. We shall also use these data in our studies below.

## X.1. *Van der Waals n-Fluid Theories*

   As mentioned in Chapter IV, there are three routes to thermodynamics in the molecular theory: (i) the pressure equation, (ii) the energy equation, and (iii) the compressibility equation. The mixing rules determined from one route are not necessarily applicable to another route. The inconsistency is due to the approximate nature of the mixing rules, not to flaws in the molecular theory. Since our emphasis is on the equations of state, we shall choose the pressure route in the discussion of mixing rules.

   When an equation of state, empirical or otherwise, developed for a pure substance is extended to mixtures, one of the important questions is the composition dependence of the new equation. This dependence, in practice, is incorporated into the equation through mixing rules applied to the reduced state variables and/or the parameters of the equation. In Chapter IV, we have given the expression for the pressure of a binary mixture of $N_A$ molecules of type A and $N_B$ molecules of type B in terms of the pair potentials $u_{ij}$ and molecular pcf $g_{ij}$ as

$$\frac{P}{\rho kT} = 1 - \frac{\beta\rho}{6} \sum_i \sum_j x_i x_j \int d\mathbf{r} \; r\frac{\partial u_{ij}}{\partial r} \; g_{ij}(r; \rho, T, \mathbf{x}) \tag{1.1}$$

We note that the composition dependence has an *explicit* part, in the double summation over the mole fractions, and an *implicit* part, contained in the arguments of the pcfs $g_{ij}$, $\mathbf{x} = (x_A, x_B)$. Attempts have been made to approximate eq. (1.1) by the van der Waals $n$-fluid theories (i.e., one-fluid, two-fluid, and three-fluid theories), where the composition dependence is simplified and the pcfs are evaluated at reduced states characterized by the energy and size parameters. For concreteness, we look at a binary mixture of spherical molecules of species A and B whose potentials of interaction are characterized by only two parameters: a size parameter $\sigma_{ij}$ and an energy parameter $\varepsilon_{ij}$. We denote the large spheres by size $\sigma_{AA}$ and small spheres by size $\sigma_{BB}$, i.e., $\sigma_{AA} > \sigma_{BB}$. The energy parameters are $\varepsilon_{AA}$ and $\varepsilon_{BB}$, respectively. The cross interaction is characterized by size $\sigma_{AB}$ ($= (\sigma_{AA} + \sigma_{BB})/2$) and $\varepsilon_{AB}$ ($= \sqrt{\varepsilon_{AA}\varepsilon_{BB}}$). These combining rules correspond to the BL relation. The pressure equation written out in the reduced form is

$$\frac{P}{\rho kT} = 1 + \frac{4\pi}{6} \sum_i \sum_j x_i x_j \frac{\rho\sigma_{ij}^3\varepsilon_{ij}}{kT} \tag{1.2}$$

$$\int dr_{ij}^* r_{ij}^{*3} \frac{\partial u_{ij}^*}{\partial r_{ij}^*} g_{ij}(\frac{r}{d_{ij}}; \rho, T, \mathbf{x}), \qquad i,j = A,B$$

where $r_{ij}^* = r/\sigma_{ij}$ and $u_{ij}^* = u_{ij}/\varepsilon_{ij}$. Equation (1.2) can be contrasted with the pure fluid equation

$$\frac{P}{\rho kT} = 1 + \frac{4\pi}{6} \frac{\rho\sigma^3\varepsilon}{kT} \int dr^* \; r^{*3}\frac{\partial u^*}{\partial r^*} g_0(\frac{r}{d}; \rho, T) \tag{1.3}$$

where $g_0$ denotes the pure fluid pcf. The approximations made in the $n$-fluid theories are on the relations between the mixture pcf $g_{ij}(r)$ and pure $g_0(r)$. For example, in the three-fluid theory (vdW3), one assumes

$$g_{AA}(\frac{r}{\sigma_{AA}}; \rho, T, \mathbf{x}) = g_0(\frac{r}{\sigma_{AA}}; \rho\sigma_{AA}^3, \frac{kT}{\varepsilon_{AA}}) \tag{1.4}$$

$$g_{BB}(\frac{r}{\sigma_{BB}}; \rho, T, \mathbf{x}) = g_0(\frac{r}{\sigma_{BB}}; \rho\sigma_{BB}^3, \frac{kT}{\varepsilon_{BB}}) \tag{1.5}$$

and

$$g_{AB}(\frac{r}{\sigma_{AB}}; \rho, T, \mathbf{x}) = g_0(\frac{r}{\sigma_{AB}}; \rho\sigma_{AB}^3, \frac{kT}{\varepsilon_{AB}}) \tag{1.6}$$

Note that in this approximation, the composition dependence in $g_{ij}$ is removed by equating $g_{ij}$ to the pure fluid (therefore composition-independent) $g_0$ whose state variables are reduced by the size and energy parameters of the $ij$ pairs. Whether this is a "good" approximation will have to be decided from comparison with exact correlation functions. We shall discuss this in the following sections.

In the two-fluid theory (vdW2), the following assumptions are made:

$$g_{AA}(\frac{r}{\sigma_{AA}}; \rho, T, \mathbf{x}) = g_0(\frac{r}{\sigma_{xA}}; \rho\sigma_{xA}^3, \frac{kT}{\varepsilon_{xA}}) \tag{1.7}$$

$$g_{BB}(\frac{r}{\sigma_{BB}}; \rho, T, \mathbf{x}) = g_0(\frac{r}{\sigma_{xB}}; \rho\sigma_{xB}^3, \frac{kT}{\varepsilon_{xB}}) \tag{1.8}$$

and

$$g_{AB}(\frac{r}{\sigma_{AB}}; \rho, T, \mathbf{x}) = \frac{g_{AA} + g_{BB}}{2} \tag{1.9}$$

where

$$\sigma_{xA}^3 = \sum_j x_j \sigma_{jA}^3 \tag{1.10}$$

$$\sigma_{xB}^3 = \sum_j x_j \sigma_{jB}^3 \tag{1.11}$$

$$\varepsilon_{xA}\sigma_{xA}^3 = \sum_j x_j \varepsilon_{jA} \sigma_{jA}^3 \tag{1.12}$$

and

$$\varepsilon_{xB}\sigma_{xB}^3 = \sum_j x_j \varepsilon_{jB} \sigma_{jB}^3 \tag{1.13}$$

Finally, the one-fluid theory (vdW1) is given by

$$g_{ij}(\frac{r}{\sigma_{ij}}; \rho, T, \mathbf{x}) \approx g_0(\frac{r}{\sigma_x}; \rho\sigma_x^3, \frac{kT}{\varepsilon_x}), \qquad i,j=A,B \tag{1.14}$$

where

$$\sigma_x^3 = \sum_i \sum_j x_i x_j \sigma_{ij}^3 \tag{1.15}$$

and

$$\varepsilon_x \sigma_x^3 = \sum_i \sum_j x_i x_j \varepsilon_{ij} \sigma_{ij}^3 \qquad (1.16)$$

The above formulation can be easily generalized to ternary and multicomponent systems. For $n$ components, it is possible to have up to $n$-fluid versions of the theory. The reason for the mixing laws for the energy parameter $\varepsilon_x$ can be seen from eq. (1.2). For the one-fluid theory, (1.2) becomes simply

$$\frac{P}{\rho kT} = 1 - \frac{4\pi}{6} \rho \sigma_x^3 \frac{\varepsilon_x}{kT} \int_0^\infty dr^* \, r^{*3} \frac{\partial u^*}{\partial r^*} \, g_0(r^*; \rho \sigma_x^3, \frac{kT}{\varepsilon_x}) \qquad (1.17)$$

This equation is the same as the one for a hypothetical *pure* fluid with molecular parameters $\sigma_x$ and $\varepsilon_x$, even though these parameters are composition dependent. This type of formulation is known in literature as the *conformal solution theory* [2]: i.e., all fluids, mixture or pure, characterized by two force constants are considered isomorphic to each other at the same reduced density $\rho^* = \rho \sigma_x^3$ and reduced temperature $T^* = kT/\varepsilon_x$. At this point, we mention two other approximations that are of historical interest: (a) the random mixture model, and (b) the average potential model. The random mixture model simple states that all the rdf $g_{ij}$ are the same

$$g_{AA}(r; \rho, T, x) = g_{AB}(r; \rho, T, x) = g_{BB}(r; \rho, T, x) \qquad (1.18)$$

whereas the average potential model assumes that

$$g_{AB}(r; \rho, T, x) = \frac{1}{2}\left[g_{AA}(r; \rho, T, x) + g_{BB}(r; \rho, T, x)\right] \qquad (1.19)$$

Equation (1.18) is patently incorrect: it does not account for the effects arising from the size and energy differences in the pair potentials $u_{ij}$. Equation (1.19) is plausible, but is valid only for extremely simple cases. We note that in these two equations no reduction of the state variables, $\rho$ and $T$, is involved. Henderson and Lennard [3] have given a thorough discussion of these models.

## X.2. *Application to Hard-Sphere Mixtures*

We can test the validity of these theories by comparing the pressure values obtained above with known simulation results. In the following, we shall examine (a) the effects of molecular sizes on mixing rules in the absence of attractive energies (e.g., in HS mixtures), then (b) the effects of attractive energies in the presence of size differences (i.e., in mixtures of LJ molecules).

Hard sphere species differ from one another only by size. We shall denote the diameters of big spheres by $d_{AA}$ and of small spheres by $d_{BB}$. The cross interaction is characterized by $d_{AB}$ ($= (d_{AA} + d_{BB})/2$). We shall assume additive diameters here. The virial equation reduces in the HS case to

$$\frac{P}{\rho kT} = 1 + \frac{4\pi}{6} \sum_i \sum_j x_i x_j \rho d_{ij}^3 \, g_{ij}(d_{ij}; \rho, x), \qquad i,j = A,B \qquad (2.1)$$

where $g_{ij}(d_{ij})$ is the contact value of the $ij$ pcf. It depends on the density and compositions of the system, but not on temperature. Equation (2.1) can be contrasted with the pressure equation for pure hard spheres:

$$\frac{P}{\rho kT} = 1 + \frac{4\pi}{6} \rho d^3 g_0(d; \rho)$$                                                      (2.2)

where $g_0$ denotes pcf for pure hard spheres. The question posed in the $n$-fluid theories is the relation between the mixture pcf $g_{ij}(r)$ and the pure pcf $g_0(r)$. For hard spheres, we need only consider the contact values.

The contact value of $g_0$ is accurately given by the Carnahan-Starling (CS) formula

$$g_0(d) = \frac{4 - 2y}{4(1 - y)^3}$$                                                                              (2.3)

Thus if we have information on mixture pcf, we could test the $n$-fluid theories. Lee and Levesque [4] have carried out Monte Carlo simulations for HS mixtures. Both MC and CS results will serve as our standards. The results are listed in Table X.1. We compare three typical mixtures:

1. Low density with diameter ratio ($d_{AA}/d_{BB} =1.9$)
2. High density with diameter ratio ($d_{AA}/d_{BB} =1.1$)
3. High density with diameter ratio ($d_{AA}/d_{BB} =3.3$)

Table X.1.  *Test of vdW n-Fluid Theories for HS Mixtures*

| | | | Compressibility Factor, $\beta P/\rho$ | | | |
|---|---|---|---|---|---|---|
| $d_{AA}/d_{BB}$ | $\rho d_{AA}^3$ | $x_A$ | CS or MC* | vdW1 | vdW2 | vdW3 |
| 1.111 | 1.0825 | 0.01 | 7.56 | 7.56 | 7.57 | 7.58 |
| | | 0.25 | 9.44 | 9.4 | 9.66 | 9.96 |
| | | 0.50 | 12.3* | 12.0 | 12.5 | 13.0 |
| | | 0.75 | 15.8 | 15.7 | 16.2 | 16.7 |
| | | 0.99 | 20.7 | 20.7 | 20.8 | 20.8 |
| 1.111 | 1.1266 | 0.01 | 8.37 | 8.37 | 8.38 | 8.4 |
| | | 0.25 | 10.58* | 10.6 | 10.9 | 11.3 |
| | | 0.50 | 13.8 | 13.7 | 14.4 | 15.1 |
| | | 0.75 | 18.4 | 18.3 | 19.0 | 19.7 |
| | | 0.99 | 24.8 | 24.8 | 24.8 | 24.9 |
| 1.905 | 0.3036 | 0.25 | 1.23 | 1.22 | 1.23 | 1.24 |
| | | 0.50 | 1.406 | 1.40 | 1.42 | 1.44 |
| | | 0.75 | 1.65 | 1.64 | 1.66 | 1.69 |
| 3.333 | 1.8225 | 0.01 | 1.13 | 1.13 | 1.14 | 3.24 |
| | | 0.25 | 2.33 | 2.11 | 3.21 | 1303 |
| | | 0.50 | 8.814* | 6.03 | 15.5 | 5209 |

*Monte Carlo simulation of Lee and Levesque [3].

For equimolar mixtures at low densities ($\rho d_{AA}^3 =0.3036$), the compressibility from the CS equation is 1.406. All three theories give comparable results, with the vdW1 giving the best prediction. At higher density ($\rho d_{AA}^3 = 1.0825$) with diameter ratio $d_{AA}/d_{BB} =1.111$, vdW1 gives a value of 11.993 as compared to the MC number 12.3.

VdW2 gives 12.48, and vdW3 gives 10.327. Since the diameters are very close, we can expect all three theories to be fairly accurate. (In the limit $d_{AA} = d_{BB}$, all three theories are exact). At a higher diameter ratio $d_{AA}/d_{BB} = 3.333$, corresponding to a molecular volume ratio of ~37, MC gives $\beta P/\rho = 8.814$ and vdW1 gives 6.027. VdW2 and vdW3 theories are totally inadequate at this density, giving 15.49 and 5205, respectively. The vdW3 theory fails to give a reasonable value due to the magnitude of the contact value of $g_{AA}(d_{AA}) = 5453$. This is caused by the high value of the reduced density $\rho d_{AA}^3 = 1.8225$. Namely, the vdW3 theory presumes in one of its components a hypothetical pure fluid composed entirely of $N$ large spheres (see eq. (1.4)). This picture is physically untenable because actually half of the molecules ($N/2$) in the mixture are small spheres that are interspersed among the large ones, thus leaving a larger than presumed "empty space" or *free volume* available for the other molecules. Consequently, the actual total pressure is considerably lower than that given by the hypothetical fluid filled with large spheres. The above analysis clearly demonstrates that the vdW1 theory is the best theory among the vdW $n$-fluid theories. VdW2 and vdW3 are inadequate for large size differences.

## X.3. *Application to Lennard-Jones Mixtures*

Recently, Hoheisel and Lucas [5] and coworkers [6] [7] have carried out extensive simulations of binary mixtures of the LJ molecules with different size ratios (up to 2) and energy ratios (up to 5). They compared the pcf obtained from the vdW1 and the mean density approximation (MDA) [8] with simulation data. Their results showed that both theories are reasonable, but the MDA theory produces better $J$-integrals (see below). The MDA asserts

$$\sigma_x^3 = \sum\sum x_i x_j \sigma_{ij}^3 \tag{3.1}$$

and

$$g_{ij}(\frac{r}{\sigma_{ij}}; \rho, T, \mathbf{x}) = g_0(\frac{r}{\sigma_x}; \rho\sigma_x^3, \frac{kT}{\varepsilon_{ij}}) \tag{3.2}$$

The vdW1 theory is given by eqs. (1.14-16). The MDA differs from vdW1 in the use of the individual energy parameters $\varepsilon_{ij}$ for calculation of the reduced temperatures. The two theories coincide when $\varepsilon_{AA} = \varepsilon_{BB}$.

The virial pressure for LJ molecules can be written in terms of the $J$-integrals as

$$\frac{\beta P}{\rho} = 1 - 16\pi \sum_i\sum_j x_i x_j \beta\varepsilon_{ij}\rho\sigma_{ij}^3 \left[ J_{ij}^{(6)} - 2J_{ij}^{(12)} \right] \tag{3.3}$$

where the $J$-integrals are defined as

$$J^{(n)} \equiv \int dr^* \, r^{*2-n} g(r^*) \tag{3.4}$$

and $r^* = r/\sigma$ is the reduced distance. Hoheisel et al. [9] have given extensive tabulation of the $J$-integrals. For pressures, Hoheisel, Deiters and Lucas [10] and later Shukla et al. [11] have made MD calculations and compared the vdW1, vdW2, vdW3, and MDA approximations. Table X.2 presents their results. The vdW2 and vdW3 theories are again found to be inadequate for large diameter ratios, and are thus not shown. The vdW1 and MDA theories are compared with the hard sphere expansion (HSE) theory of Mansoori and Leland [12], and the Weeks-Chandler-Andersen (WCA) [13], Lee-Levesque (LL) [14], and Grundke-Henderson (GH) [15] theories (WCA-LL-GH). For small diameter ratios (<1.3), and the energy ratios varying from 1.0 to 4.5, the deviations in vdW1 go from −1.2% to −3.5%. MDA sustains a deviation up to −5%, while WCA-LL-GH remains below 1.3%, improving at higher energy ratios (down to −0.2%). At the diameter ratio of 2.0, most of the theories deteriorate: the vdW1 goes from −13.7%, −17%, to −29.5% at higher energy ratios; the MDA goes from −13.7% to −32.4%; while WCA-LL-GH fluctuates between −6.8% and −9%. This comparison shows that the WCA-LL-GH theory gives the most accurate results for the LJ mixtures. Results become worsened when the diameter ratio is increased. Similar behavior is observed at a lower density isochore ($\rho^* = 0.8$).

The *n*-fluid assumptions have been tested by Hoheisel and Lucas (*Ibid.*). Figures X.1 and 2 show the predictions of eq. (1.14) (vdW1) and eq. (3.2) (MDA) for the pcfs. For equal-sized molecules, MDA is slightly better than vdW1 (Figure X.1). As the diameter ratio reaches 1.65 (Figure X.2), the first peak is well prediction by both theories. However, the packing order is incorrect from the first minimum onward: i.e., the predicted pcf oscillates out of phase with MD data. The second MD peak occurs at ~7.6 Å, while the second peak of MDA occurs prematurely at ~5.8 Å. This discrepancy is caused by the assumption of equal-sized spheres (with diameter $\sigma_x$) in the MDA and vdW1 theories, whereas in the mixture, the second-neighbor small spheres surrounding a center sphere are likely to form a "sandwich" with a large sphere in between (i.e., in the ordered concatenation: *center-large sphere-small sphere*). This incorrect packing order is also observed for the MDA or vdW1 theories in mixtures of other size ratios. Thus as theories for the liquid structure, vdW1 and MDA are poor candidates. From observations made above we draw the following conclusions:

1.  VdW2 and vdW3 are not viable theories for HS and LJ mixtures. VdW1 theory (which is identical to the MDA in the HS case) is surprisingly accurate for hard spheres with large size difference.

2.  Both MDA and vdW1 give reasonable predictions for the pressure of the LJ fluid when the differences in energy and size parameters are not too excessive. The two theories are comparable, but MDA is slightly better in *J*-integral calculations. At large size differences, neither vdW1 nor MDA provides quantitative predictions.

## X.4. *The Lattice Gas Model of Mixtures*

The LCM of solutions, in its 20 years of existence, has been shown to work well for calculation of vapor-liquid equilibria of mixtures of highly polar substances (e.g., water, acetone and methanol), and is widely used in practice. The original formulation [16] for the excess Gibbs free energy is

$$\frac{G^E}{NkT} = - x_A \ln(x_A + x_B \Lambda_{BA}) - x_B \ln(x_B + x_A \Lambda_{AB}) \tag{4.1}$$

where $\Lambda_{ij}$ are the Wilson parameters. This formula could be traced back to the earlier developments of the lattice gas model by Guggenheim [17]. In his formulation, the gas molecules of species A and B are arranged on a regular lattice and the properties of

Table X.2. The Pressure Values of Equimolar Mixtures of LJ Molecules at Various Size and Energy Ratios. ($\varepsilon_{11}/k$ =34 K, $\sigma_{11}$ =2.85 Å, $T$ =200 K) (Shukla et al., Fluid Phase Equil., 1986) [9]

| $\varepsilon_{22}/\varepsilon_{11}$ | $\sigma_{22}/\sigma_{11}$ | $n^* = n\sigma_x^3$ | $P_{MD}^{(bar)}$ | $\Delta$VDW1 (%) | $\Delta$MDA (%) | $\Delta$HSE (%) | $\Delta$WCA– LL–GH (%) |
|---|---|---|---|---|---|---|---|
| 1.00 | 1.30 | 0.9 | 3445 | − 2.5 | − 2.5 | 4.7 | 1.3 |
| 1.00 | 2.00 | 0.9 | 1549 | − 13.7 | − 13.7 | − 15.1 | − 6.8 |
| 1.50 | 1.00 | 0.9 | 5307 | − 0.2 | − 0.4 | 5.8 | 1.4 |
| 1.50 | 1.30 | 0.9 | 3563 | − 3.5 | − 3.8 | 2.0 | 1.1 |
| 1.50 | 2.00 | 0.9 | 1629 | − 17.0 | − 17.2 | − 27.5 | − 9.5 |
| 2.50 | 1.00 | 0.9 | 5271 | 2.4 | 0.8 | 5.6 | 0.78 |
| 2.50 | 1.30 | 0.9 | 3691 | − 2.7 | − 4.2 | 1.5 | − 1.0 |
| 2.50 | 2.00 | 0.9 | 1778 | − 20.8 | − 21.8 | − 16.4 | − 9.2 |
| 3.50 | 1.30 | 0.9 | 3680 | − 2.0 | − 5.0 | 1.1 | − 0.2 |
| 3.50 | 2.00 | 0.9 | 1851 | − 25.7 | − 27.7, | − 23.3 | − 8.7 |
| 4.50 | 1.00 | 0.9 | 5159 | 8.2 | 2.7 | 5.8 | 0.8 |
| 4.50 | 1.30 | 0.9 | 3551 | − 1.2 | − 6.0 | 0.6 | − 0.54 |
| 4.50 | 2.00 | 0.9 | 1905 | − 29.5 | − 32.4 | − 8.4 | − 7.5 |
| 1.50 | 1.05 | 0.8 | 3323 | − 0.5 | − 0.7 | 4.6 | 0.9 |
| 1.50 | 1.30 | 0.8 | 2347 | − 3.4 | − 3.6 | 2.3 | − 0.9 |
| 1.50 | 1.55 | 0.8 | 1765 | − 8.2 | − 8.4 | − 2.8 | − 6.1 |
| 1.50 | 2.00 | 0.8 | 1078 | − 14.9 | − 15.1 | − 12.4 | − 8.3 |
| 2.50 | 1.05 | 0.8 | 3323 | 0.8 | − 0.7 | 4.1 | − 0.1 |
| 2.50 | 1.30 | 0.8 | 2323 | − 3.3 | − 4.8 | 1.7 | − 1.1 |
| 2.50 | 1.55 | 0.8 | 1690 | − 8.9 | − 10.2 | − 3.5 | − 3.8 |
| 2.50 | 2.00 | 0.8 | 1081 | − 19.8 | − 20.8 | − 18.0 | − 9.4 |
| 3.50 | 1.05 | 0.8 | 3162 | 3.7 | 0.2 | 4.9 | 0.3 |
| 3.50 | 1.30 | 0.8 | 2213 | − 2.1 | − 5.4 | 2.4 | − 0.9 |
| 3.50 | 1.55 | 0.8 | 1716 | − 9.5 | − 12.3 | − 3.7 | − 2.9 |
| 3.50 | 2.00 | 0.8 | 1097 | − 23.2 | − 25.3 | − 21.5 | − 9.0 |
| 4.50 | 1.05 | 0.8 | 2935 | 6.1 | 0.0 | 4.9 | − 0.3 |
| 4.50 | 1.30 | 0.8 | 2092 | − 1.0 | − 6.3 | 3.0 | − 0.8 |
| 4.50 | 1.55 | 0.8 | 1568 | − 9.2 | − 13.8 | − 2.9 | − 0.7 |
| 4.50 | 2.00 | 0.8 | 1076 | − 27.1 | − 30.5 | − 25.4 | − 7.7 |
| 1.50 | 1.30 | 0.68858 | 1531 | − 3.4 | − 3.6 | 2.2 | − 1.1 |
| 1.50 | 2.00 | 0.70863 | 711 | − 12.2 | − 12.2 | − 5.3 | − 5.1 |
| 4.50 | 1.30 | 0.5000 | 444 | − 2.5 | − 6.5 | 3.2 | − 4.7 |
| 4.50 | 2.00 | 0.42518 | 109 | − 7.3 | − 9.1 | 8.3 | − 8.3 |

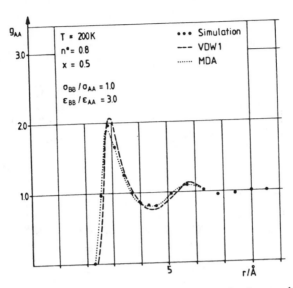

FIGURE X.1. *The pair correlation function* $g_{AA}$. *The first peak is well predicted by the vdW1 and MDA theories. Since A and B are of equal size* ($\sigma_{AA} = \sigma_{BB}$), *the predicted periods of oscillations are also in phase with MD results. (Hoheisel and Lucas, Mol. Phys., 1984)*

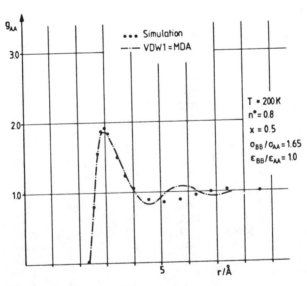

FIGURE X.2. *The pair correlation function* $g_{AA}$ *for small spheres. As* $\varepsilon_{AA} = \varepsilon_{BB}$, *MDA = vdW1. The theoretical packing order is incorrect due to the assumption in vdW1 of equal-sized* ($\sigma_x$) *molecules. The oscillations* (— · —) *are out of phase with MD results* (•). *(Hoheisel and Lucas, Mol. Phys., 1984)*

solutions are derived from combinatorial considerations. Granted that real-gas molecules are not localized on the lattice sites, as are the molecules of a crystal, the lattice gas model serves as a useful model for liquids and provides physical insights on mixture behavior. Its limitations will be subjects of discussion in the sections dealing with modern liquid theories.

Consider a regular crystalline lattice (the type of lattice structure is immaterial here) in which each lattice point is surrounded by $z$ nearest neighbors. A schematic square lattice is shown in Figure X.3. The number $z$ is called the *coordination number*. It is well defined for regular crystalline structures. For example, $z$ for the regular cubic lattice is 6, for a body-centered cubic 8, and for a face-centered cubic 12. A similar quantity is used in liquid theories. Scattering experiments on liquids have shown that $z \approx 4$ for highly polar liquids and ~8 ($\pm 2$) for nonpolar liquids. In addition, the coordination number of an A molecule, $z_A$, may differ from the coordination number of B, $z_B$. Figure X.4 shows such a case. To form a binary mixture on lattice, we start with a situation as shown in Figure X.3(a): a block of $N_A$ molecules of A bordering on a block of $N_B$ molecules of B before mixing. Mixing is carried out by interchanging an interior A molecule with an interior B molecule, and the procedure is repeated until the final mixture is obtained. We assume that the interaction energy is due solely to nearest-neighbor (n.n.) interactions, i.e., there is energy $u_{AA}$ between a pair of AA molecules, $u_{BB}$ between a pair of BB molecules, and $u_{AB}$ between a pair of AB molecules. The total energy of the lattice gas before mixing is

$$U' = \frac{z}{2}N_A u_{AA} + \frac{z}{2}N_B u_{BB} \tag{4.2}$$

(upon neglecting boundary effects). After one A molecule is interchanged with one B molecule, there is an increment of the total energy because $z$ ($z = 4$ in Figure X.3) AA bonds and $z$ BB bonds are broken while $2z$ ($2z = 8$ for the square lattice) AB bonds are formed. (We use the term "bonds" here to refer to n.n. interactions.) The change in total energy is

$$U_{\text{after mixing}} - U_{\text{before mixing}} = \Delta U' = 2zu_{AB} - zu_{AA} - zu_{BB} \equiv 2w \tag{4.3}$$

where we have introduced the interchange energy $w$. Repeated interchanges produce a final mixture with $N_{AA}$ n.n. AA bonds and $N_{BB}$ n.n. BB bonds as well as $N_{AB}$ ($=N_{BA}$) n.n. AB bonds. These numbers satisfy the overall balance equations

$$2N_{AA} + N_{BA} = zN_A \tag{4.4}$$

and

$$2N_{BB} + N_{AB} = zN_B \tag{4.5}$$

The factor 2 in front of $N_{AA}$ and $N_{BB}$ accounts for double counting of the same pair. In addition, the fraction of B molecules as neighbors to a central A molecule can be calculated as

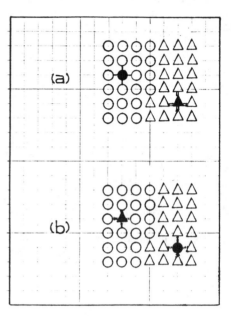

FIGURE X.3. Interchange of gas molecules A (○) and B (Δ) on a square lattice. (a) Before interchange; (b) after interchange. We note that four $u_{AA}$ and four $u_{BB}$ bonds are destroyed, while four new $u_{AB}$ bonds are formed. The displaced molecules are darkened for visualization.

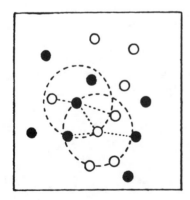

FIGURE X.4. Nearest neighbors in a liquid for a binary mixture of molecules A (○) and B (●). Note that $n_{BA} = 3$ (i.e. there are three B molecules as nearest neighbors to A), and $n_{AA} = 3$, whereas $n_{AB} = 3$ and $n_{BB} = 2$. The coordination numbers: $z_A = n_{AA} + n_{BA} = 6$; and $z_B = n_{BB} + n_{AB} = 5$. Thus the local mole fractions $x_{BA} = 3/6 = 0.5$, and $x_{AB} = 3/5 = 0.6$. (Lee and Starling, Fluid Phase Equil., 1985)

$$x_{BA} = \frac{N_{BA}}{zN_A} \tag{4.6}$$

and the fraction of A neighbors surrounding an A center is

$$x_{AA} = \frac{2N_{AA}}{zN_A} \tag{4.7}$$

That is, on the average, $z$ nearest neighbors of center A contain $zx_{AA}$ A molecules and $zx_{BA}$ B molecules. These fractions $x_{ij}$ are called the *local compositions (LC)*. They are not necessarily equal to the bulk mole fractions $x_A$ or $x_B$ due to asymmetrical forces of interaction between A molecules and B molecules. Mixtures where $x_{AA} = x_A$ and $x_{BA} = x_B$ are called *random mixtures*. They belong to a special class of ideal solutions. For nonideal solutions, the local mole fractions are different, and the mixtures are nonrandom mixtures. We show this nonrandomness by referring to Figure X.3(b) which presents a case of highly nonrandom mixing. We count 36 for $N_{AA}$, 32 for $N_{BB}$, and 15 for $N_{AB}$ ($=N_{BA}$). Thus on the average the local compositions are given by

$$x_{AA} = \frac{2N_{AA}}{zN_A} = \frac{(2)(36)}{(4)(22)} = 0.83 \tag{4.8}$$

This value is compared with $x_A = 22/42 = 0.524$. Similarly, $x_{BB} = 15/(4)(22) = 0.170$ as compared to $x_B = 20/42 = 0.476$. (Note that we have included bonds extending outward from the boundaries towards neighbors not shown. The approach is valid in the thermodynamic limit $N \to \infty$.) From relation (4.4), we also have the normalization condition

$$x_{AA} + x_{BA} = 1 \tag{4.9}$$

Similar relations can be given for the local mole fractions with center B (by interchanging the indices A and B above).

The configurational internal energy $U' (= U - U_{ideal\ gas})$ for the final mixture can be calculated as

$$U' = N_{AA}u_{AA} + N_{BB}u_{BB} + N_{AB}u_{AB} \tag{4.10}$$

Substituting the expressions of local compositions into (4.10), we have

$$U' = \frac{z}{2}N_A x_{AA}u_{AA} + \frac{z}{2}N_B x_{BB}u_{BB} + zN_A x_{BA}u_{BA} \tag{4.11}$$

$$= \frac{z}{2}N[x_A x_{AA}u_{AA} + x_B x_{BA}u_{BA}] + \frac{z}{2}N[x_B x_{BB}u_{BB} + x_A x_{AB}u_{AB}]$$

To use the notation of Guggenheim [18], we identify our $u_{AA} = $ his $-(2/z)E_A$, our $u_{BB} = $ his $-(2/z)E_B$, and our $u_{AB} = $ his $(1/z)[w - E_A - E_B]$. Substitution into (4.11) gives

$$U' = -\frac{z}{2}(N_A - N_A x_{BA})\frac{2E_A}{z} - \frac{z}{2}(N_B - N_B x_{AB})\frac{2E_B}{z} \tag{4.12}$$

$$+ zN_A x_{BA}\frac{1}{z}(w - E_A - E_B)$$

$$= -N_A E_A - N_B E_B + \frac{wN_{AB}}{z}$$

where we have used the relation $N_A x_{BA} = N_B x_{AB}$. Thus, the internal energy depends on the number of AB pairs actually formed in the mixture. When $w = 0$, i.e., the interchange energy is zero, there is no change in the total internal energy upon mixing. (This is one of the conditions for ideal solutions.)

The formation of AB pairs cannot be arbitrarily assigned. It depends on the actual interaction forces operating among the molecules as well as on the state conditions of the mixture. In statistical mechanics, the number is given by an ensemble average. Wilson [19] has proposed the formulas

$$N_{BA} = zN_A \frac{x_B \Lambda_{BA}}{x_A + x_B \Lambda_{BA}} \tag{4.13}$$

or

$$x_{BA} = \frac{x_B \Lambda_{BA}}{x_A + x_B \Lambda_{BA}} \tag{4.14}$$

and

$$x_{AA} = \frac{x_A}{x_A + x_B \Lambda_{BA}} \tag{4.15}$$

where $\Lambda_{ij}$ is a parameter defined by

$$\Lambda_{BA} \equiv \frac{V_B \exp(-J_{BA}/kT)}{V_A \exp(-J_{AA}/kT)} \tag{4.16}$$

and $J_{ij}$ are the nearest-neighbor energies of interaction. As we shall see, there is a molecular basis for these formulas.

## X.5. *A Liquid Theory of Local Compositions*

The lattice gas model discussed above imposes a crystalline structure on the gas molecules. This picture is evidently unrealistic. Molecules in gases and liquids are capable of rapid diffusional motions that destroy the latticework. This phenomenon is observed during melting of a solid. In order to generalize the local composition concepts to liquids, we undertake a complete reformulation of the theoretical basis. First, we abandon the lattice structure and introduce instead the modern liquid structural theories expounded in earlier chapters.

Consider a GCE for a binary mixture of molecules of species A and B. We do not require a crystalline structure here.  The partition function $\Xi$ is given in general by

$$\Xi = \sum_{N_A \geq 0}^{\infty} \sum_{N_B \geq 0}^{\infty} \frac{z_A^{N_A} z_B^{N_B}}{N_A! N_B!} \int d\{N\} \, \exp[-\beta V_N(\{N\})] \tag{5.1}$$

where $z_i = \exp(\beta\mu_i)/\Lambda_i^3$, $i$ = A,B, are the activities (these ought to be distinguished from the coordination numbers defined earlier). $\beta = 1/kT$, and $k$ is the Boltzmann constant.  $\Lambda_i$ is the de Broglie thermal wavelength $h/(2\pi m_i kT)^{1/2}$, $T$ is the absolute temperature, $m_i$ is the mass of molecule $i$, $h$ is Planck's constant, and $\mu_i$ is the chemical potential of species $i$.  The potential energy $V_N$ is given by

$$V_N = \sum_{i=1}^{N_A} u_A^{(1)}(i) + \sum_{j=1}^{N_B} u_B^{(1)}(j) + \sum_{\gamma=A,B} \sum_{\delta=A,B} \sum_{i=1}^{N} {\sum_{i<j}^{N}}' u_{\gamma\delta}^{(2)}(ij) \tag{5.2}$$

$$+ \sum_{\gamma,\delta,\varepsilon=A,B} {\sum_{i,j,k=1}}' u_{\gamma\delta\varepsilon}^{(3)}(ijk) + \cdots$$

where $u^{(1)}$ is the one-body external potential, $u^{(2)}$ is the pair interaction potential, $u^{(3)}$ is the triplet interaction potential, etc.  The arguments $ijk$ represent the position vectors, $\mathbf{r}_i$, $\mathbf{r}_j$, and $\mathbf{r}_k$ of the $N$ molecules as well as the Euler angles $\omega_i$, $\omega_j$, and $\omega_k$.  (The latter are necessary for interactions among polar molecules .)  To simplify matters, we consider a homogeneous system where $u^{(1)} = 0$.  If pairwise additivity can be applied, all interactions higher than $u^{(2)}$ can be neglected.  However, this is not the case for dense liquids and highly polar molecules (especially for hydrogen-bonding fluids such as water and methanol).  To minimize cluttered notation, we shall consider the pairwise additive systems first.  For fluids where $u^{(3)}$ and $u^{(4)}$, etc., are important, we recommend the effective pair potential method of Stillinger [20], i.e., recombination of all the higher-order effects into an effective pair potential $\bar{u}(ij; T, \rho)$ and treatment of the system as if it were pairwise additive.  This is achieved at the expense of rendering $\bar{u}$ temperature and density dependent.  For mixtures, $\bar{u}$ is also composition dependent.  Thus,

$$V_N(\{N\}) = \sum_{\gamma=A,B} \sum_{\delta=A,B} \sum_{i=1}^{N} {\sum_{i<j}^{N}}' \bar{u}_{\gamma\delta}(ij; T, \rho, x) \tag{5.3}$$

The pcf $g_{\gamma\delta}^{(2)}(ij)$ is given by

$$g_{\gamma\delta}^{(2)}(12) = \frac{1}{\rho_\gamma \rho_\delta} \frac{1}{\Xi} \sum_{N_A = \delta_{A\gamma} + \delta_{A\delta}}^{\infty} \sum_{N_B = \delta_{B\delta} + \delta_{B\gamma}}^{\infty} \frac{z_A^{N_A} z_B^{N_B}}{(N_A - \delta_{A\gamma} - \delta_{A\delta})! (N_B - \delta_{B\delta} - \delta_{B\gamma})!} \tag{5.4}$$

$$\cdot \int d\{N-2\} \, \exp[-\beta V_N]$$

where $\delta_{ij}$ is the Kronecker delta.  For binary mixtures, we have the combinations $g_{AA}^{(2)}(ij)$, $g_{AB}^{(2)}(ij)$, $g_{BA}^{(2)}(ij)$ and $g_{BB}^{(2)}(ij)$.  Due to combinatorial symmetry, $g_{AB}^{(2)} = g_{BA}^{(2)}$.  To simplify notation we shall omit the superscript (2) on the pcfs.  Physically, $g_{\gamma\delta}(r_{ij})$ gives the probability that one finds a molecule of type $\gamma$ surrounding a molecule of type $\delta$ at the radial distance $r_{ij}$. Figure X.4 illustrates the nearest neighbors in a liquid mixture.

The nearest-neighbor number, $n_{AA}(L)$, of A molecules surrounding a central A molecule and within a sphere of radius $L$ is

$$n_{AA}(L) = \rho_A \int_0^L dr\, 4\pi r^2 g_{AA}(r) \tag{5.5}$$

Similarly,

$$n_{AB}(L) = \rho_A \int_0^L dr\, 4\pi r^2 g_{AB}(r) \tag{5.6}$$

and

$$n_{BB}(L) = \rho_B \int_0^L dr\, 4\pi r^2 g_{BB}(r) \tag{5.7}$$

Figure IV.5 is a plot [21] of the coordination number $z(r)$ $(= N(r))$ for a pure fluid that gives the number of neighbors at distance $r$ surrounding a central molecule.

## EXPRESSIONS OF THE LOCAL COMPOSITIONS IN TERMS OF PCF

For a mixture of $N_A$ molecules of A species and $N_B$ molecules of B species, the number of nearest-neighbor AA pairs formed is denoted by $<N_{AA}>$. The angular brackets denote the ensemble average. The average number of nearest neighbor AB pairs is $<N_{AB}>$. We shall similarly define $<N_{BA}>$ and $<N_{BB}>$. If the mixture is a lattice gas, the total number of "bonds" that can be formed from the A molecules is $z_A N_A$, and that from the B molecules is $z_B N_B$. Earlier we have assumed that the coordination number $z_A$ of A is the same as $z_B$ of B, i.e., $z_A = z_B$. Generally speaking, this equality does not hold in liquids. The quasi-chemical theory (Guggenheim *Ibid.*) defines the local compositions as

$$x_{AA} = \frac{2<N_{AA}>}{z_A N_A} \tag{5.8}$$

$$x_{BA} = \frac{<N_{BA}>}{z_A N_A} \tag{5.9}$$

$$x_{AB} = \frac{<N_{AB}>}{z_B N_B} \tag{5.10}$$

$$x_{BB} = \frac{2<N_{BB}>}{z_B N_B} \tag{5.11}$$

Clearly, $x_{BA}$ is the fraction of the mean number of AB pairs actually formed in the mixture over the number of all pairs that could have been formed by the $N_A$ A molecules (including both AA and AB pairs). In terms of a single molecule, $x_{BA}$ gives the fraction of the $z_A$ bonds surrounding a central A molecule occupied by the B molecules, and thus the name *local mole fraction*. For example, for the face-centered cubic (f.c.c.) lattice, each A molecule is surrounded by 12 nearest neighbors. If, on the average, seven of these positions are occupied by the B molecules, the local fraction of B ( i.e., $x_{BA}$) is $7/12 = 0.5833$. Since the remaining positions are occupied by the A molecules, $x_{AA} = 0.4167$. The same reasoning can be applied to a B-center with fractions $x_{AB}$ and $x_{BB}$. In a liquid, the precise number of nearest neighbors that could be found in a crystal must give way to a statistical average, This is afforded by the pcfs (see eqs. (5.5, 6 and 7)). If we follow the same reasoning as for the lattice gas, the local compositions can now be defined for any fluid as

$$x_{AA} = \frac{n_{AA}(L_{AA})}{n_{AA}(L_{AA}) + n_{BA}(L_{BA})} \tag{5.12}$$

$$x_{BA} = \frac{n_{BA}(L_{BA})}{n_{AA}(L_{AA}) + n_{BA}(L_{BA})} \tag{5.13}$$

$$x_{AB} = \frac{n_{AB}(L_{AB})}{n_{AB}(L_{AB}) + n_{BB}(L_{BB})} \qquad \text{etc.} \tag{5.14}$$

Also, the coordination numbers $z_j$ are

$$z_A \equiv n_{AA}(L_{AA}) + n_{BA}(L_{BA}), \qquad\qquad z_B \equiv n_{BB}(L_{BB}) + n_{AB}(L_{AB}) \tag{5.15}$$

These definitions depend on the radii $L_{ij}$ chosen for the calculation. This ambiguity arises not from the statistical theory but from the approximations inherent in the local composition theory when applied to liquids. As we substitute the definitions of the coordination numbers into the above ratios, we obtain

$$x_{AA} = \frac{\rho_A \int_0^{L_{AA}} dr\, 4\pi r^2 g_{AA}(r)}{\rho_A \int_0^{L_{AA}} dr\, 4\pi r^2 g_{AA}(r) + \rho_B \int_0^{L_{BA}} dr\, 4\pi r^2 g_{BA}(r)} \tag{5.16}$$

$$= \frac{x_A}{x_A + x_B \int_0^{L_{BA}} dr\, 4\pi r^2 g_{BA}(r) / \int_0^{L_{AA}} dr\, 4\pi r^2 g_{AA}(r)}$$

Similarly,

$$x_{BA} = \frac{x_B \int\limits_0^{L_{BA}} dr\, 4\pi r^2 g_{BA}(r) \Big/ \int\limits_0^{L_{AA}} dr\, 4\pi r^2 g_{AA}(r)}{x_A + x_B \int\limits_0^{L_{BA}} dr\, 4\pi r^2 g_{BA}(r) \Big/ \int\limits_0^{L_{AA}} dr\, 4\pi r^2 g_{AA}(r)} \tag{5.17}$$

The local compositions $x_{AB}$ and $x_{BB}$ could be similarly defined. These equations make an interesting contrast with the equations of Wilson [22] for the local compositions; i.e.,

$$x_{AA} = \frac{x_A}{x_A + x_B \Lambda_{BA}} \tag{5.18}$$

and

$$x_{BA} = \frac{x_B \Lambda_{BA}}{x_A + x_B \Lambda_{BA}} \tag{5.19}$$

The Wilson parameters $\Lambda_{ij}$ could be identified in the liquid theory as

$$\Lambda_{BA} = \int\limits_0^{L_{BA}} dr\, 4\pi r^2 g_{BA}(r) \Big/ \int\limits_0^{L_{AA}} dr\, 4\pi r^2 g_{AA}(r) \tag{5.20}$$

Furthermore, we could rewrite the pcf in terms of the potentials of mean force [23]

$$\int\limits_0^{L_{AA}} dr\, 4\pi r^2 g_{AA}(r) = \int\limits_0^{L_{AA}} dr\, 4\pi r^2 \exp[-\beta W_{AA}(r)] \tag{5.21}$$

$$= V_{AA} \exp[-\beta \overline{W}_{AA}]$$

and

$$\int\limits_0^{L_{BA}} dr\, 4\pi r^2 g_{BA}(r) = \int\limits_0^{L_{BA}} dr\, 4\pi r^2 \exp[-\beta W_{BA}(r)] \tag{5.22}$$

$$= V_{BA} \exp[-\beta \overline{W}_{BA}]$$

where $V_{AA}$ is the spherical volume $4\pi L_{AA}^3/3$, and $V_{BA} = 4\pi L_{BA}^3/3$. The second equalities in eqs. (5.21 and 22) were obtained from the mean value theorem of calculus; i.e., the potential of mean force $W_{ij}(r)$ is evaluated at some mean location in the region of integration. Figure X.5 shows the simulated potential of mean force for an LJ fluid (argon) in the liquid state [24]. Thus,

$$\Lambda_{BA} = \frac{V_{BA}}{V_{AA}} \exp[-\beta(\overline{W}_{BA} - \overline{W}_{AA})] \tag{5.23}$$

This is precisely the form given by Wilson (*Ibid.*) for the distribution parameter $\Lambda_{BA}$.

Wilson's $J_{ij}$, in the present interpretation, are actually the mean potentials of mean force of Kirkwood. We note that in eqs. (5.21) and (5.22) the rdfs $g_{AA}(r)$, $g_{AB}(r)$, and $g_{BB}(r)$ and the potentials of mean force $W_{AA}(r)$, $W_{AB}(r)$, and $W_{BB}(r)$ are all composition dependent [25]. Therefore, our definitions of the parameters $\Lambda_{ij}$ are also composition dependent. Wilson gave $\Lambda_{ij}$ as composition-independent parameters. The limitations of his $\Lambda_{ij}$ have been commented on by Flemr [26]. The definitions given here are based on the liquid molecular theories and are thus free from these limitations [27]. Wilson's molar volume $\bar{V}_j$ corresponds in the present treatment to the volume of the first coordination shell. If we happen to choose equal radii, $L_{AA} = L_{BA} = L_{AB} = L_{BB}$, the ratio $V_{BA}/V_{AA}$ would be unity, and we would obtain the Whiting-Prausnitz [28] expressions for local compositions.

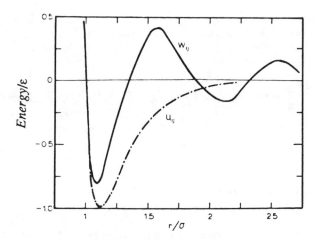

FIGURE X.5. *The potential of mean force $W_{ij}(r)$ (solid line) for a Lennard-Jones liquid obtained from computer simulation. The pair potential $u_{ij}(r)$ is given as the dash-dot curve. (Posch et al., Mol. Phys., 1981)*

## ENERGY EXPRESSIONS

As noted in eq. (4.11), the internal energy is given by an expression involving the local compositions. On the other hand, the internal energy can also be expressed in terms of the distribution functions as (Hill [29])

$$U' = \frac{1}{2} \frac{N_A(N_A-1)}{V} \int_0^\infty dr \, 4\pi r^2 g_{AA}(r)\bar{u}_{AA}(r) + \frac{1}{2} \frac{N_A N_B}{V} \int_0^\infty dr \, 4\pi r^2 \, g_{AB}(r)\bar{u}_{AB}(r) \quad (5.24)$$

$$+ \frac{1}{2} \frac{N_B N_A}{V} \int_0^\infty dr \, 4\pi r^2 g_{BA}(r)\bar{u}_{BA}(r) + \frac{1}{2} \frac{N_B(N_B-1)}{V} \int_0^\infty dr \, 4\pi r^2 g_{BB}(r)\bar{u}_{BB}(r)$$

There is a major difference between the local composition expression (4.11) and the statistical mechanical (SM) expression (5.24): (4.11) provides for *local* (or nearest-neighbor) energies only, whereas (5.24) includes long-range contributions (the upper limit of integration goes to infinity). For dense fluids, second and third neighbors contribute significantly to the total energy. The correlation functions also show prominent peaks beyond the first maxima. (See, e.g., Figure IV.1). The long-range contributions (beyond the nearest neighbors) are denoted by $U^{LR}$. Then the relation between the local and the statistical mechanical energies is

$$U_{AA} = \frac{z_A}{2} N_A x_{AA} u_{AA} + U_{AA}^{LR} \tag{5.25}$$

$$= \frac{1}{2} \frac{N_A(N_A-1)}{V} \int_0^\infty dr \, 4\pi r^2 g_{AA}(r) \bar{u}_{AA}(r)$$

where $U_{AA}$ is the AA pair contribution to the internal energy. $U_{AA}^{LR}$ is the long-range AA contribution beyond the first neighbors. If we separate the SM expression into short- and long-range parts

$$\frac{1}{2} \frac{N_A(N_A-1)}{V} \int_0^\infty dr \, 4\pi r^2 g_{AA}(r) \bar{u}_{AA}(r) = \frac{\rho_A N_A}{2} \int_0^{L_{AA}} dr \, 4\pi r^2 g_{AA}(r) \bar{u}_{AA}(r) + U_{AA}^{LR} \tag{5.26}$$

comparison with (5.25) gives

$$z_A x_{AA} u_{AA} = \rho_A \int_0^{L_{AA}} dr \, 4\pi r^2 g_{AA}(r) \bar{u}_{AA}(r) \tag{5.27}$$

This formula is consistent with the definition of $x_{AA}$ given in (5.12) and (5.21), i.e.,

$$n_{AA} = z_A x_{AA} = \rho_A V_{AA} \exp(-\beta \bar{W}_{AA}) \tag{5.28}$$

where the mean value theorem has been applied to the energy integral. Similar expressions could be derived for the other pairs. We note that the statistical mechanical formula (5.24) includes long-range interactions that propagate over the distances of several neighborhoods (including the first neighbors, the second neighbors, the third neighbors, etc.). This long-ranged nature is evident in the pcfs of liquids (see Fig. IV.1.) The 5–6 peaks shown correspond to many layers of neighbors whereas in the LCM (in the quasi-chemical form) only nearest-neighbor interactions are taken into account. The contributions of second and third neighbors to the energy of a liquid have been known to be substantial: in simulation studies (Verlet [30]) where the potentials of interaction were truncated at $2.5\sigma$ ($\sigma$ being the size parameter of the LJ potential), the long-range corrections could amount to 50% of the total energy. The first neighbor distance ends at about $1.5\sigma$ ~$1.6\sigma$ at high densities (see Verlet [31]; these distances refer to the positions of the first minima). Thus, terminating at the first neighbors would cause large errors in the thermodynamic quantities obtained according to statistical mechanics. Correctly interpreted, the LCM $x_{ij}$ and $U_{ij}$ refer only to first neighbor contributions.

## ENERGY RELATIONS

Expression (5.27) gives an interpretation of the energies $u_{ij}$ in terms of $g_{ij}$. We now want to express the energies in terms of other experimental quantities. We shall apply the MSA approximation [32]. For hard spheres,

$$g_{ij}(r) = 0, \qquad \text{for } r < \frac{1}{2}(d_i + d_j) \tag{5.29}$$

and

$$C_{ij}(r) \approx -\beta \bar{u}_{ij}(r), \qquad \text{for } r > \frac{1}{2}(d_i + d_j) \tag{5.30}$$

where $C_{ij}$ is the dcf and $d_j$ is the diameter of hard spheres of type $j$. These conditions have been generalized to soft spheres by Rosenfeld and Ashcroft [33]. They gave the condition

$$\int d\mathbf{r} \; g_{ij}(r)[C_{ij}(r) + \beta \bar{u}_{ij}(r)] = 0 \tag{5.31}$$

where $\bar{u}_{ij}(r)$ is a pair potential with soft repulsion. This approximation is called soft MSA (SMSA). Obviously, conditions (5.29 and 30) satisfy (5.31). Thus (5.31) is a generalization of (5.29 and 30) to a wider variety of potentials. Rosenfeld and Ashcroft (*Ibid.*) have studied the plasma potential, the LJ potential, and the inverse power potential, and found that the correlation functions all obey (5.31) with small errors of only a few percent. Thus we can rewrite, for example, the OZ relation

$$h_{ij}(\mathbf{r}_1, \mathbf{r}_2) = C_{ij}(\mathbf{r}_1, \mathbf{r}_2) + \sum_k \rho_k \int d\mathbf{r}_3 \; h_{ik}(\mathbf{r}_1, \mathbf{r}_3)C_{kj}(\mathbf{r}_3, \mathbf{r}_2) \tag{5.32}$$

by setting $h_{ik}(\mathbf{r}_1, \mathbf{r}_3) = g_{ik}(\mathbf{r}_1, \mathbf{r}_3) - 1$, and $C_{kj}(\mathbf{r}_3, \mathbf{r}_2) = -\beta \bar{u}_{kj}(\mathbf{r}_3, \mathbf{r}_2) + [C_{kj}(\mathbf{r}_3, \mathbf{r}_2) + \beta \bar{u}_{kj}(\mathbf{r}_3, \mathbf{r}_2)]$ to get

$$h_{ij}(\mathbf{r}_1, \mathbf{r}_2) = C_{ij}(\mathbf{r}_1, \mathbf{r}_2) - \sum_k \rho_k \int d\mathbf{r}_3 \; g_{ik}(\mathbf{r}_1, \mathbf{r}_3)\beta \bar{u}_{kj}(\mathbf{r}_3, \mathbf{r}_2) \tag{5.33}$$

$$- \sum_k \rho_k \int d\mathbf{r}_3 \; C_{kj}(\mathbf{r}_3, \mathbf{r}_2)$$

$$+ \sum_k \rho_k \int d\mathbf{r}_3 \; g_{ik}(\mathbf{r}_1, \mathbf{r}_3)[C_{kj}(\mathbf{r}_3, \mathbf{r}_2) + \beta \bar{u}_{kj}(\mathbf{r}_3, \mathbf{r}_2)]$$

Since this equation is valid for any $\mathbf{r}_1$ and $\mathbf{r}_2$, we choose $\mathbf{r}_2 = \mathbf{r}_1$. Thus, $h_{ij}(\mathbf{r}_1, \mathbf{r}_1) = h_{ij}(0) = -1$, and $C_{ij}(\mathbf{r}_1, \mathbf{r}_1) = C_{ij}(0)$, and (5.33) becomes

$$-1 = C_{ij}(0) - \sum_k \rho_k \int d\mathbf{r}_3 \; g_{ik}(\mathbf{r}_1, \mathbf{r}_3)\beta \bar{u}_{kj}(\mathbf{r}_3, \mathbf{r}_1) + \beta \left[ \frac{\partial P}{\partial \rho_j} \right]_{T,V} - 1 \tag{5.34}$$

$$+ \sum_k \rho_k \int d\mathbf{r}_3 \; g_{ik}(\mathbf{r}_1, \mathbf{r}_3)[C_{kj}(\mathbf{r}_3, \mathbf{r}_1) + \beta \bar{u}_{kj}(\mathbf{r}_3, \mathbf{r}_1)]$$

where we have used the isothermal compressibility equation [33] $\beta \partial P/\partial \rho_j =$

$1 - \sum_k \rho_k \int dr\ C_{kj}(r)$. The last term of (5.34) is zero according to (5.31). We then have

$$\frac{U_{AA}}{NkT} + \frac{U_{BA}}{NkT} = \frac{x_A}{2}\left\{C_{AA}(0) + \beta\frac{\partial P}{\partial \rho_A}\right\} \tag{5.35}$$

and

$$\frac{U_{BB}}{NkT} + \frac{U_{AB}}{NkT} = \frac{x_B}{2}\left\{C_{BB}(0) + \beta\frac{\partial P}{\partial \rho_B}\right\} \tag{5.36}$$

Thus the internal energies are related to the partial compressibilities and the dcf at zero wavelengths. For hard spheres, the energies are zero. We recover the known relations (VIII.6.5) for HS. The derivatives $\beta\partial P/\partial\rho_j$, or by Gibbs-Duhem relation $\beta\rho_j\partial\mu_i/\partial\rho_i$, can be obtained from experiments. The dcfs at zero distance can be found, e.g., from scattering experiments.

## EXCESS FREE ENERGY

The conventional expression for the configurational Helmholtz free energy $A'$ is [34]

$$-\frac{\alpha A'}{NkT} = x_A \ln\{x_A + x_B\Lambda_{BA}\} - x_A\beta\overline{W}_{AA} \tag{5.37}$$

$$+ x_B \ln\{x_B + x_A\Lambda_{AB}\} - x_B\beta\overline{W}_{BB}$$

with

$$\Lambda_{BA} = \frac{V_{BA}}{V_{AA}}\exp\{-\beta(\overline{W}_{BA} - \overline{W}_{AA})\} \tag{5.38}$$

where $\alpha$ has been introduced into the equation as a proportionality factor between the free energy and the potential of mean force $\overline{W}_{ij}$. Note that the thermodynamic relation between the Helmholtz free energy and the internal energy is the Gibbs-Helmholtz relation

$$\frac{\partial(A'/NkT)}{\partial(1/kT)} = \frac{U'}{N} \tag{5.39}$$

Thus, differentiating (5.37) with respect to $\beta = 1/kT$ should yield the internal energy

$$\frac{U'}{N} = \frac{x_A}{\alpha}\left\{x_{AA}\frac{\partial(\beta\overline{W}_{AA})}{\partial\beta} + x_{BA}\frac{\partial(\beta\overline{W}_{BA})}{\partial\beta}\right\} \tag{5.40}$$

$$+ \frac{x_B}{\alpha}\left\{x_{BB}\frac{\partial(\beta\overline{W}_{BB})}{\partial\beta} + x_{AB}\frac{\partial(\beta W_{AB})}{\partial\beta}\right\}$$

Therefore the AA pair energy is

$$\alpha\frac{U_{AA}}{N} = x_A x_{AA}\frac{\partial \beta \overline{W}_{AA}}{\partial \beta} \tag{5.41}$$

In the derivations, we have assumed that the parameter $\alpha$ is a constant of temperature. Using relation (5.25) (ignoring the long-range contribution), we obtain the equivalent relation

$$\frac{\alpha z_A}{2}u_{AA} = \frac{\partial \beta \overline{W}_{AA}}{\partial \beta} \tag{5.42}$$

At low densities, $W_{ij}(r) = \overline{u}_{ij}(r)$. Thus $\partial \beta \overline{W}_{AA}/\partial \beta$ should satisfy this boundary condition and give the pair energy $\overline{u}_{ij}$. Equating the local energy $u_{AA}$ with the pair energy, we should have

$$\alpha = \frac{2}{z_A} \tag{5.43}$$

A similar relation could be found for $z_B$. To make proper sense, we should have attached $\alpha_A$ and $\alpha_B$ separately to the concentration terms in eq. (5.37); i.e.,

$$-\frac{A'}{NkT} = \frac{x_A}{\alpha_A} \ln\{x_A + x_B\Lambda_{BA}\} - \frac{x_A}{\alpha_A}\beta\overline{W}_{AA} \tag{5.44}$$

$$+ \frac{x_B}{\alpha_B} \ln\{x_B + x_A\Lambda_{AB}\} - \frac{x_B}{\alpha_B}\beta\overline{W}_{BB}$$

Then (5.43) holds for both parameters: $\alpha_A = 2/z_A$, and $\alpha_B = 2/z_B$. We have thus shown that the proportionality constants between the Helmholtz free energy and the potentials of mean force are related to the coordination numbers $z_A$ and $z_B$. The coordination numbers in a liquid are dependent on temperature. Our earlier derivation has ignored this fact. Thus the results could only be considered as approximate. This condition was also noted by Renon and Prausnitz [35] and by Whiting and Prausnitz [36]. Since the coordination number in a dense liquid is around 8 (e.g., for LJ molecules at $\rho\sigma^3 = 0.75$ [37]), the $\alpha$ value should be approximately 0.25. This is the value used by Renon and Prausnitz (Ibid.) for nonpolar fluids. As for liquid water, the coordination number is around 4.4; thus the $\alpha$ value should be about 0.45. Renon and Prausnitz have recommended 0.47 for substances such as acetonitrile and nitromethane. For many polar liquids, the nearest-neighbor coordination number is known from scattering experiments. Thus, we have at least an estimate of $\alpha$ to use in the calculations.

For pure fluid A, (5.44) simplifies to

$$\frac{\overline{W}_{AA}^0}{\alpha_A} = \frac{A_{pure\ A}^0}{N_A} \tag{5.45}$$

Namely, the potential of mean force is related to the Helmholtz free energy per molecule by a factor of $2/z_A$. In the original work of Wilson [38], the value of $\alpha$ was taken to be 1; i.e., the coordination number $z_A = 2$. Each molecule has, on the average, two neighbor molecules. For three-dimensional fluids, this is an unlikely choice. This partly explains the failure of Wilson's formulas to predict phase separation. The actual relation between the potential of mean force and the free energy is not so simple. The potential of mean force consists of three factors according to Hill [39]: the first factor is the free energy; the second is the pair potential; and the third is energy connected with "charging" the system in the presence of the remaining $N$-2 molecules. All factors except the pair potential are dependent on density, temperature, and composition of the fluid. Thus there is no guarantee that $\alpha$ is a constant. In fact, it should also be density, temperature and composition dependent. Furthermore, the free energy formula (5.37) proposed in this section does not distinguish between long-range (SM) and short-range (LCM) contributions, although it was based on the local composition model. In order to fit data, equation (5.37) has been *empirically* generalized to account for long-range effects.

## X.6. *Distribution of Nearest Neighbors*

We have defined the local compositions in liquid mixtures via the pcfs. In this section, we shall examine the variations of the local compositions for model mixtures (as exemplified in the ratios of sizes $\sigma_{AA}/\sigma_{BB}$ and interaction energies, $\varepsilon_{AA}/\varepsilon_{BB}$). The effects of shape are also important, but will not be discussed here.

We first examine binary mixtures whose molecules interact either with the LJ potential

$$u_{ij}(r) = 4\varepsilon_{ij}\left[\left[\frac{\sigma_{ij}}{r}\right]^{12} - \left[\frac{\sigma_{ij}}{r}\right]^{6}\right] \tag{6.1}$$

or with the spherical Kihara potential

$$u_{ij}(r) = 4\varepsilon_{ij}\left[\left[\frac{\sigma_{ij}-d_{ij}}{r-d_{ij}}\right]^{12} - \left[\frac{\sigma_{ij}-d_{ij}}{r-d_{ij}}\right]^{6}\right] \tag{6.2}$$

where $i,j$=A,B and $\sigma_{ij}$, $\varepsilon_{ij}$, and $d_{ij}$ are the size, energy, and core parameters, respectively. To obtain the "exact" behavior of local compositions, simulation work should be carried out. This has recently been done by Hoheisel and Kohler [40], and Gierycz and Nakanishi [41]. On the other hand, integral equation theories, such as PY equation, can be employed for this purpose. PY is known to give reasonable results for short-ranged repulsive potentials. To check the accuracy, we have carried out PY calculations for the LJ potential at the following state conditions and compared them with simulations whenever possible:

1. $T = 119$ K, $\rho^* = \rho\sigma_{AB}^3 = 0.75$, and $x_A = 0.25$, 0.4, 0.5, 0.6, 0.75, and 0.9 for the LB-2-1 mixtures of Nakanishi et al. (*Ibid.*) (i.e., $\varepsilon_{BB}/\varepsilon_{AA} = 2.0$ and $\sigma_{BB}/\sigma_{AA} = 1.0$).

2. $T$=237 K, $\rho^*$=1.187, and $x_A = 0.25$ for the LB-2-$\sqrt{2}$ model (i.e., $\varepsilon_{AA}/\varepsilon_{BB}$= 1/2 and $\sigma_{AA}/\sigma_{BB}$ =1/2).

We note that in the LB models, the Lorentz-Berthelot rules are used to obtain cross interaction parameters; i.e.,

$$\varepsilon_{AB} = \sqrt{\varepsilon_{AA}\varepsilon_{BB}} = 119.8 \ K, \qquad \sigma_{AB} = \frac{\sigma_{AA} + \sigma_{BB}}{2} = 3.405 \ A \qquad (6.3)$$

Figure X.6 shows the dependence of the local compositions on the coordination distance $L$ for case 1 of equal sizes. As mentioned earlier, the first neighbor coordination in a liquid is not as well defined as in the crystalline solids: changing $L$ will also cause changes in LC. We observe that at $L = 1.08$ to $1.15\sigma$, a maximum departure from the bulk mole fractions occurs. Going beyond $1.2\sigma$, the local compositions tend to values of the bulk mole fractions. The trend is particularly evident in Figure X.7. The LC oscillate around the bulk compositions starting at about $2\sigma$ and finally settle down around $4\sigma$. The common choices in literature for the coordination distance were $L = 1.1\sigma$ and $1.35\sigma(\pm)$. For equal-sized molecules, the first choice corresponds to the maximum deviation from bulk concentrations (i.e., where the nonrandomness of mixing is highest). The second choice gives, however, a diminished nonrandom behavior. The arbitrariness of the choices is evident.

The agreement of PY with the MC results is satisfactory for most of the conditions compared, except at low concentrations ($x_A = 0.25$ and $0.4$) and close to $L = 1.1\sigma$. We have included the simulation data ($x_A = 0.50$) of Hoheisel and Kohler (*Ibid.*) and Gierycz and Nakanishi (*Ibid.*) for comparison. The PY calculations agree well with the simulations of Hoheisel et al. In summary, PY results are dependable at concentrations $x_A = 0.75$ and $0.60$. PY agrees with Hoheisel data at $x_A = 0.50$. At lower concentrations, there is some discrepancy near $L = 1.1\sigma$. The agreement in $x_{BB}$ is better than that in $x_{AA}$.

For unequally sized molecules, the PY results did not converge at $T = 124$ K, the temperature used by Nakanishi in MD. We present the calculations at $T = 237$ K instead. (Note: For a pure LJ fluid, the given density ($\rho^* = 1.187$) is in the solid region.) Figure X.7 also includes the MD results at $T = 124$ K. The two sets of LC show similar trends with, however, some quantitative differences. The calculations are used to determine the variations of $x_{ij}$ with coordination distance $L$. An oscillatory behavior is clearly demonstrated. Since the density is high, the behavior is explained in the stratification of successive neighbor shells. As the range of integration is extended outward away from the central molecule, the contribution to the coordination number includes at first the nearest neighbors, next the second neighbors, and then the third neighbors. Each time a new neighborhood is included, there is a jump in the local mole fractions over the bulk fractions. This effect disappears after approximately three neighborhoods for this density state.

## X.7. *Application to the Equations of State of Mixtures*

Given an equation of state for a pure fluid, the following procedure could be used to construct an equation for mixtures according to the mixture theories developed above.

First we derive the Helmholtz free energy $A$ from the given $P$-$V$-$T$ relation. By identifying the potential of mean force as the Helmholtz free energy according to (5.45), we can immediately write down the free energy for mixtures according to (5.44); i.e.,

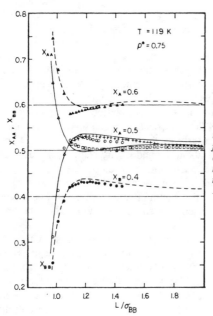

FIGURE X.6. Variations of local compositions with the coordination range L. Two LB-2-1 mixtures: an equimolar ($x_A$ =0.5) and a 60:40 ($x_A$ =0.6) mixture are shown. Symbols are from the simulation results of Gierycz and Nakanishi [40] and Hoheisel and Kohler [39]. Solid lines are from PY calculations. Upper curves are $x_{AA}$; and lower curves are $x_{BB}$. Note that at $L/\sigma_{BB} \approx 1.2$ one has maximum nonrandom mixing. At large L, the LC approaches the bulk mole fraction as expected. (Lee at al., Fluid Phase Equil., 1986)

FIGURE X.7. Variations of local compositions with the coordination range, L. A 25:75 ($x_A$ =0.25) LB-2-$\sqrt{2}$ mixture is shown (i.e., $\varepsilon_{BB}/\varepsilon_{AA}$ =2, and $\sigma_{BB}/\sigma_{AA}$ =$\sqrt{2}$). Asterisks: simulation data [40] at $T$ = 124 K and $\rho\sigma_{BB}^3$ = 1.187 (B is the larger sphere). Lines: PY calculations at $T$ = 237 K and same density. Peaks in the oscillations are due to inclusion of successive layers of neighbors in counting. (Lee at al., Fluid Phase Equil., 1986)

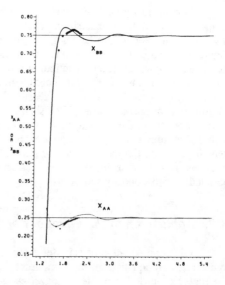

$$\frac{-A'}{NkT} = \frac{x_A}{\alpha_A} \ln\left[x_A \exp\left[\frac{-\alpha_A \hat{A}_{AA}}{kT}\right] + x_B F_{BA} \exp\left[\frac{-\alpha_A \hat{A}_{BA}}{kT}\right]\right]$$

$$+ \frac{x_B}{\alpha_B} \ln\left[x_B \exp\left[\frac{-\alpha_B \hat{A}_{BB}}{kT}\right] + x_A F_{AB} \exp\left[\frac{-\alpha_B \hat{A}_{AB}}{kT}\right]\right]$$

(7.1)

where $F_{ji}$ are the volume ratios. In this equation, approximation (5.45) is made that the potentials of mean force are individually equal to the scaled component Helmholtz free energies. Later on, empirical "fitting" of the parameters will implicitly correct for the deficiencies. To obtain the pressure, we differentiate $A'$ with respect to volume:

$$P = -\left.\frac{\partial A}{\partial V}\right|_T$$

(7.2)

We thus obtain

$$P' = x_A(x_{AA}P_{AA} + x_{BA}P_{BA}) + x_B(x_{BB}P_{BB} + x_{AB}P_{AB})$$

(7.3)

where

$$\frac{P_{ij}}{N} = -\left.\frac{\partial \hat{A}_{ij}}{\partial V}\right|_T$$

(7.4)

We note that (7.3) satisfies the explicit composition dependence of eq. (3.1) (N.B. The local composition $x_{ji}$ contains the factor $x_j$.) As to the implicit composition dependence, we shall introduce the MDA mixing rules for the state variables:

$$T_{ij}^* = \frac{kT}{\varepsilon_{ij}}$$

(7.5)

$$\rho^* = \rho \sigma_x^3$$

(7.6)

$$\sigma_x^3 = \sum_i \sum_j x_i x_j \sigma_{ij}^3$$

(7.7)

A concrete example will illustrate this development. We have chosen the Strobridge equation [42] for the Helmholtz free energy, which reads

$$\beta \hat{A}_{ij} = \left[A_1 + \frac{A_2}{T_{ij}^*} + \frac{A_3}{T_{ij}^{*2}} + \frac{A_4}{T_{ij}^{*3}} + \frac{A_5}{T_{ij}^{*5}}\right]\rho^*$$

(7.8)

$$+ \frac{1}{2}\left[A_6 + \frac{A_7}{T_{ij}^*}\right]\rho^{*2} + \frac{1}{3}A_8\rho^{*3} + \frac{1}{5}\left[\frac{A_{15}}{T_{ij}^*}\right]\rho^{*5}$$

$$+\frac{1}{2c}\left[\frac{A_9}{T_{ij}^{*3}}+\frac{A_{10}}{T_{ij}^{*4}}+\frac{A_{11}}{T_{ij}^{*5}}\right](1-\exp(-c\rho*^2))$$

$$+\frac{1}{2c^2}\left[\frac{A_{12}}{T_{ij}^{*3}}+\frac{A_{13}}{T_{ij}^{*4}}+\frac{A_{14}}{T_{ij}^{*5}}\right]\left[1-(1+c\rho*^2)\exp(-c\rho*^2)\right]$$

The constants $A_i$ could be found in [42]. This free energy is incorporated into the mixture by using eq. (7.1). It is differentiated to give the *local pressures*

$$P'=\sum_i x_i \frac{\sum_j x_j F_{ji} P_{ji}\ \exp\left[-\alpha\hat{A}_{ji}/RT\right]}{\sum_k x_k F_{ki}\ \exp\left[-\alpha\hat{A}_{ki}/RT\right]} \qquad (7.9)$$

where

$$\frac{\beta P_{ij}'}{\rho}=\lambda\ Z_{\text{conf}}^{ij(0)}+(\lambda-1)\ Z_{\text{conf}}^{ij(\rho)} \qquad (7.10)$$

and

$$-Z_{\text{conf}}^{ij}=\left[A_1+\frac{A_2}{T_{ij}^*}+\frac{A_3}{T_{ij}^{*2}}+\frac{A_4}{T_{ij}^{*3}}+\frac{A_5}{T_{ij}^{*5}}\right]\rho*+\left[A_6+\frac{A_7}{T_{ij}^*}\right]\rho*^2+A_8\rho*^3+\frac{A_{15}}{T_{ij}^*}\rho*^5 \qquad (7.11)$$

$$+\left[\frac{A_9}{T_{ij}^{*3}}+\frac{A_{10}}{T_{ij}^{*4}}+\frac{A_{11}}{T_{ij}^{*5}}\right]\rho*^2\ \exp(-c\rho*^2)+\left[\frac{A_{12}}{T_{ij}^{*3}}+\frac{A_{13}}{T_{ij}^{*4}}+\frac{A_{14}}{T_{ij}^{*5}}\right]\rho*^4\exp(-c\rho*^2)$$

where the temperature $T^*$ and density $\rho^*$ are reduced according to the MDA prescriptions (7.5) and (7.6). The result is a hybrid mixing rule based on the LCM and MDA. To utilize these mixing rules, we must define the cross parameters, $\varepsilon_{ij}$, $\sigma_{ij}$, $F_{ij}$, and $\lambda_{ij}$. These are given by the combining rules

$$\sigma_{AB}=\xi(\sigma_{AA}\ \sigma_{BB})^{1/2}, \qquad\qquad \varepsilon_{AB}=\zeta(\varepsilon_{AA}\ \varepsilon_{BB})^{1/2} \qquad (7.12)$$

$$F_{ij}=\delta^3\left[\frac{\sigma_{ii}^3}{\sigma_{jj}^3}\right]^{1/2}, \qquad\qquad \lambda_{ij}=\frac{\lambda_{ii}+\lambda_{jj}}{2}$$

where three binary interaction parameters (BIPs) ($\xi$, $\zeta$, and $\delta$) are introduced. They are obtained by fitting the equations to experimental data.

To calculate the vapor-liquid equilibria we also need the fugacities. They are derived from the mixture equation of state:

$$\ln \frac{\hat{f}_i}{x_i \rho kT} = \frac{1}{\alpha} \left[ \ln \sum_k x_k F_{ki} \exp\left[\frac{\alpha \hat{A}_{ki}}{RT}\right] - 1 + \sum_j x_j \frac{F_{ij} \exp(\alpha \hat{A}_{ij}/RT)}{\sum_k x_k F_{kj} \exp(\alpha \hat{A}_{kj}/RT)} \right] \qquad (7.13)$$

$$+ \frac{1}{\alpha} \left[ \ln \sum_k x_k F_{ki} - 1 + \sum_j x_j \frac{F_{ij}}{\sum_k x_k F_{kj}} \right] + (Z-1)[1 + \bar{R}_i]$$

where

$$\bar{R}_i = \frac{1}{\sigma_x^3} \frac{\partial \sigma_x^3}{\partial n_i}\bigg|_{T,V,n_{j \neq i}} = 2 \left[ \frac{\sum_j x_j \sigma_{ij}^3}{\sigma_x^3} - 1 \right] \qquad (7.14)$$

We could also have developed the mixture equation of state according to the CSM. The parameters for such development are given by

$$\sigma_x^3 = \sum_i \sum_j x_i x_j \, \sigma_{ij}^3 \qquad (7.15)$$

$$\varepsilon_x \sigma_x^3 = \sum_i \sum_j x_i x_j \varepsilon_{ij} \sigma_{ij}^3$$

$$\lambda_x \sigma_x^3 = \sum_i \sum_j x_i x_j \lambda_{ij} \sigma_{ij}^3$$

In the following we shall compare the performance of LCM with CSM.

### RESULTS AND DISCUSSIONS

The above equation of state has been applied to the calculation of the vapor-liquid equilibria (VLE) of some 60 binary mixtures [43], including nonpolar-nonpolar, nonpolar-polar, and polar-polar fluid mixtures. The procedure is to first fit the pure fluid data to the pure equation of state, thus determining the pure parameters $\sigma$, $\varepsilon$, $\lambda$ for the pure substances. For polar substances, it is necessary to include the polar effects by the mean potential temperature:

$$\frac{\varepsilon_{ij}}{k} = \frac{\varepsilon_{ij}^0}{k} + \frac{D_{ij}}{T}, \qquad \varepsilon_{ij}^0 = \zeta (\varepsilon_{ii}^0 \varepsilon_{jj}^0)^{1/2} \qquad (7.16)$$

and $T_{ij}^* = kT/\varepsilon_{ij} = kT/(\varepsilon_{ij}^0 + D_{ij}/T)$, where $D_{ij}$ is a polar parameter. Thus, additional temperature dependence is introduced into the equation of state to account for the polar effects. Equations (7.8 and 13) have been applied to VLE and P-V-T calculations. We note that for the polar parameter, $D_{AB} = (D_{AA} + D_{BB})/2$.

Mixtures with methane (all together 13 systems: from ethane to n-decane and from $CO_2$ to water) have been studied. The deviations are reported in terms of equilibrium $K$ ratios ($K_i = y_i/x_i$). This ratio is very sensitive to relative volatility, especially when the heavy component exists preferentially in the liquid phase. A small change in $x_i$ can cause a large change in $K_i$. For methane (1)-n-decane (2) mixtures, $K_1$ was predicted to within 4%. The results were also compared with the CSM. For most of the cases studied, the LCM performs better than the CSM.

The methane-water system was investigated to pressures of 690 bar with good agreement (see Fig. X.8). For the acetone-water system, the CSM failed to reproduce the azeotrope due to the large difference in molecular volumes. Figure X.9 gives a comparison of the two theories. The hybrid LCM worked well for this mixture. Many hydrogen-bonding fluids were tested, such as alcohols, ammonia, and water. $P$-$x$ and $T$-$x$ diagrams showed that the LCM was successful in correlating these data.

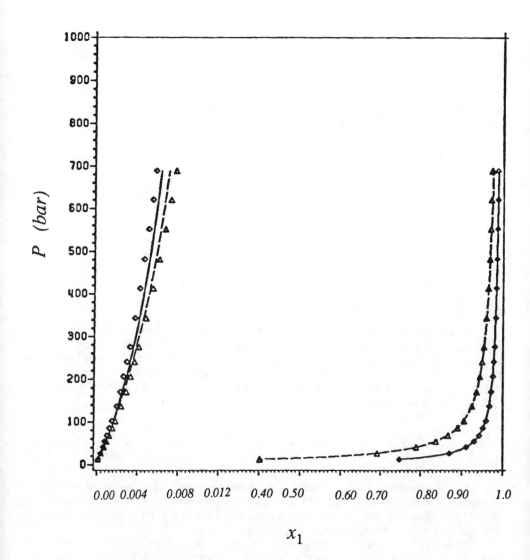

FIGURE X.8. *P-x diagram for methane(1)-water(2) mixtures. Symbols: experimental data. Solid lines: LCM results at T = 411 K. Dashed lines: LCM results at T = 444 K. Note the break of scale in the abscissa.*

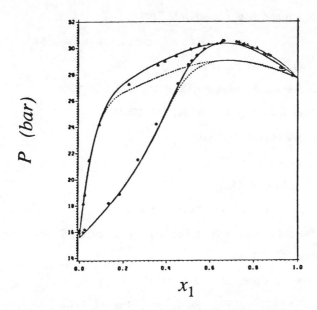

FIGURE X.9. P-x diagram for acetone(1)-water(2) mixtures. Symbols: experimental data at T = 473.15 K. Solid lines = local composition (LCM) results.
Dashed lines: conformal solution (CSM) results. The system exhibits an azeotrope which is well predicted by LCM.

## References

[1]    E.A. Guggenheim, *Mixtures*, (Oxford University Press, London, 1952).

[2]    H.C. Longuet-Higgins, Proc. Roy. Soc. (London) A205, 247 (1951).

[3]    D. Henderson and P.J. Lennard, in *Physical Chemistry: An Advanced Treatise*,
       Volume VIIIB, edited by D. Henderson (Academic Press, New York, 1971)
       pp.486ff.

[4]    L.L. Lee and D. Levesque, Mol. Phys. 26, 1351 (1973).

[5]    C. Hoheisel and K. Lucas, Mol. Phys. 53, 51 (1984).

[6]    C. Hoheisel, U. Deiters, and K. Lucas, Mol. Phys. 49, 159 (1983).

[7]    K.P. Shukla, M. Luckas, H. Marquardt, and K. Lucas, Fluid Phase Equil., 26, 129
       (1986).

[8]    G.A. Mansoori and T.W. Leland, J. Chem. Soc. Faraday Trans. II, 8, 320 (1972).

[9]    C. Hoheisel and K. Lucas, *Ibid.* (1984).

[10]   C. Hoheisel, U. Deiters, and K. Lucas, Mol. Phys. 49, 159 (1983).

[11]   K.P. Shukla, M. Luckas, H. Marquardt, and K. Lucas, Fluid Phase Equil. 26, 129
       (1986).

[12]  G.A. Mansoori and T.W. Leland, *Ibid.* (1972).

[13]  J.D. Weeks, D. Chandler, and H.C. Andersen, J. Chem. Phys. **54**, 5237 (1971).

[14]  L.L. Lee and D. Levesque, *Ibid.* (1973).

[15]  E.W. Grundke and D. Henderson, Mol. Phys. **24**, 269 (1972).

[16]  G.M. Wilson, J. Am. Chem. Soc. **86**, 127 (1964).

[17]  E.A. Guggenheim, *Ibid.* (1952).

[18]  E.A. Guggenheim, *Ibid.* (1952).

[19]  G.M. Wilson, *Ibid.* (1964).

[20]  F.H. Stillinger, J. Phys. Chem. **74**, 3677 (1970).

[21]  K. Toukubo and K. Nakanishi, J. Chem. Phys. **65**, 1937 (1976).

[22]  G.M. Wilson, *Ibid.* (1964).

[23]  J.G. Kirkwood, J. Chem. Phys. **3**, 300 (1935).

[24]  H. Posch, F. Vesely, and W.A. Steele, Mol. Phys. **44**, 241 (1981).

[25]  L.L. Lee and H.M. Hulburt, J. Chem. Phys. **58**, 44 (1973).

[26]  V. Flemr, Collect. Czech. Chem. Commun. **41**, 3347 (1976).

[27]  L.L. Lee and K.E. Starling, Fluid Phase Equil. **21**, 77 (1985).

[28]  W.B. Whiting and J.M. Prausnitz, Fluid Phase Equil. **9**, 119 (1982).

[29]  T.L. Hill, *Statistical Mechanics* (McGraw-Hill, New York, 1956).

[30]  L. Verlet, Phys. Rev. **159**, 98 (1967).

[31]  L. Verlet, Phys. Rev. **165**, 201 (1968).

[32]  J.L. Lebowitz and J.K. Percus, Phys. Rev. **144**, 251 (1966).

[33]  J.L. Lebowitz, J. Chem. Phys. **41**, 133 (1964).

[34]  W.B. Whiting and J.M. Prausnitz, *Ibid.* (1982).

[35]  H. Renon and J.M. Prausnitz, AIChE J. **14**, 135 (1968).

[36]  W.B. Whiting and J.M. Prausnitz, Fluid Phase Equil. **9**, 119 (1982).

[37]  K. Toukubo and K. Nakanishi, J. Chem. Phys. **65**, 1937 (1976).

[38]  G.M. Wilson, *Ibid.* (1964).

[39]  T.L. Hill, *Statistical Mechanics* (MaGraw-Hill, New York, 1956).

[40]  C. Hoheisel and F. Kohler, Fluid Phase Equil. **16**, 13 (1984).

[41]  P. Gierycz and K. Nakanishi, Fluid Phase Equil. **16**, 255 (1984); also K. Nakanishi, Y. Adachi, and I. Fujihara, *Ibid.*, **29**, 347 (1986).

[42] T.R. Strobridge, NBS Technical Note **129** (1962).

[43] M.H. Li, T.H. Chung, C.-K. So, L.L. Lee, and K.E. Starling, in *Equations of State: Theories and Applications,* edited by K.C. Chao and R.L. Robinson, Jr., (American Chemical Society Symposium Series No. 300, Washington D.C., 1986), pp.250ff.

## Exercises

1. Prove eq. (3.3) starting from the virial equation of state.

2. Discuss the conditions under which the lattice gas expressions will yield an ideal solution. What are the expressions for the local mole fractions? (Refer to the reference of Guggenheim [16].)

3. Discuss the validity of the Guggenheim mixing procedure as exemplified in Figure X.3. Under what conditions will the exchange energy expression (4.3) no longer be valid? What happens when the mixture is a concentrated solution? Discuss the implications of size differences on mixing lattice gases. For example, what is the spatial arrangements for mixing diatomics and monatomics?

4. Discuss the consequences of temperature-dependent $\alpha$ in formula (5.37). What then is the energy expression?

# CHAPTER XI

# THE PERTURBATION THEORIES

$\mathbf{W}$e have studied a number of model potentials: the hard spheres, the soft spheres, and the Lennard-Jones molecules. These are all idealized models for real gases. The interaction forces in real gases are more complicated. For example, most real gases are polar or multipolar: carbon dioxide is quadrupolar, and hydrogen chloride is both dipolar and quadrupolar. In addition, polyatomic molecules are nonspherical: the breadth-to-length ratio of bromine is about 0.547, and that of carbon disulfide is about 0.9. All these factors influence the physical properties of the substance. Simple potential models cannot adequately describe these effects. However, simpler models could serve as reference potentials, and the additional effects, such as quadrupolar forces, could be treated as perturbations on the reference systems. In principle, when the reference system chosen is close to the final system, one would also expect the properties produced by adding the perturbation terms to be close to the full system. This is the basis of the perturbation approach.

On the other hand, a theory for the liquid state was lacking till progress had been made for the model fluid of hard spheres. For simple repulsive potentials, a reference hard sphere fluid is capable of giving a highly accurate description of their structure in the liquid state. Thus hard spheres make a good candidate as reference for repulsive potentials. Similarly, the LJ potential is a reasonable reference for small polar molecules (e.g., for Stockmayer molecules), especially when the polar forces are weak. On the other hand, highly nonspherical and highly polar molecules would have to be approximated by nonspherical reference potentials [1]. There is as yet no satisfactory theory for hydrogen bonding fluids. The situation, nonetheless, is improving. In this chapter, we study perturbation theories for isotropic potentials and polar/multipolar (angle-dependent) potentials. The success would depend on both the type of theory chosen and the way that the full potential is separated into the reference and perturbative parts. Among the theories for liquids, the perturbation approach is by far the most successful and will continue to be useful in the foreseeable future.

## XI.1. *The Isotropic Fluids*

Two classes of simple fluids must be distinguished when the perturbation approach is applied: fluids with repulsive forces only, and those also possessing attractive interactions. The former include such (hypothetical) fluids as the hard spheres and fluids of inverse power potentials; the latter are represented by the LJ potential. The fluids with repulsive interaction behave in ways similar to hard spheres. The perturbation expansion is rapidly convergent at the zero order. For example, with hard spheres as reference, the Helmholtz free energy of the Weeks-Chandler-Andersen [2] (WCA) repulsive potential can be calculated accurately by using an equivalent HS diameter determined from either the Barker-Henderson [3] rule (giving a temperature-dependent diameter) or the WCA rule (giving a diameter dependent on both the temperature and the density). For attractive potentials, using only size correspondence is not sufficient. The first-order correction to the free energy should be retained (e.g., in the Verlet-Weis approach [4]). The situation becomes even more critical with regard to polar fluids where retention of the second- and third-order terms is insufficient sometimes, and a resummation method (e.g., the Padé approximants) must be used to give good agreement.

The perturbation techniques can be applied to the equilibrium properties of the fluid or to the structure of the fluid, the latter being a more stringent test of the perturbation theory used. For thermodynamic properties, the perturbation is applied to the Helmholtz free energy, and a series, the $\lambda$-expansion [5], is formed. The expansion parameter $\lambda$ is directly related to the perturbing potential $u_p(r)$. Since the factors $\beta u(\cdot)$ occur as a product, this type of expansion is called a temperature expansion (or $u$-expansion). The other expansion is in terms of density. For example, the PY2 expansion is based on density and could be reformulated as a perturbation expansion.

## THE $\lambda$-EXPANSION

For a fluid whose pair interaction potential $u(r)$ can be separated into $u_0(r)$, the reference part, and $u_p(r)$, the perturbing part, we can write

$$u(r; \lambda) = u_0(r) + \lambda u_p(r) \tag{1.1}$$

where $\lambda$ is a parameter varying between 0 and 1 ($0 \leq \lambda \leq 1$). This is the simplest form of the homotopic continuation. Other separations (e.g., Perram and White [6], and Melnyk and Smith [7]) are possible. When $\lambda = 0$, we recover the reference system (since $u = u_0$) and when $\lambda = 1$, we have the full system (i.e., $u = u_0 + u_p$). The partition function is

$$Z_N(\lambda) = \frac{1}{N!\Lambda^{3N}} \int dr^N \exp\left[-\beta \sum [u_0(r_{ij}) + \lambda u_p(r_{ij})]\right] \tag{1.2}$$

Since the Helmholtz free energy is related to $Z_N(\lambda)$ by $A(\lambda) = -kT \ln Z_N(\lambda)$, we can expand $A(\lambda)$ in terms of $\lambda$ in a Taylor series around the value $\lambda = 0$:

$$A(\lambda) = A(0) + \lambda \frac{\partial A(0)}{\partial \lambda} + \frac{\lambda^2}{2!} \frac{\partial^2 A(0)}{\partial \lambda^2} + \frac{\lambda^3}{3!} \frac{\partial^3 A(0)}{\partial \lambda^3} + \cdots \tag{1.3}$$

Substitution of (1.2) into (1.3) gives

$$A = A_0 + A_1 + A_2 + A_3 + \cdots \tag{1.4}$$

The term $A_0 = A(0)$, i.e., the free energy of the reference system. The term $A_1$ is given as

$$A_1 = \frac{1}{2} \lambda \rho^2 \int dr_1 \int dr_2 \, u_p(r_{12}) g_0(r_{12}) \tag{1.5}$$

$$= \frac{N}{2} \lambda \rho \int_0^\infty dr_{12} \, 4\pi r_{12}^2 \, u_p(r_{12}) g_0(r_{12})$$

Note that (1.5) can be obtained from a direct differentiation of $A(\lambda)$ with respect to $\lambda$:

$$\frac{\partial A}{\partial \lambda}\bigg|_{T,N,V} = - kT \frac{\partial}{\partial \lambda} \ln Z_N(\lambda) = - \frac{kT}{Z_N} \frac{\partial Z_N}{\partial \lambda} \tag{1.6}$$

$$= - \frac{kT}{Z_N} \int dr^N \exp[-\beta V_N][-\beta u_p(r_{12})] \frac{N(N-1)}{2}$$

$$= \frac{1}{2} \frac{N(N-1)}{Z_N} \int dr_1 \, dr_2 \, u_p(r_{12}) \int dr_3 \cdots dr_N \exp[-\beta V_N]$$

$$= \frac{\rho^2}{2} \int dr_1 \, dr_2 \, u_p(r_{12}) g_0(r_{12})$$

where the definition of the pcf $g^{(2)}$ has been used. Since the derivative $\partial A/\partial \lambda$ is evaluated at $\lambda = 0$, the pcf obtained is the reference system $g_0$. The second-order term can similarly be derived. The procedure is lengthy. We display only the result here.

$$A_2 = - \frac{\lambda^2 \beta \rho N}{4} \int dr_2 \, [u_p(r_{12})]^2 \, g_0(r_{12}) \tag{1.7}$$

$$- \frac{\lambda^2 \beta \rho^2 N}{2} \int dr_2 \, dr_3 \, u_p(r_{12}) u_p(r_{23}) g_0^{(3)}(r_{12}, r_{13}, r_{23})$$

$$- \frac{\lambda^2 \beta \rho^3 N}{8} \int dr_2 \, dr_3 \, dr_4 \, u_p(r_{12}) u_p(r_{34}) [g_0^{(4)}(r_{12}, r_{13}, r_{14}, r_{23}, r_{24}, r_{34}) - g_0(r_{12}) g_0(r_{34})]$$

$$+ \frac{\lambda^2 N}{8} \left[\frac{\partial \rho}{\partial P}\right]_0 \cdot \left\{\frac{\partial}{\partial \rho}\left[\rho^2 \int dr \, u_p(r) g_0(r)\right]\right\}^2$$

$A_3$ and higher-order terms are long expressions. The $A_3$ term will be discussed in the case of polar fluids later. In the 1970s, Weeks, Chandler, and Anderson [8] proposed another way of expanding the free energy, which for repulsive potentials converges rapidly.

## THE WEEKS-CHANDLER-ANDERSEN EXPANSION

The difference between the $\lambda$-expansion discussed above and the WCA expansion lies in the choice of the expansion functions. The $\lambda$-expansion expands, in the language of functional variations, in terms of the difference of the pair potentials: $\Delta u(r) = u(r) - u_0(r)$; the WCA expands in terms of the Boltzmann factors: $\Delta e(r) = \exp[-\beta u(r)] - \exp[-\beta u_0(r)]$. The latter quantity, for the HS reference, is also called the *blip function* because of its appearance (Figure XI.1).

The approach employed here is functional expansion. The pair potential, as before, is composed of two parts:

$$u(r) = u_0(r) + u_p(r) \tag{1.8}$$

The Boltzmann factor is defined as

$$e(r) \equiv \exp[-\beta u(r)] \tag{1.9}$$

The Helmholtz free energy $A[e(r)]$, considered as a functional of $e(r)$, is expanded into the functional Taylor series

$$A[e] = A[e_0] + \int dr\, \frac{\delta A[e_0]}{\delta e(r)} \Delta e(r) + \frac{1}{2!} \int dr \int dr'\, \frac{\delta^2 A[e_0]}{\delta e(r)\delta e(r')} \Delta e(r)\Delta e(r') + \cdots \tag{1.10}$$

The first term on the RHS is the free energy of the reference system (with $u_0$). The second term involves the first functional derivative $\delta A/\delta e(r)$, which can be evaluated from the definition of the partition function (1.2):

$$\frac{\delta A[e_0]}{\delta e(r_{12})} = -\frac{N}{2} \rho kT\, y_0(r_{12}) \tag{1.11}$$

where $y(r) = g(r)\exp[\beta u(r)]$ is the background correlation function. Similarly, the second-order derivative is given as

$$\frac{\delta A[e_0]}{\delta e(r)\delta e(r')} = -\rho^3 kT \int d\mathbf{r}_1\, d\mathbf{r}_2\, d\mathbf{r}_3\, \delta(\mathbf{r}_{12}, \mathbf{r})\delta(\mathbf{r}_{13}, \mathbf{r}')y_0(r_{12})y_0(r_{13})J_0^{(3)}(\mathbf{r}_1\mathbf{r}_2\mathbf{r}_3) \tag{1.12}$$

$$-\frac{1}{4}\, \rho^4 kT \int d\mathbf{r}_1\, d\mathbf{r}_2\, d\mathbf{r}_3\, d\mathbf{r}_4\, \delta(\mathbf{r}_{12}, \mathbf{r})\delta(\mathbf{r}_{34}, \mathbf{r}')y_0(r_{12})y_0(r_{34})J_0^{(4)}(\mathbf{r}_1\mathbf{r}_2\mathbf{r}_3\mathbf{r}_4)$$

where the $J$-functions are

$$J^{(3)}(\mathbf{r}_1\mathbf{r}_2\mathbf{r}_3) \equiv h(r_{23}) + g(r_{23})\left[ \frac{g^{(3)}(\mathbf{r}_1, \mathbf{r}_2, \mathbf{r}_3)}{g(r_{12})g(r_{13})g(r_{23})} - 1 \right] \tag{1.13}$$

and $J^{(4)}$ under the superposition approximation is

$$J^{(4)}(\mathbf{r}_1\mathbf{r}_2\mathbf{r}_3\mathbf{r}_4) = h(r_{13})h(r_{24}) + h(r_{13})h(r_{23})h(r_{24}) + h(r_{13})h(r_{14})h(r_{23})h(r_{24}) \tag{1.14}$$

$$+ h(r_{14})h(r_{23}) + h(r_{13})h(r_{14})h(r_{24})$$

$$+ h(r_{13})h(r_{14})h(r_{23}) + h(r_{14})h(r_{23})h(r_{24})$$

Higher-order derivatives are quite complicated and are not given here. Thus to second

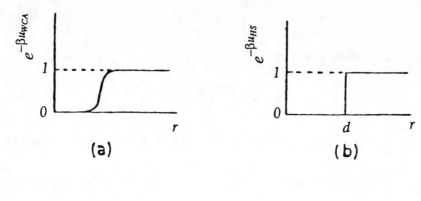

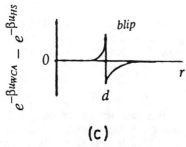

FIGURE XI.1. *The blip function as difference between the Boltzmann factors:* $\exp[-\beta u_{WCA}] - \exp[-\beta u_{HS}]$

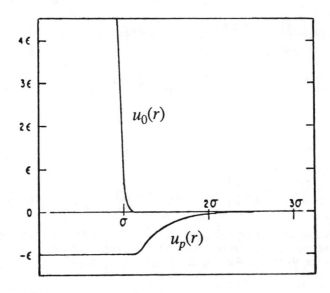

FIGURE XI.2. *The WCA separation of the Lennard-Jones potential:* $u_0(r)$ *and* $u_p(r)$.

order, the Helmholtz free energy is

$$\frac{A}{NkT} = \frac{A_0}{NkT} - \frac{\rho}{2} \int d\mathbf{r} \, y_0(r) \left[ e^{-\beta u(r)} - e^{-\beta u_0(r)} \right] \tag{1.15}$$

$$- \frac{\rho^2}{2V} \int d\mathbf{r}_1 \, d\mathbf{r}_2 \, d\mathbf{r}_3 \, d\mathbf{r} \, d\mathbf{r}' \, \delta(\mathbf{r}_{12}, \mathbf{r}) \delta(\mathbf{r}_{13}, \mathbf{r}') y_0(r_{12}) y_0(r_{13}) J_0^{(3)}(\mathbf{r}_1 \mathbf{r}_2 \mathbf{r}_3) \, \Delta e(r) \Delta e(r')$$

$$- \frac{\rho^3}{8V} \int d\mathbf{r}_1 \, d\mathbf{r}_2 \, d\mathbf{r}_3 \, d\mathbf{r}_4 \, d\mathbf{r} \, d\mathbf{r}' \, \delta(\mathbf{r}_{12}, \mathbf{r}) \delta(\mathbf{r}_{34}, \mathbf{r}') y_0(r_{12}) y_0(r_{34}) J_0^{(4)}(\mathbf{r}_1 \mathbf{r}_2 \mathbf{r}_3 \mathbf{r}_4) \Delta e(r) \Delta e(r') + \cdots$$

This expansion coupled with a well-chosen separation of the full potential $u(r)$ into its parts, $u_0(r)$ and $u_p(r)$, constitutes one of the more successful perturbation schemes for isotropic fluids. To illustrate the application, we cite the example of Verlet and Weis (*Ibid.*) for the LJ molecules.

## PERTURBATION EXPANSION FOR LJ FLUID

The procedure adopted by Verlet and Weis is a two-step perturbation scheme. The first step is a perturbation from the HS reference to the WCA-type soft-sphere potential. The second step is to perturb the WCA potential by adding on the attractive part of the LJ potential. The usual LJ potential is separated in the WCA formulation into

$$u_{LJ} = u_{WCA} + u_{ATT} \tag{1.16}$$

where

$$u_{WCA}(r) = 4\varepsilon \left[ \left[ \frac{\sigma}{r} \right]^{12} - \left[ \frac{\sigma}{r} \right]^6 \right] + \varepsilon, \qquad r \le \sqrt[6]{2}\sigma \tag{1.17}$$

$$= 0, \qquad r > \sqrt[6]{2}\,\sigma$$

and

$$u_{ATT}(r) = -\varepsilon, \qquad r \le \sqrt[6]{2}\sigma \tag{1.18}$$

$$= 4\varepsilon \left[ \left[ \frac{\sigma}{r} \right]^{12} - \left[ \frac{\sigma}{r} \right]^6 \right], \qquad r > \sqrt[6]{2}\sigma$$

Figure XI.2 shows the two potentials as continuous functions. The potential $u_{WCA}$ is purely repulsive, whereas $u_{ATT}$ is purely attractive. To get the Helmholtz free energy of the full LJ system, Verlet and Weis (*Ibid.*) chose the $\lambda$-expansion

$$\frac{A_{LJ}}{NkT} = \frac{\beta A_0}{N} + \frac{\beta A_1}{N} + \cdots = \frac{A_{WCA}}{NkT} + \frac{\rho}{2kT} \int_0^\infty dr \, 4\pi r^2 \, u_{ATT}(r) g_{WCA}(r) + \cdots \tag{1.19}$$

To obtain the reference $A_{WCA}$ and $g_{WCA}(r)$, they made a second expansion with respect to the HS quantities. This time the expansion is the WCA formula (1.15).

$$\frac{A_{WCA}}{NkT} = \frac{A_{HS}}{NkT} - \frac{\rho}{2} \int_0^\infty dr \, 4\pi r^2 y_{HS}(r) \left[ e^{-\beta u_{WCA}(r)} - e^{-\beta u_{HS}(r)} \right] + \cdots \tag{1.20}$$

where the subscript HS refers to hard-sphere quantities. If we select the hard spheres to be of such diameter $d_e$ that the second integral of (1.20) vanishes

$$\int_0^\infty dr \, 4\pi r^2 \, y_{HS}(r; d_e) \left[ e^{-\beta u_{WCA}(r)} - e^{-\beta u_{HS}(r; d_e)} \right] = 0 \tag{1.21}$$

then the free energy $A_{WCA}$, is given to second order at least by the reference hard sphere $A_{HS}$. In fact, Andersen et al. [9] have shown that $A_{WCA}$ from condition (1.21) is accurate to the fourth order. This appears to be a highly accurate approximation. Later numerical calculations have borne this out. Condition (1.21) is called the WCA rule for the equivalent HS diameter. In contrast, another rule used by Barker and Henderson [10] is

$$d_{BH} \equiv \int_0^\infty dr \left[ 1 - e^{-\beta u_{WCA}(r)} \right] \tag{1.22}$$

Thus $d_{BH}$ is dependent on temperature. $d_e$ from (1.21) is dependent on both temperature and density. Verlet and Weis expressed $d_e$ in terms of $d_{BH}$ as

$$d_e = d_{BH}(1 + f \, \delta) \tag{1.23}$$

where $f$ is a factor related to the packing fraction $y_w$:

$$f = \frac{1 - (17/4)y_w + 1.362y_w^2 - 0.8751y_w^3}{(1 - y_w)^2} \tag{1.24}$$

The packing fraction $y_w$, used in Wertheim's solution for PY, is related to the true system packing fraction $y = \pi \rho d_e^3/6$ by

$$y_w = y - \frac{y^2}{16} \tag{1.25}$$

The $\delta$ factor in (1.23) is a function of temperature only and is approximated by the formula

$$\delta = (210.31 + \frac{404.6}{T^*})^{-1} \tag{1.26}$$

where $T^* = kT/\varepsilon$. Similarly the Barker-Henderson diameter is computed as

$$d_{\mathrm{BH}} = \frac{0.3837 + 1.068/T^*}{0.4293 + 1/T^*} \tag{1.27}$$

These numerical formulas enable us to calculate the Helmholtz free energy of the WCA potential analytically based on an equivalent HS system. For example, given the temperature $T$ and density $\rho$, we can calculate $d_{\mathrm{BH}}$ and $\delta$ from (1.27 and 26) and $f$ and $d_e$ from (1.24 and 23).* Knowing $d_e$, we obtain the equivalent HS packing fraction from

$$y = \frac{\pi}{6} \rho d_e^3 \tag{1.28}$$

This $y$ can be used to obtain the compressibility of the equivalent HS system by the Carnahan-Starling (CS) equation

$$Z_{\mathrm{HS}} = \frac{1 + y + y^2 - y^3}{(1-y)^3} \tag{1.29}$$

Equation (1.29) can be integrated to give the HS free energy

$$\frac{A_{\mathrm{HS}} - A_{\mathrm{ideal}}}{NkT} = \frac{4y - 3y^2}{1-y} \tag{1.30}$$

Thus to fourth order (see Andersen et al. [11]), the WCA free energy is given by

$$\frac{A_{\mathrm{WCA}} - A_{\mathrm{ideal}}}{NkT} \approx \frac{4y - 3y^2}{(1-y)^2} \tag{1.31}$$

Ross [12] has proposed a modified variational formula for the free energy of inverse-12 potential in terms of a reference hard sphere fluid. It has also met with success. Once we have the free energy, the other thermodynamic quantities ($Z$ or $U$) can be obtained from known relations.**

Next, we need to calculate the LJ fluid properties from the reference WCA potential. Due to the difference between the repulsive potentials and those with attractive terms, the first (zero order) term of (1.19) alone is not sufficient to describe the LJ behavior. At least the first-order integral should be kept. This integral requires the pcf $g_{\mathrm{WCA}}(r)$ of the WCA fluid, which is not readily available. Weeks, Chandler, and Andersen (*Ibid.*) introduced another approximation for $g_{\mathrm{WCA}}$,

$$g_{\mathrm{WCA}}(r) \approx y_{\mathrm{HS}}(\frac{r}{d_e}) e^{-\beta u_{\mathrm{WCA}}(r)} \tag{1.32}$$

where the $y_{\mathrm{HS}}(r)$ function for hard spheres can be obtained from the Verlet-Weis construction. Using (1.21), we find the first-order correction to the free energy as

---

*Since $y$ is not yet known, some trial and error is needed.

**Note that the $Z_{\mathrm{WCA}}$ obtained from (1.31) is not the same as $Z_{\mathrm{HS}}$ of (1.29), since $P = \rho^2 \partial(A_{\mathrm{WCA}}/N)/\partial\rho$, and $y = \pi\rho d_e^3/6$ and $d_e$ is dependent on $\rho$.

$$2\pi\rho\int_{d_e}^{\infty} dr \ r^2 u_{\text{ATT}}(r) y_{\text{HS}}(\frac{r}{d_e}) \tag{1.33}$$

Verlet and Weis have made Monte Carlo simulation of the WCA system and compared with calculations from (1.32) (see Table XI.1). Good agreement is obtained. For the attractive correction (1.33), the calculated free energy is compared with known LJ data in Table XI.2. We again observe close agreement for densities from $\rho^* = \rho\sigma^3 = 0.40$ to 1.1 and temperatures $T^*$ from 0.75 to 3.5.

*Table XI.1. Properties of the WCA Soft Repulsive Potential*

| | | Compressibility, $\beta P/\rho$, and Energy, $U^* \equiv \beta U^c/N$ | | | | |
| $\rho^*$ | $T^*$ | Z(MC) | $U^*$(MC) | $d_e$ | Z(eq.1.32) | $U^*$(eq.1.32) |
|---|---|---|---|---|---|---|
| 1.1 | 3.05 | 12.67 | 5.48 | 0.9623 | 12.99 | 5.45 |
| 0.85 | 2.81 | 6.92 | 2.37 | 0.9701 | 7.04 | 2.41 |
| 0.84 | 0.75 | 10.23 | 0.73 | 1.0225 | 10.40 | 0.74 |
| 0.65 | 1.35 | 4.89 | 0.62 | 1.003 | 4.99 | 0.64 |
| 0.40 | 1.35 | 2.53 | 0.23 | 1.0047 | 2.55 | 0.24 |

*Table XI.2. Perturbation Contributions to LJ Free Energy*

| Perturbation due to Attractive Potential, $\beta A_1/N = (\beta\rho/2)\int dr \ 4\pi r^2 u_{att}(r) g_{wca}(r)$ | | | |
| $\rho^*$ | $T^*$ | MC | Eq.(1.32) |
|---|---|---|---|
| 1.1 | 3.05 | -2.83 | -2.82 |
| 0.85 | 2.81 | -2.33 | -2.34 |
| 0.84 | 0.75 | -8.82 | -8.85 |
| 0.65 | 1.35 | -3.62 | -3.63 |
| 0.40 | 1.35 | -2.05 | -2.06 |

It is also of interest to investigate the effects of different ways of separating the LJ potential into perturbation parts. The Barker-Henderson [13] separation is

$$u_0(r) = u_{\text{LJ}}(r), \qquad r \le \sigma \tag{1.34}$$

$$= 0, \qquad r > \sigma$$

and

$$u_p(r) = 0, \qquad r \le 0 \tag{1.35}$$

$$= u_{\text{LJ}}(r), \qquad r > 0$$

whereas the separation due to McQuarrie and Katz [14] is

$$u_0(r) = 4\varepsilon(\frac{\sigma}{r})^{12} \tag{1.36}$$

and

$$u_p(r) = -4\varepsilon(\frac{\sigma}{r})^6 \tag{1.37}$$

Now if we rewrite the $\lambda$-expansion of the free energy as

$$A = A_0 + A_1 + \Delta A \tag{1.38}$$

where $\Delta A = A_2 + A_3 + \cdots$, we can compare the magnitudes of the ratio $\Delta A/A_1$ for different separation schemes (1.17 and 18), and (1.34 to 37). A smaller ratio would indicate a better separation scheme, since $\Delta A/A_1$ measures directly the rate of convergence of the perturbation expansion. Verlet and Weis have carried out MC simulations for all three divisions and the comparison is given in Table XI.3. It is seen that the McQuarrie-Katz $\Delta A/A_1$ ratio is 2%, that of Barker-Henderson is 6%, and that of WCA is a small 0.4% at temperatures and densities near the triple point of argon. One explanation of the success of the WCA separation is that the resulting $u_0$ and $u_p$ are continuous and smooth functions. This renders the expansion more stable. In addition, the attractive part is bounded (not so in the McQuarrie-Katz scheme). Recently, new separation schemes have been proposed by Kang, Lee, Ree, and Ree [15]. Additional flexibility is built into eq. (1.17) when the additive constant $\varepsilon$ is replace by a function $F(r)$ to be determined.

*Table XI.3. Convergence of the $\lambda$-Expansion*

| Contributions to Free Energy, $A = A_{wca} + A_1 + \Delta A$ | | | | | |
|---|---|---|---|---|---|
| $u_p$ due to | $\rho^*$ | $T^*$ | $\beta A_1/N$ | $\beta\Delta A/N$ | $\Delta A/A_1$ |
| McQuarrie-Katz | 0.84 | 0.75 | -14.99 | -0.30 | 0.02 |
| Barker-Henderson | 0.85 | 0.72 | -7.79 | -0.45 | 0.06 |
| Weeks-Chandler-Andersen | 0.84 | 0.75 | -8.81 | -0.038 | 0.004 |

## XI.2. *Polar and Multipolar Fluids*

The major distinction between the isotropic potentials and the polar/multipolar potentials, as far as a mathematical representation is concerned, is that the latter are dependent on the orientations of the molecules in addition to the center-to-center distance. A detailed discussion of the angular coordinates was given in Chapter IV. The interactions between dipoles and quadrupoles, for example, could be expressed in these coordinates. Chapter VII has given the expressions for these interactions. For the sake of clarity, a brief discussion is given below.

### THE DIPOLE-DIPOLE INTERACTION

Roughly speaking, a molecule produces a dipole moment when the center of the positive charge distributions in the molecule does not coincide with the center of the negative charge distributions (such as in the molecules HCl, HBr, and CO). This produces a polarization, or a charge dislocation (see Figure XI.3). If we draw an arrow from the center of negative charges to that of the positive charges, we obtain a vector representing the direction of the dipole moment. The magnitude of the dipole moment is determined by the product $qL$, where $q$ is the charge of one center and $L$ is the position vector from one center to the other. The dipole moment is obtained from a Taylor series expansion of the electrostatic potential, $q/(\mathbf{r} - \mathbf{r}_i)$, as the coefficient of the $\nabla(1/r)$

term [16]. Given one dipolar molecule 1, the other dipolar molecule 2 will experience an electrostatic force due to 1. The Stockmayer potential represents this twofold interaction by a sum of the LJ potential and the ideal dipole-dipole interaction:

$$u_{\alpha\beta}(r_{12}, \theta_1, \theta_2, \phi_{12}) = 4\varepsilon_{\alpha\beta}\left[\left[\frac{\sigma_{\alpha\beta}}{r_{12}}\right]^{12} - \left[\frac{\sigma_{\alpha\beta}}{r_{12}}\right]^{6}\right] \tag{2.1}$$

$$+ \frac{\mu_\alpha\mu_\beta}{r_{12}^3}[\sin\theta_1 \sin\theta_2 \cos\phi_{12} - 2\cos\theta_1 \cos\theta_2]$$

where the angles $\theta_1$, $\theta_2$, and $\phi_{12}$ refer to the relative two-body coordinates as described earlier. The subscripts $\alpha$ and $\beta$ refer to the chemical species of the pair of molecules (e.g., $\alpha$= HCl, and $\beta$= CO). Thus (2.1) is suitably written for *mixtures*. $\mu_\alpha$ is the dipole moment of species $\alpha$, etc. Note that the dipole-dipole interaction described above is the so-called ideal dipole (or point dipole). It is obtained from the true dipole by letting the charge separation $L$ diminish to the limit zero and increasing $q$ proportionally so as to keep the original value $|qL|$ constant. The dipole-dipole interaction can also be written in the vector form

$$u_{\alpha\beta}^{DD}(r_{12}, \theta_1, \theta_2, \phi_{12}) = -\frac{\mu_\alpha\mu_\beta}{r_{12}^3}\left[\frac{3(\hat{\mathbf{L}}_1\cdot\mathbf{r}_{12})(\hat{\mathbf{L}}_2\cdot\mathbf{r}_{12})}{r_{12}^2} - \hat{\mathbf{L}}_1\cdot\hat{\mathbf{L}}_2\right] \tag{2.2}$$

where $\hat{\mathbf{L}}_i$ is the unit vector in the direction of the dipole of molecule $i$. Table VII.2 gives a listing of all alternative expressions.

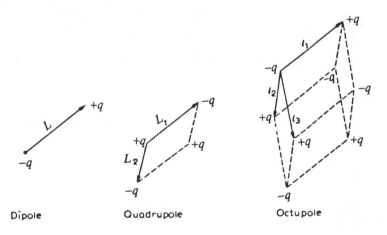

*FIGURE XI.3. Formation of multipoles due to charge separation.*

## THE QUADRUPOLE-QUADRUPOLE INTERACTION

As mentioned before, a dipole moment arises in a molecule when the center of positive charges is separated from that of the negative charges. If, furthermore, we can distinguish four local centers (Figure XI.3), two positive and two negative, in a molecule, we have a quadrupole. To represent the strength of a quadrupole, we define a quadrupole moment $\mathbf{Q}$, which is a second-order tensor given by

$$\mathbf{Q} = \frac{1}{2} \int d\mathbf{r} \; \rho_e(r)(3\mathbf{r}\mathbf{r}^T - r^2 \mathbf{I}) \tag{2.3}$$

where $\rho_e(r)$ is the charge density distribution and $\mathbf{r}$ is the position vector of the charges in the molecule with reference to some given origin, $\mathbf{I}$ is the unit tensor, and $r^2 = x^2 + y^2 + z^2$. Thus for a molecule with four centers of equal charge $q$, and with the $x$-, $y$-, and $z$-axes running along the symmetry axes of the molecule,

$$Q_{xx} = \frac{1}{2} \, q\sum_{i=1}^{4}(3x_i^2 - r_i^2), \qquad Q_{yy} = \frac{1}{2} \, q\sum_{i=1}^{4}(3y_i^2 - r_i^2), \qquad Q_{zz} = \frac{1}{2} \, q\sum_{i=1}^{4}(3z_i^2 - r_i^2) \tag{2.4}$$

For axially symmetric molecules, only one component $Q$ (say $Q_{zz}$) is independent. The quadrupole-quadrupole interaction in body-fixed coordinates is then

$$u_{\alpha\beta}^{QQ}(12) = \frac{3Q_\alpha Q_\beta}{4r_{12}^5} \Big[ 1 - 5\cos^2\theta_1 - 5\cos^2\theta_2 - 15\cos^2\theta_1 \cos^2\theta_2 \tag{2.5}$$

$$+ 2\{\sin\theta_1 \sin\theta_2 \cos\phi_{12} - 4\cos\theta_1 \cos\theta_2\}^2 \Big]$$

Refer to Table VII.2 for more detailed information. Next, we discuss the perturbation expansions for polar fluids.

## THE POPLE EXPANSION

The pair potential for molecules with polar/multipolar interactions can be written as

$$u_{\alpha\beta}(12) = u_{\alpha\beta}^0(r_{12}) + u_{\alpha\beta}^a(12) \tag{2.6}$$

where the isotropic potential $u_{\alpha\beta}^0(r)$ is an unweighted average of $u_{\alpha\beta}(12)$ over the orientations of the pair of molecules

$$u_{\alpha\beta}^0(r_{12}) = \langle u_{\alpha\beta}\rangle_{\omega_1\omega_2} \equiv \frac{1}{(4\pi)^2} \int_0^\pi d\theta_1 \sin\theta_1 \int_0^{2\pi} d\phi_1 \int_0^\pi d\theta_2 \sin\theta_2 \int_0^{2\pi} d\phi_2 \; u_{\alpha\beta}(12) \tag{2.7}$$

where the angular brackets denote the averaging over angles as defined. The anisotropic part also obeys

$$u_{\alpha\beta}^{a}(12) = u_{\alpha\beta}(12) - u_{\alpha\beta}^{0}(r_{12}) \tag{2.8}$$

From this definition, $u_{\alpha\beta}^{a}(12)$ does not include the spherical harmonic terms with $L'L''L =000$. Pople [17] gave the expansion of the free energy $A$ in terms of the aniso-tropic part $u_{\alpha\beta}^{a}/kT$:

$$A = A_0 + A_1 + A_2 + A_3 + \cdots \tag{2.9}$$

where $A_0$ is the free energy of the reference potential $u^0$. Due to the choice of the separation of the potentials, (2.6), one can show that the first-order term $A_1 =0$. Thus the first correction term from (2.9) is the $A_2$ term. Before we exhibit explicitly $A_2$, $A_3$, etc., we examine the convergence of the expansion (2.9). It has been found, from com-parisons with simulation results, that retaining the terms of the expansion up to $A_3$ is still insufficient to get good agreement. Stell et al. [18] proposed a resummation based on the $A_2$ and $A_3$ terms as

$$A = A_0 + \frac{A_2}{(1 - A_3/A_2)} \tag{2.10}$$

Equation (2.10) is called the Padé approximant of order [0, 1]. Namely, the higher-order terms $A_4$, $A_5$, etc., have been approximated by combinations of powers of $A_2$ and $A_3$. This resummation has been found to be quite successful for polar/multipolar fluids for the prediction of thermodynamic quantities. Thus it has become a standard method in the perturbation theory for polar fluids. To obtain explicit expressions for $A_2$ and $A_3$, let us assume that the isotropic $u^0$ is the LJ potential, and the polar/multipolar potentials include the quadrupole interactions:

$$u_{\alpha\beta}^{a}(12) = u_{\alpha\beta}^{112}(12) + u_{\alpha\beta}^{123}(12) + u_{\alpha\beta}^{213}(12) + u_{\alpha\beta}^{224}(12) \tag{2.11}$$

The indices 112 denote dipole-dipole interaction, 123= dipole-quadrupole, 213 = quadrupole-dipole, and 224 = quadrupole-quadrupole interactions. $A_2$ consists of terms involving pcfs, and $A_3$ includes two types: one involving the pcf and the other involving the triplet correlation functions.

$$A_2 = -\frac{1}{4kT} \sum_{\alpha}^{m} \sum_{\beta}^{m} \rho_\alpha \rho_\beta \int dr_1 \, dr_2 \, <u_{\alpha\beta}^{a}(12)^2>_{\omega_1 \omega_2} g_{\alpha\beta}^{LJ}(r_{12}) \tag{2.12}$$

where $g^{LJ}$ is the pcf of the LJ potential; or for potentials like (2.11),

$$A_2 = \sum_{\alpha\beta} \left[ A_2^{\alpha\beta}(112) + 2A_2^{\alpha\beta}(123) + A_2^{\alpha\beta}(224) \right] \tag{2.13}$$

where

$$\frac{A_2^{\alpha\beta}(112)}{NkT} = -\frac{2\pi}{3} x_\alpha x_\beta \frac{\tilde{\rho}_{\alpha\beta}}{\tilde{T}_{\alpha\beta}^2} \left[ \frac{\sigma_{\alpha\alpha}^3 \sigma_{\beta\beta}^3 \varepsilon_{\alpha\alpha}\varepsilon_{\beta\beta}}{\sigma_{\alpha\beta}^6 \varepsilon_{\alpha\beta}^2} \right] \tilde{\mu}_\alpha^2 \, \tilde{\mu}_\beta^2 \, J_{\alpha\beta}^{(6)} \tag{2.14}$$

$$\frac{A_2^{\alpha\beta}(123)}{NkT} = -\pi x_\alpha x_\beta \frac{\tilde{\rho}_{\alpha\beta}}{\tilde{T}_{\alpha\beta}^2} \left[ \frac{\sigma_{\alpha\alpha}^3 \sigma_{\beta\beta}^5 \varepsilon_{\alpha\alpha}\varepsilon_{\beta\beta}}{\sigma_{\alpha\beta}^8 \varepsilon_{\alpha\beta}^2} \right] \tilde{\mu}_\alpha^2 \, <\tilde{Q}_\beta>^2 \, J_{\alpha\beta}^{(8)}$$

$$\frac{A_2^{\alpha\beta}(224)}{NkT} = -\frac{14\pi}{5} x_\alpha x_\beta \frac{\tilde{\rho}_{\alpha\beta}}{\tilde{T}_{\alpha\beta}^2} \left[ \frac{\sigma_{\alpha\alpha}^5 \sigma_{\beta\beta}^5 \varepsilon_{\alpha\alpha}\varepsilon_{\beta\beta}}{\sigma_{\alpha\beta}^{10} \varepsilon_{\alpha\beta}^2} \right] <\tilde{Q}_\alpha>^2 <\tilde{Q}_\beta>^2 \, J_{\alpha\beta}^{(10)}$$

where $\tilde{\rho}_{\alpha\beta} = \rho\sigma_{\alpha\beta}^3$, etc., $\tilde{T}_{\alpha\beta} = kT/\varepsilon_{\alpha\beta}$, etc., $\tilde{\mu}_\alpha \equiv \mu_\alpha/(\varepsilon_{\alpha\alpha}\sigma_{\alpha\alpha}^3)^{1/2}$, $<\tilde{Q}_\alpha> \equiv <Q_\alpha>/(\varepsilon_{\alpha\alpha}\sigma_{\alpha\alpha}^5)^{1/2}$, and the $J$-integrals are defined as

$$J_{\alpha\beta}^{(n)} \equiv \int_0^\infty dr^* \, r^{*-(n-2)} g_{\alpha\beta}^{LJ}(r^*) \tag{2.15}$$

where $r^* = r/\sigma_{\alpha\beta}$. They arise from squaring the multipolar potentials. The $A_3$ term is split into two parts, $A_{3A}$ and $A_{3B}$, the former being terms involving the pcf of the reference fluid and the latter involving the triplet correlation functions. The details could be found in Flytzani-Stephanopoulos et al. [19] and Gubbins et al. [20]. We have given these theoretical formulas. Next we shall examine the performance of these equations on polar fluids.

## XI.3. *Applications to Polar Fluids*

We shall examine some model polar fluids where simulation data are available for comparison. They include hard spheres with embedded point dipoles and quadrupoles, as well as spheres with both multipoles.

### DIPOLAR HARD SPHERES

Patey and Valleau [21] [22] [23] carried out MC simulation of purely dipolar hard spheres over a range of dipole moments (from 0.25 debye to 1.75 debye). In this case, the pcf of the reference system in the perturbation equations should be replaced by the HS ones. Results at two densities were reported: $\rho/\rho_{cl.pkg.} = 0.64$ and 0.59, where $\rho_{cl.pkg.}$ is the density of close packing for hard spheres. The MC results are listed in Table XI.4 together with the Padé results (eq. (2.10)) and the MSA results. It is seen that the Padé results are satisfactory up to the highest values of the dipole moment tested (with error of 1.6% at $\mu = 1.75$ debye). On the other hand, the MSA results are already inadequate at $\mu = 0.75$ debye (with an error of $-40\%$).

Tests on other thermodynamic properties show the same trend. The configurational internal energy is compared in Table VII.1. At $\mu = 1.75$ debye, the MC value is $-4.80$, whereas the Padé approximant gives $-4.64$. On the other hand, the MSA gives $-3.21$, a gross underestimation. The comparison of the compressibility factor again shows that Padé approximant gives accurate predictions.

### QUADRUPOLAR HARD SPHERES

The quadrupole-quadrupole interaction is shorter ranged ($\sim r^{-5}$) compared to the dipolar case ($\sim r^{-3}$). Therefore one should expect faster convergence and better results for quadrupolar fluids. This is indeed the case. Patey and Valleau (*Ibid.*) made MC

*Table XI.4. Free Energy\* of Dipolar Hard Spheres at T = 298.15 K*

| $\mu$ (debye) | Free Energy, $-\beta(A - A_{HS})/N$ | | |
|---|---|---|---|
| | MC | MSA | Padé |
| 0.25 | 0.0035 | 0.0018 | 0.00304 |
| 0.75 | 0.21 | 0.124 | 0.204 |
| 1.25 | 1.14 | 0.732 | 1.168 |
| 1.75 | 3.19 | 2.12 | 3.24 |

\*At reduced density $\rho_{cl.pkg.} = 0.59$.

calculations for hard spheres with embedded axial quadrupoles. The results on the free energy are given in Table XI.5. They have also included $A_2$ and $A_3$ for comparison. Inclusion of $A_2$ in the sum $A_0 + A_2$ (i.e., the second-order theory) gives a more negative free energy than MC. Inclusion of $A_3$ (the third order theory), which is opposite in sign to $A_2$, overcorrects the free energy, i.e., giving a value less negative than MC. On the other hand, the resummation (2.10) avoids such oscillations and gives close agreement with MC. The internal energy values are given in Table VII.1. We see that the values from Padé are dependable for quadrupole moments $Q^{*2} = Q^2/kTd^5 \leq 0.714$. At $Q^{*2} = 1.67$, the Padé value is about $-10\%$ in error. However, the $A_2$-corrected and $A_3$-corrected perturbation series are worse, deviating $-100\%$ and $+100\%$ from the MC values, respectively. The improvement due to the Padé resummation is obvious.

*Table XI.5. Properties of Quadrupolar Spheres*

| $Q^{*2}$ | Free Energy, $-\beta(A - A_{HS})/N$ | | | |
|---|---|---|---|---|
| | 2nd Order | 3rd Order | Padé | MC |
| 0.02 | 0.001 | 0.001 | 0.001 | 0.0008 |
| 0.1 | 0.0249 | 0.0242 | 0.0242 | 0.0239 |
| 0.4 | 0.04 | 0.351 | 0.356 | 0.349 |
| 0.71429 | 1.273 | 0.996 | 1.045 | 1.008 |
| 1.0 | 2.495 | 1.735 | 1.912 | 1.844 |
| 1.25 | 3.898 | 2.414 | 2.823 | 2.698 |
| 1.66666 | 6.93 | 3.413 | 4.597 | 4.324 |
| $Q^{*2}$ | Configurational Energy, $-\beta U/N$ | | | |
| | 2nd Order | 3rd Order | Padé | MC |
| 0.02 | 0.002 | 0.00198 | 0.00198 | 0.002 |
| 0.4 | 0.798 | 0.652 | 0.673 | 0.655 |
| 0.71429 | 2.546 | 1.715 | 1.904 | 1.801 |
| 1.0 | 4.989 | 2.71 | 3.378 | 3.182 |
| 1.25 | 7.796 | 3.345 | 4.868 | 4.497 |
| 1.66666 | 13.86 | 3.308 | 7.645 | 6.987 |

## HARD SPHERES WITH EMBEDDED DIPOLES AND QUADRUPOLES

For hard spheres embedded with both dipoles and quadrupoles, eq. (2.11) applies. Simulations at the reduced dipole moments of $\mu^{*2} = \mu^2/kTd^3 = 1.0$ and $Q^{*2} = Q^2/kTd^5$ from 0.02 and 1.66666 are available. Table XI.6 gives the free energy and internal energy. Again Padé results are in good agreement with the MC data. At $Q^{*2} = 1.66666$, the Padé method gives $-5.617$, and simulation gives $-5.752$ ($+1\%$). For the internal energy at $\mu^* = 1.0$ and $Q^{*2} > 0.8333$, the Padé values are in error by $+5\%$. The $A_3$-correction, on the other hand, is in error by more than +40% under the same conditions.

*Table XI.6.  Properties of Dipole-Quadrupole Fluids at $\mu^*=1.0$*

| | Free Energy, $-\beta(A - A_{HS})/N$ | | | |
|---|---|---|---|---|
| $Q^{*2}$ | 2nd Order | 3rd Order | Padé | MC |
| 0.02 | 0.996 | 0.524 | 0.676 | 0.638 |
| 0.1 | 1.239 | 0.546 | 0.795 | 0.757 |
| 0.4 | 2.268 | 0.656 | 1.326 | 1.312 |
| 0.71429 | 3.829 | 0.736 | 2.118 | 2.157 |
| 1.0 | 5.675 | 0.659 | 3.012 | 3.081 |
| 1.25 | 7.624 | 0.389 | 3.912 | 3.992 |
| 1.66666 | 11.566 | -0.681 | 5.617 | 5.752 |
| | Configurational Energy, $-\beta U/N$ | | | |
| $Q^{*2}$ | 2nd Order | 3rd Order | Padé | MC |
| 0.02 | 2.081 | 0.576 | 1.134 | 1.056 |
| 0.4 | 4.537 | -0.3 | 2.101 | 2.118 |
| 0.71429 | 7.657 | -1.622 | 3.289 | 3.418 |
| 1.0 | 11.35 | -3.7 | 4.611 | 4.827 |
| 1.25 | 15.25 | -6.455 | 5.919 | 6.255 |
| 1.66666 | 23.13 | -13.61 | 8.346 | 9.075 |

From the comparisons made above, we see that the Padé approximant is quite effective in reproducing simulation results and has become an integral part in modern perturbation theories for polar fluids. Some typical polar/multipolar fluids are listed in Table XI.7 together with their LJ parameters and polar moments. The magnitudes of the moments fall within the ranges of the dipole and quadrupole moments studied above, and their behavior can thus be described by the Padé method. A full-scale description of real polar/multipolar fluids, however, still awaits future investigation.

*Table XI.7.  Potential Parameters for Polar Molecules*

| | LJ Parameters | | Electrostatic moments | | Reduced Moments | |
|---|---|---|---|---|---|---|
| | $\varepsilon/k$ (K) | $\sigma$ (Å) | $\mu \times 10^{18}$ (esu·cm) | $Q \times 10^{26}$ (esu·cm$^2$) | $\mu^*$ | $Q^*$ |
| $CO_2$ | 190 | 4.0 | 0 | -4.3 | 0 | 0.83 |
| $O_2$ | 118 | 3.46 | 0 | -0.39 | 0 | 0.137 |
| $N_2$ | 95.1 | 3.71 | 0 | -1.52 | 0 | 0.501 |
| HCl | 218 | 3.51 | 1.03 | 3.8 | 0.903 | 0.949 |

## XI.4.  *The Perturbation Theories for Correlation Functions*

The perturbation approaches discussed above relate to the thermodynamic behavior of the fluids. It is also possible to apply the same expansion techniques for correlation functions. Here again we have two major types of expansion: the density expansions (expansions in terms of the density function $\rho^{(1)}(r)$), and the temperature expansions (expansions in terms of the pair potential $u(r)/kT$ or the Boltzmann factor $\exp[-\beta u(r)]$). The Gubbins-Gray [24] expansion is in terms of $u(r)/kT$; the Andersen-Weeks-Chandler [25] expansion is in terms of $\exp(-\beta u)$; and the Perram-White [26] expansion modifies the reference part of the $\exp(-\beta u)$-expansion. Other expansion methods have been proposed— notably those due to Madden and Fitts [27], Melnyk and Smith [28], and MacGowan et al. [29] We shall discuss these methods in the following.

## GUBBINS-GRAY EXPANSION

In the Gubbins-Gray expansion, the pair potential is separated, as before, into a reference part $u_r$ and a perturbing part $u_p$

$$u(12) = u_r(12) + u_p(12) \tag{4.1}$$

In case the reference potential $u_r$ is isotropic, the expansion of the pair correlation function $g(12)$ in terms of $u/kT$ is

$$g(12) = g_0(r_{12}) + g_1(12) + g_2(12) + \cdots \tag{4.2}$$

where $g_0(r_{12})$ is the pair correlation function of the reference potential $u_r(r_{12})$ (i.e., the zero order term). $g_1$ and $g_2$ are the first-order and second-order corrections due to the perturbing potential $u_p(12)$, respectively, i.e.,

$$g_1(r_{12}\omega_1\omega_2) = - \beta u_p(r_{12}\omega_1\omega_2)g_0(r_{12}) \tag{4.3}$$

$$- \beta\rho \int dr_3 \left[ u_p(r_{13}\omega_1\omega_3)>_{\omega_3} + <u_p(r_{23}\omega_2\omega_3)>_{\omega_3} \right] g_0^{(3)}(r_1 r_2 r_3)$$

$$- \frac{1}{2}\beta\rho^2 \int r_3 r_4 <u_p(r_{34}\omega_3\omega_4)>_{\omega_3\omega_4} \left[ g_0^{(4)}(r_1 r_2 r_3 r_4) - g_0(r_{12})g_0(r_{34}) \right]$$

$$+ \beta\frac{\partial}{\partial\rho}[\rho^2 g_0(r_{12})] \left\{ \int dr_{34} <u_p(r_{34}\omega_3\omega_4)>_{\omega_3\omega_4} g_0(r_{34}) \right.$$

$$\left. + \frac{1}{2}\rho \int dr_{34} dr_{45} <u_p(r_{34}\omega_3\omega_4)>_{\omega_3\omega_4} \left[ g_0^{(3)}(r_3 r_4 r_5) - g_0(r_{34}) \right] \right\}$$

If the choice of the reference potential is $u_r = <u(12)>_{\omega_1\omega_2}$, equation (4.3) simplifies to

$$g_1(r_{12}\omega_1\omega_2) = - \beta u_p(r_{12}\omega_1\omega_2)g_0(r_{12}) \tag{4.4}$$

$$- \beta\rho \int dr_3 \left[ <u_p(r_{13}\omega_1\omega_3)>_{\omega_3} + <u_p(r_{23}\omega_2\omega_3)>_{\omega_3} \right] g_0^{(3)}(r_1 r_2 r_3)$$

Furthermore, if the anisotropic part $u_p$ is of the multipolar type (i.e., $L \neq 0$, as in the dipole-dipole and quadrupole-quadrupole interactions, etc.), the angle averages of $u_p$ are zero. We are left with

$$g_1(12) = - \beta u_p(12)g_0(r_{12}) \tag{4.5}$$

The general expression for $g_2$ involves correlation functions up to $g^{(6)}$. For $u_p$ of the multipolar type ($L \neq 0$), we need correlations up to $g^{(4)}$:

$$g_2(\mathbf{r}_{12}\omega_1\omega_2) = \frac{1}{2}\beta^2\Bigg\{ u_p(\mathbf{r}_{12}\omega_1\omega_2)^2 g_0(r_{12})$$

$$+ \rho \int d\mathbf{r}_3 <[u_p(\mathbf{r}_{13}\omega_1\omega_3) + u_p(\mathbf{r}_{23}\omega_2\omega_3)]^2>_{\omega_3} g_0^{(3)}(\mathbf{r}_1\mathbf{r}_2\mathbf{r}_3)$$

$$+ \frac{1}{2}\rho^2 \int d\mathbf{r}_3 d\mathbf{r}_4 <u_p(\mathbf{r}_{34}\omega_3\omega_4)^2>_{\omega_3\omega_4}\Big[g_0^{(4)}(\mathbf{r}_1\mathbf{r}_2\mathbf{r}_3\mathbf{r}_4) - g_0(r_{12})g_0(r_{34})\Big]$$

$$- \frac{1}{2}\frac{\partial}{\partial\rho}[\rho^2 g_0(r_{12})]\Big[2\int d\mathbf{r}_{34} <u_p(\mathbf{r}_{34}\omega_3\omega_4)^2>_{\omega_3\omega_4} g_0(r_{34})$$

$$+ \rho \int d\mathbf{r}_{34} d\mathbf{r}_{45} <u_p(\mathbf{r}_{34}\omega_3\omega_4)^2>_{\omega_3\omega_4} [g_0^{(3)}(\mathbf{r}_3\mathbf{r}_4\mathbf{r}_5) - g_0(r_{34})]\Big]\Bigg\}$$

(4.6)

The above equations (4.5) and (4.6) are applicable to the pcfs of multipolar fluids.

For quadrupolar fluids, it was found that the first order theory (inclusive of the $g_1$ term) is accurate only for small quadrupole moments (e.g., $Q^* \equiv Q/(\varepsilon\sigma^5)^{1/2} \leq 0.3$) when the isotropic Lennard-Jones potential was used as reference. Murad et al. [30] have calculated the second order terms for a quadrupolar fluid of pair interaction

$$u(12) = u_{LJ}(r_{12}) + u_{QQ}(12)$$

(4.7)

where $u_{QQ}$ is the quadrupole-quadrupole interaction. The condition studied was $\rho\sigma^3 = 0.75$, $kT/\varepsilon = 1.154$ and $Q^{*2} = 0.5$. The first 18 spherical harmonic coefficients of $g(12)$ were obtained (i.e., $L'L''L$ = 022, 202, 220, 222, 224, 242, 422, 244, 424, 442, 246, 426, 044, 404, 440, 444, 446, and 448). The results are displayed in Figures X.4 and 5 for $L'L''L$ = 224 and 222. The results from the first order perturbation ($g(12) \approx g_0 + g_1$) and the second order perturbation ($g(12) \approx g_0 + g_1 + g_2$), as well as LHNC and QHNC, are shown. The second order theory gives a good description of $g(224;r)$. However, the other two spherical harmonic coefficients (220 and 222) are not well predicted. The QHNC theory is quite satisfactory in this case. We note that the first order perturbation is

$$g(12) = g_0(r_{12}) - \beta u_p(12)g_0(r_{12})$$

(4.8)

For quadrupole-quadrupole interaction ($L'L''L$ = 224), $u_p$ consists of the term— $u^{224}$. Thus the $g(12)$ obtained from (4.8) has only one multipolar term, i.e., $g(224;r)$, while all other terms ($g(220;r)$, $g(222;r)$, etc.) are zero. Obviously, this could not be correct.

## ANDERSEN-WEEKS-CHANDLER EXPANSION

The approach adopted by Andersen, Weeks and Chandler [31] is to expand both the free energy $A$ and the background correlation function $y(r)$ in terms of the Boltzmann factor. As was noted earlier, the functional derivative of $\ln Q_N$ with respect to $e(s) \equiv \exp[-\beta(s)]$ is

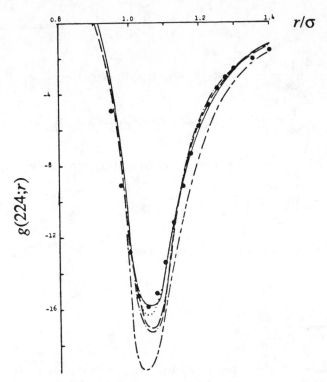

FIGURE XI.4. *Spherical harmonic coefficient* $g(224,;r)$ *of LJ+QQ fluid at T\** = 1.154, ρ\* = 0.75, *and* Q\*² =0.50. •: *MD.* ———: *second order perturbation.* · · · : *first order perturbation.* -- · -- · --: *LHNC.* - - - -: *QHNC.* *(Murad, Gubbins, and Gray, Chem. Phys. Lett., 1979).*

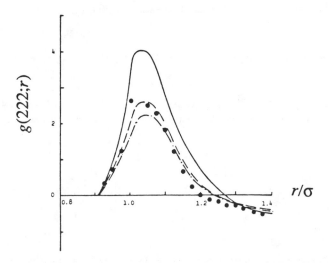

FIGURE XI.5. *Spherical harmonic coefficient* $g(222,;r)$ *of LJ+QQ fluid at T\** = 1.154, ρ\* = 0.75, *and* Q\*² =0.50. *Legend as in Fig.XI.4. First order theory gives* g(222;r) = 0. *(Murad, Gubbins, and Gray, Chem. Phys. Lett., 1979).*

$$\frac{\delta \ln Q_N}{\delta e(s)} = \frac{1}{Q_N} \frac{\delta}{\delta e(s)} \int d\mathbf{r}_1 d\mathbf{r}_2 \cdots d\mathbf{r}_N \left[ e(r_{12})e(r_{13})e(r_{14})...e(r_{N-1,N}) \right] \qquad (4.9)$$

$$= \frac{V}{Q_N} \frac{\delta}{\delta e(s)} \int d\mathbf{r}_{12} d\mathbf{r}_{13}...d\mathbf{r}_{1N} \left[ e(r_{12})e(r_{13})...e(r_{N-1,N}) \right]$$

$$= \frac{V}{Q_N} \frac{N(N-1)}{2} \int d\mathbf{r}_{12} d\mathbf{r}_{13} d\mathbf{r}_{1N} \, \delta(s, r_{12}) \left[ e(r_{13})e(r_{14}) \cdots e(r_{N-1,N}) \right]$$

$$= \frac{N(N-1)V}{2Q_N} \int d\mathbf{r}_{13} \cdots d\mathbf{r}_{1N} \left[ e(r_{13})e(r_{14})...e(r_{N-1,N}) \right]$$

$$= \frac{V}{2} e^{+\beta u(s)} \rho^{(2)}(s) = \frac{\rho^2 V}{2} g^{(2)}(s) e^{+\beta u(s)}$$

Thus

$$y(s) = \frac{2}{\rho^2 V} \frac{\delta \ln Q_N}{\delta e(s)} \qquad (4.10)$$

We have shown that the functional derivative of $\ln Q_N$ gives $y(s)$. Continued differentiations of $y(s)$ with respect to $e(s)$ give higher order functional derivatives. Thus the functional Taylor expansion of $y(r)$, to first order in $\exp(-u/kT)$, is

$$y(\mathbf{r}; u) = y_0(\mathbf{r}; u_r) \qquad (4.11)$$

$$\cdot \left[ 1 + \frac{2\rho}{V} \int d\mathbf{r}_1 d\mathbf{r}_2 d\mathbf{r}_3 \, \delta(\mathbf{r}, r_{12}) y_0(r_{13})[e(r_{13}) - e_r(r_{13})] J^{(3)}(\mathbf{r}_1, \mathbf{r}_2, \mathbf{r}_3) \right.$$

$$\left. + \frac{\rho^2}{2V} \int d\mathbf{r}_1 d\mathbf{r}_2 d\mathbf{r}_3 d\mathbf{r}_4 \, \delta(\mathbf{r}, r_{12}) y_0(r_{34})[e(r_{34}) - e_r(r_{34})] J^{(4)}(\mathbf{r}_1, \mathbf{r}_2, \mathbf{r}_3, \mathbf{r}_4) + \cdots \right]$$

where the integrals $J^{(3)}$ and $J^{(4)}$ were given in eqs. (1.13 and 14). The subscript 0 denotes reference system properties, and $e_r(r) = \exp[-\beta u_r(r)]$. The lowest (zero) order approximation is

$$y(r) \approx y_0(r) \qquad (4.12)$$

or

$$g(r)e^{+\beta u(r)} = y_0(r), \qquad g(r) = y_0(r)e^{-\beta u(r)} \qquad (4.13)$$

This is the WCA approximation mentioned earlier (see (1.32)). At high densities ($\rho^* = 0.85$), the $g(r)$ of the LJ fluid is well approximated by the $g_{WCA}(r)$ according to (4.13). The importance of the repulsive part of the potential in determining the structure, especially at high densities, is clearly in evidence.

## PERRAM-WHITE EXPANSION

The Perram-White [32] expansion of the pcf to the zeroth order resembles the WCA expansion. However, their treatment of the reference potential is different. They incorporated contributions from the long-range anisotropic potential into the isotropic reference potential. Consider a pair potential composed of an isotropic part $u_{iso}$ and an anisotropic part $u_{aniso}$

$$u(12) = u_{iso}(r_{12}) + u_{aniso}(12) \tag{4.14}$$

In perturbation theories, one has the option of choosing a reference potential $u_r$ (noting that the perturbing potential could be *defined* as the remainder $u_p \equiv u - u_r$), Perram and White constructed a $u_r$ through

$$u_r(r_{12}) \equiv u_{iso}(r_{12}) + u_{avg}(r_{12}) \tag{4.15}$$

$$u_p(12) \equiv u_{aniso}(12) - u_{avg}(r_{12}) \tag{4.16}$$

where the average potential $u_{avg}(r_{12})$ is obtained by averaging over the Boltzmann factor

$$\exp[-\beta u_{avg}(r_{12})] \equiv \frac{1}{\Omega^2} \int d\omega_1 d\omega_2 \exp[-\beta u_{aniso}(12)] \tag{4.17}$$

$\Omega$ is $4\pi$ for linear molecules and $8\pi^2$ for nonlinear molecules. Namely, $u_{avg}(r_{12})$ is the angle-averaged (via the Boltzmann factor) part of the anisotropic potential. The purpose of this averaging is to include some of the angular effects in the reference potential, so that subsequent truncation of the perturbation series even at low orders will retain sufficient anisotropic contributions. We note that $u_{avg}(r_{12})$ is isotropic. Thus the reference potential $u_r(r_{12})$ remains isotropic. To first order, the expansion of the full $g(12)$ is

$$g(12) = g_0(r_{12})\exp[-\beta u_p(12)] + \frac{\rho}{\Omega} \int dr_3 d\omega_3 \, g_0^{(3)}(r_1 \, r_2 \, r_3)[f_p(13) + f_p(23)] + \cdots \tag{4.18}$$

where $g_0(r_{12})$ is the pcf of the reference potential given by (4.15). The anisotropic Mayer factor $f_p(12)$ is defined as

$$f_p(12) \equiv \exp[-\beta u_p(12)] - 1 = \exp\left\{-\beta[u_{aniso}(12) - u_{avg}(r_{12})]\right\} - 1 \tag{4.19}$$

And $g_0^{(3)}(r_1 \, r_2 \, r_3)$ is the triplet correlation function of $u_r$. When eq. (4.18) is truncated at the zero order

$$g(12) = g_0(r_{12})\exp[-\beta u_p(12)] \tag{4.20}$$

Multiplying by $\exp[+\beta u(12)]$, we have

$$y(12) = y_0(r_{12}) \tag{4.21}$$

Superficially, this equation is similar to eq.(4.12), the WCA zero order theory. However, the reference potential selected by Perram and White was different (see eq. (4.15)). To derive (4.18), Perram and White [33] expanded the pcf $g(12)$ in terms of $\lambda$, similar to the $\lambda$-expansion discussed earlier. Here $\lambda$ occurs in

$$u(12) = u_r(r_{12}) + u_p(12; \lambda) \tag{4.22}$$

$$= u_{iso}(r_{12}) + u_{avg}(r_{12}) - kT \ln[1 + \lambda f_p(12)]$$

In the language of functionals, this expansion is in terms of the Mayer factor $f_p$. Smith and Melnyk [33] have proposed a similar expansion based on the reference averaged Mayer (RAM) function. Either the background correlation function $y(12)$ (in the theory of RAMY) or the pcf $g(12)$ (RAMG) is expanded in terms of the Mayer factor. The results are

**REFERENCE AVERAGED MAYER EXPANSION**

*RAMY*

$$y(12) = y_0(r_{12}) + \rho y_1(12) \tag{4.23}$$

or

*RAMG*

$$g(12) = y_0(r_{12})e^{-\beta u(12)} + \rho y_1(12)e^{-\beta u_r(r_{12})} \tag{4.24}$$

Equivalently,

$$y(12) = y_0(r_{12}) + \rho y_1(12)e^{-\beta u_p(12)} \tag{4.25}$$

The first order term in both cases is

$$y_1(12) \equiv y_0(r_{12}) \left\{ \frac{1}{\Omega} \int d\mathbf{r}_3 d\omega_3 h_0(r_{13}) y_0(r_{23}) \left[ e^{-\beta u(23)} - e^{-\beta u_r(r_{23})} \right] \right. \tag{4.26}$$

$$\left. + \frac{1}{\Omega} \int d\mathbf{r}_3 d\omega_3 \ h_0(r_{23}) y_0(r_{13}) \left[ e^{-\beta u(13)} - e^{-\beta u_r(r_{13})} \right] \right\}$$

We note first that the RAMY and the RAMG equations are not equivalent. The Boltzmann factor $\exp[-\beta u_p]$ does not figure in RAMY. On the other hand, the RAMG equation also differs from the Perram-White equation (4.18). The RAMG theory contains the total correlation function $h_0$ in the integrand; while upon making the superposition approximation for $g^{(3)}$, the Perram-White integrand contains $g_0$ in place of $h_0$. This difference comes from the different expressions used to approximate the three-body correlation functions $y^{(3)}(\mathbf{r}_1, \mathbf{r}_2, \mathbf{r}_3)$ and $g^{(3)}(\mathbf{r}_1, \mathbf{r}_2, \mathbf{r}_3)$.

Melnyk and Smith [34] have applied the RAMY and RAMG equations to hard dumbbell fluids. The RAMY and RAMG results are comparable in most cases, except at bond lengths $L^* = 0.4$ and 0.6. RAMG seems to perform better than RAMY.

## ANGULAR MEDIAN POTENTIAL METHOD

The angular median potential used as a spherical reference potential [34] [35] is defined as the symmetrical potential obtained from a given angular potential via the formula

$$\int d\omega_1 d\omega_2 \ \text{sgn}\left[u(R_{12}\omega_1\omega_2) - \bar{u}(R_{12})\right] \equiv 0 \tag{4.27}$$

where sgn is the sign function (= +1 for positive argument and −1 for negative argument), $R_{12}$ is the distance from the center of mass of molecule 1 to that of molecule 2, and $\omega_i$ are the Euler angles of the nonspherical molecules $i$. Thus for every fixed $R$ the median potential $\bar{u}$ assumes a value that divides equally the angular regions in which $u(12)$ falls above and below $\bar{u}$. The rationale for such a division is to annul the first-order term in the perturbation expansion. We have shown in the $\lambda$-expansion that when the potential is divided according to (1.1)

$$u(12) = \bar{u}(R_{12}) + u_p(12) \tag{4.28}$$

the first-order free energy for the anisotropic potential is

$$\frac{\partial A}{\partial \lambda} = \frac{1}{2} \int d1 d2 \ \rho_0^{(2)}(R_{12}) u_p(12) \tag{4.29}$$

$$= \frac{1}{2} \int dr_1 d\mathbf{R}_{12} \ \rho_0^{(2)}(R_{12}) \int d\omega_1 d\omega_2 \ u_p(R_{12}\omega_1\omega_2)$$

The integral could be *separated* into one over the distance $\mathbf{R}_{12}$ and the other over the angles $\omega_1$ and $\omega_2$. When we apply the definition of $\bar{u}$, the first-order term is zero. There are several advantages for this reference potential. Unlike other perturbation schemes, $\bar{u}$ is independent of the temperature and density of the system. The amount of calculations required is greatly reduced. The median potential has been used for calculating the structure and thermodynamic properties of soft and hard diatomics, and nonlinear polyatomics. Accurate thermodynamic information is obtained. However, the structure is poorly predicted [36]. Therefore, it is not useful for getting the pcfs.

As an example, let us consider the two-center LJ diatomics (2cLJ). At different bond lengths ($L^* = 0, 0.3292$, and 0.793), the corresponding median potentials have been evaluated [37]. Figure XI.6 shows the results. The potential minima move to larger separations when the bond lengths are increased. The well depths also become shallower. To calculate the free energy, a hard sphere fluid was used as reference

$$\frac{A}{NkT} = \frac{y(4-3y)}{(1-y)^2} + 2\pi\rho\beta \int_d^\infty dR \ R^2 \bar{u}(R) g^{HS-PY}(R/d;y) - \left[\frac{1}{2}y^4 + y^2 + \frac{1}{2}y\right] \tag{4.30}$$

The hard sphere diameter $d$ was found by minimizing the free energy. This variational procedure was due to Ross [38]. The last term is an empirical correction when the pcf $g^{HS}$ from the PY approximation is used. ($y = \pi\rho d^3/6$.) The pressure and energy obtained for 2cLJ are shown in Figure XI.7. The pressures were better predicted than the

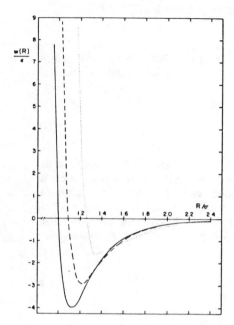

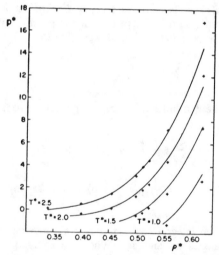

FIGURE XI.6. Calculated median potentials for LJ diatomics. ————: $L^* = 0$ (LJ spheres). - - - -: $L^* = 0.3292$. · · · : $L^* = 0.793$. (MacGowan et al., J. Chem. Phys., 1984).

FIGURE XI.7. Pressure (a) and energy (b) for LJ diatomics as calculated from median potential according to perturbation theory. (a): Reduced pressure at $L^* = 0.5471$. (b): Reduced energy at $L^* = 0.5471$. (MacGowan et al., J. Chem. Phys., 1984).

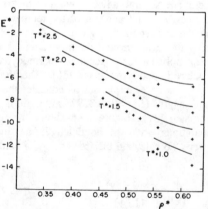

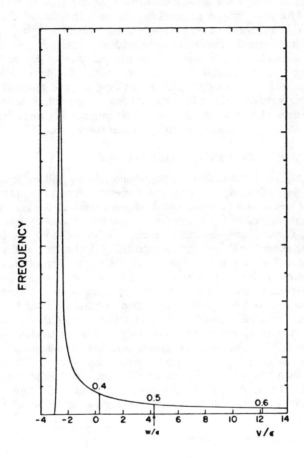

FIGURE XI.8. Frequency distribution for potential energy values of the LJ diatomic with $L^* = 0.793$. The distance $r/\sigma$ is set to 1.2. The value of the median potential for 50% division is marked by the arrow. The frequency units are arbitrary. (MacGowan et al., J. Chem. Phys. 1984).

energies. As the bond length was elongated (to $L^* = 0.793$), the predictions worsened. This was attributed to the frequency distributions of the potential energy. Figure XI.8 exhibits the case for $L^* = 0.793$. For a fixed $R/\sigma = 1.20$, the full potential $u(12)$ was negative in almost 40% of the angular domain. A large portion of negativity had not been accounted for in the median potential calculations. Thus the calculated energy was too *positive*. This indicates that the median potential tends to neglect significant local fluctuations of the potential. Furthermore, the median potential approach is not suitable for multipolar molecules. For example, the median potential for dipolar hard spheres is simply the hard sphere potential itself. Other schemes for determining the median potentials have been proposed [39], such as dividing the pair potential into different parts and making separate medians for separate parts. In general, perturbations based on median potentials work well for the thermodynamic properties of small linear molecules, but deteriorate for elongated molecules and nonlinear polyatomics [40].

## NONSPHERICAL REFERENCE POTENTIAL METHOD

For highly elongated molecules, perturbation theory will no longer be able to reproduce the thermodynamic and structural properties based on a spherical reference fluid. The next step is to use a nonspherical reference potential that can capture some of the essential anisotropies of the interaction. Most nonspherical systems studied were diatomics, e.g., Lennard-Jones diatomics, repulsive WCA diatomics, and hard dumbbells (HD). Some work on liquid crystals was reported [41]. Earlier Sandler [42], Mo and Gubbins [43], and Boublik [44] have investigated the perturbation expansions with nonspherical references. Kohler et al. [45] as well as Fischer [46] extended the blip function formalism of Weeks, Chandler, and Andersen [47] to nonspherical diatomics. Tildesley [48] and Quirke [49] have formulated perturbation expansions for the background correlation functions of site-site potentials. Lee et al. [50] have used the functional expansion method (see below) to generate perturbation series for quadrupolar fluids with multipolar fluids as references. We cite as an example the perturbation method due to Tildesley for the 2cLJ fluid based on a nonspherical hard dumbbell reference.

As in WCA, the site-site LJ potential for 2cLJ is divided into repulsive and attractive parts. The repulsive ss pcf is expanded in terms of the hard dumbbell background correlation functions with the aid of RISM approximation (see Chapter XIV):

$$g_{r\alpha\gamma}(r) = [g_{d\alpha\gamma}]_{RISM}(r) + \Delta f_{\alpha\gamma}(r)[y_{d\alpha\gamma}(r)]_{RISM} \tag{4.31}$$

$$+ \frac{1 + f_{r\alpha\gamma}}{Lr} \sum_\eta \int_{|r-L|}^{r+L} dx\, x\Delta f_{\eta\gamma}(x)[y_{d\alpha\eta}(x)]_{RISM}$$

where subscript $r$ is for the repulsive reference, and subscript $d$ for the hard dumbbells. $\Delta f_{\alpha\gamma} \equiv f_{r\alpha\gamma} - f_{d\alpha\gamma}$ is the difference in Mayer factors. $L$ is the bond length of the 2cLJ molecule and the hard dumbbell, taken to be the same here. The subscript RISM denotes quantities obtained from the RISA theories (Chapter XIV). In arriving at eq. (4.31), one recognizes that the soft repulsion ss pcf and the HD ss pcf differ in the sharp cusps that show up at $d$ and $d+L$ for hard core fluids. (They become "shoulders" for soft potentials.) Obviously $y_{r\alpha\gamma}(r) \neq y_{d\alpha\gamma}(r)$ for all $r$. However, if we subtract out all the cluster diagrams contributing to the cusps, $y^{(c)}$, i.e., diagrams with $h_r$ of $h_l$ bonds (see Chapter XIV), we obtain the "smoothed" correlation functions $y^{(s)} \equiv y - y^{(c)}$ which to lowest order

$$y_{r\alpha\gamma}^{(s)}(r) \approx y_{d\alpha\gamma}^{(s)}(r) \tag{4.32}$$

Equation (4.31) came from (4.32). The *effective* hard dumbbell diameter $d^*_{\alpha\gamma}$ is determined by annulling a blip function

$$\int dr \ r^2 y_{d\alpha\gamma}\left[e^{-\beta u_{r\alpha\gamma}(r)} - e^{-\beta u_{d\alpha\gamma}(r\,;d^*_{\alpha\gamma})}\right] = 0 \qquad (4.33)$$

where $u_{r\alpha\gamma}(r)$ is the site-site repulsive WCA potential. The full site-site pcf (for 2cLJ) is, to zero order, given by

$$g_{\alpha\gamma}(r) \approx g_{r\alpha\gamma}(r) \qquad (4.34)$$

We observe that there have been many approximations made in order to arrive at (4.34). The predictions for the site-site pcf of 2cLJ fluids (here used to represent nitrogen, fluorine, chlorine, and carbon dioxide) are reasonable, but with sizable differences at $r < L + \sigma$. This was attributed to the approximations used in the RISA theories. It is conceivable that when simulation data for hard dumbbells are used, agreement should improve.

For Lennard-Jones spheres with embedded ideal quadrupoles, Lee et al. [51] found that using the anisotropic apcf at the quadrupolar strength $Q^* = 0.5$ as reference one could achieve predictions of the harmonic coefficients of the apcf at $Q^* = 1.0$ better than using a spherical reference (see below).

## XI.5. *The Method of Functional Expansions*

Similar to the functional expansion methods for integral equations, we can choose a generating functional and expand this functional in terms of an independent function (such as the Mayer factor or the singlet density) to generate a perturbation series. The WCA and Smith-Melnyk approaches discussed above belong to this category. Here we make use of the formalism of inhomogeneous systems as described by Lebowitz and Percus [52] to generate the perturbation series.

### EXPANSIONS IN TERMS OF THE BOLTZMANN FACTOR

Let us consider an inhomogeneous system described by the grand canonical ensemble under the influence of an external potential $w$ (i.e., the source of the inhomogeneity). The partition function is given by

$$\Xi[w] \equiv \sum_{N \geq 0} \frac{z^N}{N!} \int d\{N\} \ \exp[-\beta V_N(\{N\})]\exp\left[-\beta\sum_{i=1}^{N}w(i)\right] \qquad (5.1)$$

where $z$ is the activity and $V_N(\{N\})$ is the (internal) N-body potential energy. The singlet density $\rho^{(1)}(1;w)$ is, correspondingly

$$\rho^{(1)}(1; w) = \frac{1}{\Xi}[w] \sum_{N \geq 0} \frac{z^N}{(N-1)!} \int d\{N-1\}\exp[-\beta V_N(\{N\})]\exp\left[-\beta\sum_{i=1}^{N} w(i)\right] \qquad (5.2)$$

Percus [53] has given the functional derivatives of in $\Xi$ with respect to the Boltzmann factor $e(i) = \exp[-\beta w(i)]$ as

$$\frac{\delta \Xi[w]}{\delta e(1)} = e^+(1)\rho^{(1)}(1; w) \tag{5.3}$$

$$\frac{\delta^2 \Xi[w]}{\delta e(1)\delta e(2)} = e^+(1)e^+(2)F_2(1,2;w) \tag{5.4}$$

$$= e^+(1)e^+(2)\Big[\rho^{(2)}(1,2;w) - \rho^{(1)}(1;w)\rho^{(1)}(2;w)\Big]$$

$$\frac{\delta^3 \Xi[w]}{\delta e(1)\delta e(2)\delta e(3)} = e^+(1)e^+(2)e^+(3)F_3(1,2,3;w) \tag{5.5}$$

with

$$e^+(i) \equiv e^{+\beta w(i)} \tag{5.6}$$

$F_n(\{n\})$ are the nth order Ursell functions:

$$F_2(1,2) = \rho^{(1)}(1,2) - \rho^{(1)}\rho^{(1)}(2) \tag{5.7}$$

$$F_3(1,2,3) = \rho^{(3)}(1,2,3) - \rho^{(2)}(1,2)\rho^{(1)}(3) + 2\rho^{(1)}(1)\rho^{(1)}(2)\rho^{(1)}(3) \tag{5.8}$$

$$- \rho^{(2)}(2,3)\rho^{(1)}(1)$$

$$- \rho^{(2)}(3,1)\rho^{(1)}(2)$$

etc. Now if we consider $e^+(i)\rho^{(1)}(i;w)$ as the (PY) generating functional dependent on the Boltzmann factor $e(i)$, we have already explicitly shown its first two functional derivatives. Thus we can carry out the functional Taylor expansion with respect to a reference potential $w_0$

$$\rho^{(1)}(1;w)e^{+\beta w(1)} = \rho^{(1)}(1;w_0)e^{+\beta w_0(1)} \tag{5.9}$$

$$+ \int d3\, e_0^+(1)e_0^+(3)F_2(1,3;w_0)[e(3) - e_0(3)]$$

$$+ \frac{1}{2} \int d3d4\ e_0^+(1)e_0^+(3)e_0^+(4)\ F_3(1,3,4;w_0)[e(3) - e_0(3)]\cdot[e(4) - e_0(4)]$$

$$+ \cdots$$

where the subscript 0 denotes the quantity arising from the reference external potential $w_0$. Now the initial $w_0$ is caused by a test particle of type $b$ at location $r_0$ (or 0)

$$w(i) = u_{ab}(i, 0) \tag{5.10}$$

where $u_{ab}$ is the pair potential between the particle (of type $a$) at $r_i$ in the bath and the test particle (of type $b$) at $r_0$. The final external potential $w$ is caused by a new test particle also of type $a$ (the same as the bath particles) at $r_0$. The change (or turning-on) of test particle causes changes in the generating functional $e^+\rho^{(1)}$ according to (5.9). The effects on the probability densities are known to be [54]

$$\rho^{(1)}(1;w) = \frac{\rho^{(2)}(1,0)}{\rho^{(1)}(0)} \tag{5.11}$$

$$\rho^{(2)}(1,2;w) = \frac{\rho^{(3)}(1,2,0)}{\rho^{(1)}(0)} \tag{5.12}$$

where $\rho^{(2)}(1,0)$ and $\rho^{(3)}(1,2,0)$ are the pair and triplet correlation functions of a homogeneous system. Then (5.9) becomes, in the presence of the test particle,

$$\frac{y(12)}{y_0(12)} - 1 = \rho \int d3 \left[ \frac{g_0^{(3)}(1,2,3)}{g_0(12)} - g_0(32) \right] e_0(32)[e(32) - e_0(32)] \tag{5.13}$$

$$+ \frac{\rho^2}{2} \int d3d4 \; G(1234)e_0(32)e_0(42)[e(32) - e_0(32)][e(42) - e_0(42)] + \cdots$$

where

$$G(1234) \equiv \frac{g_0^{(4)}(1234)}{g_0(12)} + 2g_0(32)g_0(24) \tag{5.14}$$

$$- \frac{g_0^{(3)}(123)g_0(42)}{g_0(12)} - \frac{g_0^{(3)}(124)g_0(32)}{g_0(12)} - g_0^{(3)}(234)$$

The subscripted variables are derived from $u_{ab}$, and the unsubscripted variables from $u_{aa}$. The integration $\int d3$ corresponds to the expression

$$\int d3 \equiv \frac{1}{\Omega} \int d\mathbf{r}_3 \int d\omega_3 = \frac{1}{8\pi^2} \int\limits_{-\infty}^{+\infty} dx_3 \int\limits_{-\infty}^{+\infty} dy_3 \int\limits_{-\infty}^{+\infty} dz_3 \int\limits_{0}^{\pi} d\theta_3 \sin\theta_3 \int\limits_{0}^{2\pi} d\phi_3 \int\limits_{0}^{2\pi} d\chi_3 \tag{5.15}$$

Upon making the superposition approximation

$$g_0^{(3)}(1,2,3) \approx g_0(12)g_0(23)g(13) \tag{5.16}$$

we have from (5.13)

$$\frac{y(12)}{y_0(12)} - 1 = \frac{\rho}{\Omega} \int d\mathbf{r}_3 d\omega_3 [g(13) - 1]y_0(32)[e(32) - e_0(32)] \tag{5.17}$$

We note that $g(13)$ is for the full potential $u_{aa}$. To understand this, we consider a bath of molecules of species $a$ containing tagged particles 1 and 3. While the system is charged, the test particle 2 at $\mathbf{r}_0$ switches from species $b$ to $a$. Particles 1 and 3 are in the bath, thus interacting with each other by $u_{aa}$. The interactions between 1 and 2 and between 3 and 2 are of the $u_{ab}$ type since 2 is initially a $b$-particle. This explains the terms in the superposition (5.16). Comparison of (5.17) with the RAMY expansion of Melnyk et al. shows that they differ at least in two places: (1) in (5.17), $g(13) - 1 =$

$h(13)$ is the total correlation function of the full system, while in (4.23) $h_0(13)$ is for the reference system; and (2) equation (4.24) contains two integrals interchanging the roles of $(r_1, r_3)$ and $(r_3, r_2)$, whereas equation (5.17) contains only one integral. The source of the second difference arises from the fact that in (4.22), the expansion was in terms of the *pair* potential $u(12)$ and the combinatorial factor was $N(N-1)/2$; whereas in deriving (5.13), the expansion was in terms of the *external* potential $w(i)$, with a combinatorial factor of $N$. This point can be seen from a similar expansion given by Percus [55] for the case of zero reference potential $w_0 = 0$. He gave

$$\rho^{(1)}(1;w)e^{\beta w(1)} = \rho^{(1)}(1) + \int d2 F_2(1,2)[e^{-\beta w(2)} - 1] \qquad (5.18)$$

$$+ \frac{1}{2} \int d2 d3 \ F_3(1,2,3)[e^{-\beta U(2)} - 1][e^{-\beta U(3)} - 1] + \cdots$$

Our equation (5.9) would reduce to (5.18) when $w_0$ was set to zero. Thus (5.9) is a generalization of Percus expression to a perturbation form. Equation (5.9) is also applicable to angle-dependent reference potentials. Equation (5.17) has been applied to the (generalized) Stockmayer potentials with quadrupolar [56] and dipolar interactions [57]. Figure XI.9 shows the spherical harmonic coefficients for a quadrupolar fluid with $Q^*=0.5$ ($T^* = 1.277$ and $\rho^* = 0.85$). The LJ potential was again taken as the reference. Close agreement was obtained. However, for higher quadrupole moments (e.g., $Q^* = 1$), the perturbation expansion gave poor results. This was attributed to the spherical LJ reference, which was unable to describe the strong angular quadrupolar forces. We thus used a nonspherical reference, i.e., the LJ+QQ potential at $Q^* = 0.5$ as reference. The improvements in the results were quite substantial (see Figure XI.10). Similar application to LJ+$\mu\mu$ potential has been carried out. Figure XI.11 shows the $g^{000}$ term for $\mu^* = 1.0$ ($T^* = 1.0$, and $\rho^* = 0.85$). LJ potential was taken as the reference. The perturbation (5.17) gave reasonable predictions. However, the structures $h^{110}$ and $h^{112}$ were poorly predicted. This may be caused by the superposition approximation and/or the truncation of the perturbation expansion. We note also that (5.17) is asymmetrical in its arguments. The symmetry properties were discussed in details by Lee and Chung [58].

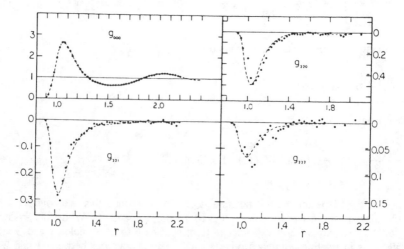

*FIGURE XI.9. The spherical harmonic coefficient $g_{LL'M}(r)$ of the apcf in an LJ+QQ liquid at $T^* = 1.277$, $\rho^* = 0.85$, and $Q^* = 0.5$. •: MD. - - - -: first order perturbation with LJ reference. (Lee et al., Physica, 1982)*

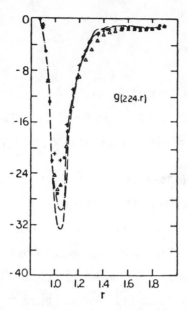

FIGURE XI.10. The spherical harmonic coefficient $g(224;r)$ for LJ+QQ liquid at $T^* = 1.294$, $\rho^* = 0.85$, and $Q^* = 1.0$. •: MD. +++: first order perturbation with LJ reference. Δ: first oreder perturbation with nonspherical reference (LJ+QQ at $Q^* = 0.5$). - - - -: QHNC. — · — · —: LHNC. (Lee et al. Physica, 1982.)

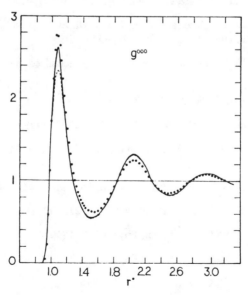

FIGURE XI.11. The angle-averaged pair correlation function $g^{000}(r)$ for LJ+μμ liquid at $T^* = 1.0$, $\rho^* = 0.85$, and $\mu^* =1.0$. •: MC data (Adams and Adams, Mol. Phys. 1981.) ———: first order perturbation theory. - - - -: modified LHNC theory. (Lee and Chung, J. Chem. Phys., 1983.)

# References

[1]   D.J. Tildesley, Mol. Phys. **41**, 341 (1980).

[2]   J.D. Weeks, D. Chandler, and H.C. Andersen, J. Chem. Phys. **54**, 5237 (1971).

[3]   J.A. Barker and D. Henderson, J. Chem. Phys. **47**, 2856 (1967); and *Ibid.* **47**, 4714 (1967).

[4]   L. Verlet and J.-J. Weis, Phys. Rev. A**5**, 939 (1972).

[5]   R.W. Zwanzig, J. Chem. Phys. **22**, 1420 (1954).

[6]   J.W. Perram and L.R. White, Mol. Phys. **28**, 527 (1974).

[7]   T.W. Melnyk and W.R. Smith, Mol. Phys. **40**, 317 (1980).

[8]   J.D. Weeks, D. Chandler, and H.C. Andersen, J. Chem. Phys. **54**, 5237 (1971).

[9]   H.C. Andersen, D. Chandler, and J.D. Weeks, J. Chem. Phys. **56**, 3812 (1972).

[10]  J.A. Barker and D. Henderson, *Ibid.* (1967).

[11]  H.C. Andersen, D. Chandler, and J.D. Weeks, J. Chem. Phys. **56**, 3812 (1972).

[12]  M. Ross, J. Chem. Phys. **71**, 1567 (1979).

[13]  J.A. Barker and D. Henderson, J. Chem. Phys. **52**, 2315 (1970).

[14]  D.A. McQuarrie and J.L. Katz, J. Chem. Phys. **44**, 2393 (1966).

[15]  H.S. Kang, C.S. Lee, T. Ree, and F.H. Ree, J. Chem. Phys. **82**, 414 (1985); also H.S. Kang, T. Ree, and F.H. Ree, *Ibid.*, **84**, 4547 (1986).

[16]  C.J.F. Böttcher, *Theory of Electric Polarization*, second edition, vol. 1 (Elsevier, New York, 1973).

[17]  J.A. Pople, Proc. Royal Soc. A**221**, 508 (1954).

[18]  G. Stell, J.C. Rasaiah, and H. Narang, Mol. Phys. **23**, 393 (1972).

[19]  M. Flytzani-Stephanopoulos, K.E. Gubbins, and C.G. Gray, Mol. Phys. **30**, 1649 (1975).

[20]  K.E. Gubbins, C.G. Gray, and J.R.S. Machado, Mol. Phys. **42**, 817 (1981).

[21]  G.N. Patey and J.P. Valleau, Chem. Phys. Letters **21**, 297 (1973).

[22]  G.N. Patey and J.P. Valleau, J. Chem. Phys. **61**, 534 (1974).

[23]  G.N. Patey and J.P. Valleau, J. Chem. Phys. **64**, 170 (1976).

[24]  K.E. Gubbins and C.G. Gray, Mol. Phys. **23**, 187 (1972).

[25]  H.C. Andersen, J.D. Weeks, and D. Chandler, Phys. Rev. A **4**, 1597 (1971).

[26]  J.W. Perram and L.R. White, Mol. Phys. **28**, 527 (1974).

[27]  W.G. Madden and D.D. Fitts, Chem. Phys. Letters **28**, 427 (1974); Mol. Phys. **28**, 1095 (1974); Mol. Phys. **30**, 809 (1975); Mol. Phys. **31**, 1923 (1976); W.G. Madden, D.D. Fitts, and W.R. Smith, Mol. Phys. **35**, 1017 (1978).

[28] W.R. Smith, Can. J. Phys. **52**, 2022 (1974); W.R. Smith, I. Nezbeda, T.W. Melnyk, and D.D. Fitts, Discuss. Faraday Soc. **66**, 130 (1976); T.W. Melnyk and W.R. Smith, Mol. Phys. **40**, 317 (1980).

[29] D. MacGowan, E.M. Waisman, J.L. Lebowitz, and J.K. Percus, J. Chem. Phys. **80**, 2719 (1984); D. MacGowan, J. Chem. Phys. **81**, 3224 (1984).

[30] S. Murad, K.E. Gubbins, and C.G. Gray, Chem. Phys. Letters **65**, 187 (1979); Chem. Phys. **81**, 87 (1983).

[31] H.C. Andersen, J.D. Weeks, and D. Chandler, *Ibid.* (1971).

[32] J.W. Perram and L.R. White, *Ibid.* (1974).

[33] T.W. Melnyk and W.R. Smith, Mol. Phys. **40**, 317 (1980).

[34] M.S. Shaw, J.D. Johnson, and B.L. Holian, Phys. Rev. Letters **50**, 1141 (1983).

[35] J.L. Lebowitz and J.K. Percus, J. Chem. Phys. **79**, 443 (1983).

[36] D. MacGowan, J.D. Johnson, and M.S. Shaw, J. Chem. Phys. **82**, 3765 (1985); I. Nezbeda, W.R. Smith, and S. Labik, J. Chem. Phys. **81**, 935 (1984).

[37] D. MacGowan, E.M. Waisman, J.L. Lebowitz, and J.K. Percus, J. Chem. Phys. **80**, 2719 (1984).

[38] M. Ross, J. Chem. Phys. **71**, 1567 (1979).

[39] D. MacGowan, J. Chem. Phys. **81**, 3224 (1984).

[40] D. MacGowan, D.B. Nicolaides, J.L. Lebowitz, and C.K. Choi, Mol. Phys. **58**, 131 (1986).

[41] S. Singh and Y. Singh, Mol. Cryst. Liq. Cryst. **87**, 211 (1982).

[42] S.I. Sandler, Mol. Phys. **28**, 1207 (1974); J.O. Valderrama, S.I. Sandler, and M. Flinger, Mol. Phys. **42**, 1041 (1981).

[43] K.C. Mo and K.E. Gubbins, J. Chem. Phys. **63**, 1490 (1975).

[44] T. Boublik, Coll. Czech. Chem. Commun. **39**, 2333 (1974).

[45] F. Kohler, N. Quirke, and J.W. Perram, J. Chem. Phys. **71**, 4128 (1979).

[46] J. Fischer, J. Chem. Phys. **72**, 5371 (1980).

[47] J.D. Weeks, D. Chandler, and H.C. Andersen, J. Chem. Phys. **54**, 5237 (1971); H.C. Andersen, J.D. Weeks, and D. Chandler, Phys. Rev. **A4**, 1597 (1971).

[48] D.J. Tildesley, Mol. Phys. **41**, 341 (1980).

[49] N. Quirke and D.J. Tildesley, J. Phys. Chem. **87**, 1972 (1983).

[50] L.L. Lee, E. Assad, H.A. Kwong, T.H. Chung, and J.M. Haile, Physica **110A**, 235 (1982).

[51] L.L. Lee, E. Assad, H.A. Kwong, T.H. Chung, and J.M. Haile, *Ibid.* (1982).

[52] J.L. Lebowitz and J.K. Percus, Phys. Rev. **122**, 1675 (1961).

[53]   J.K. Percus, in *The Equilibrium Theory of Classical Fluids*, edited by H.L. Frisch and J.L. Lebowitz (Benjamin, New York, 1964).

[54]   J.K. Percus, in *The Equilibrium Theory of Classical Fluids*, edited by H.L. Frisch and J.L. Lebowitz (Benjamin, New York, 1964).

[55]   J.K. Percus, in *The Equilibrium Theory of Classical Fluids*,  Ibid. (1964).

[56]   L.L. Lee, E. Assad, H.A. Kwong, T.H. Chung, and J.M. Haile, Physica **110A**, 235 (1982).

[57]   L.L. Lee and T.H. Chung, J. Chem. Phys. **78**, 4712 (1983).

[58]   L.L. Lee and T.H. Chung, *Ibid.* (1983).

## Exercises

1.   Use functional differentiation to show the relations (1.11 and 12).  Discuss the change of coordinate systems, if any.

2.   Reproduce the entries in Table XI.2 for LJ fluids, using the Verlet-Weis perturbation approach.

3.   Discuss alternative methods of separating the full potential into its perturbative parts $u_p$ according to Perram and White (ref. [5]), and the RAMY theory of Melnyk et al. (ref. [6]).  Compare their theories with the WCA approach.

4.   Apply Padé resummation to calculating the carbon dioxide properties, using the parameters given in Table XI.7.  Compare the results with published experimental data on the energy and pressure of $CO_2$.

# CHAPTER XII

# ELECTROLYTE SOLUTIONS

Electrolyte solutions historically set themselves apart from ordinary solutions in the study of mixtures. For many years, standard texts on solutions were written for *nonelectrolytes* (e.g., Hildebrand and Scott [1]). This fact bespeaks the difficulty in treating electrically charged species. One of the earliest efforts in describing dissolved salts was undertaken by Debye and Hückel [2]. However, it was not until the 1960s and 1970s when liquid state theories had well developed that a common approach to electrolyte and nonelectrolyte solutions became possible. The common basis is the molecular distribution functions. (For a historical review, see Falkenhagen *Electrolytes* [3]). In this chapter, we present one of the statistical mechanical models for electrolyte solutions, namely the *charged hard spheres*. This model, though simple, represents an important advance in the study of concentrated electrolyte solutions.

Statistical mechanics treats the ionized species in the same way as it treats ordinary fluids, i.e., by considering the Hamiltonian of an N-body system consisting of cations, anions, and solvent molecules. The power of the statistical methods lies in their ability of treating particles of dissimilar nature on the same basis. The electrostatic Coulomb forces developed for macroscopic bodies are assumed to operate on the molecular level. Statistical mechanics is used to provide the laws governing the distribution of particles. For example, the Boltzmann distribution used in the Debye-Hückel (DH) theory is a special case of statistical distributions. As we shall see, this distribution is uniquely determined by the Hamiltonian of the system.

Debye and Hückel introduced a model for electrolyte solutions where the ions are treated as *point electric charges* (i.e., charges having *no excluded volume*) obeying classical electrostatic principles (the Poisson equation). The charge density is given by an exponential distribution law (Boltzmann distribution). It is successful in describing dilute solution properties. As such, the DH theory was considered a milestone in electrolyte theories. However, for concentrated solutions, the point charge model is no longer valid. Large deviations in osmotic and mean activity coefficients are observed. Thus DH is not quantitative at high ionic strengths. Obviously, new and more accurate theories need be developed.

In the intervening years, progress was made due to efforts by Mayer [4], Bjerrum [5], Guggenheim [6], and others. In the 1970s, an integral equation theory, the mean spherical model (MSA), was solved for charged hard spheres with finite size. This success represents another step forward in modeling ionic solutions, because it takes the sizes of ions into account. The solution given by MSA reveals an intricate interplay between ion size and charge strength. Further developments along the liquid distribution function approach produced the Percus-Yevick (PY) and hypernetted chain (HNC) version for electrolyte solutions. Meanwhile, Monte Carlo (MC) simulations were performed for the so-called primitive model of ionic solutions (see below). These developments are instrumental in the establishment of a quantitative theory.

In this chapter, we study a model electrolyte, i.e., mixtures of charged hard spheres of finite sizes. The solvent molecules, on the other hand, are not explicitly treated. For example, the water molecules are removed and replaced by a dielectric continuum. This idealization of the ionic solutions is called the *primitive model*. For aqueous solutions, the dielectric medium will have the permittivity of water. As a consequence, the electrostatic forces of interaction, which are Coulombic in nature, retain their values in water. Such treatment of solvents was earlier considered by McMillan and Mayer [7] and is called the McMillan-Mayer theory of solutions.

We note a number of limitations of the primitive model. The water molecules, being of similar dimension as the ions, are actually *granular*, i.e., *visible* to the charged ions through hydration or simple exclusion. Recent studies have examined the effects of the *granularity* by injecting dipolar spheres (simulating water molecules) into the primitive model. However, a full-scale treatment of water molecules in terms of more sophisticated potential models (such as the Matsuoka-Clementi-Yoshimine potential[8]) is at present a formidable task. There are major theoretical difficulties in treating aqueous ionic solutions. Second, the repulsive interactions among ions are not so harsh as suggested by the hard sphere model. Keeping these points in mind, we should consider the primitive model as an approximation for real electrolyte solutions. Attempts have been made to apply the MSA model to describe real salt solutions. Due to these limitations, it has been found that an adjustable ionic diameter or a state-dependent dielectric constant has to be used in order to achieve quantitative agreement.

In the 1970s, Waisman and Lebowitz [9] solved the MSA for the restricted primitive model. A rapid succession of new results on more sophisticated models of ionic solutions ensued. The MSA equations have been solved for asymmetrical ions, mixtures of charged hard spheres and dipolar spheres, and charged particles near a wall. The MSA itself has been improved in many ways by the techniques of cluster resummations (e.g. in the optimized random phase approximation (ORPA) [9] and the Γ-ordering theory [10]). Other integral equations have been tested and shown promise, such as the PY-type equation (e.g., Allnatt [11]) and the HNC [12]. Due to progress in numerical transform techniques (e.g., the fast Fourier transform), the HNC equation has recently been applied to charged soft spheres. Computer simulations on the Coulombic and *screened* Coulombic systems were also carried out. New features of the structure of ionic solutions have been discovered. All these developments add considerably to our understanding of electrolyte solutions.

Since the DH theory has played a major role at the early stages of solution theories, we devote sections XII.3 and 4 to a full account. The DH theory is limited to dilute solutions where only long-range electrostatic forces are at work and short-range interactions are absent. For concentrated solutions, the short-range electrostatic and exclusion forces become more important. Thus the primitive model postulates a finite impenetrable core for the ions. The MSA results for this model are presented in section XII.5. In comparison with simulation data, the energy values predicted by MSA are accurate but the structure given by MSA is rather poor. In section XII.6, asymmetrical ions are considered. It is found that the HNC equation is by far the most accurate theory for ionic potentials. Section XII.7 is devoted to a study of HNC. We also show results for solvent effects using the LHNC and QHNC equations. The moment conditions due to Stillinger and Lovett are discussed. In addition, we give an account of the simulation results for Coulomb and screened Coulomb potentials.

## XII.1.  *Review of Electrostatics*

In order to discuss the forces and potentials for charged molecules, we need to review the fundamental laws in classical electrostatics. An important point to keep in mind is that for classical systems the laws governing macroscopic bodies also apply to molecules. First we discuss Coulomb's law.

### COULOMB'S LAW

Coulomb's law simply states that for two point charges $q_1$ and $q_2$ separated by a distance $r$, the force **F** exerted along the line joining the two charges is given by

$$\mathbf{F}_{12} = \frac{1}{\varepsilon_m} \frac{q_1 q_2}{r^2} \mathbf{u}_{12} \tag{1.1}$$

where $\varepsilon_m$ is the permittivity of the medium where the two charges are embedded, $\mathbf{u}_{12}$ is the unit vector in the direction of $\mathbf{r}_{12} = \mathbf{r}_2 - \mathbf{r}_1$, and $r = |\mathbf{r}_{12}|$. Note that here we have adopted the convention omitting the factor $4\pi$ from the denominator in the definition of the permittivity ($4\pi$ accounts for the solid angle in spherical geometry). For free space (i.e., vacuum), $\varepsilon_0 = 111.2 \times 10^{-12}$ Coulomb$^2$/N m$^2$ or $(\varepsilon_0)^{-1} = 9 \times 10^9$ N–m$^2$/Coulomb$^2$. In a medium with permittivity $\varepsilon_m$ the dielectric constant $\varepsilon$ is defined as the ratio

$$\varepsilon \equiv \frac{\varepsilon_m}{\varepsilon_0} \tag{1.2}$$

For vacuum, the dielectric constant is 1. In literature, the nomenclature for $\varepsilon$ and $\varepsilon_m$ is not always distinct. The statement that $\varepsilon$ *is a dielectric constant* may mean either the permittivity $\varepsilon_m$ or the dielectric constant $\varepsilon$ (a ratio). One must decide from the context if $\varepsilon$ stands for permittivity or otherwise. Here, we adhere to distinct notations.

### GAUSS'S LAW

Gauss's law deals with the electric field on the surface of a body enclosing a distribution of charges. It is expressed in terms of an integral

$$\iint_{CS} \mathbf{E} \cdot d\mathbf{S} = \sum_{i=1}^{k} \frac{4\pi q_i}{\varepsilon_m} \tag{1.3}$$

where **E** is the electric field given by

$$\mathbf{E}_1 = \frac{\mathbf{F}_{12}}{q_1} \tag{1.4}$$

where $\mathbf{F}_{12}$ is the force given by Coulomb's law (1.1); i.e., $\mathbf{E}_1$ is the electric force per unit charge at location $\mathbf{r}_1$. In eq. (1.3), the integration is over a closed (regular) surface (CS) enclosing a volume that contains $k$ separate centers of charges $q_i$ ($i=1,...,k$). The integrand is the dot (inner) product of the electric field **E** with surface element $d\mathbf{S}$ whose magnitude is the surface area and whose direction is that of the outward pointed normal **n**. Simply stated, (1.3) states that the projection of the mean electric field on the surface is equal to the sum of the charges inside the body (divided by the permittivity $\varepsilon_m$ of the medium). When the charge distribution inside the body is continuous, the sum of charges should be replaced by an integral; i.e.,

$$\oiint_{CS} E \cdot dS = \frac{4\pi}{\varepsilon_m} \iiint_{V_b} \rho_e \, dV \tag{1.5}$$

where $\rho_e$ is the electrical charge density (charges per volume), and $V_b$ is the volume of the body. We have stated these theorems without proof. One should consult texts [13] on electromagnetism for details.

## POISSON'S EQUATION

The Poisson equation can be easily derived from Gauss's law. In mathematics, Green's divergence theorem states

$$\oiint_{CS} dS \, \nabla\phi \cdot n = \iiint_{V_b} dV \, \nabla \cdot \nabla\phi \tag{1.6}$$

where CS encloses the volume $V_b$. $\nabla$ is the gradient operator, and $\phi$ is some scalar potential. Equation (1.6) says that the accumulation of the normal component of the gradient of $\phi$ on the surface is equal to the volume integral of the Laplacian of $\phi$. Now if the electric field E is expressed as the gradient of some potential function $\phi$, i.e., $E = -\nabla\phi$, we have immediately from (1.5)

$$-\iiint_{V_b} dV \, \nabla \cdot \nabla\phi = \frac{4\pi}{\varepsilon_m} \iiint_{V_b} dV \, \rho_e \tag{1.7}$$

Since the volume $V_b$ is arbitrary, the integrands on both sides of (1.7) must be equal:

$$\Delta\phi = -\frac{4\pi}{\varepsilon_m} \rho_e \tag{1.8}$$

where the Laplacian $\Delta = \nabla \cdot \nabla$. This is Poisson's equation.

## XII.2. *The McMillan-Mayer Theory of Solutions*

A number of solution theories in use make the simplifying assumption that the solvent molecules are *smeared* out, forming a uniform background, called the *dielectric* or *neutralizing continuum* (e.g., in the primitive model of electrolyte solutions or in the one-component plasmas, respectively). This *smoothing* operation has the advantage of reducing considerably the level of difficulty since the solvent molecules, the hydrogen-bonding water, no longer figure explicitly in the formulation. What remains of the water is its permittivity, because $\varepsilon_m$ of the medium is set equal to that of water. Such a viewpoint is employed in the McMillan-Mayer (MM) theory of solutions. The important task is to find a way of *averaging* out the solvent molecules while preserving their effects.

We start by considering an experiment of osmosis: a pure solvent $w$ (e.g., water) is placed in partition I of a container and separated by a semipermeable membrane from partition II containing a solution with solute $a$ and solvent $w$. The membrane allows passage of water molecules but not solute molecules. At equilibrium, the system is isothermal, and the activity of water, $z_w^I$ in I must be equal to $z_w^{II}$ in II. However, in partition II, $z_w^{II}$ would be less than its pure component counterpart, given the same pressure and temperature, because $w$ is mixed with $a$ in partition II. To increase its activity to the level in I, we must increase the system pressure on II, $P^{II}$, above $P^I$. When equality $z_w^I = z_w^{II}$ has eventually established, we find $P^{II} > P^I$. The difference $P_{osm} = P^{II} - P^I$ is called the *osmotic pressure*. As in the case of fluids, $P_{osm}$ is a function of the concentration of the solute as well as the temperature and the pressure of the system.

Next, we present the MM theory of solutions. Since we are dealing with an open system, we shall use the grand canonical ensemble (GCE). In order to distinguish between the exact and MM approaches, we use one asterisk (*) to denote the exact ensemble quantities and two asterisks (**) to denote the averaged quantities where the solvent (e.g., water) molecules have been *smoothed* out. The averaging is applied to partition II.

In the GCE the partition function $\Xi$ is related to the thermodynamic pressure by

$$\Xi^{*I} = \exp(\beta^I P^I V^I) \qquad \text{(region I)} \qquad (2.1)$$

$$\Xi^{*II} = \exp(\beta^{II} P^{II} V^{II}) \qquad \text{(region II)} \qquad (2.2)$$

where the superscripts I and II denote the properties in regions I and II, respectively. The MM partition function for region II is defined as

$$\Xi^{**II} \equiv \frac{\Xi^{*II}}{\Xi^{*I}} = \exp(\beta P_{osm} V) \qquad (2.3)$$

where $P_{osm} = P^{II} - P^I$. We have set $V^I = V^{II}$. Also $\beta^I = \beta^{II}$ at equilibrium.

Next we want to find the expression for $\Xi^{**}$ in terms of an exact description. First we rewrite the definitions of $\Xi^*$ and the $N+M$-tuplet correlation function as

$$\Xi^* = \sum_{N \geq 0}^{\infty} \sum_{M \geq 0}^{\infty} \frac{z_a^N z_w^M}{N! M!} Q_{N+M}^* \qquad (2.4)$$

and

$$g^{*(n+m)}(1_a, 2_a, \ldots, n_a; 1_w, 2_w, \ldots, m_w; z_a, z_w) \qquad (2.5)$$

$$= \rho_a^{-n} \rho_w^{-m} \Xi^{*-1} \sum_{N \geq n}^{\infty} \sum_{M \geq m}^{\infty} \frac{z_a^N z_w^M}{(N-n)!(M-m)!}$$

$$\cdot \int d(n+1)_a \cdots dN_a d(m+1)_w \cdots dM_w \, e^{-\beta V_{N+M}}$$

where

$$Q_{N+M}^* = \int d1_a \, d2_a \cdots dN_a \, d1_w \, d2_w \cdots dM_w \, \exp(-\beta V_{N+M}) \qquad (2.6)$$

and $V_{N+M}$ is the total potential energy of $N$ solute molecules and $M$ solvent molecules. The above definitions are for a mixture of one solute species $a$ and one solvent species $w$. For multicomponent mixtures, the expressions can be generalized. In order to examine the details, we write out the series for $\Xi^*$:

$$\Xi^{*\mathrm{I}} = 1 + \frac{z_w}{1!} Q^*_{0+1} + \frac{z_w^2}{2!} Q^*_{0+2} + \frac{z_w^3}{3!} Q^*_{0+3} + \cdots \tag{2.7}$$

and

$$\Xi^{*\mathrm{II}} = 1 + \frac{z_a}{1!0!} Q^*_{1+0} + \frac{z_w}{0!1!} Q^*_{0+1} + \frac{z_a z_w}{1!1!} Q^*_{1+1} + \frac{z_a^2}{2!0!} Q^*_{2+0} + \frac{z_w^2}{0!2!} Q^*_{0+2} + \cdots \tag{2.8}$$

$$= \left[ 1 + \frac{z_w}{1!} Q^*_{0+1} + \frac{z_w^2}{2!} Q^*_{0+2} + \cdots \right] + \frac{z_a}{1!} \left[ Q^*_{1+0} + \frac{z_w}{1!} Q^*_{1+1} + \frac{z_w^2}{2!} Q^*_{1+2} + \cdots \right]$$

$$+ \frac{z_a^2}{2!} \left[ Q^*_{2+1} + \frac{z_w}{1!} Q^*_{1+1} + \frac{z_w^2}{2!} Q^*_{2+2} + \cdots \right] + \cdots$$

The first pair of parentheses of (2.8) is simply $\Xi^{*\mathrm{I}}$. Thus division of $\Xi^{*\mathrm{II}}$ by $\Xi^{*\mathrm{I}}$ gives

$$\frac{\Xi^{*\mathrm{II}}}{\Xi^{*\mathrm{I}}} = 1 + (\Xi^{*\mathrm{I}})^{-1} \frac{z_a}{1!} \left[ \sum_{M \geq 0}^{\infty} \frac{z_w^M}{M!} Q^*_{1+M} \right] \tag{2.9}$$

$$+ (\Xi^{*\mathrm{I}})^{-1} \frac{z_a^2}{2!} \left[ \sum_{M \geq 0}^{\infty} \frac{z_w^M}{M!} Q^*_{2+M} \right] + \cdots$$

Thus the MM partition function has the form

$$\Xi^{**\mathrm{II}} = \sum_{N \geq 0}^{\infty} \frac{z_a^N}{N!} k_a^N Q^{**}_N = 1 + \frac{z_a}{1!} k_a Q^{**}_1 + \frac{z_a^2}{2!} k_a^2 Q^{**}_2 + \cdots \tag{2.10}$$

where $k_a$ and $Q^{**}_1, Q^{**}_2, \ldots$, etc., are quantities to be determined by comparison with the exact expression (2.9). If we equate coefficients of same powers in activity $z_a$, we have

$$k_a^N Q^{**}_N = \frac{1}{\Xi^{*\mathrm{I}}} \sum_{M=0}^{\infty} \frac{z_w^M}{M!} Q^*_{N+M} \tag{2.11}$$

There is some arbitrariness in assigning values to $k_a$ and $Q^{**}_N$. The choice made by Friedman et al. [14] is to consider first the related quantity

$$\int d1_a \, d2_a \cdots dN_a \, g^{*(N+0)}(1_a, 2_a, \ldots, N_a; z_a, z_w) \tag{2.12}$$

$$= \frac{1}{\rho_a^N} \frac{1}{\Xi^{*\mathrm{II}}} \left\{ z_a^N Q_{(N+0)}^* + \frac{z_a^{N+1}}{1!0!} Q_{(N+1)+0}^* + \frac{z_a^{N+1} z_w}{1!1!} Q_{(N+1)+1}^* + \frac{z_a^N z_w}{0!1!} Q_{N+1}^* + \cdots \right\}$$

This is not quite the quantity yet we wanted in (2.11). We note that in the limit $z_a \to 0$, $z_a/\rho_a = 1$, and

$$\lim_{z_a \to 0} \Xi^{*\mathrm{II}} = \Xi^{*\mathrm{I}} \tag{2.13}$$

Thus

$$\lim_{z_a \to 0} \int d1_a \cdots dN_a \, g^{*(N+0)}(1_a, \ldots, N_a; z_a = 0, z_w) \tag{2.14}$$

$$= \left[ \lim_{z_a \to 0} \left[ \frac{z_a^N}{\rho_a^N} \right] \right] \frac{1}{\Xi^{*\mathrm{I}}} \left\{ Q_{N+0}^* + \frac{z_w}{1!} Q_{N+1}^* + \frac{z_w^2}{2!} Q_{N+2}^* + \cdots \right\}$$

The term in the braces is precisely the summation in (2.11). Thus we can write

$$k_a^N Q_N^{**} = \lim_{z_a \to 0} \frac{\rho_a^N}{z_a^N} \int d1_a \cdots dN_a \, g^{*(N+0)}(1_a, \ldots, N_a; z_a = 0, z_w) \tag{2.15}$$

Friedman et al. (*Ibid.*) chose

$$k_a \equiv \lim_{z_a \to 0} \frac{\rho_a}{z_a} \tag{2.16}$$

and

$$Q_N^{**} \equiv \lim_{z_a \to 0} \int d1_a \cdots dN_a \, g^{*(N+0)}(1_a, \ldots; N_a; z_a = 0, z_w) \tag{2.17}$$

If we define the potential of mean force $W^{*(N+M)}$ by

$$W^{*(N+0)}(1_a, \ldots, N_a; z_a, z_w) \equiv -kT \ln g^{*(N+0)}(1_a, \ldots, N_a; z_a, z_w) \tag{2.18}$$

the MM configurational partition function can be written in the familiar form

$$Q_N^{**} = \lim_{z_a \to 0} \int d1_a \cdots dN_a \exp\left[ -\beta W^{*(N+0)}(1_a, \ldots, N_a; z_a, z_w) \right] \tag{2.19}$$

Finally,

$$\Xi^{**\mathrm{II}} = \sum_{N \geq 0}^{\infty} \frac{z_a^{**N}}{N!} \int d1_a \cdots dN_a \exp\left[ -\beta \{ \lim_{z_a \to 0} W^{*(N+0)}(1_a, \ldots, N_a, z_a, z_w) \} \right] \tag{2.20}$$

with

$$z_a^{**} \equiv z_a k_a = z_a \left[ \lim_{z_a \to 0} \frac{\rho_a}{z_a} \right] \tag{2.21}$$

We have succeeded in expressing the MM partition function in terms of the potential of mean force $W^{*(N+0)}$ of the exact ensemble. Since the potential of mean force depends on the solvent activity $z_w$ and the temperature through (2.18) (note that the $N$-tuplet correlation function $g^{*(N+0)}$ depends on these variables), the potential energy of the ionic system (after excluding the solvent) is no longer constant and is state-dependent, a price to pay for simplification. Note that the limit $z_a \to 0$ denotes infinite dilution quantities, i.e., when the solute molecules disappear from the solution.

In the following, we shall incorporate the MM theory in the primitive model of ionic solution. In section XII.7, we shall discuss the effects of solvent *granularity* whereby the solvent molecules are treated explicitly. As eq. (2.20) retains the form of a partition function, we could derive the thermodynamic properties as usual. Also from the definition of $\Xi^{**}$ (2.3), the pressure we obtain from the MM theory is the osmotic pressure, not the system pressure $P$.

### XII.3. *The Debye-Hückel Theory*

The DH theory is of historical significance in electrolyte solutions. It is now recognized as the limiting law at infinite dilution. In this theory, the ions are point charges and the solvent molecules are replaced by a dielectric continuum with the permittivity $\varepsilon_{H_2O}$ of water. For charged hard spheres, the potential between ion 1 and ion 2 is given by the Coulomb interaction subject to hard core repulsion

$$u(r_{12}) = +\infty, \qquad\qquad r \le d \tag{3.1}$$

$$= \frac{z_1 z_2 e^2}{\varepsilon r_{12}}, \qquad\qquad r > d$$

For DH, the diameter $d$ is taken to be zero (no hard core). $z_j$, $(j = 1,2)$, is the valence of the charged ion $j$ (for example, in sodium sulfate $Na_2SO_4$, $z_+ = +1$ for the cation and $z_- = -2$ for the anion); $e$ is the charge of an electron, i.e., $-1.602 \times 10^{-19}$ Coulomb or, in electrostatic units, $-4.803 \times 10^{-10}$ esu. We usually take the absolute value $|e|$ in what follows.

When salts such as NaCl dissolve in water, the ions $Na^+$ and $Cl^-$ dissociate and separate from each other. However, the attractive Coulomb force between oppositely charged ions should recombine $Na^+$ with $Cl^-$. What prevents them from coalescing into a neutral molecule? The answer is to be found in the large values of the dielectric constant of water, 80.176 at 20°C. The high permittivity accounts for the existence of separate ions in aqueous solutions, because it reduces the forces of attraction to such an extent that thermal motions of the medium could easily overcome the forces of recombination. To make a quantitative comparison, the work required to separate $Na^+$ and $Cl^-$ ions from their closest approach (roughly equal to the sum of the radius of hydrated $Na^+$ ion, 2.4 Å, and the Pauling radius of $Cl^-$, 1.81 Å, 2.4+1.81 = 4.21 Å) to infinity is (using esu units)

$$\text{Work} = \int_d^\infty dr \, \frac{e_+ e_-}{\varepsilon r^2} = - \frac{e_+ e_-}{\varepsilon d} \tag{3.2}$$

$$= \frac{(4.8 \times 10^{-10})^2}{80.176 \times 4.2 \times 10^{-8}} = 6.4 \times 10^{-14} \text{ erg}$$

This value can be compared with the thermal fluctuations of the ions; i.e.,

$$\text{KE} = \frac{3}{2} kT = \frac{3}{2} \, 1.38 \times 10^{-16} \times 293 \, K = 6 \times 10^{-14} \text{ erg} \tag{3.3}$$

The energies of thermal motion of the ions are of the same order of magnitude as the work required to separate the two ions.

Thus it is highly likely that ions will dissociate in aqueous environment. The dissociation is enhanced further by the interaction with neighboring ions. However, in a solvent with smaller permittivity (such as methanol), the ionic dissociation is reduced. In air, the dielectric constant =1, and the radius of $Na^+$ is smaller (about 0.95 Å) (in the absence of hydration). The closest approach is approximately 2.8 Å. The energy required to separate the ions to infinity is

$$\frac{(4.8 \times 10^{-10})^2}{2.8 \times 10^{-8}} = 8.2 \times 10^{-12} \text{ erg} \tag{3.4}$$

which is 130 times greater than in water. As a consequence, in air very few salt molecules exist in a dissociated state.

In DH, one calculates the *average electrostatic potential* (AEP) $\Psi_i(r)$ surrounding a given ion $i$. This central ion possesses a charge $ez_i$. Although the direct interaction between ion $i$ and another ion $j$ is given by, say, (3.1), the work required to bring a unit charge from infinity to a distance $r$ from $i$ is influenced by other ions surrounding $i$. Oppositely charged ions tend to "aggregate" around the center $i$, forming an ion atmosphere (called the *cosphere*). This cosphere intervenes between ions $i$ and $j$ and, as a consequence, *screens* the interaction (3.1). The resulting AEP is a sum of the bare electrostatic potential (3.1) and its cosphere potential, called the Debye screened potential. The theoretical basis for this case is the Poisson equation (1.8)

$$\nabla^2 \Psi_i(r) = - \frac{4\pi}{\varepsilon_m} q_{(i)}(r) \tag{3.5}$$

where $r$ is the distance from the central charge $i$, and $q_{(i)}(r)$ the charge density (number of charges per unit volume) at distance r measured from center i. This density is related to the pair (number) densities $\rho_{j/i}$ via

$$q_{(i)}(r) = e \sum_j z_j \rho_{j/i}(r) \tag{3.6}$$

The summation is over all ionic species in the solution (e.g., for a mixture of NaCl and $CaBr_2$, $j = Na^+$, $Ca^{++}$, $Cl^-$ and $Br^-$). The local pair densities of ions $j$ surrounding the ion $i$ can be expressed in terms of the pcfs

$$\rho_{j/i}(r) = \rho_j g_{ji}(r) \tag{3.7}$$

We have here introduced the pcfs $g_{ji}$. In the following we shall make approximations in order to solve for the DH theory. The pcf could be expressed in terms of the Kirkwood potential of mean force (PMF), $W_{ji}(r)$,

$$g_{ji}(r) = \exp[-\beta W_{ji}(r)] \tag{3.8}$$

The PMF should include all pair interactions, the electrostatic as well as excluded volume forces. The first approximation to be made in DH is to replace $W_{ji}$ by the AEP

$$W_{ji}(r) \approx ez_j \Psi_i(r) \qquad \text{(Approximation 1)} \tag{3.9}$$

Since $\Psi_i$ does not contain the core interaction, (3.9) ignores the excluded volume effects. The pcf is then

$$g_{ji}(r) = \exp[-\beta ez_j \Psi_i(r)] \tag{3.10}$$

This approximation to pcf is reminiscent of the *Boltzmann* factor used for low density pcfs. Substituting (3.6) into the Poisson equation we have the Poisson-Boltzmann equation

$$\Delta\Psi_i(r) = -\frac{4\pi}{\varepsilon_m} e \sum_j z_j \rho_j \exp[-\beta ez_j \Psi_i(r)] \tag{3.11}$$

where $\Delta$ is the Laplacian. The major deficiency of this equation is the absence of the excluded volume. The linearization to be incorporated later will also introduce errors into the equation. However, the errors are not as serious as the neglect of core repulsion. Debye has corrected the lack of core term to the extent only for the central ion $i$ and not for ions in the ionic atmosphere. Outhwaite [15] formulated a modified Poisson-Boltzmann (MPB) equation incorporating repulsion in $W_{ji}(r)$. The resulting equation is a more successful theory than DH.

To solve (3.11), DH linearizes the exponential term by retaining only the first-order terms in the expansion. Note that the neutrality of the whole solution requires balancing the positive and negative charges

$$\sum_j \rho_j z_j = 0 \tag{3.12}$$

This is the *electroneutrality* condition. Thus

$$\sum_j z_j \rho_j [1 - \beta ez_j \Psi_i(r)] \approx \sum_j z_j \rho_j - \sum_j \beta \rho_j z_j^2 e\Psi_i(r) = -\sum_j \beta \rho_j z_j^2 e\Psi_i(r) \tag{3.13}$$

(Approximation 2)

The linearized Poisson-Boltzmann equation is

$$\Delta \Psi_i(r) = \kappa^2 \Psi_i(r) \qquad (3.14)$$

where $\kappa$ is called the inverse Debye length,

$$\kappa^2 \equiv \frac{4\pi e^2}{\varepsilon_m kT} \sum_j \rho_j z_j^2 \qquad (3.15)$$

$\kappa$ has the units of inverse length. Note that

$$I \equiv \frac{1}{2} \sum_j \rho_j (z_j e)^2 \qquad (3.16)$$

is the ionic strength. Thus $\kappa^2 = 8\pi I/(\varepsilon_m kT)$. In spherical coordinates, the Laplacian is

$$\Delta = \frac{1}{r^2} \frac{\partial}{\partial r} \left[ r^2 \frac{\partial}{\partial r} \right] + \frac{1}{r^2 \sin \theta} \frac{\partial}{\partial \theta} \left[ \sin \theta \frac{\partial}{\partial \theta} \right] + \frac{1}{r^2 \sin^2 \theta} \left[ \frac{\partial^2}{\partial \phi^2} \right] \qquad (3.17)$$

Since the AEP $\Psi$ is independent of $\theta$ and $\phi$, we have

$$\frac{1}{r^2} \frac{d}{dr} \left[ r^2 \frac{d\Psi_i}{dr} \right] = \kappa^2 \Psi_i \qquad (3.18)$$

The general solution of this equation is

$$\Psi_i(r) = \frac{A}{r} e^{-\kappa r} + \frac{B}{r} e^{+\kappa r} \qquad (3.19)$$

We need two boundary conditions to determine the integration constants $A$ and $B$. First, we note that as $r \to \infty$ (i.e., the position $r$ is far removed from the center), $\Psi_i(r)$ is zero. Thus $B = 0$ in (3.19). To determine the second constant, there are several choices. For point ions, there is no proper boundary condition, because the potential diverges at the center. Nonetheless, we could require that $A$ be such that the singularity has the same strength (or mathematically, "the pole has the same residue") as that of the free ion. This is clearly the case when the two ions approach each other: the divergence of the interaction overwhelms the effect of the cosphere:

$$A = \frac{z_i e}{\varepsilon_m} \qquad (3.20)$$

Thus

$$\Psi_i(r) = \frac{z_i e}{\varepsilon_m r} e^{-\kappa r} \qquad (3.21)$$

The AEP turns out to be the Debye screened potential. It gives the average electrical potential experienced by a unit charge at $r$ due to interactions with the central ion $i$ and its cosphere. Naturally, for each species $k$ in the ionic solution, one could write down an equation of the form (3.21). As a consequence, DH theory implies that the PMF is of the form

$$W_{ji}(r) = ez_j\Psi_i(r) \tag{3.22}$$

and the pcf

$$g_{ji}(r) = \exp[-\beta ez_j\Psi_i(r)] \approx 1 - \frac{e^2 z_j z_i}{\varepsilon_m kT}\ \frac{e^{-\kappa r}}{r} \tag{3.23}$$

upon linearizing the exponential.

We recall that the DH theory contains two approximations: (i) neglect of the excluded volume, and (ii) linearization of the Boltzmann factor. The physical basis of DH is the classical Poisson equation. However, it is also possible to derive the screened potential from statistical mechanics, in this case from the Ornstein-Zernike (OZ) relation for correlation functions. Simplifying assumptions will have to be made. This is done next.

## XII.4. *Derivation from Statistical Mechanics*

Let us consider, for simplicity, a single salt dissolved in water. In the MM picture, the solvent is replaced by the dielectric background. The OZ relation for the ion species can be written as

$$h_{ij}(\mathbf{rr'}) - C_{ij}(\mathbf{rr'}) = \sum_l \rho_l \int ds\ C_{il}(\mathbf{rs})h_{lj}(\mathbf{sr'}) \tag{4.1}$$

In order to reproduce the DH results, we assume low salt concentrations. Thus the limiting law cluster expansions of $h_{ij}$ and $C_{ij}$ could be used; i.e.,

$$C_{ij} = f_{ij} = e^{-\beta u_{ij}} - 1 \approx -\frac{u_{ij}}{kT} \tag{4.2}$$

(This approximation is similar to MSA.)

$$h_{ij} = e^{-\beta W_{ij}} - 1 \approx -\frac{W_{ij}}{kT} \tag{4.3}$$

where $W_{ij}$ is again the potential of mean force. The second equality is valid only at high temperatures. The interaction potential $u_{ij}$ is given by (3.1). Thus (4.1) becomes

$$-W_{ij}(r) = -\frac{z_i z_j e^2}{\varepsilon_m r} + \sum_l \beta\rho_l \int ds\ \frac{z_i z_l e^2}{\varepsilon_m s}\ W_{lj}(|r-s|) \tag{4.4}$$

We assume the solution of $W_{ij}$ to be of the form

$$W_{ij}(r) = \frac{z_i z_j e^2}{\varepsilon_m} w(r) \tag{4.5}$$

where $w(r)$, a function to be determined, is independent of species:

$$- w(r) = - \frac{1}{r} + \sum_l \frac{\rho_l z_l^2 e^2}{\varepsilon_m kT} \int ds \, \frac{w(|\mathbf{r} - \mathbf{s}|)}{s} \tag{4.6}$$

$$= - \frac{1}{r} + \frac{\kappa^2}{4\pi} \int ds \, \frac{w(|\mathbf{r} - \mathbf{s}|)}{s}$$

The integral could be viewed as the result of the convolution of two functions:

$$v(r) \equiv \frac{1}{r} \qquad \text{and} \qquad w(r) \tag{4.7}$$

Fourier transforming (4.6), we have

$$- \tilde{w}(k) = - \tilde{v}(k) + \frac{\kappa^2}{4\pi} \tilde{v}(k) \, \tilde{w}(k) \tag{4.8}$$

where $\tilde{\ }$ denotes the Fourier transformed function. Thus

$$\tilde{w}(k) = \frac{\tilde{v}(k)}{1 + (\kappa^2/4\pi) \, \tilde{v}(k)} \tag{4.9}$$

To find $\tilde{v}(k)$, we note that

$$\tilde{v}(k) = \frac{4\pi}{k} \int_0^\infty dr \, \frac{1}{r} \, r \sin(kr) = \frac{4\pi}{k} \lim_{p \to 0} \int dr \, e^{-pr} \sin(kr) \tag{4.10}$$

$$= \frac{4\pi}{k} \lim_{p \to 0} \frac{k}{p^2 + k^2} = \frac{4\pi}{k^2}$$

where we have introduced the factor $e^{-pr}$ to maintain convergence. The second equality is the Laplace transform of $\sin(kr)$. Thus

$$\tilde{w}(k) = \frac{4\pi}{k^2 + \kappa^2} \tag{4.11}$$

The inverse transform is

$$w(r) = \frac{4\pi}{8\pi^3 r} \int_0^\infty dk \, \frac{4\pi}{k^2 + \kappa^2} \, k \sin(kr) = \frac{e^{-\kappa r}}{r} \tag{4.12}$$

Thus

$$W_{ij}(r) = \frac{z_i z_j e^2}{\varepsilon_m} \frac{e^{-\kappa r}}{r}$$

(4.13)

This result is the same as eq. (3.22). We have reproduced the DH results using the OZ equation.

The derivation of Debye's results from statistical mechanics serves at least two purposes: (i) determination of the conditions under which the DH theory is valid, and (ii) improvements of DH by incorporating corrections that were missing from DH. It is evident that DH depends on assumptions on the correlation functions $h_{\alpha\gamma}$ and $C_{\alpha\gamma}$ and is more likely to be valid at low concentrations and high temperatures. Recent studies [16] show that for 2-2 electrolytes at as low as $10^{-5}$ M, the DH results are inadequate in both energy and osmotic coefficient calculations. Another imprecision becomes clear: the OZ equation (4.4) was solved as if the correlation functions had the Coulombic form inside the hard core. The excluded volume effects were not properly accounted for. Kirkwood and Poirier [17] have corrected this deficiency and obtained an oscillatory $w(r)$. In the following, we shall present the thermodynamic functions of the DH theory.

## DEBYE-HÜCKEL THERMODYNAMICS

The internal energy of the DH system could be derived from Green's expression

$$\frac{U^c}{V} = \frac{1}{2} \frac{e^2}{\varepsilon_m} \sum_i \sum_j \rho_i \rho_j z_i z_j \int dr \frac{1}{r} [4\pi r^2 g_{ij}(r)]$$

(4.14)

where $U^c = U - U_{\text{ideal gas}}$. Substituting the Debye pcf into (4.14), we get

$$U^c = -\frac{\kappa^3}{8\pi\beta} = -\sum_{j=+,-} \frac{\rho_j (z_j e)^2}{2\varepsilon_m/\kappa}$$

(4.15)

This DH results could be explained as the dielectric energy. In terms of dielectrics, the point charges with their cospheres could be treated as a collection of *nonoverlapping* spherical capacitors with radius $1/\kappa$, where $\kappa$ is the Debye inverse screening length. Each of these capacitors has a capacitance of $\varepsilon_m/\kappa$ (i.e., the same capacitance as in electrostatics for a sphere of radius $1/\kappa$). Since the capacitor energy is given by

$$\text{Energy} = \frac{Q^2}{2C}$$

(4.16)

with $Q = z_k e$ and $C = \varepsilon_m/\kappa$, $2N$ such capacitors will yield an energy

$$U^c = \sum_{k=1}^{2N} \frac{(z_k e)^2}{2\varepsilon_m/\kappa}$$

(4.17)

$$\frac{U^c}{V} = \sum_{j=+,-} \frac{\rho_j (z_j e)^2}{2\varepsilon_m/\kappa} \tag{4.18}$$

(Note that the subscript $k$ counts the number of *capacitors*, in this case, $N^+ + N^-$ capacitors; while $j = +, -$ counts the number of ionic species.) In terms of the inverse screening length, we recover (4.15), i.e.

$$\frac{U^c}{V} = -\frac{\kappa^3}{8\pi\beta} \tag{4.19}$$

In deriving (4.15), we have assumed that the dielectric constant $\varepsilon$ is not a function of temperature. Actually, it is a function of temperature. To correct for the temperature effects, the energy calculated from (4.15) should be augmented by the derivative $\partial \ln \varepsilon / \partial \ln T$:

$$U = U^c \left[ 1 + \frac{\partial \ln \varepsilon_m}{\partial \ln T} \right] \tag{4.20}$$

This is called the Born-Bjerrum correction (Born [18], Bjerrum [19]). In comparing the results to actual experimental data, one should include this correction.

Using the Gibbs-Helmholtz relation, we can obtain the Helmholtz free energy $A$ from the internal energy

$$\frac{A^c}{V} = -\frac{\kappa^3}{12\pi\beta} \tag{4.21}$$

The osmotic pressure $P_{osm}$ or the osmotic coefficient $\phi = P_{osm}/\rho kT$ is obtained from the derivative $-P_{osm} = \partial A/\partial V$:

$$\phi - 1 = -\frac{\kappa^3}{24\pi \sum_i \rho_i} \tag{4.22}$$

Other thermodynamic quantities can be obtained from the usual relations. For example, the excess Gibbs free energy is

$$G = A + P / \sum_i \rho_i \tag{4.23}$$

The chemical potential $\mu_j$ could be obtained as

$$\frac{\mu_j}{kT} = \ln(\rho_j \Lambda_j^3) - \frac{1}{2} \frac{\kappa z_j^2 e^2}{\varepsilon_m kT} \tag{4.24}$$

The activity coefficient $\ln \gamma_j$ is defined as the correction to the ideal-solution behavior

$$kT \ln (\rho_j \Lambda_j^3 \gamma_j) \equiv \mu_j \tag{4.25}$$

Thus

$$\ln \gamma_j = -\frac{1}{2} \frac{\kappa z_j^2 e^2}{\varepsilon_m kT} \tag{4.26}$$

Since $\gamma_j$ could not be measured experimentally, the quantity in use is the mean activity coefficient, defined as

$$\ln \gamma_\pm \equiv \frac{v_+ \ln \gamma_+ + v_- \ln \gamma_-}{v_+ + v_-} \tag{4.27}$$

where $v_+$ and $v_-$ are stoichiometric coefficients of dissociation

$$C_{v_+} A_{v_-} \rightarrow v_+ C^{z_+} + v_- A^{z_-} \tag{4.28}$$

where $C$ is the cation and $A$ is the anion. For example, $Na_3PO_4 \rightarrow 3Na^+ + PO_4^{-3}$, $v_+ = 3$ and $v_- = 1$. $\gamma_\pm$ is a stoichiometrically averaged activity coefficient. From (4.26), we have

$$\ln \gamma_\pm = \beta G^c = -\frac{\kappa}{2} \frac{\beta e^2}{\varepsilon_m} |z_+ z_-| \tag{4.29}$$

Thus the mean activity coefficient is proportional to the square root of the ionic strength $I$. A typical activity coefficient curve is shown in Figure XII.1 for NaCl up to saturation (at 5 ~6 M). The logarithmic curve first drops below zero then changes sign at higher concentrations. The linear Debye-Hückel region is confined to low $I$ (roughly less than 0.01 M). It is grossly inadequate at higher concentrations.

## XII.5. *Mean Spherical Approximation in the Restricted Primitive Model*

We have discussed the DH theory at considerable length. However, the utility of the theory is limited. Significant advances occurred in the early 1970s when the MSA was solved for the restricted primitive model (RPM) of electrolytes. The anions and cations in RPM are charged hard spheres of equal size. Thus the pair potential is

$$u_{ij}(r) = +\infty, \qquad r < d \tag{5.1}$$

$$= \frac{q_i q_j}{\varepsilon_m r}, \qquad r > d$$

where $i,j = +,-$, $q_j = z_j e$, and $d$ is the common diameter of the hard spheres. The MSA assumes that

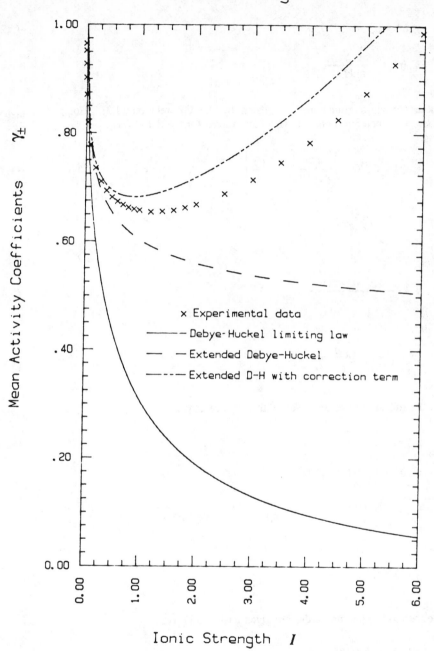

FIGURE XII.1. *The mean activity coefficients of aqueous NaCl solution.* ———: *Debye-Hückel eq. (4.29).* -----: *from* $\ln \gamma_\pm = -A|z_+z_-|\sqrt{I}$ $/(1+\beta a\sqrt{I}.$ — -- —: *from* $\ln \gamma_\pm = -A|z_+z_-|\sqrt{I}$ $/(1+\beta a\sqrt{I} + CI.$ *(Zemaitis et al., AIChE, 1986)*

$$g_{ij}(r) = 0, \qquad\qquad r < d \qquad\qquad (5.2)$$

$$C_{ij}(r) \approx -\frac{q_i q_j}{\varepsilon_m kT} \frac{1}{r}, \qquad\qquad r > d$$

These correlation functions are related by the OZ relation (4.1). Going to Laplace space, Waisman and Lebowitz [20] have solved for the dcf $C_{ij}(r)$

$$C_{ij}(r) = C_{ij}^{HS}(r) - \frac{q_i q_j}{\varepsilon_m kTd}\left[2\tau - \frac{\tau^2 r}{d}\right], \qquad\qquad r < d \qquad\qquad (5.3)$$

$$= -\frac{q_i q_j}{\varepsilon_m kTr}, \qquad\qquad r > d$$

where

$$\tau \equiv \frac{x^2 + x - x(1+2x)^{1/2}}{x^2} \qquad\qquad (5.4)$$

and

$$x^2 \equiv \kappa^2 d^2 = \frac{4\pi d^2}{\varepsilon_m kT}\sum_{j=+,-} \rho_j q_j^2 \qquad\qquad (5.5)$$

For the rdf, the results are given in the Laplace space

$$g_{++}(r) = g_{--}(r) \qquad\qquad (5.6)$$

$$g_{++}(r) + g_{+-}(r) = 2g^{HS}(r)$$

$$\tilde{g}_D(s) = \tilde{g}_{++}(s) - \tilde{g}_{+-}(s) = -\frac{2p^2 s}{\pi\rho} \frac{1}{(s^2 + 2ps + 2p^2)e^{sd} - 2p^2}$$

where

$$p \equiv \frac{(1+2\kappa d)^{1/2} - 1}{2d}$$

The thermodynamic properties are given analytically as

**INTERNAL ENERGY**

$$\frac{U - U^{HS}}{VkT} = -\frac{x^2 + x - x(1+2x)^{1/2}}{4\pi d^3} \qquad\qquad (5.7)$$

## OSMOTIC COEFFICIENT

$$\phi = \phi^{HS} + \frac{3x + 3x(1+2x)^{1/2} - 2(1+2x)^{3/2} + 2}{72y} \tag{5.8}$$

(This pressure is obtained from the energy route.) $\phi^{HS}$ is the hard-sphere compressibility and $y = \pi d^3(\rho_+ + \rho_-)/6$. It is also possible to obtain $\phi$ from the virial theorem. MSA gives the contact values of $g_{\alpha\beta}(r=d^+)$ as

$$g_{ij}(d^+) = g^{HS}(d^+) - \frac{q_i q_j}{\varepsilon kTd}(1 - \tau^2) \tag{5.9}$$

Thus,

$$\phi_{(virial)} = \phi^{HS} + \frac{1}{3\rho kT}\frac{U - U^{HS}}{V} \tag{5.10}$$

In (5.10), $\phi^{HS}$ is to be calculated from the virial pressure equation.

## HELMHOLTZ FREE ENERGY

$$\frac{A - A^{HS}}{VkT} = -\frac{3x^2 + 6x + 2 - 2(1+2x)^{3/2}}{12\pi d^3} \tag{5.11}$$

It is also clear that the MSA results reduce to the DH values as the diameter $d \rightarrow 0$. For example, as $d \rightarrow 0$ the quantity, $p$ approaches $\kappa/2$ and the structure

$$\lim_{d \rightarrow 0} \tilde{g}_D(s) = -\frac{\kappa^2}{2\pi\rho(s+\kappa)} \tag{5.12}$$

This corresponds to the DH rdf of (3.23). Table XII.1 compares the thermodynamic properties of a 1-1 electrolyte solution with MC data of Card and Valleau [21] at low concentrations (up to 2 M). MSA gives satisfactory results. The energy values obtained from the MSA are especially accurate. Table XII.2 [22] compares the excess energy $U^e/(NkT)$ with MC results at high Bjerrum length. The MSA energy is again shown to be satisfactory at high densities and ionic strength. For the correlation functions, Hirata and Arakawa [23] as well as Henderson and Smith [24] have calculated $g_{ij}(r)$ in real space using zonal representations. However, comparison with MC results gave poor agreement. MSA $g_{++}(r)$ turned negative at high ionic strengths, an unphysical behavior. Thus MSA is not reliable for structural calculations.

## XII.6. *Mean Spherical Approximation in the Primitive Model*

For charged hard spheres of unequal diameters (the primitive model or PM), the pair potential is

*Table XII.1. Osmotic Coefficients and Energies of 1-1 Electrolytes*

| Conc. | $\phi = \beta P_{osm}/\rho$ | | | | $\beta U/\rho$ | | | |
|-------|------|-------|-------|-------|-------|-------|-------|-------|
| (moles/l) | MC | MSA | PY | HNC | MC | MSA | PY | HNC |
| 0.00911 | 0.970 | 0.969 | 0.970 | 0.970 | 0.103 | 0.099 | 0.101 | 0.101 |
| 0.10376 | 0.945 | 0.931 | 0.946 | 0.946 | 0.274 | 0.268 | 0.271 | 0.271 |
| 0.425 | 0.977 | 0.945 | 0.984 | 0.980 | 0.434 | 0.426 | 0.429 | 0.429 |
| 1.00 | 1.094 | 1.039 | 1.108 | 1.091 | 0.552 | 0.541 | 0.542 | 0.545 |
| 1.968 | 1.346 | 1.276 | 1.386 | 1.340 | 0.651 | 0.636 | 0.638 | 0.646 |

*Table XII.2.  Energies at High Bjerrum Lengths for RPM Salts*

| $\rho^*$ | $B/d_{--}$ | Excess Energy, $-U^c/VkT$ | | |
|-------|---------|-------|-------|-------|
| | | MC | MSA | HNC |
| 0.2861 | 1.8823 | 0.839 | 0.804 | 0.832 |
| | 9.4111 | 5.465 | 5.278 | 5.431 |
| | 47.055 | 33.62 | 31.87 | - |
| 0.4788 | 1.5873 | 0.783 | 0.725 | 0.767 |
| | 7.9368 | 4.876 | 4.664 | 4.859 |
| | 39.684 | 28.61 | 27.75 | 28.66 |
| 0.669 | 1.4191 | 0.756 | 0.675 | 0.731 |
| | 7.0952 | 4.601 | 4.288 | 4.540 |
| | 35.476 | 26.54 | 25.28 | 26.20 |
| 0.7534 | 1.3634 | 0.711 | 0.657 | 0.719 |
| | 6.8172 | 4.511 | 4.16 | 4.343 |
| | 34.0859 | 25.71 | 24.45 | 25.46 |

$$u_{ij} = \infty, \qquad\qquad r < d_{ij} \tag{6.1}$$

$$= \frac{q_i q_j}{\varepsilon_m r}, \qquad\qquad r > d_{ij}$$

where $i,j = +,-$ and the HS diameters $d_{++} \neq d_{--}$. For the cross interaction, $d_{+-}$ is assumed to be *additive*, i.e., $d_{+-} = (d_{++}+d_{--})/2$. The MSA assumptions are

$$g_{ij}(r) = 0, \qquad\qquad r < d_{ij} \tag{6.2}$$

$$C_{ij}(r) \approx - \frac{q_i q_j}{\varepsilon_m kTr}, \qquad\qquad r > d_{ij}$$

Blum [25] has solved the MSA equations for PM. As in the DH theory, a characteristic length, the shielding parameter $\Gamma$, appears in this theory:

$$2\Gamma = \alpha \left\{ \sum_{i=1}^{n} \rho_i \left[ \frac{z_i - (\pi/2\Delta)d_i^2 P_n}{1 + \Gamma d_i} \right]^2 \right\}^{1/2} \tag{6.3}$$

where we have used $d_i$ for the HS diameter of *ii* interaction, and $d_{ij} = (d_i+d_j)/2$. Also the symbols are defined as follows

$$P_n \equiv \frac{1}{\Omega} \sum_k \frac{\rho_k d_k z_k}{1 + \Gamma d_k}, \qquad \Omega \equiv 1 + \frac{\pi}{2\Delta} \sum_k \frac{\rho_k d_k^3}{1 + \Gamma d_k} \qquad (6.4)$$

$$\zeta_n \equiv \sum_k \rho_k (d_k)^n, \quad \Delta \equiv 1 - \frac{\pi \zeta_3}{6}, \qquad \alpha^2 \equiv \frac{4\pi e^2}{\varepsilon_m kT}$$

As the size $d_i \to 0$, we see that $\lim_{d_i \to 0} 2\Gamma = \kappa$, the Debye inverse length. On the other hand, when $d_{++} = d_{--}$,

$$4\Gamma^2 = \frac{\kappa^2}{(1 + \Gamma d)^2} \qquad (6.5)$$

The root is

$$-2\Gamma d = 1 - (1 + 2x)^{1/2} \qquad (6.6)$$

where $x$ is the Waisman-Lebowitz $\kappa d$. The result is consistent with their formula.

**INTERNAL ENERGY**

$$-\frac{U - U^{HS}}{VkT} = \frac{e^2}{\varepsilon_m kT} \left\{ \Gamma \sum_{i=1}^{n} \frac{\rho_i z_i^2}{1 + \Gamma d_i} + \frac{\pi}{2\Delta} \Omega P_n^2 \right\} \qquad (6.7)$$

**HELMHOLTZ FREE ENERGY**

$$\frac{A - A^{HS}}{VkT} = \frac{U - U^{HS}}{VkT} + \frac{\Gamma^3}{3\pi} \qquad (6.8)$$

**OSMOTIC COEFFICIENT**

$$\phi - \phi^{HS} = -\frac{\Gamma^3}{3\pi\rho} - \frac{\alpha^2}{8\rho} \left[ \frac{P_n}{\Delta} \right]^2 \qquad (6.9)$$

**MEAN ACTIVITY COEFFICIENT**

$$\ln \gamma_\pm - \ln \gamma_\pm^{HS} = \frac{U - U^{HS}}{NkT} - \frac{\alpha^2}{8\rho} \left[ \frac{P_n}{\Delta} \right]^2 \qquad (6.10)$$

## INVERSE ISOTHERMAL COMPRESSIBILITY

$$\beta \frac{\partial P}{\partial \rho}\bigg|_T = \frac{1}{4\zeta_0 \pi^2} \sum_{j=1}^{n} \rho_j Q_j^2 \tag{6.11}$$

where

$$Q_j = \frac{2\pi}{\Delta}\left[1 + \zeta_2 d_j\left[\frac{\pi}{2\Delta}\right] + \frac{1}{2}a_j P_n\right] \tag{6.12}$$

and

$$a_j = \frac{\alpha^2}{2\Gamma(1 + \Gamma d_j)}\left[z_j - P_n d_j^2 \frac{\pi}{2\Delta}\right] \tag{6.13}$$

## DIRECT CORRELATION FUNCTIONS

For $r > (d_i + d_j)/2$, $C_{ij}(r)$ is given by the MSA assumption (6.2). For $r < (d_i + d_j)/2$, we divide the region into two parts, "A": $0 \le r \le (d_j - d_i)/2 \equiv x$, and "B": $(d_j - d_i)/2 \le r \le (d_i + d_j)/2 \equiv y$, (assuming that $d_i < d_j$).

```
  |  ───────────  |  ───────────  |
  0                x                y
      Region "A"       Region "B"
```

For $C_{++}$ and $C_{--}$, region "A" is nonexistent. In region "A",

$$C_{ij}(r) = C_{ij}^{(0)}(r) - \frac{2e^2}{\varepsilon_m kT}\left[-z_i N_j + X_i(N_i + \Gamma X_i) - \frac{d_i}{3}(N_i + \Gamma X_i)^2\right] \tag{6.14}$$

where $C_{ij}^{(0)}(r)$ are the dcfs in the HS mixture. The new constants are

$$X_i \equiv \frac{z_i - (\pi/2\Delta) d_i^2 P_n}{1 + \Gamma d_i} \tag{6.15}$$

and

$$-N_i \equiv \frac{z_i \Gamma + (\pi/2\Delta) d_i P_n}{1 + \Gamma d_i} \tag{6.16}$$

Thus

$$N_i + \Gamma X_i = -\frac{\pi}{2\Delta} d_i P_n \tag{6.17}$$

In region "B", the dcfs are given by

$$rC_{ij}(r) - rC_{ij}^{(0)}(r) = \frac{e^2}{\varepsilon_m kT} \cdot \tag{6.18}$$

$$\left[ (d_i - d_j) \left\{ \frac{X_i + X_j}{4} [(N_i + \Gamma X_i) - (N_j + \Gamma X_j)] - \frac{d_i - d_j}{16} [(N_i + \Gamma X_i + N_j + \Gamma X_j)^2 - 4N_i N_j] \right\} \right.$$

$$- \left\{ (X_i - X_j)(N_i - N_j) + (X_i^2 + X_j^2)\Gamma + (d_i + d_j)N_i N_j - \frac{1}{2} [d_i(N_i + \Gamma X_i)^2 + d_j(N_j + \Gamma X_j)^2] \right\} r$$

$$+ \left\{ \frac{X_i}{d_i}(N_i + \Gamma X_i) + \frac{X_i}{d_j}(N_j + \Gamma X_i) + N_i N_j - \frac{1}{2} [(N_i + \Gamma X_i)^2 + (N_i + \Gamma X_j)^2] \right\} r^2$$

$$\left. + \left\{ (6d_i^2)^{-1}(N_i + \Gamma X_i)^2 + (6d_j^2)^{-1}(N_j + \Gamma X_j)^2 \right\} r^4 \right]$$

## RADIAL DISTRIBUTION FUNCTIONS

For the rdf, the contact values are given by

$$d_{ij} g_{ij}(d_{ij}) = \frac{d_{ij}}{1 - \xi_3} + \frac{3}{2} \frac{\xi_2 d_i d_j}{(1 - \xi_3)^2} - \frac{e^2 X_i X_j}{\varepsilon_m kT} \tag{6.19}$$

For the rdf at $r > (d_i + d_j)/2$, the expressions are given in the Laplace space by Blum and Høye [26]. It is of the form

$$\tilde{g}_{ij}(s) = \frac{D_0 D_\pm}{D_T} \tilde{g}_{ij}^0(s) - \frac{D_0 \Gamma^2}{s\alpha^2 \pi D_T} e^{-s\sigma_{ij}} a_i a_j + \Delta \tilde{g}_{ij}(s) \tag{6.20}$$

Here, the first term, aside from a factor, gives essentially the PY solution for a pure HS mixture, the second term is purely electrostatic in nature, and the third term is derived from the cross interaction between the hard core and the electric charges. The algebraic expressions of these terms have been given by Blum and Høye (*Ibid.*) and will not be repeated here. However, for PM at low concentrations, Blum showed that as $P_n \approx 0$ (this happens when $d_i = d_j$ or when $\rho d_j^3 \leq 0.1$), the Laplace transform of $g_{ij}$ is

$$\tilde{g}_{ij}(s) \approx \tilde{g}_{ij}^{HS}(s) - A_{ij} e^{-sd_{ij}} \left[ s^2 + 2\Gamma s + 2\Gamma^2 - \frac{2\Gamma^2}{\alpha^2} \sum_m \rho_m a_m^2 e^{-sd_m} \right]^{-1} \tag{6.21}$$

where $a_j$ has been defined in eq. (6.13), and

$$A_{ij} = \frac{z_i z_j \alpha^2}{4\pi(1 + \Gamma d_i)(1 + \Gamma d_j)} \tag{6.22}$$

Inverse Laplace transformation gives

$$g_{ij}(r) = g_{ij}^{\text{HS}}(r) - \frac{A_{ij}}{r} \sum_{m=0}^{\infty} \frac{\Gamma^{m+1}}{m!\,\alpha^{2m}} \tag{6.23}$$

$$\times \left[ \sum_{\{l_t\}} \left[ \prod_{t=1}^{m} \rho_{l_t} a_{l_t}^2 \right] F_m(r - d_{ij} - \sum_t d_{l_t}) \right]$$

where $F_m(r) = 0$ for $r < 0$, and

$$F_m(r) = r^{m+1} e^{\Gamma r} \left[ j_{m-1}(\Gamma r) - j_m(\Gamma r) \right] \tag{6.24}$$

The sum is over all partitions $\{l_t\} = l_1, .., l_m$; $j_m(x)$ is the spherical Bessel function with $j_{-1}(x) = \cos(x)/x$.

Strictly speaking, both RPM and PM are idealized models. However, they are improvements over the DH theory in that they treat the hard-core repulsion more realistically. Waisman and Lebowitz (*Ibid.*) have shown that if one lets the diameter $d$ go to zero, the RPM formulas reduce to the DH ones. Thus attempts have been made to calculate properties of real electrolyte solutions by RPM or PM formulas [27]. Since PM is more realistic than RPM, we shall present below an application of the PM method to sea water properties [28].

### *Example 1: Sea water properties*

Sea water consists essentially of dissolved NaCl salt. The concentrations vary from 0.1 to 0.5 M (depending on locations). Traces of other salts are also present (see Table XII.3). Properties such as the osmotic pressure, the mean activity coefficient, etc., are of interest and have been extensively measured. To apply the PM method, we need to know first the diameters of the cation and the anion, $d_{++}$ and $d_{--}$. Pauling has given the crystal radii of a number of ions (see Table XII.4). This is a good starting point. Since real ions are not hard spheres, some softness is present. In perturbation theory, this is reflected in a temperature-dependent HS diameter. The temperature dependence can be determined either from theoretical considerations, as discussed in Chapter XI, or, for engineering purposes, from fitting $d$ to reproduce experimental information. The latter step is taken here. In addition, the effective ion sizes should reflect the hydration of ions in aqueous environment.

The next problem is to obtain the value of the shielding parameter $\Gamma$. It is given

*Table XII.3. Major Constituents of Sea Water*

| Constituent | molality |
|---|---|
| $Na^+$ | 0.4756 |
| $Mg^{++}$ | 0.0542 |
| $Ca^{++}$ | 0.0103 |
| $K^+$ | 0.01007 |
| $Cl^-$ | 0.5544 |
| $SO_4^{-2}$ | 0.0286 |
| $HCO_3^-$ | 0.0024 |
| $Br^-$ | 0.00084 |
| Total Dissolved solids | 1.1364 |
| Specific gravity (20°C) | 1.0243 |

*Table XII.4. Pauling Ionic Radii*

| Ionic Radius (in Angstroms) | | | | | |
|------|------|------|------|------|------|
| Li | 0.60 | Na | 0.95 | K | 1.33 |
| Cu | 0.96 | Rb | 1.48 | Ag | 1.26 |
| Cs | 1.69 | Au | 1.37 | Mg | 0.65 |
| Ca | 0.99 | Zn | 0.74 | Sr | 1.13 |
| Cd | 0.97 | Ba | 1.35 | Hg | 1.10 |
| B | 0.20 | Al | 0.50 | Ga | 0.62 |
| Y | 0.93 | In | 0.81 | La | 1.15 |
| Zr | 0.80 | Sn | 0.71 | Ce | 1.01 |
| Pb | 0.84 | N | 0.11 | P | 0.34 |
| O | 1.40 | S | 0.29 | Cr | 0.52 |
| Se | 0.42 | Mo | 0.62 | F | 1.36 |
| Cl | 1.81 | Br | 1.95 | I | 2.16 |

by eq. (6.3) once we have the concentrations ($\rho_+$ and $\rho_-$), the temperature (used in $\alpha^2$), and the dielectric constant $\varepsilon$ (for water at 25°C, $\varepsilon = 78.358$) (see Table XII.5). Equation (6.3) is actually of sixth degree in $\Gamma$ for a binary salt. Thus in principle, there are six roots for $\Gamma$. Only one of the six roots is physically meaningful. We calculate first $x/2$ from the RPM eq. (5.5), then iterate through eq. (6.3) to obtain a $\Gamma_{trial}$, obtained by setting $P_n = 0$ (Table XII.6). Next we use this trial value in further iterations (for $P_n \neq 0$) until a convergent answer $\Gamma$ is obtained. Table XII.6 shows the variations of $\Gamma$ at different temperatures and concentrations for a 1-1 salt.

The effect of $P_n$ on $\Gamma$ is not appreciable. For the most stringent cases studied (at $B/d_- = 0.8515$, $\rho^* = 3.0975$, and $d_{++}/d_{--} = 0.2$), the error in $\Gamma_{trial}$ is about +6%. $|P_n|$ increases with decreasing diameter ratio $d_{++}/d_{--}$ and decreases with increasing Bjerrum length $B$, which is defined as

$$B \equiv \frac{|z_+ z_-| e^2}{\varepsilon_m kT} \tag{6.25}$$

The Pauling radii for a number of ions are given in Table XII.4. Using the Pauling crystal diameters, we are able to calculate the osmotic coefficients of the NaCl aqueous solution from 0.9 M to 5 M according to (6.9). The results are tabulated in Table XII.7. The overall deviation is 3.9%. This result is obtained without any fitting of parameters. Discrepancies occurred at low temperatures. The calculated values are consistently too low. This may indicate that the Pauling crystal radii, which were determined from crystal measurements, do not describe accurately the sizes of the ions in an aqueous environment. If we choose instead, the sizes $d_+ = 2.190$ Å for $Na^+$ close to its

*Table XII.5. The Dielectric Constant of Water*

| T°C | $\varepsilon$ | $\partial \ln \varepsilon / \partial \ln T$ |
|-----|--------|---------|
| 20 | 80.176 | -1.3459 |
| 25 | 78.358 | -1.3679 |
| 30 | 76.581 | -1.3900 |

Table XII.6. The Shielding Parameter, $\Gamma$.

| $d_+/d_-$ | The Shielding Parameter at Different Coupling Strengths, $P_n$ | | | |
|---|---|---|---|---|
| | $B/d_-$ | $\rho^*$ | $\Gamma(P_n = 0)$ | $\Gamma(P_n \neq 0)$ |
| 0.8 | 1.2772 | 0.9178 | 1.0069($P_n = 0$) | 1.0096($P_n = -0.013$) |
| | 6.3857 | 0.9178 | 1.7024 (0) | 1.7047 (−0.009) |
| | 31.928 | 0.9178 | 2.7701 (0) | 2.7717 (−0.005) |
| 0.6 | 1.1353 | 1.3068 | 1.1342 (0) | 1.1493 (−0.035) |
| | 5.6762 | 1.3068 | 1.9391 (0) | 1.9517 (−0.023) |
| | 28.3808 | 1.3068 | 3.1813 (0) | 3.1902 (−0.014) |
| 0.4 | 0.9934 | 1.9504 | 1.3073 (0) | 1.3531 (−0.063) |
| | 4.9666 | 1.9504 | 2.2939 (0) | 2.3328 (−0.043) |
| | 24.8332 | 1.9504 | 3.8381 (0) | 3.8661 (−0.027) |
| 0.2 | 0.8515 | 3.0975 | 1.5966 (0) | 1.6968 (−0.057) |
| | 4.2571 | 3.0975 | 2.956 (0) | 3.0444 (−0.044) |
| | 21.2856 | 3.0975 | 5.1668 (0) | 5.2334 (−0.030) |

hydrated diameter, and $d_- = 3.618$ Å for Cl⁻, we could reduce the deviations down to 1.7%. Since the ions are not hard spheres, their effective sizes can be represented by temperature-dependent diameters. Using the experimental information as a guide, the following relations are obtained:

For Na⁺:
$$d_+ = -0.78132 + \frac{6141.4}{T} - \frac{9.3052 \times 10^5}{T^2} \tag{6.26}$$

For Cl⁻:
$$d_- = 3.3428 + \frac{183.13}{T} - \frac{3.0233 \times 10^4}{T^2}$$

The temperatures are in Kelvin and the diameters are in Å. With these choices, the deviations in osmotic coefficient predictions for the NaCl solutions are reduced to 0.7%. Note that in the above calculations, the scale conversion [29] between the MM picture of electrolytes and the Lewis-Randall practical units used in experiments has been implicitly accounted for by fitting the parameters to data.

Table XII.7. Osmotic Coefficients Obtained from MSA with Pauling Diameters*

| T K | Osmotic Coefficient, $\phi = \beta P_{osm}/\rho$ | | | |
|---|---|---|---|---|
| | Molarity, M | $\phi$ (expt.) | $\phi$ (calc.) | Error% |
| 298.15 | 1.0144 | 0.9391 | 0.8949 | -4.8% |
| 310.65 | 2.3367 | 1.0229 | 0.9689 | -5.3% |
| 298.15 | 3.3756 | 1.0887 | 1.0297 | -5.4% |
| 348.15 | 4.3895 | 1.1871 | 1.1421 | -3.8% |
| 298.15 | 5.2390 | 1.2730 | 1.1819 | -7.2% |

*Pauling cation size, $d_{Na+} = 1.90$ Å. Anion size, $d_{Cl-} = 3.62$ Å.

This example demonstrates the capability of the PM equations in predicting the osmotic pressures of a number of electrolyte solutions of the 1-1 and 1-2 types. Using the Pauling radii alone, we could obtain an order of magnitude estimation. Note that Blum's solution for the MSA contains the geometric factors (e.g., $\Delta$), electrostatic factors (e.g., $\alpha$), and the cross terms (e.g., $P_n$, $\Omega$ and $\Gamma$). These functional relationships become possible only through MSA. More extensive correlations for over 100 industrial salt solutions have been developed recently [30].

The PM and RPM methods can be equally well applied to molten salts, i.e., salts existing at high temperatures and thus in the liquid state. In this case, there is no dielectric background except that generated by the salt. Normally, the dielectric constant is taken to be unity (see e.g., Larsen [31]). Studies have shown that the MSA gives satisfactory energy predictions for the PM and the RPM, even at high reciprocal temperatures, and Bjerrum length. Other properties, such as the osmotic and activity coefficients, are less satisfactory, unless they are derived from the energy relation. On the other hand, the structure $(g_{++}(r), g_{+-},$ and $g_{--}(r))$ from MSA is poor. At high $B$, $g_{++}(r)$ may become negative at small r, an unphysical behavior [32]. Thus the good agreement of the energy equation in MSA must be due to fortuitous cancellations of errors in the integrals.

## XII.7. *Hypernetted Chain Equation*

The integral equations, such as PY and HNC, have been applied to the calculations of the structure of the electrolyte solutions. Rasaiah et al. [33] have shown that the results from HNC are more dependable than the PY results (see, e.g. Table XII.1). Further studies confirmed the superiority of the HNC method. Recall that the HNC (for pure species) rests on the relation

$$C(12) = h(12) - \ln y(12) \tag{7.1}$$

Equation (7.1) can be written as

$$C(12) = h(12) - \ln g(12) - \beta u(12) \tag{7.2}$$

$$= h(12) - \ln[1 + h(12)] - \beta u(12)$$

$$= h(12) - h(12) + \frac{1}{2} h(12)^2 - \frac{1}{3} h(12)^3 + \cdots - \beta u(12)$$

$$= -\beta u(12) + \frac{1}{2} h(12)^2 - \frac{1}{3} h(12)^3 - \cdots$$

Thus for small $h(12)$, the HNC reduces to the MSA prescription (at $r > d$). In light of this theoretical relation, HNC is a correction to the MSA. As will be seen, HNC also performs better than MSA in practice. Equation (7.1) coupled with the OZ relation can be solved numerically for, e.g., the restricted primitive model (Carley [34], Rasaiah and Friedman [35], Larsen [36]). The numerical procedure can be made more efficient by using the fast Fourier transform technique. Lately, the same technique is applied to the case of a soft repulsive core potential plus the Coulomb interaction [37].

The internal energy, and pressure, obtained from the HNC are listed in Tables XII.1, 2 and 8. The conditions range from dilute 1-1 electrolyte solutions to molten salts (e.g., at $\rho^*=0.669$ and $B =35.476$). The HNC theory is compared with some other theories, such as the MSA(E), and the reference HNC (RHNC [38]) equation

$$\ln y(r) - g(r) + C(r) = \ln y_0(r) - g_0(r) + C_0(r) \tag{7.3}$$

where the subscript 0 denotes the reference potential quantities. In this case, the reference is taken to be the HS potential. And since the PY equation works well for hard spheres, the PY values for $y_0$, $g_0$, and $C_0$ are used in the RHNC. HNC gives good energy predictions from low $B$ to high $B$. On the other hand, the MSA values (obtained from the energy route) are also reasonable, though less accurate than the HNC values. For pressures, the agreement is less satisfactory for all theories. For HNC and RHNC, the pressures were calculated from the virial pressure equation.

*Table XII.8. The Osmotic Pressures in RPM*

| $\rho^*$ | $B/d$ | Osmotic Pressures, $P_{osm}/\rho kT$ | | | |
|---|---|---|---|---|---|
| | | MC | MSA(E)* | HNC | RHNC |
| 0.2861 | 1.8823 | 1.62 | 1.75 | 1.76 | 1.70 |
| | 9.4111 | 0.79 | 1.13 | 1.05 | 0.96 |
| | 47.055 | -0.79 | -1.53 | - | - |
| 0.4788 | 1.5873 | 2.95 | 3.14 | 2.83 | |
| | 7.9368 | 1.99 | 2.44 | 2.45 | 2.08 |
| | 39.684 | -0.65 | 0.30 | 0.03 | -0.71 |
| 0.669 | 1.4191 | 5.04 | 5.10 | 5.86 | 4.71 |
| | 7.0952 | 4.48 | 4.65 | 5.18 | 3.94 |
| | 35.476 | 1.49 | 2.79 | 2.46 | 0.68 |
| 0.7534 | 1.3634 | 6.71 | 6.59 | 7.88 | 5.95 |
| | 6.8172 | 6.55 | 6.17 | 7.21 | 5.17 |
| | 34.0859 | 2.61 | 4.40 | 4.51 | 1.85 |

*MSA(E) is the pressure calculated based on the MSA energy route.

The structures $g_{++}(r)$ $(=g_{--}(r)$ for RPM) and $g_{+-}(r)$ are given in Figure XII.2. The dcfs are given in Figure XII.3. The low Bjerrum value ($B = 4.707$) refers to the electrolyte solution regime and the high $B = 35.476$ refers to the molten salt state. It is seen that for low $B$, the HNC and RHNC values are quite satisfactory. For high $B$, both theories are qualitatively correct, even though some finer structural details are missing in HNC and RHNC (e.g., the first peak of $g_{++}$ from MC at $B = 35.476$ is sharper than those from the theories).

The HNC and RHNC equations satisfy the Stillinger-Lovett [39] moment conditions. The latter are exact conditions that the correlation functions of the ionic solutions must satisfy. Let us define the moments (the zeroth moment and second moment) of the correlation function $h_{ij}(r)$ as

$$M_{ij}^{(0)} \equiv \int d\mathbf{r}\ h_{ij}(r) = 4\pi \int_0^\infty dr\ r^2 h_{ij}(r) \tag{7.4}$$

and

$$M_{ij}^{(2)} \equiv \int d\mathbf{r}\ r^2\ h_{ij}(r) = 4\pi \int_0^\infty dr\ r^4 h_{ij}(r) \tag{7.5}$$

where $M^{(0)}$ and $M^{(2)}$ are the zeroth and second moments of the total correlation function, respectively. The Stillinger-Lovett zeroth moment condition is equivalent to the electroneutrality condition: that the total electrolyte solution must be neutral. In terms of $M^{(0)}$ (or equivalently of $h_{ij}$), this condition for a single-salt solution is

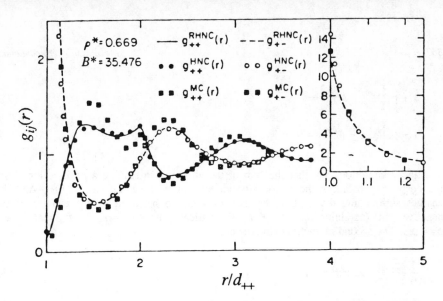

FIGURE XII.2. *Pair correlation functions for the restricted primitive model: Comparison with MC simulation on a molten salt. Density* $\rho^* = \rho d_{++}^3 = 0.669$. *Bjerrum length* $B^* = |z_+ z_-| e^2 / \varepsilon k T d_{++} = 35.476$. *The first peak of* $g_{++}$ *is not well predicted by the theories. (Larsen, J. Chem. Phys., 1978)*

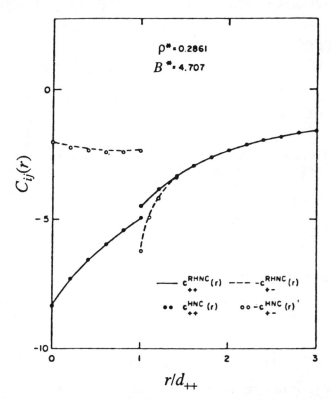

FIGURE XII.3. *Direct correlation functions for the RPM in a concentrated electrolyte solution. Lines: RHNC. Symbols: HNC. (Larsen, J. Chem. Phys., 1978)*

$$1 + \rho_+\left[M_{++}^{(0)} - M_{+-}^{(0)}\right] = 0 \tag{7.6}$$

$$1 + \rho_-\left[M_{--}^{(0)} - M_{-+}^{(0)}\right] = 0 \tag{7.7}$$

The above conditions say that the sum of the charges surrounding a positive ion (from the cospheres) balances the charges of the central positive ion (eq. (7.6)). And the charges surrounding the central negative ion are also balanced by the charges of the negative ion (condition (7.7)). For symmetrical ions $(h_{++} = h_{--})$, $M_{++}^{(0)} = M_{--}^{(0)}$ and $\rho_+ = \rho_-$. The second moment conditions are

$$\frac{1}{6}\frac{\kappa^2}{z_i e}\sum_j \rho_j z_j e M_{ij}^{(2)} = -1 \tag{7.8}$$

where

$$\kappa^2 = \frac{4\pi}{\varepsilon_m kT}\sum_j \rho_j z_j^2 e^2 \tag{7.9}$$

For a single-salt solution, we have

$$\frac{1}{6}\kappa^2\rho + (M_{++}^{(2)} - M_{+-}^{(2)}) = -1 \tag{7.10}$$

and

$$\frac{1}{6}\kappa^2\rho - (M_{--}^{(2)} - M_{-+}^{(2)}) = -1 \tag{7.11}$$

Rasaiah and Friedman [40] have demonstrated numerically that both moment conditions were satisfied by the HNC equation. For RPM electrolytes, these conditions also say

### THE FIRST MOMENT CONDITION

$$\lim_{k\to 0}\frac{\rho[\tilde{h}_{++}(k) - \tilde{h}_{+-}(k)]}{2} = -1 \tag{7.12}$$

### THE SECOND MOMENT CONDITION

$$\lim_{k\to 0}\frac{\partial^2}{(\partial kd)^2}\frac{\rho[\tilde{h}_{++}(k) - \tilde{h}_{+-}(k)]}{2} = \frac{2}{x^2} \tag{7.13}$$

where $x^2$ is the Waisman-Lebowitz $x$ (see eq.(5.5)). $\tilde{h}_{ij}(k)$ is the Fourier transform of $h_{ij}(r)$.

Recently, the HNC equation has been solved for a soft reference potential [41]

$$u_{ij}(r) = B_{ij}\left[\frac{\sigma_i + \sigma_j}{r}\right]^n + \frac{q_i q_j}{\varepsilon r} \tag{7.14}$$

*n* takes on the value 9. The anion ion size is chosen to be the same as the cation ion size (symmetric ions). The constants $\sigma_+ = \sigma_- = 1.4214$ Å, and

$$B_{ij} = kT \frac{5377.75}{T} \frac{|z_i z_j|}{\sigma_i + \sigma_j} \tag{7.15}$$

The structure, $g_{++}(r)$ and $g_{+-}(r)$, for a 2:2 electrolyte are displayed in Figure XII.4. These are the low density states ($C_{st} = 0.005$ M). The $g_{+-}(r)$ from HNC is reasonable. The $g_{++}(r)$ from HNC gives an exaggerated peak at $r = 2d$. Rossky et al. [42] examined the cause and determined that this exaggeration was due to the missing bridge diagram ( ) in the HNC approximation. When the bridge clusters were added, the discrepancy was removed. Recent work [43] based on the RHNC with asymmetric ions further showed the importance of the bridge functions. Overall, the HNC results are quite accurate (after excluding the low concentration range: $C_{st} < 0.2$ M). The $g_{+-}(r)$ distribution from HNC is particularly accurate for the states studied. A new integral equation, the HMSA, of Zerah and Hansen [44] has recently been applied to molten salts. This equation is essentially an *interpolation* between the MSA (for short ranges) and the HNC (for long ranges), since the HNC theory is well suited for long-range potentials (in particular, Coulombic systems) and MSA performs well for harshly repulsive short-range forces (e.g., hard spheres). Thus first we separate the pair potential into short-range repulsive $u^r$ and long-range attractive $u^a$ parts (such as in the WCA division)

$$u_{ij}(r) = u_{ij}^r(r) + u_{ij}^a(r) \tag{7.16}$$

The HMSA closure is then

$$g_{ij}(r)e^{\beta u_{ij}^r(r)} - 1 = f_{ij}(r)^{-1}\left[\exp\left\{f_{ij}(r)[h_{ij}(r) - C_{ij}(r) - \beta u_{ij}^a(r)]\right\} - 1\right] \tag{7.17}$$

where $f_{ij}(r)$ is a switching function, that must be zero at $r = 0$ and unity at $r = \infty$. For simplicity, the following form has been chosen by Rogers and Young [45]

$$f_{ij}(r) = 1 - e^{-\alpha_{ij}r} \tag{7.18}$$

Other choices are also permissible. The parameters $\alpha_{ij}$ are chosen such that the pressure consistency condition (i.e., $P^c = P^v$) is satisfied. Zerah and Hansen (*Ibid.*) have applied the HMSA to several molten salts and found good agreement with MC data. (See Figure XII.5 for the case of molten KCl). The pressures calculated from HMSA are also closer to MC values than HNC.

## SOLVENT GRANULARITY

The primitive model of electrolyte solutions replaces the solvent molecules by a dielectric continuum having the same dielectric constant $\varepsilon$. Whether this assumption is valid is of interest, particularly in studying solvation, where the "granularity" of the solvent particles could not be ignored. Levesque et al. [46] have solved the LHNC and QHNC equations for a mixture of cations, anions, and dipolar hard spheres simulating the solvent molecules. The potential model is

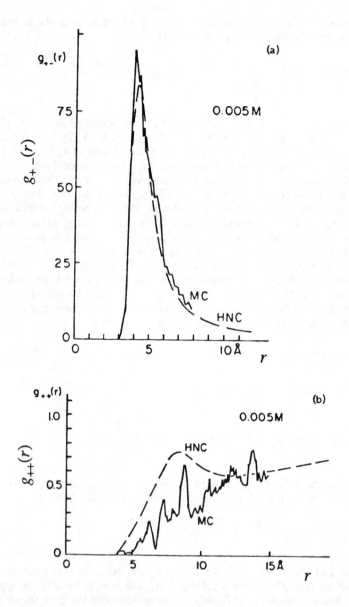

FIGURE XII.4. Comparison of HNC and MC results for the pair correlation functions of charged soft spheres at 0.005M (64 ion samples; 10000 passes). (Rossky et al., J. Chem. Phys., 1980)

$$u_{\alpha\beta}(12) = u_{\alpha\beta}^{000}(r)\phi^{000}(12) + u_{\alpha\beta}^{101}(r)\phi^{101}(12) + u_{\alpha\beta}^{011}(r)\phi^{011}(12) + u_{\alpha\beta}^{112}(r)\phi^{112}(12) \quad (7.19)$$

where $\phi^{LMN}$ are the rotational invariants defined earlier. For the present model, the subscripts $\alpha$, $\beta$ = +,–, or $\mu$ ($\mu$ being the dipolar spheres). The coefficients are

$$u_{\alpha\beta}^{000}(r) = u_{\alpha\beta}^{HS}(r) + \frac{q_\alpha q_\beta}{r}, \qquad u_{\alpha\beta}^{101}(r) = \frac{\mu_\alpha q_\beta}{r^2} \qquad (7.20)$$

$$u_{\alpha\beta}^{011}(r) = -\frac{q_\alpha \mu_\beta}{r^2}, \qquad u_{\alpha\beta}^{112}(r) = -\frac{\mu_\alpha \mu_\beta}{r^3}$$

In the following, we shall consider a 1:1 electrolyte with equal-sized ($d_{++} = d_{--}$) cations and anions. The size of the dipolar hard spheres, $d_{\mu\mu}$, is allowed to vary (from $1.0d_{++}$ to $0.68d_{++}$). The conditions studied are given in Table XII.9. The concentrations vary from 0.05 M to 1 M for an electrolyte solution with sizes $=d_{++}=d_{--}= 4$ Å. The low Bjerrum values reported here are due to the nonconvergence of the QHNC method for $B$ greater than 40. This study compares the dipolar solvent model with the MM assumption where the solvent molecules are smoothed out. The difference is revealed in the structure $g_{+-}(r)$ and $g_{++}(r)$ as shown in Figures XII.6 and 7. The primitive model results are shown as dashed lines. The existence of the solvent molecules causes rapid oscillations in $g_{+-}$ (e.g., a second peak appears at $r= 2.0d_{++}$), signifying that the cations and anions are separated by solvent molecules, for example, in chains of

$$\oplus \quad \ominus \quad \textcircled{\mu} \quad \ominus$$

or

$$\ominus \quad \oplus \quad \textcircled{\mu} \quad \oplus$$

In Figure XII.6, the second peak is higher than the first peak, indicating that solvent-separated ion pairs are more common than ion pairs in contact. The oscillatory pair correlation function $g_{++}$ over the baseline PM curve gives indication of formation of ion-solvent complexes: hydration of the ions. However, the pcf alone is not a sufficient indicator. Higher-order distribution functions are needed for confirmation. The ion-dipole and dipole-dipole correlation functions are shown in Figure XII.8. Again the order of oscillations is determined by the solvent size.

It is also interesting to observe the behavior of the "effective" potential. Adelman [47] carried out averaging on the mixture OZ equations, transforming them into an effective OZ equation. The results are shown in Figure XII.9. Surprisingly, the effective potentials $w_{ij}^{\text{eff}}(r)$ do not correspond to the PM results. Evidently, the MM picture and Adelman's formalism represent different procedures of averaging. Levesque et al. (*Ibid.*) concluded that the solvent effects are not correctly accounted for in the PM approach. To study the solvation of ions, we must consider the granular nature of the solvent

*Table XII.9. The State Conditions for Ion-Dipole Mixtures*

| Theory | $\rho_\mu^*$ | $\mu^{*2}$ | $q^{*2}$ | $\varepsilon$ (solvent) | $d_\mu/d_+$ |
|--------|------|-------|------|-------------|---------|
| QHNC | 0.6 | 2.5 | 40 | 26.12 | 1 |
| QHNC | 0.6 | 2.5 | 40 | 26.12 | 0.68 |
| QHNC | 0.6 | 2.5 | 136 | 26.12 | 0.68 |
| LHNC | 0.8 | 2.25 | 86 | 78.5 | 0.68 |

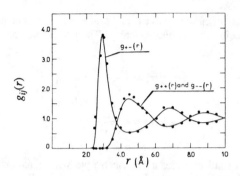

FIGURE XII.5. Pair correlation functions for molten KCl at 1043 K. Lines: HMSA results. Filled circles: MC data of Dixon and Gillan. (Zerah and Hansen, J. Chem. Phys., 1986)

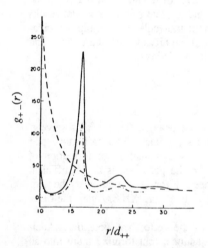

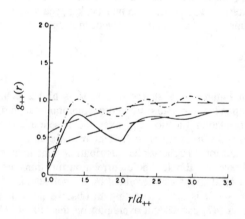

FIGURE XII.6. Solvation of ions for a mixture of hard ions and dipolar hard spheres. The unlike pair correlation function $g_{+-}(r)$ is at a reduced dipole moment $\mu^{*2} = 2.5$, $\rho d_{\mu}^3 = 0.6$, $\beta q^2/d_{++} = 136$, and $d_{\mu}/d_{++} = 0.68$. ——: 0.05M. - - -: PM at 0.05M - · · -: 0.4M. (Levesque, Weis, and Patey, J. Chem. Phys., 1980)

FIGURE XII.7. Solvation of ions for a mixture of hard ions and dipolar hard spheres. The like pair correlation function $g_{++}(r)$ is at a reduced dipole moment $\mu^{*2} = 2.5$, $\rho d_{\mu}^3 = 0.6$, $\beta q^2/d_{++} = 40$, and $d_{\mu}/d_{++} = 1$. ——: 0.1M. - · · -: 1 M. - - -: the corresponding PM results. (Levesque, Weis, and Patey, J. Chem. Phys., 1980)

molecules. Blum and coworkers [48] have solved the MSA equations for mixtures of hard ions and dipolar spheres. The results serve as first steps toward understanding the solvent granularity question.

## XII.8. *Simulation Results*

MC simulations of the primitive model of electrolyte solutions have appeared since the 1960s, e.g., Vorontsov-Vel'yaminov (1966) [49], Card and Valleau (1970) [50], Rasaiah et al. (1972) [51], Larsen (1974) [52], Patey and Valleau (1975) [53], Larsen (1976) [54], and Larsen and Rogde (1980) [55]. The simulation results proved invaluable in discriminating competing theories and revealing new structural features of electrolyte solutions. We shall select some of these works for discussion in the following.

Card and Valleau (1970) have carried out simulations for the 1:1 electrolytes at low concentrations (with the stoichiometric concentrations $C_{st}$ = 0.00911, 0.1038, 0.425, 1.0, and 1.968 M). The results are given in Table XII.1. They include the internal energy, osmotic coefficient, constant volume heat capacity, and contact values of the pcfs, $g_{++}$ and $g_{+-}$. (Note that since $d_{++} = d_{--}$, $g_{--} = g_{++}$). Simulations of the pcf were also given. Since they were for low concentrations, no special features were discovered.

Larsen [56] published a MC study of the RPM including the molten salt region. The results are summarized in Table XII.10. The quantity

$$\text{CONTACT} = 2y\,[g_{++}(d_{++}) + g_{+-}(d_{++})] \qquad (8.1)$$

where $y = \pi\rho d_{++}^3/6$, ($d_{++} = d_{--}$ and $d_{+-} = (d_{++} + d_{--})/2$). The Bjerrum length varies from the electrolyte solution regime ($B/d_{++}$ = 1.8823) to the molten salt regime ($B/d_{++}$ = 94.1103). One interesting behavior is that the like and unlike pcfs at high $B$ show formation of clusters (triplets and quadruplets, etc.) larger than the pairs (see Figure XII.10). At $r/d_{++}$ =2, $g_{++}(r)$ shows a narrow peak of size 1.21. This indicates the formation of the triplets + − +.

*Table XII.10. MC Values for High Bjerrum Lengths (Molten Salts)*

| y | Γ | −U/NkT | $g_A(d+)$ | $g_B(d+)$ | CONTACT | PV/NkT |
|---|---|---|---|---|---|---|
| 0.1498 | 2.0 | 0.839 | 0.64 | 2.36 | 0.90 | 1.62 |
| | 5.0 | 2.467 | 0.46 | 3.72 | 1.25 | 1.43 |
| | 49.98 | 33.62 | 0.09 | 31.32 | 9.41 | -0.79 |
| | 99.96 | 70.16 | 0.0 | 63.46 | 18.99 | -3.40 |
| 0.2507 | 2.0 | 0.783 | 1.42 | 3.14 | 2.28 | 3.02 |
| | 5.0 | 2.226 | 0.68 | 4.20 | 2.44 | 2.70 |
| | 50.04 | 28.61 | 0.03 | 15.70 | 7.89 | -0.65 |
| | 100.09 | 59.91 | 0.01 | 31.76 | 15.93 | -3.04 |
| 0.3503 | 2.0 | 0.756 | 2.38 | 3.75 | 4.30 | 5.04 |
| | 5.0 | 2.114 | 1.40 | 4.87 | 4.39 | 4.69 |
| | 50.02 | 26.54 | 0.09 | 13.24 | 9.34 | 1.49 |
| | 100.03 | 54.93 | 0.03 | 22.34 | 15.67 | -1.64 |
| 0.3945 | 2.0 | 0.711 | 2.60 | 4.94 | 5.94 | 6.71 |
| | 5.0 | 2.067 | 1.94 | 5.84 | 6.13 | 6.44 |
| | 49.998 | 25.71 | 0.21 | 12.70 | 10.18 | 2.61 |
| | 99.996 | 53.26 | 0.10 | 23.28 | 18.44 | 1.69 |

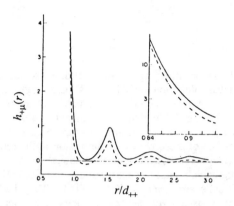

FIGURE XII.8. Solvation of ions for a mixture of hard ions and dipolar hard spheres. The ion-dipole correlation function $h_{+\mu}(r)$ is at a reduced dipole moment $\mu^{*2} = 2.25$, $\rho d_\mu^3 = 0.8$, $\beta q^2/d_{++} = 86$, and $d_\mu/d_{++} = 0.68$. ———: 0.1M. - - -: 1 M. (Levesque, Weis, and Patey, J. Chem. Phys., 1980)

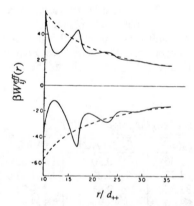

FIGURE XII.9. Effective ion-ion pair potential $\beta W_{ij}^{eff}$ at a reduced dipole moment $\mu^{*2} = 2.5$, $\rho d_\mu^3 = 0.6$, $\beta q^2/d_{++} = 136$, and $d_\mu/d_{++} = 0.68$. ———: $\beta W_{ij}^{eff}$ of Adelman (see text). - - -: McMillan-Mayer primitive model results. (Levesque, Weis, and Patey, J. Chem. Phys., 1980)

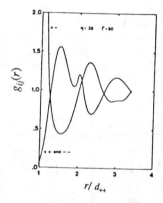

FIGURE XII.10. Formation of ion clusters in an PRM molten salt at packing fraction $y = 0.35$ and $B = 50$. The shoulder in $g_{sb++}$ at $r \approx 2.1d_{++}$ indicates the triplet structure $+ - +$. (Larsen, Chem. Phys. Letters, 1974)

The correlation functions for the 2:2 electrolytes are given in Figures XII.11 [57]. Various theoretical results (e.g., MSX, HNC, MSA, and YBG equations) are compared. The MSX equation is due to Outhwaite [58], i.e.,

$$g_{ij}(r) = g_{ij}^{HS}(r)\exp\left[\overline{C}_{ij}/g_{ij}^{HS}(r)\right] \tag{8.2}$$

where $\overline{C}_{ij}$ is the chain sum [59]. This function is a graph-theoretical entity, i.e., a sum of all graphs that have at least one $\phi$-bond and that are chains of $h_0$ and $\phi$, where $\phi(r) \equiv -\beta u_p(r)$. For the DH case, it is simply

$$\overline{C}_{ij}(r) = -\frac{z_i z_j e^2}{\varepsilon_m kT}\frac{e^{-\kappa r}}{r} \tag{8.3}$$

Figure XII.11 shows a peak of $g_{++}$ at $r = 8.4$ Å (about twice the ionic diameter $d_{++} = 4.2$ Å ). This is indication of triplet formation. It was estimated that at 0.0625 M, about 8% of the ions are in triplet form. The energy required for the formation of various ion clusters is presented in Table XII.11. Note that the energies required to form clusters larger than the triplets (e.g., quadruplets) are considerably higher.

## SCREENED COULOMB POTENTIAL

In the study of the DH theory, we have derived the averaged ("screened") Coulomb potential. The pair potential among the ions is of the Yukawa type:

$$u_{ij}(r) = \infty, \qquad\qquad r \le d_{ij} \tag{8.4}$$

$$= \frac{q_i q_j}{\varepsilon_m r_{ij}}\, e^{-z(r-d_{ij})}, \qquad\qquad r > d_{ij}$$

where $z$ is a parameter determining the range of interaction. So as $z= 0$, we shall recover the *bare* Coulomb interaction. For nonzero $z$ values, the Yukawa potential is short-ranged. The reasons for studying the Yukawa potential on the microscopic level are that

(i)    MC simulation can be done accurately in the absence of long-range forces.

(ii)   Coulomb interaction results can be recovered as $z \rightarrow 0$, thus the simulations have significance for ionic solutions.

(iii)  As shall become evident shortly, the performance of different theoretical approaches is altered (some getting better, some getting worse) as the range of intermolecular forces is shortened. For example, in the case of MSA, results deteriorate as $z$ is increased or when the range of interaction is curtailed.

Larsen and Rogde (1980) carried out MC simulations of the Yukawa potential for RPM. The energy values at three state conditions are presented in Table XII.12. We observe that the MSA (energy route) values are not satisfactory, whereas the theory called TT2A [60], gives much better results. These states are at $z = 1.5075$, 3.0, and 4.0. Thus as the pair potential becomes shorter-ranged, MSA(E) also diminishes in accuracy. The excess compressibility (over the HS value) is exhibited in Table XII.12. MSA(E) gives, for the most part, the best results (except at $B = 1.4191$, $z = 1.5075$, $\rho^* = 0.669$, where the virial pressure result is better). Figure XII.12 compares the like ($g_{++} = g_{--}$) and unlike ($g_{+-}$) pcfs. Both the MSA and EXP approximations fail as the $B$

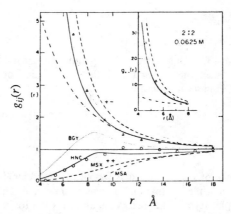

*FIGURE XII.11.* Pair correlation functions $g_{+-}(r)$ and $g_{++}(r) = g_{--}(r)$ for a 2:2 RPM electrolyte at 0.0625 M. Symbols: MC. ———: HNC. · · · : YBG. - · -: MSA. - - -: MSX (see text). (Valleau, Cohen, and Card, J. Chem. Phys., 1980)

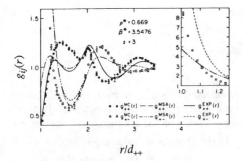

*FIGURE XII.12.* Pair correlation functions for the RPM with the screened Coulomb potential at $\rho^* = 0.669$, $B = 3.5476$, and $z = 3$. The first peak in $g_{++}$ is not well predicted by theories. (Larsen and Rogde, J. Chem. Phys., 1980)

Table XII.11. Relative electrostatic energies of ion clusters in the RPM. (Numbers are ratios of the energies over that of two like ions in contact.) (Valleau, Cohen, and Card, J. Chem. Phys., 1980)

| Cluster configuration | Relative stabilization |
|---|---|
| ⊕⊖ | −1.000 |
| ⊕⊖⊕ | −1.500 |
| ⊕⊖ ⊕ | −1.293 |
| ⊕⊕⊖⊖ | −2.333 |
| ⊕⊖ ⊕⊖ | −2.586 |
| ⊕⊖⊕⊖⊕ | −2.917 |
| ⊕⊖ ⊕⊕⊖ | −2.918 |
| ⊕⊕⊖⊕⊕⊖ | −3.700 |
| ⊖⊕⊖ ⊕⊕⊖ | −4.036 |
| ⊕⊕⊖ ⊕⊕⊖ | −4.066 |

is increased. The degree of disagreement is similar to the case of unscreened Coulomb potentials (i.e., the distinct double peaks of the $g_{++}(r)$ are not predicted by any of the theories). The EXP theory [61] gives the pcfs as

$$g_{ij}(r) = g_{ij}^{HS}(r)\exp[\overline{C}_{ij}(r)] \tag{8.5}$$

Table XII.12. Osmotic Pressure and Energy for the Screened Coulomb Potential

| z | B/d_ | Osmotic Pressures, $P_{osm}/\rho kT$ | | | Internal Energy, $-U/NkT$ | | |
|---|---|---|---|---|---|---|---|
| | | MC | MSA(E) | MSA(V) | MC | MSA(E) | TT2A |
| 1.5075 | 1.4191 | 0.8 | 0.24 | 0.64 | 0.935 | 0.741 | 0.889 |
| 3.0 | 3.5476 | 1.8 | 0.95 | 3.77 | 3.61 | 2.78 | 3.39 |
| 4.0 | 5.3197 | 3.2 | 1.79 | 8.59 | 6.57 | 5.08 | 6.22 |

Larsen and Rogde attempted to find a corresponding states principle between the Yukawa case and the unscreened Coulomb case without setting $z = 0$. However, no unique guideline could be found to match the correlations functions of the two potentials. By visually matching the structure ($g_{++}$ and $g_{+-}$) between $r = 1.0d_{++}$ and $r = 3.4d_{++}$, they determined the equivalent Bjerrum lengths in the Coulomb potential case. For example, for Yukawa potential at $B = 1.4191$, $\rho^* = 0.669$, and $z = 1.5057$, the correlation functions ($g_{++}$ and $g_{+-}$) match those of the Coulomb potential at $B = 3.5476$ and $\rho^* = 0.669$ ($z=0$).

## References

[1]   J.H. Hildebrand and R.L. Scott, The Solubility of Nonelectrolytes, Third edition (Reinhold, New York, 1924).

[2]   P. Debye and E. Hückel, Z. Physik 24, 195 and 305 (1923).

[3]   H. Falkenhagen, Electrolytes (Oxford University Press, London, 1934).

[4]   E. Mayer, J. Chem. Phys. 5, 67 (1937); and Ibid. 18, 1426 (1950).

[5]   N. Bjerrum, K. dan. Vidensk. Selsk. 7, No. 9 (1926); and Selected Papers, (Munskgaard, Copenhagen, 1949).

[6]   E.A. Guggenheim, Phil. Mag. 19, 588 (1926); and E.A. Guggenheim and J.C. Turgeon, Trans. Faraday Soc. 51, 747 (1955).

[7]   W.G. McMillan and J.E. Mayer, J. Chem. Phys. 13, 276 (1945).

[8]   O. Matsuoka, E. Clementi, and M. Yoshimine, J. Chem. Phys. 64, 1351 (1976).

[9]   H.C. Andersen and D. Chandler, J. Chem. Phys. 55, 1497 (1971); and H.C. Andersen, D. Chandler, and J.D. Weeks, Ibid., 56, 3812 (1972).

[10]  G. Stell, in Statistical Mechanics: Equilibrium Techniques, Part A, edited by B.J. Berne (Plenum, New York, 1977), pp.47-84.

[11]  A.R. Allnatt, Mol. Phys. 8, 533 (1964).

[12]  D.D. Carley, J. Chem. Phys. 46, 3783 (1967); also J.C. Rasaiah and H.L. Friedman, J. Chem. Phys. 48, 2742 (1968).

[13]  D.R. Corson and P. Lorrain, *Introduction to Electromagnetic Fields and Waves* (W.H. Freeman, San Francisco, 1962).

[14]  H.L. Friedman and W.D.T. Dale, in *Statistical Mechanics: Equilibrium Techniques*, Part A, edited by B. J. Berne, (Plenum, New York, 1977), pp.85-135.

[15]  C.W. Outhwaite, in *Statistical Mechanics*, edited by K. Singer, (Chemical Society, London, 1975), p.188.

[16]  P.J. Rossky, J.B. Dudowicz, B.L. Tembe, and H.L. Friedman, J. Chem. Phys. **73**, 3372 (1980).

[17]  J.G. Kirkwood and J.C. Poirier, J. Phys. Chem. **58**, 591 (1954).

[18]  M. Born, Z. Phys. **1**, 45 (1920).

[19]  N. Bjerrum, *Selected Papers*, (Munskgaard, Copenhagen, 1949).

[20]  E. Waisman and J.L. Lebowitz, J. Chem. Phys. **52**, 4037 (1970); also *Ibid.*, **56**, 3086 and 3093 (1972).

[21]  D.N. Card and J.P. Valleau, J. Chem. Phys. **52**, 6232 (1970).

[22]  G. Stell and B. Larsen, J. Chem. Phys. **70**, 361 (1979).

[23]  F. Hirata and K. Arakawa, Bull. Chem. Soc. Japan **48**, 2139 (1975).

[24]  D. Henderson and W.R. Smith, J. Stat. Phys. **19**, 191 (1978).

[25]  L. Blum, Mol. Phys. **30**, 1529 (1975).

[26]  L. Blum and J.S. Høye, J. Phys. Chem. **81**, 1311 (1977).

[27]  See, e.g., R. Triolo, J.R. Grigera, and L. Blum, J. Phys. Chem. **80**, 1858 (1976); R. Triolo, L. Blum, and M.A. Floriano, J. Chem. Phys. **67**, 5956 (1977); and J. Phys. Chem. **82**, 1368 (1978).

[28]  S. Watanasiri, M.R. Brulé, and L.L. Lee, J. Phys. Chem. **86**, 292 (1982).

[29]  See, e.g., H.L. Friedman, J. Solution Chem. **1**, 387 (1972).

[30]  L.H. Landis, Ph.D. Thesis, *Mixed Salt Electrolyte Solutions: Accurate Correlation for Osmotic Coefficients Based on Molecular Distribution Functions*, University of Oklahoma (1984).

[31]  B. Larsen, J. Chem. Phys. **65**, 3431 (1976).

[32]  L.L. Lee, J. Chem. Phys. **78**, 5270 (1983).

[33]  J.C. Rasaiah, D.N. Card, and J.P. Valleau, J. Chem. Phys. **56**, 248 (1972).

[34]  D.D. Carley, J. Chem. Phys. **46**, 3783 (1967).

[35]  J.C. Rasaiah and H.L. Friedman, J. Chem. Phys. **48**, 2742 (1968).

[36]  B. Larsen, J. Chem. Phys. **68**, 4511 (1978).

[37]  P.J. Rossky, J.B. Dudowicz, B.L. Tembe, and H.L. Friedman, J. Chem. Phys. **73**, 3372 (1980); and R. Bacquet and P.J. Rossky, J. Chem. Phys. **79**, 1419 (1983).

[38]   B. Larsen, J. Chem. Phys. **68**, 4511 (1978).

[39]   F.H. Stillinger and R. Lovett, J. Chem. Phys. **48**, 3858 and 3869 (1968).

[40]   J.C. Rasaiah and H.L. Friedman, J. Chem. Phys. **50**, 3965 (1969).

[41]   P.J. Rossky, J.B. Dudowicz, B.L. Tembe, and H.L. Friedman, J. Chem. Phys. **73**, 3372 (1980); and R. Bacquet and P.J. Rossky, J. Chem. Phys. **79**, 1419 (1983).

[42]   R. Bacquet and P.J. Rossky, *Ibid.* (1983).

[43]   C. Caccamo, G. Malescio, and L. Reatto, J. Chem. Phys. **81**, 4093 (1984).

[44]   J.P. Hansen and G. Zerah, Phys. Lett. **108A**, 277 (1985); also G. Zerah and J.P. Hansen, J. Chem. Phys. **84**, 2336 (1986).

[45]   F.J Rogers and D.A. Young, Phys. Rev. **A30**, 999 (1984).

[46]   D. Levesque, J.-J. Weis, and G.N. Patey, J. Chem. Phys. **72**, 1887 (1980).

[47]   S.A. Adelman and J.H. Chen, J. Chem. Phys. **70**, 4291 (1979).

[48]   L. Blum, J. Stat. Phys. **18**, 451 (1978); F. Vericat and L. Blum, *Ibid.* **22**, 593 (1980); and Mol. Phys. **45**, 1067 (1982).

[49]   P.N. Vorontsov-Vel'yaminov and V.P. Chasovskikh, High Temperatures (USSR) **13**, 1071 (1975); P.N. Vorontsov-Vel'yaminov, H.M. El'yashevich, and A.K. Kron, Elektrokhimiya **2**, 708 (1966); also P.N. Vorontsov-Vel'yaminov and H.M. El'yashevich, *Ibid.* **4**, 1430 (1968).

[50]   D.N. Card and J.P. Valleau, J. Chem. Phys. **52**, 6323 (1970).

[51]   J.C. Rasaiah, D.N. Card, and J.P. Valleau, J. Chem. Phys. **56**, 248 (1972).

[52]   B. Larsen, Chem. Phys. Letters **27**, 47 (1974).

[53]   G.N. Patey and J.P. Valleau, J. Chem. Phys. **63**, 2334 (1975).

[54]   B. Larsen, J. Chem. Phys. **65**, 3431 (1976).

[55]   B. Larsen and S.A. Rogde, J. Chem. Phys. **72**, 2578 (1980).

[56]   B. Larsen, *Ibid.* (1974).

[57]   J.P. Valleau, L.K. Cohen, and D.N. Card, J. Chem. Phys. **72**, 5942 (1980).

[58]   C.W. Outhwaite, Chem. Phys. Lett. **37**, 383 (1976).

[59]   See e.g., H.C. Andersen, in *Statistical Mechanics: Equilibrium Techniques*, Part A, edited by B.J. Berne (Plenum, New York, 1977)

[60]   B. Larsen, G. Stell, and K.C. Wu, J. Chem. Phys. **7**, 530 (1977).

[61]   H.C. Andersen and D. Chandler, J. Chem. Phys. **57**, 1918 (1972).

[62]   R.A. Robinson and R.H. Stokes, *Electrolyte Solutions*, 2nd Edition (Butterworth, London, 1970).

## Exercises

1.  Derive the Debye-Hückel thermodynamic quantities, such as energy, osmotic pressure, and chemical potential by using the radial distribution function $g_{ij}$ (3.23) in the energy and pressure equations from Chapter V. Check your results against those listed in Section XII.4.

2.  Plot the mean activity coefficient $\ln \gamma_\pm$ at 25°C as a function of ionic strength for NaCl from 0 to 5M, using DH formula. Compare with experimental data (see e.g., Robinson and Stokes [62]). Discuss the deficiencies of the DH theory at high concentrations.

3.  Use the fast Fourier transform technique to solve the HNC equation for NaCl solution at the same conditions. You may use the PM model and the Pauling radii for $Na^+$ and $Cl^-$. Calculate the osmotic pressure from the virial theorem and compare with experimental data. State explicitly the assumptions made and discuss the results.

4.  Two new integral equations, the RHNC and the HMSA, have been proposed. Apply these equations to the NaCl solution at 25°C and a number of molarities. Compare with the results from the HNC equation.

5.  Survey the literature to find alternative theories for the mean activity coefficients of 2:2 electrolytes. Discuss their merits and weaknesses.

6.  The hard ion models presented here could equally be applied to molten salts and two-component plasmas. Consult the literature to find applications of the present theories to these fluids. Discuss the similarities to and differences from the electrolyte solutions.

# CHAPTER XIII

# MOLECULAR DYNAMICS

In statistical mechanics, we learn that the measured *gross* properties of a many-particle system are averages over many microstates consistent with the constraints imposed on the system. For atomic fluids, these microstates take place on the time scale of picoseconds (e.g., for argon, the relaxation time is on the order of $10^{-13}$ seconds). The molecular dynamics (MD) method is designed to generate these microstates. Statistical averages could then be taken over these states to obtain equilibrium and dynamic properties. In particular, given $N$ particles (typically, $N= 100 \sim 1000$), Newton's equations of motion are solved to generate the time trajectories in the phase space. Numerical methods (such as the predictor-corrector methods) are used to solve the coupled dynamic equations. At a given $t$, the solution gives an instantaneous picture of the system (the spatial configuration and the velocity distribution). This picture constitutes a microstate. The situation is like a motion picture. The phase trajectory, ordered in time, is constituted of thousands of microstates and gives the time evolution of the system. The difference between the molecular dynamics and the Monte Carlo (MC) method is that the connection between successive microstates in MD is given by Newton's equations of motion (i.e., obeying classical force laws), while in MC there is no prescribed order in the sequence of events (i.e., random walks). In MD, normally 2000 to 20000 time steps are generated on a computer. The positions and velocities of $N$ particles are stored in memory. Later, the trajectories are analyzed, much like at the census bureau, by taking averages of properties such as energy, pressure, diffusion rates, neighbor distributions, etc., on the population. Time averaging is usually adopted as the means of obtaining statistical averages.

We have described the general outlines of the MD method. There are specific considerations when the method is implemented. First the force laws have to be selected by the user. One wishes to apply a law that closely describes the substance to be simulated, e.g., argon or chlorine. Next, one would fix the system at certain physical state, such as at constant temperature and pressure. This is done by choosing a representative ensemble. Early MD simulations were carried out in the microcanonical ensemble (at constant $N$, $V$, and $E$) primarily because this ensemble is easiest to generate when one uses Hamilton's equations. In the 1980s, new algorithms appeared that could produce trajectories in the canonical ($NVT$) and isothermal-isobaric ($NPT$) ensembles. In addition, nonequilibrium states could be simulated when, for example, a constant shear rate is imposed over the element cell. This is called the NEMD (nonequilibrium molecular dynamics) method [1]. Aspects of these new methods are not elementary and will not be covered in this text. We shall discuss the microcanonical method due to Rahman [2] and the isothermal ensemble due to Nosé [3] in the following.

## XIII.1. *Time Averages and Ensemble Averages: Ergodicity*

We define a dynamic variable $A$ of an $N$-body system as a physical quantity that depends on the positions $\mathbf{r}^N$ and momenta $\mathbf{p}^N$ of the $N$ particles. For example, the total energy and total momentum are dynamic variables. The value of an experimentally measured property in our statistical approach is the result of averaging many microscopic states, or fluctuations. We have already discussed the ensemble average in previous chapters. For a canonical ensemble, the probability of the occupancy of a cell $i$ in the phase space is related to its energy value $\varepsilon_i$

$$dp_i^{(N)}(\mathbf{r}^N,\mathbf{p}^N) = \frac{1}{Z_N N! h^{3N}} \; \exp\left[-\beta H_N(\mathbf{r}^N,\mathbf{p}^N)\right]_{\text{cell } i} \, d\mathbf{r}^N \, d\mathbf{p}^N \tag{1.1}$$

with $(\mathbf{r}^N, \mathbf{p}^N)$ in cell $i$. $H_N$ is the $N$-body Hamiltonian having the energy value $\varepsilon_i$ in cell $i$. The ensemble average $<A>$ is defined as

$$<A> \equiv \frac{1}{Z_N N! h^{3N}} \int d\mathbf{r}^N \, d\mathbf{p}^N \, A(\mathbf{r}^N,\mathbf{p}^N) \exp\left[-\beta H_N(\mathbf{r}^N,\mathbf{p}^N)\right] \tag{1.2}$$

Since we are able to follow the time evolution of the $N$-body system, i.e., to track the instantaneous positions $\mathbf{r}^N(t)$ and momenta $\mathbf{p}^N(t)$ of $N$ particles with time, we could also arrive at a *time average* $<A>_t$ of A:

$$<A>_t \equiv \lim_{T \to \infty} \frac{1}{T} \int_0^T dt \; A(\mathbf{r}^N(t),\mathbf{p}^N(t)) \tag{1.3}$$

The question arises as to what the relation is between $<A>$ and $<A>_t$. Conceptually, the fluctuations of microstates averaged over time should yield the mean value of a dynamic variable that is consistent with averaging over all possible distributions of these states. Thus we propose the *ergodic condition* [4]: $<A> = <A>_t$. Not all systems are ergodic [5]. Here we assume that the systems we study satisfy *ergodicity*, and time averages coincide with ensemble averages. We use $<\cdot>$ for both $<\cdot>$ and $<\cdot>_t$.

## XIII.2. *Equations of Motion*

We have already derived the equations of motion of an $N$-body system with the Hamiltonian $H_N(\mathbf{r}^N, \mathbf{p}^N)$ in Chapter I:

$$\dot{\mathbf{r}}_i = \frac{\partial H_N}{\partial \mathbf{p}_i}, \qquad \dot{\mathbf{p}}_i = -\frac{\partial H_N}{\partial \mathbf{r}_i} \tag{2.1}$$

With initial conditions $\mathbf{r}_i(0)=\mathbf{r}_i^0$, $\mathbf{p}_i(0)=\mathbf{p}_i^0$, $i=1,...,N$, eqs. (2.1) give a unique trajectory through the phase space describing the evolution of the $N$-body system. We have also shown that for a Hamiltonian with pairwise additive potential,

$$H_N(\mathbf{r}^N, \mathbf{p}^N) = \sum_i^N \frac{p_i^2}{2m} + \sum_{i<j}^N \sum^N u(r_{ij}), \qquad r_{ij} = |\mathbf{r}_j - \mathbf{r}_i| \tag{2.2}$$

Equations (2.1) become

$$\dot{\mathbf{r}}_i = \frac{\mathbf{p}_i}{m}, \qquad \dot{\mathbf{p}}_i = - \sum_{j \neq i} \nabla u(r_{ij}) \tag{2.3}$$

Combining these two equations gives the usual Newton equation of motion

$$\ddot{\mathbf{r}}_i(t) = - \frac{1}{m} \sum_{j \neq i} \nabla u(r_{ij}) \tag{2.4}$$

This is the basis of molecular dynamics simulation. Once we have solved eq. (2.4), the momenta or the velocities $v_i(t)$ can be obtained from

$$\mathbf{v}_i(t) = \dot{\mathbf{r}}_i(t) \tag{2.5}$$

Any dynamic variable $A(\mathbf{r}^N(t), \mathbf{p}^N(t))$ can then be averaged in time:

$$<A> = \lim_{T \to \infty} \frac{1}{T} \int_0^T dt\, A(\mathbf{r}^N(t),\, \mathbf{p}^N(t)) \tag{2.6}$$

For example, the average of the kinetic energy $KE = \sum_i p_i^2/2m$ is

$$<KE> = \lim_{T \to \infty} \frac{1}{T} \int_0^T dt \sum_i \frac{p_i^2(t)}{2m} \tag{2.7}$$

Since time average is equal to ensemble average in ergodic systems, we get

$$<KE> = \frac{1}{Z_N N! h^{3N}} \int d\mathbf{r}^N\, d\mathbf{p}^N \sum_i \frac{p_i^2}{2m} \exp\left\{ -\beta \sum_j \frac{p_j^2}{2m} - \beta V_N(\mathbf{r}^N) \right\} \tag{2.8}$$

$$= \frac{3}{2} NkT$$

Therefore, by carrying out the averaging in (2.7), we actually obtain the temperature $T$ of the system. We note also that (2.8) gives the kinetic contribution to the internal energy.

## XIII.3. *Algorithms of Molecular Dynamics*

Equation (2.4) shows that for an $N$-body system, its dynamics is determined by the pair potential $u(r)$. Once we are given a $u(r)$, the time evolution of the $N$ particles (the positions $r^N(t)$ and momenta $p^N(t)$) is completely determined. The task now is to solve (2.4) by numerical methods. There are two common procedures in use: (i) the predictor-corrector method due to Rahman (*Ibid.*); (ii) the leap-frog method due to Verlet [6]. We discuss them separately in the following.

The predictor-corrector method is standard in numerical analysis [7] and has many versions. First, we know the positions $r_i(t_{n-1})$ at time $t_{n-1}$, and the positions $r_i(t_n)$, velocities $v_i(t_n)$, and accelerations $a_i(t_n)$ at time $t_n = t_{n-1} + \Delta t$. The predictor formula for the positions at time $t_{n+1} = t_n + \Delta t$ is

$$\hat{r}_i(t_{n+1}) = r_i(t_{n-1}) + 2\Delta t\, v_i(t_n) \tag{3.1}$$

The new positions $\hat{r}_i(t_{n+1})$ permit the calculation of the interaction forces on particle $i$ from the given potential $u(r)$, and therefore also the accelerations $\hat{a}_i(t_{n+1})$. The corrector formulas are then

$$v_i(t_{n+1}) = v_i(t_n) + \frac{1}{2}\,\Delta t[\hat{a}_i(t_{n+1}) + a_i(t_n)] \tag{3.2}$$

$$r_i(t_{n+1}) = r_i(t_n) + \frac{1}{2}\,\Delta t[v_i(t_{n+1}) + v_i(t_n)]$$

This process can be repeated until the predicted and the corrected values of $r_i(t_{n+1})$ differ by less than a prescribed value. Choosing the time step $\Delta t = 10^{-14}$ second for argon, Rahman [8] showed that for 864 particles in a cubic box the results after one iteration through the predictor-corrector formulas did not differ appreciably from the results after two iterations.

Verlet [9] on the other hand used a second-order difference formula which bypassed the need for evaluating the velocities:

$$r_i(t_{n+1}) = -r_i(t_{n-1}) + 2r_i(t_n) + \frac{\Delta t^2}{m} F_i(t_n) \tag{3.3}$$

where $F_i(t_n)$ is the force on the $i$th particle at time $t_n$. The derivation of (3.3) is left as an exercise. For pairwise additive potentials,

$$F_i(t_n) = -\sum_{j \neq i} \nabla_i u(r_{ij}) \tag{3.4}$$

The truncation error of formula (3.3) is of the order of $(\Delta t)^3$. To obtain the velocities, we use the Lagrange three-point formula

$$v_i(t_n) = \frac{1}{2\Delta t}[r_i(t_{n+1}) - r_i(t_{n-1})] \tag{3.5}$$

This method is very efficient, since no iteration is required.

## XIII.4. *Formulas for Equilibrium Properties*

Macroscopic properties are obtained as time averages of microscopic quantities. The following quantities are readily calculated in a molecular dynamics simulation.

**TEMPERATURE**

Let us arbitrarily select one particle with tag "1". Its velocity is $v_1(t)$. The temperature of the system is obtained through the formula

$$\lim_{T' \to \infty} \frac{1}{T'} \int_0^{T'} dt \ \mathbf{v}_1(t) \cdot \mathbf{v}_1(t) = \frac{3kT}{m} \tag{4.1}$$

where $k$ is the Boltzmann constant and $m$ is the mass of the particle. We can easily prove this by changing to ensemble averaging (i.e., using ergodic properties)

$$\frac{1}{Z_N N! h^{3N}} \int d\mathbf{r}^N \ d\mathbf{p}^N \ (\mathbf{v}_1 \cdot \mathbf{v}_1) \exp\left[ -\beta \sum_i \frac{p_i^2}{2m} - \beta \sum_{i<} \sum_j u(r_{ij}) \right] = \frac{3kT}{m} \tag{4.2}$$

**INTERNAL ENERGY**

The instantaneous total energy TE($t$) of the system is composed of two parts: the kinetic energy KE($t$) $= \sum_i m v_i(t)^2/2$ and the potential energy PE($t$) $= \sum_i \sum_j u(r_{ij}(t))$ (for pairwise additive potentials). Thus, the time average is given by

$$<U> = <TE>_t = <KE>_t + <PE>_t \tag{4.3}$$

$$= <\frac{1}{2} \sum_i^N m v_i(t)^2>_t + <\sum_{i<}^N \sum_j^N u(r_{ij}(t))>_t$$

We have already shown that the time average of the kinetic part is equal to $3NkT/2$, i.e., the ideal-gas contribution. The potential (configurational) part of the internal energy $<U^c>$ can be calculated as

$$<U^c> \equiv <\sum_{i<} \sum_j u(r_{ij}(t))>_t = \lim_{T' \to \infty} \frac{1}{T'} \int_0^{T'} dt \ \sum_{i<} \sum_j u(|\mathbf{r}_j(t) - \mathbf{r}_i(t)|) \tag{4.4}$$

**PRESSURE**

The pressure is calculated from the virial theorem

$$\frac{\beta P}{\rho} = 1 - \frac{1}{6NkT} \left[ \lim_{T' \to \infty} \frac{1}{T'} \int_0^{T'} dt \ \sum_{i<}^N \sum_j^N r_{ij}(t) u'(r_{ij}(t)) \right] \tag{4.5}$$

where

$$u'(r) = \frac{du}{dr}, \qquad r_{ij}(t) = |\mathbf{r}_j(t) - \mathbf{r}_i(t)|$$

Again, simplified statistical formulas could be used.

## RADIAL DISTRIBUTION FUNCTIONS

To calculate $g(r)$, we use the interpretation that $\rho g(r_{12})4\pi r_{12}^2\, dr_{12}$ gives the number of molecules surrounding a central molecule 1 in a spherical shell of thickness $dr$ and at a distance $r_{12}$ from the center:

$$\rho g(r)4\pi r^2\, \Delta r = \lim_{T'\to\infty}\frac{1}{T'}\int_0^{T'}dt\, \Delta N(r,t) \tag{4.6}$$

where $\Delta N(r,t)$ is the instantaneous number of molecular centers found in the shell $4\pi r^2\, \Delta r$ at a distance $r$ from molecule 1.

Several thermodynamic properties, such as entropy and Helmholtz free energy, are not averages of dynamic variables. For example, entropy is a phase space property, related to the phase volume (or probability distribution). It does not depend on $r^N$ and $p^N$ directly. Thus it could not be calculated by molecular dynamics. However, entropy could be evaluated from pressure or energy through the relations of thermodynamics.

# XIII.5. *Calculation of Transport Properties*

One remarkable feature of molecular dynamics simulation is that it allows, at least in principle, the calculation of transport properties of an $N$-body system. Conventional approach used the Green-Kubo formulas [10] in linear response theory. Recently, the NEMD methods have been used. The transport properties of an $N$-body system are completely determined by its Hamiltonian. In the following, we shall give the Kubo formulas for the self-diffusion coefficient, viscosities (shear viscosity and bulk viscosity), and thermal conductivity. We will not give the derivations, as they could be found in standard texts [11].

## SELF DIFFUSION COEFFICIENT D

The self-diffusion coefficient is given by the integral of the velocity autocorrelation function as

$$D = \frac{1}{3}\int_0^\infty dt\, <\mathbf{v}_1(0)\cdot\mathbf{v}_1(t)> \tag{5.1}$$

where the velocity autocorrelation function is defined as

$$<\mathbf{v}_1(0)\cdot\mathbf{v}_1(t)> \equiv \lim_{T'\to\infty}\frac{1}{T'}\int_0^{T'}ds\, \mathbf{v}_1(s)\cdot\mathbf{v}_1(s+t) \tag{5.2}$$

It represents the time average of the correlation of the velocities of particle 1 at two different times, i.e., $s_0$ and $s_1$, separated by an interval $t = s_1 - s_0$. In other words, it is a measure of the *memory* of a particle of its previous velocity. Thus it is called the velocity autocorrelation function.

## SHEAR VISCOSITY $\eta$

The Kubo formula for shear viscosity is

$$\eta = \frac{1}{kTV} \int_0^\infty dt \, <J_v(0)J_v(t)> \tag{5.3}$$

where $J_v$ is the momentum flux

$$J_v = \sum_j \left[ \frac{p_j^x p_j^y}{m} + F_j^x r_j^y \right] \tag{5.4}$$

The summation is over all $N$ molecules in the system and the superscripts $x$ and $y$ denote the $x$-component and $y$-component of the vectors $\mathbf{p}$ (momentum), $\mathbf{F}$ (force), and $\mathbf{r}$ (position). For example, $p_j^x$ is the $x$-component of the momentum of the $j$th particle, $F_j^x$, the $y$-component of the force acting on the $j$th particle, and $r_j^y$, the $y$-coordinate of the position of the $j$th particle. Recall also that the force $\mathbf{F}_j$ on the $j$th particle can be obtained from the pair potential by

$$\mathbf{F}_j = -\sum_{i \neq j} \nabla_j u(r_{ji}) \tag{5.5}$$

## BULK VISCOSITY $\zeta$

$$\zeta = \frac{1}{9kTV} \int_0^\infty dt \, <J_b(0)J_b(t)> \tag{5.6}$$

where the momentum flux $J_b$ is given by

$$J_b = \mathbf{Tr} \left\{ \sum_j^N \left[ \frac{\mathbf{p}_j \mathbf{p}_j}{m} + \mathbf{F}_j \mathbf{r}_j \right] - \mathbf{I} \left[ PV + \left[ \frac{\partial PV}{\partial <E>} \right]_T (TE - <U>) \right] \right\} \tag{5.7}$$

where, again, $TE$ is the instantaneous total energy, $TE = \sum_j mv_j^2/2 + \sum_i \sum_j u(r_{ij})$, $\mathbf{I}$ is the unit tensor, $\mathbf{Tr}$ stands for the trace, and $<U>$ is the internal energy.

## THERMAL CONDUCTIVITY $\kappa$

$$\kappa = \frac{1}{3kT^2V} \int_0^\infty dt \, <J_T(0)J_T(t)> \tag{5.8}$$

where the energy flux $J_T$ is

$$J_T = \sum_i \frac{p_i^2}{2m} \frac{\mathbf{p}_i}{m} + \sum_{i<}\sum_j (\mathbf{I}u(r_{ij}) + \mathbf{F}_{ij}\mathbf{r}_{ij}) \cdot \frac{\mathbf{p}_i}{m} - \sum_i h_i \frac{\mathbf{p}_i}{m} \qquad (5.9)$$

where $h_i$ is the equilibrium enthalpy per particle for the $i$th particle, and $\mathbf{F}_{ij}\mathbf{r}_{ij}$ is a matrix with diagonal elements $F_{ij}^x x_{ij}$, $F_{ij}^y y_{ij}$, and $F_{ij}^z z_{ij}$ [11]. We see that in all Kubo formulas, an autocorrelation function of the form $<A(0)A(t)>$ is used. Therefore, the study of auto-correlations is an important part of the statistical mechanics of transport properties.

# XIII.6. *Techniques of Computer Simulation*

The system of eqs. (2.1) can be solved numerically on a computer using the procedures outlined in XIII.3. In carrying out the simulation, numerical techniques for saving time and simplifying formulas become important.

### INITIAL CONDITIONS

Typically few hundred to few thousand particles are employed in computer simulation. The choice is dictated by the statistics desired, the size of the memory core, and the execution time on the computer. On the other hand, such relatively small number of particles is already sufficient in simulating most real gas properties. After deciding upon the pair potential $u(r)$, we write down the Hamiltonian $H_N$:

$$H_N(\mathbf{r}^N, \mathbf{p}^N, \cdots) = \sum_{i=1}^n \frac{p_i^2}{2m} + \sum_{i<}^N\sum_j u(r_{ij}) + \text{(other } KE) + \text{(other } PE) \qquad (6.1)$$

The other kinetic energy terms involve the rotational energy (RE) and vibrational energy (VE). For example, RE is needed in diatomics simulation. If there are three-body forces, they should be included also.

Next we select the state condition to be studied: for example, in $NVE$ ensemble, the density and the velocity distribution. We choose a particle number, say, $N=500$. The density $\rho$ is related to $N$ and the volume by $\rho = N/V$. We take a cube of sides $L$, $L^3 = V = N/\rho$. For example, for argon at $\rho\sigma^3 = 0.65$, $L/\sigma = V^{1/3}/\sigma = 9.16$. The temperature cannot be preset in MD simulation, although in practice it is often adjusted during execution through velocity scaling. This scaling was the impetus of Nosé's Hamiltonian formulation. We shall have more to say on Nosé's method later when discussing isothermal ensembles.

### INITIAL CONFIGURATION

The $N$ (=500) particles are distributed at positions $\mathbf{r}_1, \mathbf{r}_2,...,\mathbf{r}_N$ in a cubic box of, say, $L = 9.16\sigma$ in such a way as to give a large value of the Boltzmann factor $\exp[-\beta V_N(\mathbf{r}_1,...,\mathbf{r}_N)]$; i.e., the system is at a very probable state in the phase space. In practice, however, one often arranges the $N$ particles on a face-centered cubic (f.c.c.) lattice. Inside the cubic box, we fill with unit cells each containing four molecules located at

$$\mathbf{r}_1 = (0, 0, 0), \quad \mathbf{r}_2 = (0, \frac{b}{2}, \frac{b}{2}), \quad \mathbf{r}_3 = (\frac{b}{2}, 0, \frac{b}{2}), \quad \mathbf{r}_4 = (\frac{b}{2}, \frac{b}{2}, 0) \qquad (6.2)$$

where $b$ is the length of the unit cell. In the example $N=500$, there are 125 such cells, with five cells aligned along the the $x, y$, and $z$-directions. Thus the length $b$ is $L/5 \approx 1.83\sigma$.

The velocities $v_1, \ldots, v_N$ are assigned to the N particles with a Maxwellian distribution; i.e., for any particle $i$, the probability of having the velocity value $v$ is given by

$$f_0(\mathbf{v}) = \rho \left[ \frac{m}{2\pi kT} \right] \exp \left[ \frac{m}{2kT} (v - v_o)^2 \right] \tag{6.3}$$

Since $v = (v_x, v_y, v_z)$, all three components obey the same type of distribution. As motion is started, the lattice begins to "melt." Since the initial configuration, the f.c.c. lattice, is not an equilibrium state, one should not collect statistical information right away. Typically, $10^2 \sim 10^4$ initial time steps are discarded. The system is then tested for equilibrium (e.g., through the constancy of temperature). After ascertainment of equilibration, property statistics are then accumulated.

## PERIODIC BOUNDARY CONDITION

Sooner or later, the motion will gradually but surely take the particles outside the cubic box (a leaking box at that). To make up for this leakage, we use periodic boundary conditions. The periodicity also minimizes surface effects. For small systems with walls, the surface influence, if not controlled, would be inordinately large. For any particle at position $(x,y,z)$ in the box, we construct 26 images at

$$(x \pm L, y, z) \quad (x \pm L, y \pm L, z) \quad (x \pm L, y \pm L, z \pm L) \tag{6.4}$$

$$(x, y \pm L, z) \quad (x \pm L, y, z \pm L)$$

$$(x, y, z \pm L) \quad (x, y \pm L, z \pm L)$$

We have a total of 27 boxes, with the one in the center as the standard. When a particle leaves the standard box from one surface, its image re-enters the opposite surface. A two-dimensional case is depicted in Figure XIII.1. This setup conserves the number of particles in the standard box at all times. The statistics are taken only with regard to the center box. It effective simulates an infinite system. The surface effects are eliminated. On the other hand, it introduces some other limitations:

1.  Because the number of particles in the box is constant, it is impossible to study the thermodynamic states in which thermal fluctuations have correlation lengths of the order of the box size, $L$. Thus critical phenomena are excluded.

2.  The time scale is limited to times shorter than $L/c$, where $c$ is the velocity of sound. For times longer than this value, disturbances which arise in one region of the box can traverse the box and re-enter through the opposite surface, leading to spurious recurrence contributions.

3.  The collective properties that can be studied are limited to discrete wave vectors; i.e.,

$$k = \frac{2\pi}{L} n, \qquad n = 1, 2, \ldots \tag{6.5}$$

with minimum wave number

$$k_0 = \frac{2\pi}{L} \tag{6.6}$$

This is due to the finite dimension $L$ of the system. For potentials such as the LJ potential, the periodic images introduce questions on the truncation of the range of interaction. At least two conventions are used [12]:

1.  *The Minimum Image Convention*: Given a particle situated anywhere in the standard box, an *imaginary* box of the same size as the standard box is formed centered on the given particle (see Figure XIII.1). The interaction forces are evaluated between this particle and the $N-1$ particles inside this imaginary box, including some that are image particles. This is equivalent to truncating the potential at the surface of the box. Exactly one image of each of the other particles is allowed to interact. The procedure is applied to all particles in the standard box.

2.  *The Spherical Cutoff Convention*: In this method, one takes into account only the particles inside a sphere of cutoff radius $r_c$ centered on the given particle (see Figure XIII.1). The diameter of the sphere is commonly taken to be the length of the standard box. This ensures that at most one image of each particle is used in evaluating the force. Smaller distances are used when the sample size is large, thus reducing the expense of simulation.

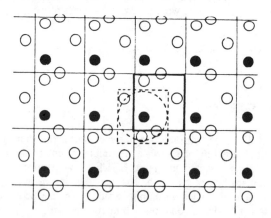

FIGURE XIII.1. *Periodic boundary conditions in two dimensions. The "standard" box is lined with heavy boundaries. Molecules in the image boxes have exactly the same spatial and velocity distribution as in the standard box. In the minimum image convention, the neighbors in the dotted boundaries are counted. In the spherical cutoff convention, only those within the dotted circle are counted. (Valleau and Whittington, Plenum, 1977)*

## THE TIME INTERVAL

The time interval $\Delta t$ between successive dynamic states is determined by the type of molecule as well as the specific relaxation process (e.g., translation or rotation) being studied. For example, for argon, the LJ parameters are $\sigma = 3.405$ Å, and $\varepsilon/k = 119.8$ $K$. Equation (3.3) can be written in the dimensionless form (with $r^* = r/\sigma$)

$$\mathbf{r}_i^*(t+\Delta t) = -\mathbf{r}_i^*(t-\Delta t) + 2\mathbf{r}_i^*(t) \tag{6.7}$$

$$-(\Delta t)^2 \left[\frac{48\varepsilon}{m\sigma^2}\right] \frac{\mathbf{r}_{ij}^*}{r_{ij}^*} \sum_{j \neq i} \left[(r_{ij}^*)^{-13} - \frac{1}{2}(r_{ij}^*)^{-7}\right]$$

Therefore,

$$\tau_0 = \left[\frac{m\sigma^2}{48\varepsilon}\right]^{1/2} \tag{6.8}$$

appears as a natural unit of time. Inserting the values of $\sigma$, $\varepsilon/k$, and $m=6.63 \times 10^{-23}$ g for argon, we obtain $\tau_0 = 3.112 \times 10^{-13}$ s. The choice of $\Delta t$ is usually made on the basis of numerical stability. A reduced time step (in units of $\tau_0$) $\Delta t^* = 0.03$ or $\Delta t = 10^{-14}$ s is used for argon. If rotational motion is present, the relaxation occurs at a time scale an order of magnitude faster. Shorter time steps must be used, on the order of $10^{-15}$ s.

## CHECKS ON ACCURACY

The success of a molecular dynamics simulation rests on minimizing the numerical errors that might accumulate with time. Several methods are used to check the numerical accuracy:

### Time Reversal

At intermediate time, say $t_n$, the velocities of the particles, $v_i(t_n)$, are changed to $-v_i(t_j)$, $\forall\ i$, and the calculation proceeds backwards for $n$ steps. The final result should repeat exactly the initial values of positions $r_i$ and velocities $v_i$ (the is called microscopic reversibility). Discrepancies reflect numerical roundoff errors.

### Constant Energy and Momentum

The total energy and total momentum are conserved quantities in a microcanonical trajectory. The constancy of energy could be demonstrated for the Hamiltonian by using the Hamilton's equations of motion. The total time derivative $dH_N(t)/dt$ for $H_N(t) = H_N(\mathbf{r}^N(t), \mathbf{p}^N(t))$ is

$$\frac{dH_N(t)}{dt} = \sum_i \left[\frac{\partial H_N}{\partial \mathbf{p}_i}\frac{d\mathbf{p}_i}{dt} + \frac{\partial H_N}{\partial \mathbf{r}_i}\frac{d\mathbf{r}_i}{dt}\right] \tag{6.9}$$

$$= \sum_i \left[\dot{\mathbf{r}}_i\dot{\mathbf{p}}_i - \dot{\mathbf{p}}_i\dot{\mathbf{r}}_i\right] = 0$$

where we have used the relation (2.1). Thus energy (represented by the Hamiltonian) is a constant of motion in Hamiltonian dynamics. The simulation is in a microcanonical ensemble. Variations in energy and momentum with time indicate inaccuracies or mistreatment of the problem. For example, if the cutoff distance $r_c$ for the potential function is too small, there will be a drain (or gain) of the energy.

## Constant Temperature

The mean kinetic energy per particle is computed and its variations with time monitored. The value should fluctuate about the desired temperature. A systematic drift in this quantity may reflect, among other things, that the system is not at equilibrium.

## Order Parameter

An order parameter $\zeta_k(t)$ is defined [13] as

$$\zeta_k(t) \equiv \frac{1}{N} \sum_{1=i\neq j}^{N} \cos(kr_{ij}) \tag{6.10}$$

where $k$ is the reciprocal vector for the corresponding lattice sites, e.g., in the f.c.c. lattice discussed above. For an ordered system, $\zeta_k(t)$ fluctuates around $N-1$; for liquids, $\zeta_k(t)$ fluctuates around zero.

## BULK PROPERTIES

To evaluate the time average of a dynamic variable $A(t)$

$$<A>_t = \lim_{T\to\infty} \frac{1}{T} \int_0^T dt\, A(t) \tag{6.11}$$

We collect the values of $A(t_j)$ at $M$ successive times, $t_1, t_2,..., t_M$, $M$ being as large as feasible (or $M$ being so large that any collection of $A(t_j)$ larger than $M$ values will not alter the average value significantly). Then the average is approximated by

$$<A>_t \approx \frac{1}{M} \sum_{j=0}^{M} A(t_j) \tag{6.12}$$

As the time correlation functions $<A(0)A(t)>$ are useful quantities, they are obtained as

$$<A(0)A(t)> = \frac{1}{M-n} \sum_{j=0}^{M-n} A(t_j)A(t_{j+n}) \tag{6.13}$$

where $t=n\Delta t$.

## TIME SAVING TECHNIQUES

### Neighbor Listing

This is a bookkeeping method first proposed by Verlet [14]. For a given interaction potential $u(r)$, one does not calculate the interaction forces for $r \to \infty$ for all particles, since this would require enormous computer time. Usually a cutoff distance $r_c$ is judiciously chosen beyond which no account of interaction is taken. For LJ potential, the common practice is $r_c \approx 2.5\sigma$. Verlet proposed a second radius, $r_v > r_c$. Then for each molecule $i$, we make a list $L_i$ of all other molecules inside a sphere of radius $r_v$ centered around molecule $i$. For $N$ molecules, e.g., $N=500$, we have 500 lists, $L_1, L_2, \cdots, L_{500}$. We note that in making the lists, particles in the 26 images, if falling within the radius $r_v$ of the border molecule, are also counted. However, we do not make lists for particles inside the image boxes. In calculating the interaction force, $\mathbf{F}_i$, on molecule $i$ (eq. (3.3)), we only take molecules in the list $L_i$ to form pairs and ignore all other molecules outside the sphere of $r_v$. This is the spherical cutoff convention mentioned above. For LJ interaction at density $\rho\sigma^3 = 0.8$, the typical number of molecules within $r_c = 2.5\sigma$ is ~24. Depending on the size of $r_v$, say $3.3\sigma$, about twice as many molecules are included in $L_i$, still considerably less than $N=500$. The saving on time is proportional to $N^2$ ($N(N-1)/2$). Since the major part (close to 95%) of calculation time is spent on evaluating the forces due to pairs, the saving is significant.

The lists $L_i$ need be updated after $n$ time steps. The value $n$ is determined by the size $r_v$. If the thickness of the shell between the sphere of radius $r_c$ and that of $r_v$ is $\theta$, i.e., $\theta = r_v - r_c$, and the maximum velocity of particles within the sphere of $r_v$ is $v_{max}$, it will take approximately $t' = \theta/v_{max}$ for a particle outside the sphere of $r_v$ to get into the cutoff sphere of $r_c$. Since meantime the center molecule is also moving, a safe time period is half of $t'$. Therefore $n = t'/2\Delta t$. Typically, $n$ is chosen to be 10 (Verlet), corresponding to $r_v = 3.3\sigma$.

An alternative bookkeeping scheme has been proposed by Berne et al.[15]. The standard box is subdivided into a number of small cells. At the beginning of each time step, the contents of each cell are listed. The cells are then scanned and relative distances are calculated only for pairs of molecules in neighboring cells. In this method, the time required at each step is always proportional to $N$. Thus as the population of particles in the system increases, this method becomes faster than the Verlet bookkeeping method.

## MULTI-TIMESTEP METHOD

In simulating several relaxation processes that take place on different time scales, e.g., segmental and rotational relaxations in polymer and biomolecules, it is more efficient to use a multi-timestep method proposed by Streett et al. [16]. The region surrounding each molecule is separated into a sphere of fast varying forces and a shell of slowly varying forces, (see Figure XIII.2). As an example, let us consider the LJ molecules

$$u(r_{ij}) = 4\varepsilon\left[\left[\frac{\sigma}{r_{ij}}\right]^{12} - \left[\frac{\sigma}{r_{ij}}\right]^6\right] \qquad (6.14)$$

The instantaneous force on molecule $i$ is given by

$$\mathbf{F}_i = \sum_{j\neq i} -\left[\frac{\mathbf{r}_{ij}}{r_{ij}}\frac{du(r_{ij})}{dr_{ij}}\right]_{r_{ij}\leq r_c} \qquad (6.15)$$

where again $r_c$ is the cutoff distance. This $\mathbf{F}_i$ can be divided into two parts, $\mathbf{F}_p$ and $\mathbf{F}_s$, given by

$$\mathbf{F}_i = \mathbf{F}_p + \mathbf{F}_s = \sum_{j\neq i}\left[\frac{-\mathbf{r}_{ij}}{r_{ij}}\frac{du}{dr_{ij}}\right]_{r_{ij}\leq r_a} + \sum_{j\neq i}\left[\frac{-\mathbf{r}_{ij}}{r_{ij}}\frac{du}{dr_{ij}}\right]_{r_a\leq r_{ij}\leq r_c} \qquad (6.16)$$

where $r_a$ is a dividing radius. It is found that the primary force $\mathbf{F}_p$ due to interactions within the sphere of $r_a$ is fast varying, whereas the secondary force $\mathbf{F}_s$ from outside the sphere of $r_a$ is slowly varying. For LJ molecules at liquid densities, $r_c$ is about $2.5\sigma$, and $r_a$ is about $1.5\sigma$. For these choices, it is found that $|\mathbf{F}_p| > |\mathbf{F}_s|$; i.e., $\mathbf{F}_p$ is more important than $\mathbf{F}_s$. The motion of the molecule is dominated by a rapidly changing primary force resulting from collisions within a cage of primary neighbors. The long-range secondary force is smaller, and changes more slowly with time. Under these circumstances, we could first evaluate the force and derivatives of the force $\mathbf{F}_s$ at the initial time $t_0$, namely $\mathbf{F}_s(t_0)$, $\mathbf{F}_s'(t_0)$, $\mathbf{F}_s''(t_0)$,..., etc. For subsequent times, instead of calculating the pair interactions, which is time-consuming, we the Taylor expansion

$$\mathbf{F}_s(t_0+k\Delta t) = \mathbf{F}_s(t_0) + \mathbf{F}_s'(t_0)(k\Delta t) + \mathbf{F}_s''(t_0)\frac{(k\Delta t)^2}{2!} + \cdots \qquad (6.17)$$

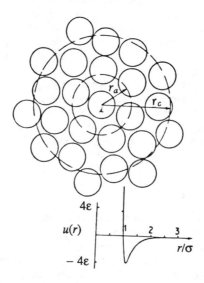

FIGURE XIII.2. *Primary and secondary neighbors of molecule i. The primary sphere has radius $r_a = 1.5\sigma$, the secondary sphere, $r_c = 2.5\sigma$. The LJ potential is drawn to scale to indicate the magnitude of interation. (Streett, Tildesley, and Saville, Mol. Phys., 1978)*

$k=1,2,...,n-1$. The expansion is terminated after the $m$th derivative $\mathbf{F}_s^{(m)}$, and we call this approximation an $m$th-order algorithm. Since $\mathbf{F}_s$ is slowly varying, the small time expansion can also yield accurate results. For $\mathbf{F}_p$, the usual summation of pair interactions is done, in this case in a sphere of small radius $r_a$. After $n$ time steps, the list of primary and secondary neighbors is recompiled, and the primary and secondary forces are recalculated, together with $\mathbf{F}_s'$, $\mathbf{F}_s''$, ... The cycle is repeated for the next $n-1$ steps. Streett et al. [17] showed that for the LJ potential, $m=3$, $n=10$, $r_c=2.5\sigma$, and $r_a=1.1\sigma$. At a reduced density of $\rho\sigma^3=0.8$, the average numbers of primary and secondary neighbors are 1.4 and 24; at $\rho\sigma^3=1.05$, the numbers are 3.1 and 29, respectively. In a system of 256 molecules, the multi-timestep method is faster, by a factor of 3 to 5, than Verlet's list method. A run is considered successful if the resulting particle trajectories are essentially the same as those obtained in a conventional MD run with the same starting point. In most cases, the difference in positions and velocities is about one part in $10^5$ after 1000 time steps. Teleman and Jönsson [18] has given an updated version of the multi-timestep method.

## XIII.7. *Simulation in Isothermal Ensembles: The Nosé Method*

Nosé [19] in 1984 proposed a method for simulating isothermal ensembles. It was based on the observation that in earlier MD simulations, one attempted to "control" the temperature of the *NVE* ensemble by scaling the velocities of the particles between time steps. This practice had become the "norm" in *NVE* ensemble simulation. To find a theoretical justification for this procedure, Nosé discovered that it is possible to *construct* a new Hamiltonian in the *NVE* ensemble in such a way that it actually produces trajectories in the *canonical* ensemble, i.e., in the *isothermal* ensemble. This new Hamiltonian is a *construction*, i.e., it does not necessarily correspond to a physically real system (or, for that metter, it does not matter whether the system represented by the new Hamiltonian is realizable or not). Nosé calls this system the *extended system* (ES). The original *NVE* ensemble had the configuration $\mathbf{r} \equiv \mathbf{r}^N$ and the momentum distribution $\mathbf{p} \equiv \mathbf{p}^N$ with real time $t$. (N.B., we have simply used $\mathbf{r}$ and $\mathbf{p}$ to represent the $N$-vectors). In the ES, we define a new set of *virtual* coordinates: the configuration $\mathbf{q} \equiv \mathbf{q}^N$, the momentum distribution $\mathbf{b} = \mathbf{b}^N$, and the *virtual* time $\tau$ by

$$\mathbf{r}_i \equiv \mathbf{q}_i \tag{7.1}$$

$$\mathbf{p}_i \equiv \frac{\mathbf{b}_i}{s}$$

$$t = \int_0^\tau \frac{d\tau}{s}$$

where a scaling parameter $s$ is introduced to readjust the time $t$. Thus the real velocity $d\mathbf{r}_i/dt$ is related to the virtual variables by

$$\frac{d\mathbf{r}_i}{dt} = s \frac{d\mathbf{r}_i}{d\tau} = s \frac{d\mathbf{q}_i}{d\tau} \tag{7.2}$$

A simple explanation of these transformations is that the time is scaled by $dt = d\tau/s$. As will be seen shortly, when time is scaled, momenta are scaled. And when momenta are scaled, temperature is scaled. The scaling factor $s$ provides one additional degree of freedom and acts like an external system that imposes a constraint on the $N$-particle system, in this case, the constant temperature condition. The new Hamiltonian of ES is assumed to have the form

$$H_N = \sum_i \frac{b_i^2}{2ms^2} + \frac{p_s^2}{2Q} + V_N(\mathbf{q}) + gkT \ln s \tag{7.3}$$

The Hamiltonian, as before, is composed of the *KE* and *PE* terms, in this case with a new kinetic energy term $p_s^2/2Q$ and a new potential energy term $gkT \ln s$. $p_s$ is the generalized momentum (having units *energy·time*) conjugate to $s$ (dimensionless) the time scaling factor, and $Q$ (having units *energy·time$^2$*) behaves as the *mass* for the motion of $s$. $T$ in eq. (7.3) is the externally fixed temperature (i.e., the constant temperature *thermal bath*). $g$, as will be shown, is essentially the number of degrees of freedom of the $N$-particle system. The curious choice of the new *PE* as a logarithm, i.e., $gkT \ln s$, is found by hindsight to be essential for producing the distribution laws in the canonical ensemble. Other choices of *PE* have been proposed by other researchers [20]. These alternatives have been reviewed by Nosé [21]. Next, we assume that Hamilton's equations of motion apply to the new Hamiltonian (7.3) in terms of the virtual variables. The set of equations (2.1) now becomes

$$\frac{dq_i}{d\tau} = \frac{\partial H_N}{\partial b_i} = \frac{b_i}{ms^2} \tag{7.4}$$

$$\frac{db_i}{d\tau} = -\frac{\partial H_N}{\partial q_i} = -\frac{\partial V_N}{\partial q_i}$$

$$\frac{ds}{d\tau} = \frac{\partial H_N}{\partial p_s} = \frac{p_s}{Q}$$

$$\frac{dp_s}{d\tau} = -\frac{\partial H_N}{\partial s} = \frac{1}{s}\left[\sum_i \frac{b_i^2}{ms^2} - gkT\right]$$

The conserved quantities of motion are the new energy $E$, the total momentum $\Sigma_i b_i$ and the angular momentum $\Sigma_i q_i \times b_i$. As before, for the energy

$$\frac{dH_N(\tau)}{d\tau} = \sum_i \left[\frac{\partial H_N}{\partial b_i}\dot{b}_i + \frac{\partial H_N}{\partial q_i}\dot{q}_i\right] + \frac{\partial H_N}{\partial p_s}\dot{p}_s + \frac{\partial H_N}{\partial s}\dot{s} = 0 \tag{7.5}$$

The original *real* Hamiltonian $H_N^0$ is written as

$$H_N^0(\mathbf{r}, \mathbf{p}) \equiv \sum_i \frac{p_i^2}{2m} + V_N(\mathbf{r}) \tag{7.6}$$

The new Hamiltonian eq. (7.3) could be expressed in terms of $H_N^0$ as

$$H_N(\mathbf{q}, \mathbf{b}) = H_N^0(\mathbf{q}, \mathbf{b}/s) + \frac{p_s^2}{2Q} + gkT \ln s \tag{7.7}$$

Next, we shall establish the relation between the computer simulation carried out in the virtual variables according to (7.4) and the distribution functions in the isothermal canonical ensemble. The results to be anticipated are: the distribution functions obtained by simulating the extended system are precisely the canonical distribution functions. Since, for the new Hamiltonian (7.3) and the new equations of motion (7.4), the energy is conserved, the simulation could be done in the style of a microcanonical ensemble, however with consequent results in the canonical ensemble. This is the *idée fixe* of Nosé. To show this, we first examine the partition function $Z_N$. By definition,

$$Z_N \equiv \int dp_s \int ds \int db \int dq \; \delta_D(H_N^0(\mathbf{q}, \mathbf{b}/s) + p_s^2/2Q + gkT \ln s - E) \tag{7.8}$$

where $\delta_D$ is the Dirac delta function and $E$ is the constant energy. We transform from the virtual (simulation) variables back to the real (physical) variables: $\mathbf{p}_i = \mathbf{b}_i/s$, and

$\mathbf{q}_i = \mathbf{r}_i$. Hence

$$Z_N = \int dp_s \int d\mathbf{p} \int d\mathbf{r} \int ds \; s^{3N} \delta_D(H_N^0(\mathbf{r}, \mathbf{p}) + p_s^2/2Q + gkT \ln s - E) \tag{7.9}$$

We note that, with respect to the variable $s$, the argument in the Dirac delta has one and only one root. We use the mathematical relation that

$$\delta_D(f(t)) = \frac{\delta_D(t-t_0)}{f'(t_0)} \tag{7.10}$$

where $f'(t) \equiv df/dt$ and $t_0$ is a root of $f(t)$, i.e., $f(t_0) = 0$. This identity could be demonstrated, provided $f$ is a univalent function, i.e., for any given values $f$ and $t$, $f = f(t)$ if and only if $t = t(f)$ and the derivatives $df/dt = 1/(dt/df)$ exist. The Dirac delta is known in mathematics as a *distribution* [22]. For an arbitrary function $\phi(t)$, the integral gives

$$\int dt \; \delta_D(f(t)) \; \phi(t) = \int df \; \frac{dt}{df} \delta_D(f) \; \phi(t=t(f)) = \frac{dt}{df}\bigg|_{f=0} \cdot \phi(t_0) = \frac{\phi(t_0)}{df(t_0)/dt} \tag{7.11}$$

Since $\phi(t)$ is arbitrary, this proves the property of the distribution (7.10).

$$Z_N = \frac{1}{gkT} \int dp_s \int d\mathbf{p} \int d\mathbf{r} \int ds \; s^{3N+1} \delta_D(s - e^{-[H_N^0(\mathbf{r}, \mathbf{p}) + p_s^2/2Q - E]/gkT}) \tag{7.12}$$

$$= \frac{1}{gkT} e^{[(3N+1)/g]E/kT} \int dp_s \; e^{-[(3N+1)/g]p_s^2/2QkT} \int d\mathbf{p} \int d\mathbf{r} \; e^{-[(3N+1)/g]H_N^0(\mathbf{r}, \mathbf{p})/kT}$$

If we choose $g = 3N+1$, the partition function of the ES (used in the simulation work) is equivalent (aside from a multiplicative constant) to that of the canonical ensemble:

$$Z_N = C \int d\mathbf{p} \int d\mathbf{r} \; \exp\left[-\frac{H_N^0(\mathbf{r}, \mathbf{p})}{kT}\right] \tag{7.13}$$

Thus the distribution function

$$\rho^{(N)}(\mathbf{r}, \mathbf{p}) \sim \exp\left[-\frac{H_N^0(\mathbf{r}, \mathbf{p})}{kT}\right] \tag{7.14}$$

This is the canonical ensemble distribution, since $T$ is set externally. (We have neglected a constant factor for $\rho^{(N)}$ here.) Any dynamic variable $A$ calculated in ES according to eqs. (7.4) with arguments $\mathbf{q}$ and $\mathbf{b}/s$ corresponds precisely to those in a canonical ensemble:

$$\lim_{T' \to \infty} \frac{1}{T'} \int_0^{T'} d\tau \; A(\mathbf{q}, \mathbf{b}/s) = \langle A(\mathbf{q}, \mathbf{b}/s) \rangle_{ES} = \langle A(\mathbf{r}, \mathbf{p}) \rangle_{CE} \tag{7.15}$$

where $< \cdots >_{ES}$ and $< \cdots >_E$ denote ensemble averages in the extended system and in the canonical ensemble, respectively. The first equivalence in (7.15) is obtained by sampling the dynamic variable at *equal* intervals $\Delta \tau$ in virtual time $\tau$. This is called *virtual time sampling*. Since $\tau = \tau(t)$ is a function of real time, the real time interval $\Delta t$ of each time step is unequal. The relation is expressed by eq. (7.2) where $t = \int dt/s$, $s$ being a function of time according to (7.4).

## AN APPLICATION

Nosé [23] tested the method on a system of 108 argon atoms (mass 39.9 g/mol), interacting with the LJ potential ($\epsilon = 1.039$ kJ/mol, $\sigma = 3.446$ Å), which was truncated at 8.5 Å. The MD cell was a cube of length 17.5 Å, corresponding to 29.88 cm³/mol. The periodic boundary condition was imposed. A fifth oder predictor-corrector algorithm was employed for the integration of the equations of motion (7.4). The time step $\Delta \tau$ was chosen to be $2.5 \times 10^{-15}$ s. Each run consisted of 2500 time steps, with the first 500 steps discarded before taking averages. The real time $t$ was obtained by multiplying the time of simulation $\tau$ by the factor $<s^{-1}>$, thus ignoring any fluctuations in the unequal real time intervals.

The detailed dynamics depends on the value of $Q$ chosen. Simulations were carried out with three different $Q$ values: 1, 10, and 100 (ps)² kJ/mol in order to examine the effects. Two isotherms $T = 100$ K and 150 K were studied. The results are listed in Table XIII.1. In Nosé dynamics, temperature is allowed to fluctuate around $T_{eq}$, the temperature in equilibrium with the thermal bath. Figure XIII.3 shows the fluctuations for $Q = 1$ (in given units). The fluctuations $\Delta T^2 \equiv <(\delta T)^2>$ seem to increase with large $Q$ (see Table XIII.1).

The same spirit of *scaling* could be used to generate statistics in constant pressure ensembles. Andersen [24] introduced the *NPH* ensemble by scaling the MD cell since volume is conjugate to the pressure. Works by Parrinello and Rahman [25], Haile and Gupta [26], and Hoover et al. [27] have further advanced the scaling techniques. In particular, Hoover's Hamiltonian involved dissipative forces in terms of a friction coefficient. However, the method has been shown to be equivalent to the Nosé technique [28].

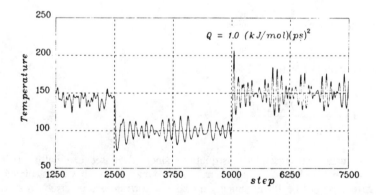

FIGURE XIII.3. *Change of Temperature. The first 1250-2500 steps were carried out by conventional NVE MD. At step 2500, the simulation is switched to the Nosé isothermal ensemble with* $Q = 1.0$ (ps)² *kJ/mol at temperature* $T_{eq} = 100K$. *At step 5000,* $T_{eq}$ *is changed to 150 K. (Nosé, Mol. Phys., 1984)*

Table XIII.1. *Molecular Dynamics Simulation Using Nosé's Method for Lennard-Jones Molecules at V = 29.88 cm³/mol.*

| Initial condition | $T_{eq}$ K | $Q$ (kJ mol$^{-1}$)(ps)$^2$ | $\langle T \rangle$ K | $\Delta T$ K | $\langle \phi \rangle$ kJ mol$^{-1}$ | $P$ GPa | $\langle s \rangle$ | $\langle \frac{1}{s} \rangle$ | $c_v$ R | $D$ $10^{-5}$ cm$^2$ s$^{-1}$ |
|---|---|---|---|---|---|---|---|---|---|---|
| f.c.c. | standard MD | | 143.3 | 6.8 | −5.188 | 0.123 | 1.0 | 1.0 | | 4.1 |
| run 1 | 100 | 1 | 99.6 | 8.3 | −5.520 | 0.049 | 1.419 | 0.706 | 2.54 | 1.6 |
| run 2 | 150 | 1 | 149.3 | 12.4 | −5.115 | 0.137 | 1.071 | 0.936 | 2.36 | 4.9 |
| run 2 | 100 | 1 | 99.9 | 9.0 | −5.525 | 0.045 | 1.418 | 0.707 | 2.82 | 1.6 |
| run 2 | 100 | 10 | 99.3 | 6.9 | −5.549 | 0.039 | 1.437 | 0.697 | 1.89 | 2.3 |
| run 2 | 100 | 100 | 100.2 | 7.4 | −5.537 | 0.042 | 1.425 | 0.703 | 2.04 | 1.9 |
| run 3 | 150 | 1 | 149.8 | 12.7 | −5.094 | 0.140 | 1.064 | 0.943 | 2.42 | 4.2 |
| run 3 | 150 | 10 | 149.1 | 12.0 | −5.151 | 0.132 | 1.101 | 0.910 | 2.63 | 3.3 |
| run 3 | 150 | 100 | 148.7 | 25.9 | −5.099 | 0.140 | 1.118 | 0.908 | 9.25 | 3.5 |

*$Q$ is the psudo-mass conjugate to $P_s$. $\Delta T$ is root-mean-squared temperature fluctuations. $\langle s \rangle$ is the averaged time scaling factor. $c_v$ is the constant volume heat capacity. $D$ is the diffusion coefficient. (Nosé, Mol. Phys., 1984)

# References

[1]   D.J. Evans, Phys. Lett. A**74**, 229 (1979); J. Stat. Phys. **22**, 81 (1980); Mol. Phys. **37**, 1745 (1979); and Phys. Rev. A**23**, 1988 (1981).

[2]   A. Rahman, Phys. Rev. A**136**, 405 (1964).

[3]   S. Nosé, Mol. Phys. **52**, 255 (1984); and J. Chem. Phys. **81**, 511 (1984)

[4]   I.E. Farquhar, *Ergodic Theory in Statistical Mechanics* (John Wiley, New York, 1964).

[5]   A.S. Wightman, in *Statistical Mechanics at the Turn of the Decade*, edited by E.G.D. Cohen (Marcel Dekker, New York, 1971).

[6]   L. Verlet, Phys. Rev. **159**, 98 (1967).

[7]   C.W. Gear, *Numerical Initial Value Problems in Ordinary Differential Equations* (Prentice-Hall, Englewood Cliffs, New Jersey, 1971).

[8]   A. Rahman, *Ibid.* (1964).

[9]   L. Verlet, *Ibid.* (1967).

[10]  M.S. Green, J. Chem. Phys. **20**, 1281 (1952); *Ibid.* **22**, 398 (1954); and R. Kubo, M. Yokota, and S. Nakajima, J. Phys. Soc. Japan **12**, 1203 (1957); also R. Kubo, in *Lectures in Theoretical Physics*, edited by W.E. Britten and L.G. Dunham, Volume 1 (Interscience, New York, 1958).

[11]  H.T. Davis, in Adv. Chem. Phys. **24**, 257, edited by I Prigogine and S.A. Rice (Wiley, New York, 1973).

[12]  J.P. Valleau and S.G. Wellington, in *Statistical Mechanics: Equilibrium Techniques*, Part A, edited by B.J. Berne (Plenum, New York, 1977).

[13]  J. Kushik and B.J. Berne, in *Statistical Mechanics: Time-Dependent Processes*, Part B, edited by B.J. Berne (Plenum, New York, 1977).

[14]  L. Verlet, Phys. Rev. **159**, 98 (1967).

[15]  B.J. Berne and G.D. Harp, *Advances in Chemical Physics*, **17**, 63, edited by I. Prigogine and S.A. Rice (Wiley, New York, 1970); also B.J. Berne and D. Forster, in *Annual Review of Physical Chemistry*, vol. 22, edited by H. Eyring (Annual Review Inc., Palo Alto, California, 1971) pp.563-596.

[16]  W.B. Streett, D.J. Tildesley, and G. Saville, Mol. Phys. **35**, 639 (1978).

[17]  W.B. Streett, D. Tildesley, and G. Saville, *Ibid.* (1978).

[18]  O. Teleman and B. Jösson, J. Comp. Chem. **7**, 58 (1986); and Mol. Phys. **60**, 193 (1987).

[19]  S. Nosé, *Ibid.* (1984).

[20]  J.M. Haile and S. Gupta, J. Chem. Phys. **79**, 3067 (1983).

[21]  S. Nosé, J. Chem. Phys. **81**, 511 (1984).

[22]  L. Schwartz, *Théories des Distributions*, Tomes 1 and 2, Actualités Scientifiques et Industrielles (Hermann et Cie., Paris, 1957).

[23]  S. Nosé, Mol. Phys. **52**, 255 (1984).

[24]  H.C. Andersen, J. Chem. Phys. **72**, 2384 (1980).

[25]  M. Parrinello and A. Rahman, Phys. Rev. Lett. **45**, 1196 (1980); and J. appl. Phys. **52**, 7182 (1981).

[26]  J.M. Haile and S. Gupta, J. Chem. Phys. **79**, 3067 (1983).

[27]  W.G. Hoover, A.J.C. Ladd, and B. Moran, Phys. Rev. Lett. **48**, 1818 (1982); A.J.C. Ladd and W.G. Hoover, Phys. Rev. **B28**, 1756 (1983); D.J. Evans, J. Chem. Phys. **78**, 3297 (1983); and D.J. Evans, W.G. Hoover, B.H. Failor, B. Moran, and A.J.C. Ladd, Phys. Rev. **A28**, 1016 (1983).

[28]  D.J. Evans and B.L. Holian, J. Chem. Phys. **83**, 4069 (1985).

## Exercises

1.  Prove Verlet's numerical formula (3.3) by using Taylor's expansions for increments $+\Delta t$ and $-\Delta t$. Use this algorithm to solve the equations of motion (2.1) for LJ molecules simulating argon at $T=$ 100 K and 29.88 $cm^3/mol$. Find the energy, pressure, and pair correlation function. Compare the results with Rahman's predictor-corrector method.

2.  For the above problem, evaluate the velocity autocorrelation function.

3.  Show that eq. (5.1) for the diffusion coefficient is equivalent to the Einstein formula

$$<|\mathbf{r}_i(t) - \mathbf{r}_i(0)|^2> = 6Dt$$

The quantity in the angular brackets gives the *mean squared displacement*.

4.  Survey the literature on the evaluation of the transport properties $\eta$ and $\kappa$. What are the existing techniques? And what are the limitations, if any?

5.  Explain the *NPH* ensemble method of Andersen. Carry out a simulation for the fluid of Problem 1 by this method. (Using the pressure value of Problem 1 as input.)

6.  Discuss the similarities and differences between the MC and MD methods. What properties are evaluated by these methods?

7.  Survey the literature to find out the systems (polar fluids, plasmas, surfactants, etc.) and properties of the solids, liquids, and gases that are simulated by the MD method. Speculate on future applications of the MD methods.

8.  Use Nosé's method to simulate the fluid of Problem 1.

# CHAPTER XIV

# INTERACTION SITE MODELS FOR POLYATOMICS

$S$o far we have limited our discussions to monatomic molecules (i.e., each molecule consists of a single atom). This picture is appropriate only for noble gases (e.g., argon, krypton, and xenon). Real molecules are usually polyatomic (e.g., methane, carbon dioxide, and hydrogen chloride); several atoms are held together by chemical bonds. To treat these molecules properly in a theoretical framework, a different type of correlation function must be devised. This objective can be achieved in two ways: (1) by use of angular pair correlation functions (apcf) $g(r_{12}, \omega_1, \omega_2)$; (2) by site-site pair correlation functions (ss pcf) $g_{\alpha\gamma}(r_{12}^{\alpha\gamma})$. The sites refer to geometrical locations in a given molecule. The latter approach is referred to as the *interaction site model* (ISM) of molecules. The correlation functions are formulated as probability distributions of pairs of sites belonging to different molecules. These sites may be chosen as centers of atoms composing the molecule, or they may be centers of charges or centers of mass. In some formulations, one considers *auxiliary* sites, or imaginary positions in the molecule, not corresponding to any real atoms. There is no theoretical limitation on the choices except for the requirement that the sites chosen should result in a realistic representation of the intermolecular interaction forces. There are several points to be made regarding ISM. First, the interaction energy between a pair of nonspherical molecules is *assumed* to be the sum of site-site interactions [1] (e.g., for two HF molecules, there are four pairs, H–H, H–F, F–H, and F–F, plus pairs arising from any necessary auxiliary sites). This assumption is good at short ranges for overlapping forces in real molecules. However, the site-site potential is not reliable at long ranges with regard to multipolar and induction forces. It neglects the effects of bonding charge density and is not applicable to nonlocalized electronic interactions (such as $\pi$ electrons) [2]. Second, the site-site correlation functions $g_{\alpha\gamma}$ are *defined* probabilistically. This is perfectly legitimate, since we could inquire about the chances of finding one site $\alpha$ at one location, and simultaneously finding another site $\gamma$ at a second location. The definition could be made exact. Third, approximate theories are formulated to calculate these ss pcfs. These theories are called the interaction site approximations (ISA). They are based on the Ornstein-Zernike formula generalized to site-site models. There are a number of alternative forms of OZ in existence. We distinguish, for example, here the approximate Ornstein-Zernike-like (OZ-RISM) formula proposed by Chandler and Andersen [3] and the Ornstein-Zernike inverse matrix (OZ-IM) expression [4], which is mathematically an *exact* relation. Most work in literature was carried out using the OZ-RISM method.

Some typical ss pcf are shown in Figures IV.15 and 16 for ammonia and water. The site-site correlations in $NH_3$ consist of the pairs N-H, N-N and H-H. (Note that for all pairs, including the the H-H pair, the members must belong to different $NH_3$ molecules). For water we have the O-O, O-H, and H-H pairs. The ISA theories, applied to simple polyatomic models, reproduce well the structure factor from scattering experiments. Thus increased attention has been paid to this approach. In the following we shall discuss a number of ISM theories. Applications to various model fluids will also be examined.

## XIV.1. *The Site-Site Potentials*

A variety of potential functions have been used to fit the structural data and other properties of polyatomic liquids. Most are simple models, many of which have been discussed earlier. Prominent among these models are the site-site Lennard-Jones potential, the exp-6 potential, and the distributed-charge Coulomb potentials. The fluids studied include nitrogen, methane, ethane, ethylene, chlorine, hydrogen fluoride, benzene, neopentane, and many others. We examine some representative potentials below.

### NITROGEN

One of the common potentials used to represent nitrogen is the site-site LJ potential of Cheung and Powles [5]. Other more accurate potentials [6] have been proposed. Ling and Rigby [7] have given a review. The Cheung-Powles potential is simpler and easier to use in simulation work. The interaction between a nitrogen atom in molecule 1 and another in molecule 2 is given by

$$u(r) = 4\varepsilon \left[ \left[ \frac{\sigma}{r} \right]^{12} - \left[ \frac{\sigma}{r} \right]^{6} \right], \qquad r = |\mathbf{r}_{\gamma_2} - \mathbf{r}_{\alpha_1}| \qquad (1.1)$$

where $\alpha_1 = N$, the nitrogen atom in molecule 1, and $\gamma_2 = N$, the nitrogen atom in molecule 2. The potential parameters are listed in Table XIV.1. The bond length between the N-N atoms is $L=1.09$ Å. The potential curve is depicted in Figure XIV.1. This potential yields correctly the second virial coefficient, the thermodynamic, and transport properties of nitrogen [8]. However, in order to describe the x-ray diffraction data, one needs a slightly modified potential. Narten et al. [9] proposed a Williams exp-6 potential for nitrogen:

$$u(r) = \varepsilon \frac{\lambda}{b} \left\{ \left[ \frac{6}{\lambda} \right] e^{\lambda(1 - r/r_0)} - \left[ \frac{r_0}{r} \right]^{6} \right\} \qquad (1.2)$$

The potential parameters are also given in Table XIV.1. This potential is specifically designed for use with the ISA theory. Since ISA is an approximate theory, (1.2) is not necessarily the true nitrogen potential. The two potentials are compared in Figure XIV.1. The exp-6 curve is less repulsive at small $r$. The results obtained by using both ss LJ and ss exp-6 in the ISA calculation are shown in Figure XIV.2. Quite accurate fit to experimental data was obtained (except at the second maximum, where ISA gives a

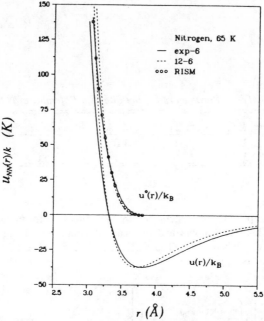

FIGURE XIV.1.   The site-site potential for nitrogen.  $u_{NN}(r)/k$ is in Kelvin. Dotted line: site-site Lennard-Jones potential.  Solid line: site-site exp-6 potential.  Also shown are the repulsive reference potentials, including the one derived from RISM calculation (see [8]).  (Narten et al., J. Chem. Phys., 1980)

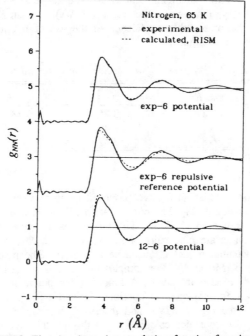

FIGURE XIV.2. The site-site pair correlation function for nitrogen at 65 K.  $\cdots$ : from neutron scattering experiments.  ———: RISA calculation. The results from exp-6,  the repulsive part of exp-6, and LJ 12-6 potential are shown.  The exp-6 curves are displaced by +2 and +4 along the ordinate. (Narten et al., J. Chem. Phys., 1980) [8]

higher peak).

*Table XIV.1. Parameters for the Site-Site Potential of Nitrogen*

| Potential | $\varepsilon/k$, $K$ | $\sigma$, Å | $L$, Å | $\lambda$ | b | $r_0$, Å |
|-----------|-----------|-----------|-----------|-----------|-----------|-----------|
| ss LJ | 37.3 | 3.31 | 1.09 | - | - | - |
| ss exp-6 | 37.3 | - | 1.1 | 11.55 | 5.55 | 3.80 |

## METHANE

The site-site potential has been used by Williams [10] for methane. It consists of five centers: four hydrogen atoms and one carbon atom. The bond length of C–H is taken to be 1.107 Å (Habenschuss et al. [11]), and is assumed to be rigid. The site-site exp-6 potential used is

$$u(r) = \varepsilon \left[ \frac{h}{h-6} \right] \left\{ \left[ \frac{6}{h} \right] e^{h(1 - r/r_0)} - \left[ \frac{r_0}{r} \right]^6 \right\} \tag{1.3}$$

The potential parameters are given in Table XIV.2. Habenschuss et al. [12] have carried out x-ray diffraction studies of methane at 92 K. The above site-site potential predicts, through the ISA theory, the x-ray structure factor $S(k)$ surprisingly well. (See Figure XIV.3.) One now has reasonable confidence in the use of interaction site models to represent simple molecular liquids.

*Table XIV.2: Parameters for the Site-Site Potential of Methane*

| Potential | $\varepsilon/k$, $K$ | $\sigma$, Å | h | $r_0$, Å |
|-----------|-----------|-----------|-----------|-----------|
| C–C | 48.8 | 3.25 | 13.6 | 3.78 |
| C–H | 20.7 | 2.80 | 12.59 | 3.43 |
| H–H | 6.83 | 2.42 | 12.21 | 3.26 |

## CARBON DISULFIDE

Tildesley and Madden [13] have proposed a three-center Lennard-Jones (3cLJ) site-site potential for liquid $CS_2$. The configuration of $CS_2$ is linear (S=C=S). The bond length of C=S is taken to be 1.57 Å. The three sites are taken to coincide with the atomic positions. There are nine pairs of interaction for each configuration of the two molecules. (The potential model is the same as (1.1).) The potential parameters are given in Table XIV.3 (Model A). For comparison, another set of parameters (Model B) given by Haile [14] is also included. A graph of the potential contours of the 3cLJ interaction is shown for coplanar interactions derived from rotations of molecule 2 around molecule 1 (see Figure XIV.4 and inset). The attractive regions are in the T orientations. Molecular dynamics calculations of this model have been carried out and used to generate the pressure, internal energy, specific heat, structure factor, and orientation parameter $G_2$. Comparison with experiments shows favorable agreement.

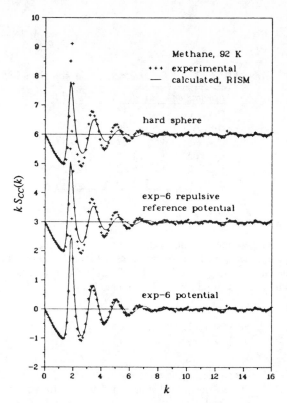

FIGURE XIV.3. *The structure factor S(k) of methane.* +++: *from x-ray scattering (Habenschuss et al. [11]). Solid lines: from RISA calculation. The results from hard-sphere, exp-6 repulsion, and exp-6 potentials are shown for comparison. The exp-6 results are much better than the exp-6 repulsive results, indicating that attractive forces contribute significantly toward the structure. (Habenschuss et al., J. Chem. Phys., 1981)*

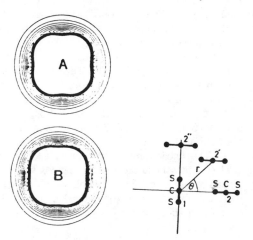

FIGURE XIV.4. *The potential contour of* $u(r_{12}\theta_1\theta_2\phi_{12})$ *for* $CS_2$ *interaction. The angle* $\theta$ *is depicted in the inset, where two* $CS_2$ *molecules are co-planar and one rotates around the other to generate the energy contours shown. Molecules 2 and 2″ give the T-orientation. Both model A and model B are displayed. (Tildesley and Madden, Mol. Phys., 1981)*

*Table XIV.3.  3c-LJ Parameters for Carbon Disulfide*

| Parameter | Model A | Model B |
|---|---|---|
| $\sigma_{CC}$, Å | 3.35 | 3.58 |
| $\sigma_{CS}$, Å | 3.44 | 3.47 |
| $\sigma_{SS}$, Å | 3.52 | 3.37 |
| $\varepsilon_{CC}/k$, $K$ | 51.20 | 79.73 |
| $\varepsilon_{CS}/k$, $K$ | 96.80 | 123.21 |
| $\varepsilon_{SS}/k$, $K$ | 183.00 | 190.70 |

**HYDROGEN FLUORIDE**

Lately, hydrogen-bonding fluids have been modeled using site-site potentials. Cournoyer and Jorgensen [15] have proposed a distributed-charge Coulomb potential to represent the interaction energies of hydrogen fluoride molecules. An LJ 12:6 potential plus Coulomb interaction is used for the F-F interaction

$$u(r) = \frac{A}{R^{12}} - \frac{C}{R^6} + \sum_\alpha \sum_\gamma \frac{q_\alpha q_\gamma}{r_{\alpha\gamma}} \tag{1.4}$$

where $R$ is the F–F distance, and $r_{\alpha\gamma}$ is the site-site distance from site $\alpha$ of molecule 1 to site $\gamma$ of molecule 2. Three sites are distinguished in the HF molecule, with two coinciding with the nuclei (H and F) and a third at a distance 0.166 Å from F along the linear bond F-H. The total bond length F-H is taken to be fixed at 0.917 Å

```
 site      site          site
   F-----O-------------H
  +Q     -2Q           +Q
```

Charges are distributed on sites, with $+Q$ on F and H and $-2Q$ at the third site to preserve charge neutrality. In the literature, this model is also called the TIPS model (Jorgensen [16]). Three adjustable parameters, $A$, $C$, and $Q$, are determined from fit to liquid properties and to gas phase dimers. The values are: $A = 300000$ kcal Å$^{12}$/mol, $C = 425$ kcal Å$^6$/mol, and $Q = 0.725$ (N.B. The electronic charge $e^2 = 332.18$ kcal Å /mol). This charge distribution gives a dipole moment of 2.04 Debye (cf. experimental 1.82 Debye) and a quadrupole moment of $2.5 \times 10^{-26}$ esu cm$^2$ (cf. experimental $2.6 \times 10^{-26}$ esu cm$^2$). The density, enthalpy of vaporization, heat capacity, isothermal compressibility, and coefficient of thermal expansion were calculated by the Monte Carlo method for liquid HF at 0°C and −70°C under 1 atm. They compared fairly well with experimental data. The linear dimers show lower energy than the cyclic dimers. Hydrogen bonding is clearly present in the FH correlations with a peak at 1.75 Å. The calculated hydrogen bond strength is 5.83 kcal/mol, which compares well with experimental estimates. There are ubiquitous hydrogen-bonded chains. This makes this model an interesting one for studying HF liquids.

## XIV.2.  *Transformation of Coordinates*

The site-site pair correlation function $g_{\alpha\gamma}(r_{\alpha_1\gamma_2})$ is a probability measure, as discussed previously (see Chapter IV), of finding a site $\alpha$ in molecule 1 at $r_\alpha$, and another site $\gamma$ in molecule 2 at $r_\gamma$ ($r_{\alpha_1\gamma_2} \equiv r_{\gamma_2} - r_{\alpha_1}$) irrespective of the orientations of the pair of molecules 1 and 2. The sites $\alpha$ and $\gamma$ may coincide with the atoms (centers of mass, or centers of charges) in the molecule. On the other hand, they may not (i.e., they may represent some fictitious but well-defined points in the molecule). The probability measure is related to the familiar apcf $g(r_{12}, \omega_1, \omega_2)$ through a geometric transformation. To clarify this relation, we first define the cc apcf in a center-of-mass-to-center-of-mass (CC) frame for, say, diatomics (see Figure XIV.5). Let the vector $\mathbf{R}_{12}$ be the CC distance. Next, we use a two-body fixed (TBF) relative (or the intermolecular) coordinate system with the $z$-axis coinciding with the vector direction $\mathbf{R}_{12}$. The Euler angles of diatomics 1 and 2 are given as $\omega_1$ and $\omega_2$, respectively, with respect to this $z$-axis, and $\omega_i = (\theta_i, \phi_i, \chi_i)$, where $\theta_i$ is the polar, $\phi_i$ the azimuthal, and $\chi_i$ the rotational angle. These are angles for the cc apcf $g(\mathbf{R}_{12}, \omega_1, \omega_2)$. Of course, it is also possible to define $g(\mathbf{r}_1, \mathbf{r}_2, \Omega_1, \Omega_2)$ in a space-fixed (SF) frame with distances $r_1, r_2$ and angles $\Omega_1, \Omega_2$ referring to a laboratory-fixed frame.

On the other hand, we define an equivalent site-to-site (SS) angular pcf (ss apcf) $G_{\alpha\gamma}(r_{\alpha_1\gamma_2}, \omega'_1, \omega'_2)$. This is achieved by moving the TBF $z$-axis from the CC axis to the SS axis, the latter vector pointing from site $\alpha_1$ to site $\gamma_2$. The new ss apcf is then dependent on the sites $\alpha$ and $\gamma$ chosen as the defining direction (thus the notation, $G_{\alpha\gamma}$). The Euler angles $\omega'_1$ and $\omega'_2$ now refer to the new SS frame. (See Figure XIV.5.) The lengths and angles between CC and SS frames are related by the geometric relations (for simplicity of notation, we shall denote $R_{12}$ by $R$, $r_{\alpha_1\gamma_2}$ by $\alpha\gamma$, and $r_{\gamma_1\alpha_2}$ by $\gamma\alpha$, etc.)

$$R^2 = \frac{1}{4}\left[ \alpha\alpha^2 + \alpha\gamma^2 + \gamma\alpha^2 + \gamma\gamma^2 - L_1^2 - L_2^2 \right] \tag{2.1}$$

$$\cos\theta_1 = \frac{1}{4L_1R}\left[ -\alpha\alpha^2 - \alpha\gamma^2 + \gamma\alpha^2 + \gamma\gamma^2 \right]$$

$$\cos\theta_2 = \frac{1}{4L_2R}\left[ \alpha\alpha^2 - \alpha\gamma^2 + \gamma\alpha^2 - \gamma\gamma^2 \right]$$

$$\cos\phi_{12} = \frac{-\alpha\alpha^2 + \alpha\gamma^2 + \gamma\alpha^2 - \gamma\gamma^2 - \cos\theta_1\cos\theta_2}{2L_1L_2\sin\theta_1\sin\theta_2}$$

where $L_1$ is the bond length of diatomic 1 (i.e., from site $\alpha_1$ to $\gamma_1$, $\alpha\neq\gamma$) and $L_2$ that of diatomic 2. In addition, the following angular relations hold:

$$c_1c_2 + s_1s_2c = c'_1c'_2 + s'_1s'_2c' = \cos\psi \tag{2.2}$$

where $s_i = \sin\theta_i$, $c_i = \cos\theta_i$, $c = \cos\phi_{12}$, $c'_i = \cos\theta_i^{\alpha\gamma}$, $s'_i = \sin\theta_i^{\alpha\gamma}$, and $c' = \cos\phi_{12}^{\alpha\gamma}$. Angle $\theta_1^\alpha$ is the polar angle of diatomic 1 in the SS frame going from site $\alpha_1$ to SS. (Similar definitions hold for the other angles.) $\cos\psi$ is the inner product of the unit bond vectors of the two diatomics; i.e., $\cos\psi \equiv \mathbf{L}_1\cdot\mathbf{L}_2 /|\mathbf{L}_1\mathbf{L}_2|$. Furthermore the following relations hold:

$$\alpha\gamma^2 = R^2 - \left[ (-1)^\alpha \frac{c_1L_1}{2} - (-1)^\gamma \frac{c_2L_2}{2} \right] \cdot 2R \tag{2.3}$$

$$+ \frac{L_1^2}{4} + \frac{L_2^2}{4} - \frac{1}{2}L_1L_2(-1)^{\alpha+\gamma}\cos\psi$$

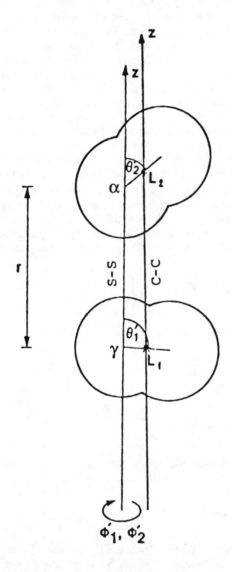

FIGURE XIV.5. *The center of mass-center of mass (CC) axis and the site-site (SS) axis for diatomics. Only the new angles (for SS axis) $\theta_1'$, $\theta_2'$, $\phi_1'$, and $\phi_2'$ are shown. $L_1$ and $L_2$ are the bond lengths of the diatomics.*

$$\alpha\gamma^2(s'_1)^2 = R^2s_1^2 + \frac{L_2^2}{4}(1 - \cos^2\psi) + (-1)^{\gamma}L_2R(c_2 - c_1\cos\psi)$$

From relations (2.1-3) one can relate the arguments of the apcfs cc apcf $g(\cdot)$ and ss apcf $G_{\alpha\gamma}(\cdot)$, and transform one function into the other. In formulas, ss pcf $g_{\alpha\gamma}$ is related to $G_{\alpha\gamma}$ by simple angle averaging:

$$g_{\alpha\gamma}(r_{\alpha_1\gamma_2}) = \int \frac{d\omega'_1}{4\pi} \int \frac{d\omega'_2}{4\pi} \, G_{\alpha\gamma}(r_{\alpha_1\gamma_2}, \omega'_1, \omega'_2) \tag{2.4}$$

With these transformations at hand we can easily convert one correlation function into the other [17]. Note that the apcf contains *more* structural information than the ss pcf. Thus the transformation from apcf to ss pcf will involve a net *loss* of information. Part of the statistics on angular distribution is lost. The transformation is not reversible. Next we examine the macroscopic properties that can be obtained from the ss pcfs.

## XIV.3 *Thermodynamic Properties*

The usefulness of the ss pcf lies partly in its direct access to a number of equilibrium properties, such as the internal energy, the mean squared force, and the isothermal compressibility. It also furnishes information on the structure factor measured by scattering experiments and the Kirkwood factors $G_n$. On the other hand, $g_{\alpha\gamma}$ alone does not contain sufficient information to determine uniquely the virial pressure and the mean squared torque. With an additional harmonic coefficient, $G_{100}$, one can calculate the virial pressure; with two coefficients, $G_{100}$ and $G_{200}$, one can calculate the mean squared torque. We first discuss the energy and the compressibility.

### INTERNAL ENERGY

The internal energy is given directly from the average of the interactions between the pairs

$$\frac{U}{NkT} = \frac{KE}{NkT} + \frac{\beta\rho}{2} \sum_{\alpha}^{m}\sum_{\gamma}^{m} \int dr_{\alpha_1\gamma_2} 4\pi(r_{\alpha_1\gamma_2})^2 \, u_{\alpha\gamma}(r_{\alpha_1\gamma_2}) g_{\alpha\gamma}(r_{\alpha_1\gamma_2}) \tag{3.1}$$

where $u_{\alpha\gamma}$ is the site-site pair potential; $r_{\alpha_1\gamma_2} = |r_{\gamma_2} - r_{\alpha_1}|$ is the scalar distance between positions $r_{\alpha_1}$ of site $\alpha$ in molecule 1, and $r_{\gamma_2}$ of site $\gamma$ in molecule 2. There are $m$ sites in each molecule.

### ISOTHERMAL COMPRESSIBILITY

The isothermal compressibility is given by [18]

$$kT\frac{\partial\rho}{\partial p}\bigg|_{T,V} = 1 + \frac{N}{V}\int dr \, h_{\alpha\gamma}(r) \tag{3.2}$$

where $h_{\alpha\gamma}$ is the site-site total correlation function $g_{\alpha\gamma} - 1$ for any pair of sites $\alpha$ and $\gamma$, and $N$ is the number of molecules in the volume $V$. The compressibility is independent of the pair $\alpha\gamma$ chosen, since in ISM the following zeroth moment condition holds for all pairs $\alpha\gamma$ and $\eta\zeta$:

$$\frac{N}{V} \int d\mathbf{r}\ h_{\alpha\gamma}(r) = \frac{N}{V} \int d\mathbf{r}\ h_{\eta\zeta}(r) \tag{3.3}$$

## MEAN SQUARED FORCE

The mean squared force (MSF) $<f^2>$ for pairwise additive potentials is given by

$$<f^2> = \frac{\rho kT}{16\pi^3} \int d\mathbf{R}_{12}\ d\omega_1\ d\omega_2\ R_{12}^2\ \Delta u(R_{12}\omega_1\omega_2) g(R_{12}\omega_1\omega_2) \tag{3.4}$$

where $\Delta u$ is the Laplacian of the pair potential $u$ with respect to the CC distance $R_{12}$. This equation can be changed into the SS frame via a reformulation of the Laplacian (Tildesley and Streett [19]). The result is

$$<f^2> = 4\pi\rho kT \sum_\alpha \sum_\gamma \int dr_{\alpha\gamma}\ r_{\alpha\gamma}^2 \Delta' u_{\alpha\gamma}(r_{\alpha\gamma}) g_{\alpha\gamma}(r_{\alpha\gamma}) \tag{3.5}$$

where $\Delta' u_{\alpha\gamma}$ is the Laplacian of the site-site pair potential $u_{\alpha\gamma}$ with respect to the SS distance $r_{\alpha\gamma}$. To obtain the virial pressure and the mean squared torque, information in addition to the ss pcf is needed. In the case of pressure, the harmonic coefficient $G_{100}$ must be known.

## PRESSURE

The virial theorem says that the pressure of a fluid is the ensemble average

$$P = \rho kT - \frac{N\rho}{6} <\sum_\alpha \sum_\gamma R_{12} \frac{\partial u_{\alpha\gamma}}{\partial R_{12}} (r_{\alpha_1\gamma_2})> \tag{3.6}$$

where, as before, $\mathbf{R}_{12}$ is the center-to-center vector and $\mathbf{r}_{\alpha_1\gamma_2}$ is the site-to-site vector. Carrying out the transformation between the CC and SS frames, we get

$$P = \rho kT - \frac{\rho^2}{6} \sum_\alpha \sum_\gamma \int dr_{\alpha\gamma} \frac{\partial u_{\alpha\gamma}}{\partial r_{\alpha\gamma}}\ g_{\alpha\gamma}(r_{\alpha\gamma}) < \hat{\mathbf{r}}_{\alpha\gamma} \cdot \mathbf{R}_{12}>_\omega \tag{3.7}$$

where $\hat{\mathbf{r}}_{\alpha\gamma}$ is the unit vector in the direction of $\mathbf{r}_{\alpha\gamma}$, and the angular average $< \cdots >_\omega$ is taken with respect to the SS frame:

$$<\hat{\mathbf{r}}_{\alpha\gamma} \cdot \mathbf{R}_{12}>_\omega \equiv \frac{\int d\omega' \int d\omega'\ G_{\alpha\gamma}(r_{\alpha\gamma},\ \omega'_1,\ \omega'_2)(\hat{\mathbf{r}}_{\alpha\gamma} \cdot \mathbf{R}_{12})}{\int d\omega'_1 \int d\omega'_2\ G_{\alpha\gamma}(r_{\alpha\gamma},\ \omega'_1,\ \omega'_2)} \tag{3.8}$$

with $\mathbf{R}_{12} = \mathbf{r}_{\alpha\gamma} - \mathbf{L}_\gamma + \mathbf{L}_\alpha$. Thus, going over to a spherical harmonic expansion, we have

$$\frac{\beta p}{\rho} = 1 - \frac{4\pi\beta\rho}{6} \sum_\alpha \sum_\gamma \int dr_{\alpha\gamma} \, r_{\alpha\gamma}^2 \, \frac{du_{\alpha\gamma}}{dr_{\alpha\gamma}} \tag{3.9}$$

$$\cdot \left\{ r_{\alpha\gamma} g_{\alpha\gamma}(r_{\alpha\gamma}) - \frac{1}{\sqrt{3}} \left[ L_\alpha G_{100}^{\alpha\gamma}(r_{\alpha\gamma}) + L_\gamma \, G_{100}^{\gamma\alpha}(r_{\gamma\alpha}) \right] \right\}$$

where $G_{100}^{\alpha\gamma}$ is the 100 spherical harmonic coefficient of the ss apcf $G(r_{12},\omega'_1,\omega'_2)$

$$G_{100}^{\alpha\gamma}(r) = (4\pi)^{-1} \int d\omega'_1 \int d\omega'_2 \, G_{\alpha\gamma}(r, \, \omega'_1, \, \omega'_2) Y_{10}(\omega_1) Y_{00}(\omega_2) \tag{3.10}$$

$$G_{100}^{\gamma\alpha}(r) = (4\pi)^{-1} \int d\omega'_1 \int d\omega'_2 \, G_{\alpha\gamma}(r, \, \omega'_1, \, \omega'_2) Y_{00}(\omega'_1) Y_{10}(\omega'_2) \tag{3.11}$$

In general, $G_{100}^{\alpha\gamma} \neq G_{100}^{\gamma\alpha}$. $L_\alpha$ is the bond radius from the center of mass to site $\alpha$. Formula (3.9) is valid for linear molecules only. Otherwise, the spherical harmonics from (3.7) and (3.8) would contain the index $\nu$ for the rotational angle $\chi$.

**MEAN SQUARED TORQUE**

Theoretically the mean squared torque (MST) is given by

$$\langle \Gamma^2 \rangle = \frac{4\pi\rho kT}{16\pi^2} \int dR_{12} d\omega_1 d\omega_2 \, R_{12}^2 \, g(R_{12}\omega_1\omega_2) Q_1^2 \, u(R_{12}\omega_1\omega_2) \tag{3.12}$$

where $Q_1^2$ is the angular Laplacian, defined by

$$Q_1^2(\cdot) \equiv \frac{c_1}{s_1} \frac{\partial(\cdot)}{\partial\theta_1} + \frac{\partial^2(\cdot)}{\partial\theta_1^2} + \frac{1}{s_1^2} \frac{\partial^2(\cdot)}{\partial\phi_1^2} \tag{3.13}$$

where $s_1 = \sin\theta_1$ and $c_1 = \cos\theta_1$. To carry out this integration, we must assume a specific potential model. In the following, we assume the site-site LJ potential. Upon transforming from CC to SS coordinates, we obtain

$$\langle \Gamma^2 \rangle = 4\pi\rho kT \sum_\alpha \sum_\gamma \left\{ \frac{1}{6} \int dr_{\alpha\gamma} \, r_{\alpha\gamma}^2 \, \Delta u(r_{\alpha\gamma}) L^2 g_{\alpha\gamma}(r_{\alpha\gamma}) \right. \tag{3.14}$$

$$+ (2\sqrt{3})^{-1} \int dr_{\alpha\gamma} \, r_{\alpha\gamma} \, A u(r_{\alpha\gamma}) L G_{100}^{\alpha\gamma}(r_{\alpha\gamma})$$

$$\left. + (6\sqrt{5})^{-1} \int dr_{\alpha\gamma} \, r_{\alpha\gamma} \, B u(r_{\alpha\gamma}) L^2 G_{200}^{\alpha\gamma}(r_{\alpha\gamma}) \right\}$$

where $A$ and $B$ are operators contained in the angular Laplacian $Q_1^2$. Operator $B$ contains the second-order derivatives, and operator $A$ contains the first-order derivatives. The results given are for homonuclear diatomics. For other geometries, one should consult the original derivation (Tildesley et al. [20]). $G_{100}$ and $G_{200}$ are the harmonic coefficients of the ss apcf $G(r_{12}\omega'_1\omega'_2)$.

## XIV.4. *The Ornstein-Zernike Relation Generalized*

The interaction site theories on structure are relations, usually in the form of integral equations, between the site-site correlation functions and the site-site potentials. Most existing theories are based on one of several versions of the Ornstein-Zernike relation which have been generalized from the monatomic to the polyatomic case. We recall that the OZ relation in the monatomic case was used to *define* the direct correlation function. For ISM, the generalized OZ serves the same purpose. However, there is a major difference here: *in the polyatomic case, there is no uniquely defined site-site direct correlation function (ss dcf)*. Or, in other words, while the site-site total correlation function (ss tcf) is well defined in terms of probability, the ss dcf contained in the matrix inverse to the tcf depends on the intramolecular components one extracts from the OZ relation. (The sense of the statement will be made precise in the following.) For example, one could use the angle-dependent tcf of molecular fluids in a generalized OZ equation (called *angular* OZ)

$$h(\mathbf{R}_{12}\omega_1\omega_2) \equiv C(\mathbf{R}_{12}\omega_1\omega_2) + \frac{\rho}{4\pi} \int d\mathbf{R}_3 \, d\omega_3 \, C(\mathbf{R}_{13}\omega_1\omega_3)h(\mathbf{R}_{32}\omega_3\omega_2) \qquad (4.1)$$

whereby an angular dcf $C(\mathbf{R}_{12}\omega_1\omega_2)$ is defined (Gray and Gubbins [21]). Next, the ss dcf is assumed to be the Euler angle average of the angular dcf

$$C_{\alpha\gamma}(r_{\alpha_1\gamma_2}) \equiv \langle C(\mathbf{R}_{12} + \mathbf{L}_\gamma - \mathbf{L}_\alpha, \, \omega_1\omega_2)\rangle_{\omega_1\omega_2} \qquad (4.2)$$

The angular average is carried out at fixed distance $r_{\alpha\gamma}$ from site $\alpha$ in 1 to site $\gamma$ in 2. On the other hand, one could *postulate* a *reasonable* OZ-like equation (OZ-RISM), as was done by Chandler and Anderson [22], then separate out an *intra*molecular structure factor $\omega$ from the compressibility matrix (see below). The remainder term is designated as the ss dcf. The resulting equation is the well-known OZ of the *reference interaction site model* or RISM:

$$\left[\left[\frac{\delta}{\rho^{(1)}}\right]\omega^{-1}(rr') - C_{\mathrm{RISM}}(rr')\right]_{\alpha\zeta} \left[(\rho^{(1)}\delta)\omega(r'r'') + F(r'r'')\right]_{\zeta\gamma} = \delta_D(rr'')\delta_{\alpha\gamma} \qquad (4.3)$$

All terms here are matrices: $\delta$ is the unit matrix, and $(\rho^{(1)}\delta)$ is the matrix with site densities as diagonal elements. The ss dcf in RISM is not equivalent to that in (4.2). Now we have at least two versions of the ss dcf. Inspired by the Kirkwood-Buff [23] formulation for mixtures, we offer a third possibility here. A *rigorous* formulation of the OZ relation is possible based on the matrix inverse relations of the two functional derivatives: the *fluctuation derivative* and the *compressibility derivative* (Lebowitz [24]). The ss dcf is contained in the compressibility derivative. However, this matrix alone does not determine a unique ss dcf. An additional condition is required. We propose the microscopic force balance embodied in the first member of the BBGKY hierarchy as the required condition. Application of BBGKY to the compressibility derivative uniquely determines an ss dcf consistent with the balance of linear momenta. We shall call this OZ the *consistent OZ* (consistent with the BBGKY relation) or the OZ *inverse matrix* (OZ-IM) relation.

It has been shown by Lebowitz [25] that for monatomics (MA) the OZ relation is equivalent to the inverse matrix multiplication of two functional derivatives: namely the fluctuation derivative $\delta\rho^{(1)}/\delta[-\beta w]$ and the compressibility derivative $\delta[-\beta w]/\delta\rho^{(1)}$, where $\rho^{(1)}$ is the nonuniform singlet density, and $w(\cdot)$ is an external potential generating the nonuniformity.

$$\frac{\delta(-\beta w(r))}{\delta \rho^{(1)}(r')} \cdot \frac{\delta \rho^{(1)}(r')}{\delta(-\beta w(r''))} = \delta_D(rr'') \qquad (4.4)$$

where $\delta_D$ is the Dirac delta, and the repeated arguments are integrated. (In this case, the argument $r'$ is integrated, $\int dr' \, (\cdot)$.) Relation (4.4) is exact since it expresses the mathematical definition of inverses. The physical interpretation of the compressibility derivative $\delta(-\beta w)/\delta \rho^{(1)}$ derives from the Gibbs-Duhem relation $\rho \partial \mu / \partial \rho = \partial P / \partial \rho$ upon noting that $w \iff -\mu$. The same relation could be generalized to polyatomics (PL) by interpreting the terms in (4.4) as matrices with species as indices. Meanwhile we recall the thermodynamic derivatives in a uniform mixture:

$$\frac{\partial \beta \mu_i}{\partial \rho_j} = \frac{\delta_{ij}}{\rho_i} - \int dr \, C_{ij}(r) \qquad (4.5)$$

$$\frac{\partial \rho_j}{\partial \beta \mu_k} = \rho_k \delta_{jk} + \rho_j \rho_k \int dr \, h_{jk}(r)$$

These equations form the basis of the Kirkwood-Buff (KB) theory of solutions. The two derivatives in (4.5) are reciprocals of each other. Since the external potential $w(\cdot)$ is thermodynamically equivalent to the chemical potential $-\mu$, the difference being the nonuniformity, equation (4.4) could be interpreted as the nonuniform counterpart of the uniform derivatives (4.5) (one of which contains the dcf). In fact, the OZ relation is a nonuniform restatement of the product of the uniform thermodynamic chemical potential and density derivatives. The total correlation function derives its physical meaning from concentration fluctuations in an open system. This corresponds to the derivative of the density $\partial \rho / \partial \mu$. The direct correlation function gives the compressibility of the system. This corresponds to the derivative of the chemical potential $\partial \mu / \partial \rho$ (see the Gibbs-Duhem relation). These two derivatives are thus tied together by the reciprocity relation. The result is the KB formulas (4.5). This connection with thermodynamics will be the *fil meneur* of our approach in generalizing the OZ relation to polyatomics.

Next, for polyatomics, we assert that a mixture of monatomics will coalesce into polyatomics when we allow the atoms to *fuse* together. This connection was recognized recently by Labik et al. [26] in deriving interesting relations between the chemical potentials of dumbbells and hard spheres by treating dumbbells as fused hard spheres. Thus the site-site approach for polyatomics could be viewed as the atom-atom interactions in a mixture of *fused* monatomics. This representation derives from the chemical physics point of view (study of the physical foundations of chemistry). The KB equations (4.5) then form the basis of the new polyatomic site-site OZ relations. We consider $N$ polyatomic molecules in an external potential field $w(\cdot)$. We adopt the atomic picture, i.e. the polyatomics are viewed as fused atoms. The statistical mechanics is formulated with respect to these atoms. Without loss of generality, let these molecules be diatomics of distinct atomic species $a$ and $b$ (i.e., heteronuclear $a$-$b$ diatomics). Note that the total number of molecules $N = N_a = N_b$. The potential energy is the sum of *inter*molecular $u(a_i b_j)$ and *intra*molecular interactions $\bar{u}(a_i b_i)$:

$$V_N + W_N = \sum_{i<j}^{N} \sum_{\alpha\gamma} u(\alpha_i\gamma_j) + \sum_{k}^{N} \bar{u}(a_kb_k) + \sum_{m}^{N} \sum_{\eta} w(\eta_m) \qquad (4.6)$$

where $\eta_m$ represents the position vector of site $\eta$ in a molecule labeled m: $\eta = a, b, \ 1 \le m \le N_a$ or $N_b$. $W_N$ is the sum of external potentials $w(\cdot)$. The grand canonical ensemble partition function is given by

$$\Xi[w] \equiv \sum_{N_a \ge 0}^{\infty}\Bigg|_{N_b = N_a} \int_{\Gamma} d\{N_a\}\ d\{N_b\} \exp[-\beta V_N - \beta W_N] \qquad (4.7)$$

The limits of integration $\Gamma$ are confined to a region of phase space reserved for non-decomposed diatomics. This formulation is based on an atomic picture. However, decomposition of the diatomics is not allowed. Therefore the summation always maintains $N_a = N_b$. The singlet density is

$$\rho_a^{(1)}(a_1;w) \equiv \frac{1}{\Xi[w]} \sum_{N_a \ge 1}^{\infty}\Bigg|_{N_b = N_a} \frac{z_a^{N_a} z_b^{N_b}}{(N_a - 1)!\ N_b!} \qquad (4.8)$$

$$\cdot \int_{\Gamma} d\{N_a - 1\} d\{N_b\} \exp[-\beta V_N - \beta W_N]$$

The pair densities are differentiated by inter- and intramolecular types; i.e.,

$$\rho_{ab}^{(2)}(a_1b_2;w) = \frac{1}{\Xi[w]} \sum_{N_a \ge 1}^{\infty}\Bigg|_{N_b = N_a} \frac{z_a^{N_a} z_b^{N_b}}{(N_a - 1)! N_b(N_b - 2)!} \qquad (4.9)$$

$$\cdot \int_{\Gamma} d\{N_a - 1\} d\{N_b - 1\} \exp[-\beta V_N - \beta W_N]$$

$$\bar{\rho}_{ab}^{(2)}(a_1b_1;w) = \frac{1}{\Xi[w]} \sum_{N_a \ge 1}^{\infty}\Bigg|_{N_b = N_a} \frac{z_a^{N_a} z_b^{N_b}}{(N_a - 1)! N_b!} \qquad (4.10)$$

$$\cdot \int_{\Gamma} d\{N_a - 1\} d\{N_b - 1\} \exp[-\beta V_N - \beta W_N]$$

where the overbar indicates *intra*molecular correlations. Note that the Boltzmann counting in the definitions reflects the diatomic nature of the molecules. We now return to eq. (4.4) and evaluate the functional derivatives for a mixture of $a$ and $b$. We shall use the shorthand notations

$$e \equiv \exp[-\beta V_N], \qquad F \equiv \frac{z_a^{N_a} z_b^{N_b}}{N_a! N_b!} \qquad (4.11)$$

$$e_a \equiv \exp[-\beta\sum_k^{N_a} w_a(k)], \qquad e_b \equiv \exp[-\beta\sum_l^{N_b} w_b(l)]$$

The functional derivative $\delta\rho^{(1)}/\delta[-\beta w]$ could be obtained from the singlet density (4.8) as

$$\frac{\delta\rho_a^{(1)}(a_1)}{\delta[-\beta w_a(a_x)]} = \frac{1}{\Xi} \sum_{N_a \geq 1}^{\infty} N_a \cdot F \cdot \int d\{N_a - 1\} d\{N_b\} \; e \cdot e_a \cdot e_b \left[ \delta_D(a_1 a_x) + \sum_{k=2}^{N_a} \delta_D(a_k a_x) \right] \quad (4.12)$$

$$- \frac{1}{\Xi} \rho_a^{(1)}(a_1) \cdot \sum F \cdot \int d\{N_a\} d\{N_b\} \; e \cdot e_a \cdot e_b \left[ \sum_{k=1}^{N_a} \delta_D(a_k a_x) \right]$$

$$= \rho_a^{(1)}(a_1) \delta_D(a_1, a_x) + \frac{1}{\Xi} \sum N_a(N_a - 1) \cdot F \cdot \int d\{N_a - 2\} d\{N_b\} \; e \cdot e_a \cdot e_b$$

$$- \rho_a^{(1)}(a_1)\rho_a^{(1)}(a_x)$$

$$= \rho_a^{(1)}(a_1)\delta_D(a_1 a_x) + \rho_{aa}^{(2)}(a_1 a_x) - \rho_a^{(1)}(a_1)\rho_a^{(1)}(a_x)$$

Similarly,

$$\frac{\delta\rho_a^{(1)}(a_1)}{\delta[-\beta w_b(b_x)]} = \frac{1}{\Xi} \sum N_a \cdot F \cdot \int d\{N_a - 1\} d\{N_b\} \; e \cdot e_a \cdot e_b \left[ \delta_D(b_1 b_x) + \sum_{l=2}^{N_b} \delta_D(b_l b_x) \right] \quad (4.13)$$

$$- \rho_a^{(1)}(a_1)\rho_b^{(1)}(b_x)$$

$$= \frac{1}{\Xi} \sum N_a \cdot F \cdot \int d\{N_a - r_{a1}\} d\{N_b - r_{b1}\} \; e \cdot e_a \cdot e_b$$

$$+ \frac{1}{\Xi} \sum N_a(N_b - 1) \cdot F \cdot \int d\{N_a - r_{a1}\} d\{N_b - r_{b2}\} \; e \cdot e_a \cdot e_b$$

$$- \rho_a^{(1)}(a_1)\rho_b^{(1)}(b_x)$$

$$= \overline{\rho}_{ab}^{(2)}(a_1 b_x) + \rho_{ab}^{(2)}(a_1 b_x) - \rho_a^{(1)}(a_1)\rho_b^{(1)}(b_x)$$

In matrix form

$$\begin{bmatrix} \dfrac{\delta\rho_a^{(1)}(a_1)}{\delta[-\beta w_a(a_x)]} & \dfrac{\delta\rho_a^{(1)}(a_1)}{\delta[-\beta w_b(b_x)]} \\[4mm] \dfrac{\delta\rho_b^{(1)}(b_1)}{\delta[-\beta w_a(a_x)]} & \dfrac{\delta\rho_b^{(1)}(b_1)}{\delta[-\beta w_b(b_x)]} \end{bmatrix} \qquad (4.14)$$

$$= \begin{bmatrix} \rho_a^{(1)}(a_1)\delta_D(a_1a_x) + F_{aa}(a_1b_x) & \bar{\rho}_{ab}^{(2)}(a_1b_x) + F_{ab}(a_1b_x) \\ \bar{\rho}_{ba}^{(2)}(b_1a_x) + F_{ba}(b_1a_x) & \rho_b^{(1)}(b_1)\delta_D(b_1b_x) + F_{bb}(b_1b_x) \end{bmatrix}$$

$$= \begin{bmatrix} \rho_a^{(1)}(a_1)\omega_{aa}(a_1a_x) + F_{aa}(a_1b_x) & \rho_a^{(1)}(a_1)\omega_{ab}(a_1b_x) + F_{ab}(a_1b_x) \\ \rho_b^{(1)}(b_1)\omega_{ba}(b_1a_x) + F_{ba}(b_1a_x) & \rho_b^{(1)}(b_1)\omega_{bb}(b_1b_x) + F_{bb}(b_1b_x) \end{bmatrix}$$

where $F_{ab} \equiv \rho_{ab}^{(2)} - \rho_a^{(1)}\rho_b^{(1)}$ is the Ursell truncated correlation function. We have introduced the *intramolecular structure* $\omega_{\alpha\gamma}$, defined as $\sum_\eta \rho_\alpha^{(1)}\delta_{\alpha\eta}\omega_{\eta\gamma} \equiv \rho_\alpha^{(1)}\delta_D(\alpha\gamma)\delta_{\alpha\gamma} + \bar{\rho}_{\alpha\gamma}^{(2)}(1 - \delta_{\alpha\gamma})$. To complete the OZ-IM relation, we must find the inverse matrix of (4.14). We could start by examining the singlet dcf for the diatomic fluid. As was done before in MA, the singlet dcf could be obtained by extracting factors from the singlet density function (4.8):

$$\rho_a^{(1)}(a_1) = z_a e^{-\beta w_a(a_1)}$$ (4.15)

$$\cdot \frac{1}{\Xi} \sum_{N_a \geq 1} \frac{z_a^{N_a-1} z_b^{N_b}}{(N_a-1)!N_b!} \int d\{N_a-1\}d\{N_b\}\, e \cdot e_b \cdot \exp\left[-\beta \sum_{k=2}^{N_a} w_a(k)\right]$$

Rearrangement gives

$$\exp[C_a^{(1)}(a_1)] \equiv \rho_a^{(1)}(a_1)\Lambda_a^3\, e^{\beta w_a(a_1) - \beta\mu_a}$$ (4.16)

$$= \frac{1}{\Xi} \sum_{N_a \geq 1} \frac{z_a^{N_a-1} z_b^{N_b}}{(N_a-1)!N_b!} \int d\{N_a-1\}d\{N_b\}\, e \cdot e_b \cdot \exp\left[-\beta \sum_{k=2}^{N_a} w_a(k)\right]$$

Therefore the functional derivative $\delta(-\beta w)/\delta\rho^{(1)}$ is given by

$$\frac{\delta(-\beta w_a(a_1))}{\delta\rho_b^{(1)}(b_x)} = \frac{\delta_{ab}\,\delta_D(a_1b_x)}{\rho_a^{(1)}(a_1)} - \frac{\delta C_a^{(1)}(a_1)}{\delta\rho_b^{(1)}(b_x)}$$ (4.17)

In the monatomic case, the second term on the RHS, i.e., the functional derivative $\delta C_a^{(1)}/\delta\rho_a^{(1)}$, is identified as the pair dcf $C_{ab}^{(2)}$. However, should we make the same choice for polyatomics? There is no simple answer to this question. If again we *defined* the pair dcf $C^{(2)}$ to be this derivative, it is plain to see from eq. (4.16) that this $C^{(2)}$ would contain intra- as well as intermolecular correlations, or this $C^{(2)}$ is not an *inter*molecular quantity. What Chandler and Andersen [27] did was to extract from $\delta C^{(1)}/\delta\rho^{(1)}$ the intramolecular factor $\omega_{ab}^{-1}$ and call the remainder the RISM site-site dcf $C_{RISM}$, i.e.,

$$C_{ab,\,RISM}(a_1b_x) \equiv \frac{\delta C_a^{(1)}(a_1)}{\delta\rho_b^{(1)}(b_x)} - \left[\frac{\delta_{ab}\delta_D(a_1b_x)}{\rho_a^{(1)}(a_1)} - \frac{\omega_{ab}^{-1}(a_1b_x)}{\rho_a^{(1)}(a_1)}\right]$$ (4.18)

However, it is not obvious that this separation is "net," i.e., a *clean* separation of the intra- and intermolecular correlations. In fact, there is no fixed formula for separating neatly the inter- from the intra-part, because in the cluster diagrams for $h_{\alpha\gamma}$ the *intra*molecular bonds are interconnected with the *inter*molecular bonds. It has recently become evident that the Chandler-Andersen *division* has problems with certain molecular conformations (e.g., linear triatomics [28]), where $C_{RISM}$ exhibits divergent long-range behavior.

In the following we shall examine another way of *dividing* the intra- and intermolecular contributions. We start with the first member of the BBGKY hierarchy for site-site density functions (see next section for derivation)

$$\nabla_1[-\beta w_a(a_1)] = \nabla_1 \ln \rho_a^{(1)}(a_1) \tag{4.19}$$

$$+ \int db_1 \, \nabla_1 \beta \bar{u}(a_1 b_1) \, \bar{\rho}_{ab}^{(2)}(a_1 b_1)/ \, \rho_a^{(1)}(a_1)$$

$$+ \sum_{\eta = a,b} \int d\eta_2 \nabla_1 \beta u(a_1 \eta_2) \, \rho_{a\eta}^{(2)}(a_1 \eta_2)/ \, \rho_a^{(1)}(a_1)$$

where $\nabla_1 \equiv \partial/\partial \mathbf{r}_{a_1}$. Clearly, the gradient of the external potential $w_a$ is related to two distinct contributions, one intramolecular and the other intermolecular. Assuming the interchangeability of the gradient and functional differential operators (see Percus [29]), we obtain the compressibility derivative of $w_a$ *inside* the gradient operator:

$$\nabla_1 \left[ \frac{\delta - \beta w_a(a_1)}{\delta \rho_b^{(1)}(b_x)} \right] = \nabla_1 \frac{\delta \ln \rho^{(1)}}{\delta \rho^{(1)}} + \int db_1 \, \nabla_1 \beta \bar{u} \frac{\delta}{\delta \rho^{(1)}} \left[ \frac{\bar{\rho}^{(2)}}{\rho^{(1)}} \right] \tag{4.20}$$

$$+ \sum_{\eta} \int d\eta_2 \nabla_1 \beta u \frac{\delta}{\delta \rho^{(1)}} \left[ \frac{\rho^{(2)}}{\rho^{(1)}} \right]$$

where we have suppressed the subscripts and arguments. The RHS consists of three terms: the ideal gas (translational energy) term, the intramolecular contribution, and the intermolecular contribution. It is natural, in this context, to identify the *intermolecular* part as corresponding to the *intermolecular* ss dcf; i.e.,

$$\nabla_1 C_{a\eta}^{(2)}(\alpha_1 \eta_x) \equiv - \sum_{\xi} \int d\xi_2 \, \nabla_1 \beta u(\alpha_1 \xi_2) \frac{\delta}{\delta \rho_\eta^{(1)}(\eta_x)} \left[ \frac{\rho_{\alpha\xi}^{(2)}(\alpha_1 \xi_2)}{\rho_a^{(1)}(\alpha_1)} \right] \tag{4.21}$$

This equation defines the ss dcf by an integrodifferential equation. The *intra*molecular part shall be called $Q_{ab}$:

$$\nabla_1 Q_{a\eta}(a_1 \eta_x) \equiv \nabla_1 \frac{\delta \ln \rho_a^{(1)}(a_1)}{\delta \rho_\eta^{(1)}(\eta_x)} \tag{4.22}$$

$$+ \int db_1 \, \nabla_1 \beta \bar{u}(a_1 b_1) \frac{\delta}{\delta \rho_\eta^{(1)}(\eta_x)} \left[ \frac{\bar{\rho}_{ab}^{(2)}(a_1 b_1)}{\rho_a^{(1)}(a_1)} \right]$$

Therefore the *Ornstein-Zernike inverse matrix* (OZ-IM) relation reads

$$\left[\varrho_{\alpha\eta}(\alpha_1\eta_x) - C^{(2)}_{\alpha\eta}(\alpha_1\eta_x)\right]\left[\rho_\eta^{(1)}(\eta_x)\delta_{\eta\xi}\ \omega_{\xi\gamma}(\xi_x\gamma_2) + F_{\eta\gamma}(\eta_x\gamma_2)\right] = \delta_{\alpha\gamma}\delta_D(\alpha_1\gamma_2) \qquad (4.23)$$

where repeated indices (e.g., $\eta = a$, $b$) are summed, and repeated arguments (e.g., $\mathbf{r}_{\eta_x}$) are integrated. It is a simple matter to show that this division into intra- and intermolecular parts is different from the RISM formulation of Chandler and Andersen (see Lee [30] [31]). One of the theoretical advantages of the OZ-IM division is that it is consistent with microscopic force balances, as given by the BBGKY equation, while the OZ-RISM (4.3) does not satisfy the BBGKY equation [32]. Note that OZ-RISM and OZ-IM share the same fluctuation derivative $\delta\rho^{(1)}/\ \delta(-\beta w)$. This ought to be so, since the derivative could be evaluated exactly (see eq. (4.13)). The difference lies in the division of terms in the compressibility derivative $\delta(-\beta w)/\ \delta\rho^{(1)}$. In other words, even though

$$\varrho_{\alpha\eta} - C^{(2)}_{\alpha\eta} = (\delta_{\alpha\eta}/\rho_\alpha^{(1)})\omega^{-1}_{\alpha\eta} - C_{\alpha\eta,\ \text{RISM}} \qquad (4.24)$$

(expressing the same result as (4.13)), but due to the fact

$$\varrho_{\alpha\eta} \neq (\delta_{\alpha\eta}/\rho_\alpha^{(1)})\ \omega^{-1}_{\alpha\eta} \qquad (4.25)$$

thus

$$C^{(2)}_{\alpha\eta,OZ-IM} \neq C^{(2)}_{\alpha\eta,RISM} \qquad (4.26)$$

The two ss dcfs are not the same. The two OZ equations are therefore not equivalent. So far there have been no applications of OZ-IM. On the other hand, OZ-RISM has been extensively used for calculating polyatomic fluid structure.

Next, we discuss the properties of the intramolecular structure $\omega_{\alpha\eta}$. For rigid polyatomic molecules (i.e., molecules with rigid bond lengths and angles), the intramolecular structure is a delta function with the range of a bond length $L$; e.g., for a diatomic molecule

$$\omega_{\alpha\gamma}(r) = \delta_{\alpha\gamma}\delta_D(r,\ 0) + \frac{\delta_D(r,\ L)}{4\pi L^2}(1 -\delta_{\alpha\gamma}) \qquad (4.27)$$

The Fourier transform of (4.27) is simply

$$\hat{\omega}(k) = \begin{bmatrix} 1 & j_o \\ j_o & 1 \end{bmatrix} \qquad (4.28)$$

where $j_o = \sin(kL)/(kL)$ is the spherical Bessel function.

## XIV.5. *Reference Interaction Site Theories*

Most calculations in literature employed the *reference interaction site approximation (RISA)* based on OZ-RISM. In our terminology, the reference interaction site model comprises two components: (1) the OZ-RISM relation (4.3), and (2) the closure relations on $C_{RISM}$ to be discussed below. The theory was originally proposed to treat hard-core fluids, such as fused hard spheres. Recently, it has been applied to *soft*-core fluids. A remarkable degree of success was achieved for dense fluids. Calculations have been made for fluids such as benzene [33], methane [34], methanol [35], and ammonia [36]. A closure condition is needed for the ss dcf. For hard-core fluids, Chandler and Andersen [37] proposed

$$C_{\alpha\gamma, \text{RISM}}(r) \approx 0, \quad \text{for } r > d_{\alpha\gamma} \tag{5.1}$$

This condition is an approximation. On the other hand, the ss pcf $g_{\alpha\gamma}(r)$ obeys the equation (an exact condition for hard cores)

$$g_{\alpha\gamma}(r) = 0, \quad \text{for } r < d_{\alpha\gamma} \tag{5.2}$$

where $d_{\alpha\gamma}$ is the collision diameter of the sites $\alpha$ and $\gamma$. Its value depends, in general, on the overall geometry of the polyatomics. For example, the carbon atom in carbon tetrachloride compared to the carbon in carbon dioxide will have a different value for the distance of closest approach simply because of steric interferences. The combined equations (4.3), (5.1) and (5.2) form the RISA theory— namely, with a given molecular geometry, these three equations form a complete set and can be solved to give the ss pcf. For soft potentials, there are several choices for the closure relation. One choice is the PY closure

$$C_{\alpha\gamma}(r) \approx y_{\alpha\gamma}(r) f_{\alpha\gamma}(r) \tag{5.3}$$

where $y_{\alpha\gamma}$ is the site-site background function, and $f_{\alpha\gamma}$ is the site-site Mayer factor. Recently, an HNC type closure [38] was used:

$$C_{\alpha\gamma}(r) \approx \exp\left[-\beta u_{\alpha\gamma}(r) + h_{\alpha\gamma}(r) - C_{\alpha\gamma}(r)\right] - h_{\alpha\gamma}(r) + C_{\alpha\gamma}(r) - 1 \tag{5.4}$$

The advantage of (5.4) is that the long-range behavior of the $C_{\text{RISM}}$ can be easily incorporated in this equation by adjusting the potential term $u$ in the exponent. We shall refer to (4.3), (5.1), and (5.2) as the original RISA, and addition of (5.3) or (5.4) for soft potentials as the soft RISA. The application of RISA to hard diatomics has been discussed in Chapter VIII for hard-core fluids. In the following we shall look at some soft potentials.

## XIV.6. *The Soft ISM*

The RISA method has been applied to polyatomics interacting with soft potentials. In this application, a closure condition on the ss dcf other than eq. (5.1) is needed. We have mentioned the PY and HNC closures. Most applications utilized these theories. Example applications to molecular models are discussed below.

### NITROGEN

Johnson et al. [39] have solved the PY closure (5.3) in RISA for liquid nitrogen. The site-site potential is a two-center LJ potential (2cLJ) with bond length 1.1 Å. The LJ parameters are $\sigma = 3.341$ Å and $\varepsilon = 0.6067 \times 10^{-14}$ erg. The condition studied is $\rho\sigma^3 =$

0.696 and $kT/\varepsilon = 4.03$. The N–N site-site pcf is shown in Figure XIV.6. It is compared with the MD results of Barojas et al. [40]. The agreement is reasonable except at the shoulder, where the RISA curve fails to show this behavior. Also included is the result from the repulsive part of the 2cLJ (with the WCA separation). The latter gives slightly poorer results.

### BENZENE

The model for benzene is taken to be a regular hexagon with an interaction site on each vertex [41]. The radius of the circle circumscribing this hexagon is chosen to be 1.756 Å. The sites coincide with the carbon atoms. A site-site LJ potential is used with parameters $\sigma = 3.5$ Å, and $\varepsilon/k = 77$ K. The temperature is 327 K, and the density $\rho = 0.00633$ Å$^{-3}$. The PY closure results for $g_{\text{site-site}}(r)$ are shown in Figure XIV.7. Comparison with simulation data (Evans and Watts [42]) shows that the qualitative features are correct. The errors occur at $r = 6.5$ Å, the third shoulder by 7%.

Fast solution algorithms for soft RISA have been recently developed (Monson [43], Enciso [44]) based on Gillan's method [45] for monatomics. This makes the solution numerically more efficient. Improved RISA, which attempts to obviate the divergent long-range behavior of the ss dcf, has been proposed by Rossky and Chiles [46]. This will be discussed further in the following sections.

## XIV.7. *The BBGKY Hierarchy for Polyatomics*

The distribution functions for polyatomics also obey the BBGKY equations as in the MA case. The derivation follows the same line. Denoting the gradient operator with respect to $\mathbf{r}_{a_1}$ as $\nabla_1$, we could differentiate the singlet density defined in (4.8) as

$$\nabla_1 \rho_a^{(1)}(a_1) = \frac{1}{\Xi} \sum N_a \cdot F \cdot \int d\{N_a - 1\} d\{N_b\} \tag{7.1}$$

$$\cdot \begin{bmatrix} - \nabla_1 \beta \overline{u}(a_1 b_1) \exp[-\beta \overline{u}(a_1 b_1)] \cdots \\ - \sum_\eta (N_\eta - 1) \nabla_1 \beta u(a_1 \eta_2) \exp[-\beta u(a_1 \eta_2)] \cdots \\ - \nabla_1 \beta w(a_1) \exp[-\beta w(a_1)] \cdots \end{bmatrix}$$

where we have exhibited only the Boltzmann factors of interest (the remainder terms are indicated by $\cdots$). Applying the definitions of the correlation functions, we obtain

$$\nabla_1 \rho_a^{(1)}(a_1) = - \int db_1 \nabla_1 \beta \overline{u}(a_1 b_1) \cdot \overline{\rho}_{ab}^{(2)}(a_1 b_1) \tag{7.2}$$

$$- \sum_\eta \int d\eta_2 \, \nabla_1 \beta u(a_1 \eta_2) \cdot \rho_{a\eta}^{(2)}(a_1 \eta_2) - \rho_a^{(1)}(a_1) \cdot \nabla_1 \beta w(a_1)$$

This is the *BBGKY relation* for the singlet density in a diatomic fluid. We observe the appearance of the intramolecular correlation $\overline{\rho}_{ab}^{(2)}$ as well as the intermolecular correlation $\rho_{ab}^{(2)}$. This distinguishes the site-site model from the the monatomic case. Higher-order hierarchies could be similarly derived. The results could be extended to other types of polyatomics.

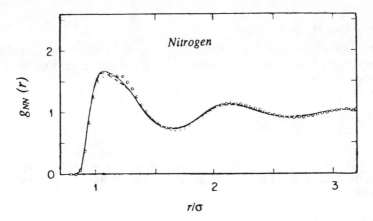

FIGURE XIV.6. The site-site pair correlation function for nitrogen at $\rho\sigma^3 = 0.696$ and $kT/\varepsilon = 4.03$. The site-site Lennard-Jones potential is used. Circles: computer simulation results. Solid line: RISA results. Dashed line: RISA results for the LJ repulsion only. (Johnson and Hazoumé, J. Chem. Phys., 1979)

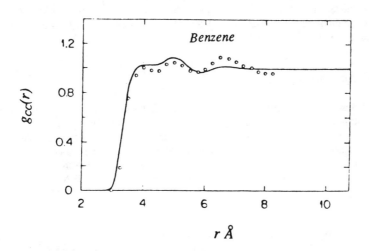

FIGURE XIV.7. The site-site pair correlation function for benzene at $\rho = 0.00633$ Å$^{-3}$ and $T = 328$ K. Circles: computer simulation results. Solid line: RISA calculations. (Johnson and Hazoumé, J. Chem. Phys., 1979)

THE PRESSURE TENSOR

It has been shown that the BBGKY equation is equivalent to the condition of linear momentum balance [47]. The result of the linear momentum equilibrium is the hydrostatic condition

$$\sum_{\nu} \nabla_{\nu} P^{\mu\nu}(\mathbf{r}) = - \sum_{\alpha} \rho_{\alpha}^{(1)}(\mathbf{r}) \nabla_{\mu} w(\mathbf{r}) \tag{7.3}$$

where $P^{\mu\nu}$ is the $\mu\nu$ component of the pressure tensor, $\mu,\nu = x,y,z$. Equating the gradients of the external potential in (7.2) and (7.3) gives the following expression for the pressure tensor:

$$\sum_{\nu} \nabla_{\nu} \cdot \beta P^{\mu\nu}(\mathbf{r}) = \sum_{\alpha} \nabla_{\mu} \rho_{\alpha}^{(1)}(\mathbf{r}) \tag{7.4}$$

$$+ \sum_{\alpha} \sum_{\eta} \int d\eta_1 \, \nabla_{\mu} \beta \overline{u}(r\eta_1) \cdot \overline{\rho}_{\alpha\eta}^{(2)}(r\eta_1)(1 - \delta_{\alpha\eta})$$

$$+ \sum_{\alpha} \sum_{\eta} \int d\eta_2 \, \nabla_{\mu} \beta u(r\eta_2) \cdot \rho_{\alpha\eta}^{(2)}(r\eta_2)$$

We have obtained a site-site expression for the pressure tensor of polyatomics. The first term on the RHS is the kinetic (ideal-gas) contribution, the second is due to *intra*molecular forces $-\nabla\beta\overline{u}$, and the third is from the usual *inter*molecular forces $-\nabla\beta u$. The pressure tensor itself is known to be not uniquely defined on the microscopic level [48]. However, the gradient of the tensor is well defined. Equation (7.4) gives the definition of $P^{\mu\nu}$ modulo a divergenceless factor.

# XIV.8. *Modifications of RISM*

As we have seen by several examples, the RISA theory works reasonably well for hard polyatomics. It is also accurate for dense fluids of soft molecules. However, it performs poorly at low densities (in fact, it does not approach the second virial limit at low densities). Recent evidences have shown that RISA is an approximate theory, especially with regard to the long-wavelength properties (e.g., orientational correlations). One such example [49] is the dielectric constant, which is given in RISA by

$$\varepsilon = 1 + 3\gamma \tag{8.1}$$

where $\gamma = 4\pi\beta\mu^2\rho/9$. This is merely the ideal-gas limit. Another example is that for linear triatomics, the ss dcf from RISA is long-ranged [50]:

$$\lim_{r \to \infty} C_{RISM}(r) \sim r \tag{8.2}$$

This is contradictory to the RISA assumption on the short-rangedness of the hard-core dcf eq. (5.1). Other points of discrepancy include the facts that (1) the inverse of the intramolecular structure $\omega$ for fused hard spheres is not defined as $k \to 0$ [51]. Namely, at long-wavelengths all the matrix elements of $\omega$ approach unity. Thus $\omega$ is singular and cannot be inverted. (2) The expansions of the ss dcf based on the assumption OZ-RISM are long-ranged. For heteronuclear diatomics, Cummings and Stell [52] have shown that the $r$-expansion for $C_{RISM}(r)$ should contain a term in reciprocal $r$, i.e., $1/r$, at long wavelengths. Equation (8.2) bespeaks the same divergence. In other words, $C_{RISM}$ is long-ranged and, in some cases, unbounded. These inherent anomalies are due to the approximate nature of the OZ-RISM relation (4.3).

To correct the inadequacies in conventional RISM, several attempts have been made. Chandler, Silbey, and Ladanyi [50] proposed, from analyses of the cluster series, a *corrected* intramolecular structure $\Omega$, such that

$$[\Omega^{-1} - \rho C_0][\omega + \rho h] \equiv \delta_D, \qquad\qquad \Omega \equiv \omega + \Delta\Omega \qquad (8.3)$$

where $\Delta\Omega$ is a correction to the intramolecular structure $\omega$ used earlier in OZ-RISM. In other words, additional cluster diagrams in $C_{RISM}$ are taken out and put into $\omega$. This shows that the original division of the compressibility derivative in OZ-RISM (into $\omega^{-1}$ and $C_{RISM}$) is *improper*. The *proper* way is (8.3). $\Omega$ is found to be related to the dcf clusters $C_l$, $C_b$, and $C_r$ by [53]

$$\Omega = (I - \rho C_l)^{-1}(\omega + \rho C_b)(I - \rho C_r)^{-1} \qquad (8.4)$$

Before explaining these clusters, we first examine the cluster series of the site-site tcf

$$h_{\alpha\gamma} = h_0 + h_l + h_r + h_b \qquad (8.5)$$

where the *left (l)*, *right (r)*, and *both (b)* clusters of $h$ are defined as diagrams with inceptive $s$-bonds (where $\rho s \equiv \overline{\rho}^{(2)}$ as defined in (4.10)) attached to the left, right, and both sides of the labeled roots, respectively; subscript 0 indicates diagrams whose root points do not intersect with the $s$-bonds. Some examples are shown in Figure XIV.8. From the OZ relation, $h_{\alpha\gamma}$ can be expressed as a chain of the $s$-bonds and bonds representing the dcfs (i.e., dcfs are topological reductions of diagrams constituting $h$ provided that they do not contain nodal circles). As a consequence, one could distinguish four types of $C$-bonds— $C_0$, $C_l$, $C_r$, and $C_b$— with similar definitions. Figure XIV.9 gives a few examples of these $C$-clusters. There is an important connectivity condition in this cluster development, namely, in any *allowed* diagram no circle (field or root point) is to be intersected by two or more $s$-bonds. It was discovered that in OZ-RISM this connectivity condition is violated. It gives rise to *unallowed* diagrams where two or more $s$-bonds share one vertex, and this is the source of anomalies in the $C_{RISM}$ function. Equation (8.3) is apparently similar to the OZ-IM relation (4.23). It remains as an interesting exercise to show the equivalence of the two definitions.

Rossky and Chiles [54] have examined the so-called unallowed cluster diagrams of Chandler et al. and proposed a *properly connected* matrix for the OZ equation:

$$h = (1 + S\rho)C[1 - (\hat{\rho} + \rho S\rho)C]^{-1}(1 + \rho S) \qquad (8.6)$$

where $\hat{\rho}$ is a 2×2 matrix for each given site connecting the cluster diagrams for 0, $r$, $l$ and $b$:

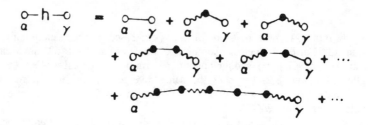

FIGURE XIV.8. Cluster series of the site-site total correlation function $h_{\alpha\gamma}$. The bond o——o represents the Mayer factor $f_{\alpha\gamma}$. The wavy line represents the intramolecular function $s_{\alpha\gamma}$. The first diagram belongs to $h_0$; the second belongs to $h_1$; the third belongs to $h_r$; the fourth belongs to $h_b$. (Chandler and Richardson, J. Phys. Chem., 1983)

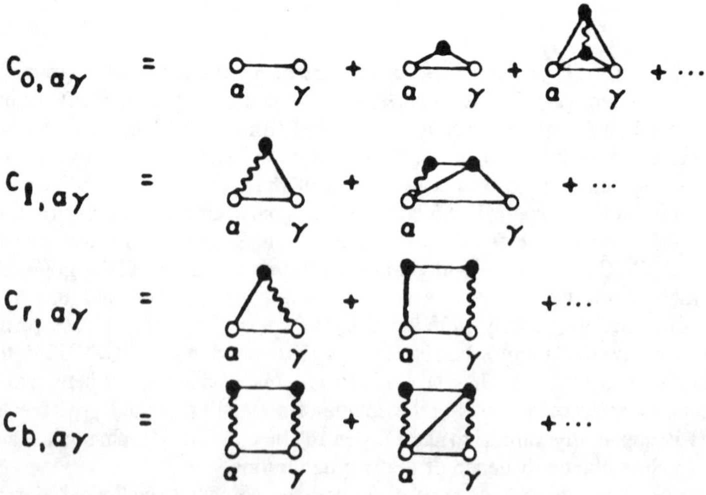

FIGURE XIV.9. Cluster series of the site-site direct correlation function $C_{\alpha\gamma}$. The bonds have the same meaning as in Figure XIV.8. These diagrams are obtained from a topological reduction of the diagrams for $h_{\alpha\gamma}$. (Chandler and Richardson, J. Phys. Chem., 1983)

$$\hat{\rho} = \begin{bmatrix} \rho & \rho \\ \rho & 0 \end{bmatrix} \tag{8.7}$$

This matrix is for use with the four types of clusters arranged in a matrix:

$$\begin{bmatrix} \text{Cluster}_0 & \text{Cluster}_r \\ \text{Cluster}_l & \text{Cluster}_b \end{bmatrix} \tag{8.8}$$

For example, in diatomics, the clusters of $h_{\alpha\gamma}$ are written as

$$\begin{bmatrix} h_{aa} & h_{ab} \\ h_{ba} & h_{bb} \end{bmatrix} = \begin{bmatrix} h_{aa,0} & h_{aa,r} & h_{ab,0} & h_{ab,r} \\ h_{aa,l} & h_{aa,b} & h_{ab,l} & h_{ab,b} \\ h_{ba,0} & h_{ba,r} & h_{bb,0} & h_{bb,r} \\ h_{ba,l} & h_{ba,b} & h_{bb,l} & h_{bb,b} \end{bmatrix} \tag{8.9}$$

Correspondingly, the $\hat{\rho}$ matrix is written as a block diagonal of size 2×2:

$$\begin{bmatrix} \rho & \rho & 0 & 0 \\ \rho & 0 & 0 & 0 \\ 0 & 0 & \rho & \rho \\ 0 & 0 & \rho & 0 \end{bmatrix} \tag{8.10}$$

The lower diagonal zero is to avoid the doubling of the $s$-bonds on a single vertex, because these are not *allowed* in a proper topological reduction of the $h$-clusters. The original OZ-RISM, which is not properly connected, produces such doubling $s$-bonds. On the other hand, the OZ-RISM is obtainable from (8.6) when one makes the (erroneous) substitution for $\hat{\rho}$ in eq. (8.7) by a matrix with full entries in $\rho$:

$$\begin{bmatrix} \rho & \rho \\ \rho & \rho \end{bmatrix} \tag{8.11}$$

However, applications of the new and properly connected OZ relation (8.6) to soft ISM (using HNC and PY closures) did not produce improvements over the original RISA. Further research is needed in order to clarify, as well as to justify, these alternative OZ relations.

## References

[1]   J.R. Sweet and W.A. Steele, J. Chem. Phys. **47**, 3022 (1967).

[2]   C.G. Gray and K.E. Gubbins, *Theory of Molecular Fluids*, vol. 1 (Clarendon Press, Oxford, 1984), p.34.

[3]   D. Chandler and H.C. Andersen, J. Chem. Phys. **57**, 1930 (1972).

[4]  L.L. Lee, Bull. Chem. Soc. Japan **58**, 710 (1985); and J. Chin. I. Ch. E. **16**, 103 (1985).

[5]  P.S.Y. Cheung and J.G. Powles, Mol. Phys. **30**, 921 (1975).

[6]  J.C. Raich and N.S. Gillis, J. Chem. Phys. **66**, 846 (1977).

[7]  M.S.H. Ling and M. Rigby, Mol. Phys. **51**, 855 (1984).

[8]  P.S.Y. Cheung and J.G. Powles, *Ibid.* (1975).

[9]  A.H. Narten, E. Johnson, and A. Habenschuss, J. Chem. Phys. **73**, 1248 (1980).

[10]  D.E. Williams, J. Chem. Phys. **47**, 4680 (1967).

[11]  A. Habenschuss and A.H. Narten, J. Chem. Phys. **74**, 5234 (1981).

[12]  A. Habenschuss and A.H. Narten, *Ibid.* (1981).

[13]  D.J. Tildesley and P.A. Madden, Mol. Phys. **42**, 1137 (1981).

[14]  J.M. Haile, private communication.

[15]  M.E. Cournoyer and W.L. Jorgensen, Mol. Phys. **51**, 119 (1984).

[16]  W.L.Jorgensen, J. Chem. Phys. **79**, 926 (1983).

[17]  See, e.g., J. Downs, K.E. Gubbins, S. Murad, and C.G. Gray, Mol. Phys. **37**, 129 (1979).

[18]  P.T. Cummings and G. Stell, Mol. Phys. **46**, 383 (1982).

[19]  D.J. Tildesley and W.B. Streett, Mol. Phys. **39**, 1169 (1980).

[20]  D.J. Tildesley, W. B. Streett and W.A. Steele, Mol. Phys. **39**, 1169 (1980).

[21]  C.G. Gray and K.E. Gubbins, *Theory of Molecular Fluids*, vol. 1 (Clarendon Press, Oxford, 1984).

[22]  D. Chandler and H.C. Andersen, J. Chem. Phys. **57**, 1930 (1972).

[23]  J.G. Kirkwood and F.P. Buff, J. Chem. Phys. **19**, 774 (1951).

[24]  J.L. Lebowitz and J.K. Percus, J. Math. Phys. **4**, 116 (1963); and J.L. Lebowitz, Phys. Rev. **133**, A895 (1964); see also F.J. Pearson and G.S. Rushbrooke, Proc. Roy. Soc. Edinburgh, A**64**, 305 (1957).

[25]  J.L. Lebowitz, *Ibid.* (1963).

[26]  S. Labik, A. Malijevsky, and I. Nezbeda, Mol. Phys. **60**, 1107 (1987).

[27]  D. Chandler and H.C. Andersen, *Ibid.* (1972).

[28]  P.T. Cummings and D.E. Sullivan, Mol. Phys. **46**, 665 (1982).

[29]  J.K. Percus, in *The Equilibrium Theory of Classical Fluids*, edited by H.L Frisch and J.L. Lebowitz (Benjamin, New York, 1964).

[30]  L.L. Lee, Bull. Chem. Soc. Japan **58**, 710 (1985).

[31]  L.L. Lee, J. Chin. Inst. Chem. Eng. **16**, 103 (1985).

[32]  L.L. Lee, Bull. Chem. Soc. Japan **58**, 710 (1985).

[33]  D.J. Evans and R.O. Watts, Mol. Phys. **32**, 93 (1976); A.H. Narten, J. Chem. Phys. **67**, 2102 (1977); and E. Johnson and R.P. Hazoumé, J. Chem. Phys. **70**, 1599 (1979).

[34]  H.J.M. Hanley and R.O. Watts, Mol. Phys. **29**, 1907 (1975); S. Murad, D.J. Evans, K.E. Gubbins, W.B. Streett, and D.J. Tildesley, Mol. Phys. **37**, 725 (1979); and A. Habenschuss and A.H. Narten, J. Chem. Phys. **74**, 5234 (1981).

[35]  W.L. Jorgensen, JACS **103**, 335 and 341 (1981); B.M. Pettitt and P.J. Rossky, J. Chem. Phys. **78**, 7296 (1983); and A.H. Narten and A. Habenschuss, J. Chem. Phys. **80**, 3387 (1984).

[36]  I.R. McDonald and M.E. Klein, J. Chem. Phys. **64**, 4790 (1976); and A.H. Narten, J. Chem. Phys. **66**, 3117 (1977).

[37]  D. Chandler and H.C. Andersen, *Ibid.* (1972).

[38]  See for example, P.J. Rossky, B.M. Pettitt and G. Stell, Mol. Phys. **50**, 1263 (1983).

[39]  E. Johnson and R.P. Hazoumé, J. Chem. Phys.**70**, 1599 (1979).

[40]  J. Barojas, D. Levesque and B. Quentrec, Phys. Rev. A**7**, 1092 (1973).

[41]  E. Johnson and R.P. Hazoumé, *Ibid.* (1979).

[42]  D.J. Evans and R.O. Watts, Mol. Phys. **32**, 93 (1976).

[43]  P.A. Monson, Mol. Phys. **47**, 435 (1982).

[44]  E. Enciso, Mol. Phys. **56**, 129 (1985).

[45]  M.J. Gillan, Mol. Phys. **38**, 1781 (1979).

[46]  P.J. Rossky and R.A. Chiles, Mol. Phys. **51**, 661 (1984).

[47]  P. Schofield and J.R. Henderson, Proc. R. Soc. London, A**379**, 231 (1982).

[48]  P. Schofield and J.R. Henderson, *Ibid.* (1982).

[49]  P.T. Cummings and G. Stell, Mol. Phys. **44**, 529 (1981).

[50]  P.T. Cummings and D.E. Sullivan, Mol. Phys. **46**, 665 (1982).

[51]  D. Chandler, R. Silbey, and B.M. Ladanyi, Mol. Phys. **46**, 1335 (1982).

[52]  P.T. Cummings and G. Stell, Mol. Phys. **46**, 383 (1982).

[53]  D. Chandler, C.G. Joslin and J.M. Deutch, Mol. Phys. **47**, 871 (1982).

[54]  P. Rossky and R.A. Chiles, Mol. Phys. **51**, 661 (1984).

## Exercises

1.  Solve the RISA equation for a low density state of nitrogen, using the site-site potential of Cheung and Powles, eq. (1.1). Compare with the known low density results for nitrogen.

2.   Show that the angular ss dcf from eq. (4.2) is different from the RISM ss dcf in (4.3).

3.   Show that the fluctuation derivative $\partial \rho^{(1)}/\partial(-\beta w)$ reduces to the uniform derivative $\partial \rho/\partial \beta \mu$ for constant density $\rho^{(1)} = \rho$.

4.   Clarify the relation between the OZ-IM relation (4.23) and the *proper* OZ (8.3) by using low density approximations for the clusters. Are the two formulations consistent with each other? Do they yield the same ss dcf?

5.   Find in the literature the site-site potential for *n*-butane and solve the RISA for this substance. Compare with scattering data at the same condition.

# CHAPTER XV

# ADSORPTION: THE SOLID-FLUID INTERFACE

In previous chapters, we have discussed fluid states where densities and temperatures were uniform throughout the physical content. In this chapter we shall introduce spatial inhomogeneities (in density and/or composition) into an otherwise uniform bath of fluid. The inhomogeneities result from an externally imposed potential, such as the ones generated by an interacting wall (the *adsorbent phase*). The bulk fluid responds by stratifications in density and composition at the interfaces (the *adsorbate phase*). We differentiate qualitatively two types of adsorption: physical adsorption (or *physisorption*) and chemical adsorption (or *chemisorption*). The two are distinguished by the formation (or absence) of chemical bonds between the surface and the adsorbed molecules. If chemical bonds are established between the adsorbate molecules and the surfaces, we call the adsorption chemisorption. Otherwise the adsorption is physisorption. The heat released during physisorption is on the order of 0.5 ~ 5 kcal/gmole. For chemisorption the heat of adsorption is on the order from 5 to 100 kcal/gmole. Table XV.1 gives the heat of adsorption, the average residence time $\tau$ of an adsorbed molecule on the surface, and the surface concentration $\Gamma$ of the adsorbate in physisorption as well as in chemisorption [1]. The demarcation between physisorption and chemisorption is not always well defined since the transition is gradual and the difference is a matter of degree.

Physisorption predominantly occurs at low temperatures (from the melting point to the critical point). It is characterized by (i) a rapid attainment of equilibrium, (ii) reversibility between adsorption and desorption, and (iii) a low energy of adsorption. Chemisorption occurs at high temperatures (with few exceptions such as the hydrogenation of ethylene on nickel at −78°C). The rate of adsorption may vary, from slow in some cases to fast in others. Once the gas is adsorbed, it is difficult to remove. Desorption is accompanied by chemical changes. For example, oxygen adsorbed on carbon is held strongly; on heating it evolves as mixtures of CO and $CO_2$. The process is not reversible. The major difference for chemisorption lies in its large energy of adsorption, which suggests establishment of chemical bonds between the gas molecules and the surface. In many cases, the adsorbed molecules undergo changes and lose their integrity. For example, ethylene molecules adsorbed on the rhodium (or platinum) (111) surface change to ethylidyne (−CH−CH$_3$) and a vinyl species (=CH−CH$_2$). Also, for chemisorption with high activation energies, the adsorption first occurs as physisorption, and then, more slowly, enters into chemical bonding with the surface. In this chapter, we shall confine our discussions to physisorption only. We are primarily interested in the *local structure* at the interfaces, i.e., on the scale from 3 to 10 Å near the boundaries. The phenomena occurring at the mesoscale, defined by de Gennes [2] as from 30 to 300 Å, though interesting in their own right, will not be discussed here. Thus we forego such phenomena as *wetting, prewetting, dry spreading,* and *contact-angle hysteresis.* Instead, we shall delineate the local structure, enthalpy of adsorption, and surface tension (specific surface free energy). These quantities are closely associated with molecular interactions at the interfaces. Recent reviews ([3] [4] [5] [6] [7] [8] [9] and [10]) on adsorption should be consulted for more detailed discussions.

Table XV.1.  Characteristics of Physisorption and Chemisorption

| Heat of Adsorption, $Q$ kcal/mole | Residence Time, $\tau$ s (25°C) | Surface Conc. $\Gamma$ moles/cm$^2$ | Comments |
|---|---|---|---|
| 0.1 | $10^{-13}$ | 0 | Adsorption nil. |
| 1.5 | $10^{-12}$ | 0 | Physisorption |
| 3.5 | $4 \times 10^{-11}$ | $10^{-12}$ | " |
| 9.0 | $4 \times 10^{-7}$ | $10^{-8}$ | " |
| 20.0 | 100 | | Chemisorption |
| 40.0 | $10^{17}$ | | " |

## XV.1. *The Surface Potentials*

In physisorption, the forces of interaction between the gas molecules and the molecules of the solid can be treated classically. For a crystalline solid, the orientation of the interfacial plane across the crystalline lattice determines the configuration of the adsorption sites on the surface. Figure XV.1 shows the (100) plane and the (111) plane of simple cubic lattices. On the (100) surface the unit cell is a square with four solid atoms (A) at the four corners and the adsorption site (S) in the center. The sides of such a cell are called saddle points (SP). On the (111) surface, the basic cell is rhombic with four atoms (A) at the four vertices, two adsorption sites (S), and five saddle points (SP). If we choose the coordinates $x$ and $y$ to lie on the interfacial plane such that the solid occupies the region $z < 0$, a gas molecule at a vertical distance $z > 0$ above the surface will interact with all the atoms of the solid (on the surface as well as below). Let the forces of interaction be given by the familiar Lennard-Jones 12-6 potential

$$w^{LJ}(r) = 4\varepsilon_{gs}\left[\left[\frac{\sigma_{gs}}{r}\right]^{12} - \left[\frac{\sigma_{gs}}{r}\right]^{6}\right] \tag{1.1}$$

where the subscript gs denotes gas-solid interaction, and $r$ is the distance between a gas molecule ($z > 0$) and a solid atom ($z \leq 0$). Due to the granularity of the surface, the sum of forces derived from all solid atoms varies periodically, depending on whether the gas molecule is directly above S, A, or SP. The attractive forces at S will be a maximum, and thus the probability of adsorption is greatest at S. If the adsorbed molecules are localized on the surface, they will be localized at the S points. Furthermore, if the molecules migrate from one S point to another S point, they will pass over the saddle point SP, since the barriers to dislocation are lowest across the SP points. Finally the A points are the locations where adsorption is least likely to occur because the solid atoms occupy those sites. Note that

$$r = \sqrt{x^2 + y^2 + z^2} \tag{1.2}$$

For fixed x-y coordinates, the forces on a gas molecule at any $z > 0$ due to the wall are the sum of pairs of interaction between the gas molecule and all solid atoms below

$$w(z)\big]_{x,y} = \sum_{\text{atom } i} w^{LJ}(r_i)\bigg]_{x,y} \tag{1.3}$$

where the subscripts x and y denote the fixed values x and y for the gas molecule, and $r_i$ is the distance between the solid atom i below and the gas molecule above. This is the pairwise additivity condition. Certainly, in addition to the wall forces (1.3), the bath particles also exert forces on the single gas molecule. Whether a particle is adsorbed on the surface or not is the result of the competition of opposing forces. If the wall forces are stronger, the bath particles will *wet* the wall. On the other hand, if attraction among particles is stronger, the wall will not be wetted and there will be a rarefaction of gases against the wall. At the S sites, the sum of the wall forces (1.3) will give a potential curve as shown in Figure XV.2 (the solid line marked S). The sharp potential well indicates strong attractive forces. Above the SP points, the sum (1.3) gives the curve marked SP. The well is less deep than that for S. Finally, at the A points the curve is the shallowest of the three. This reconfirms our previous characterizations of the S, SP, and A points.

Since the unit cell described above repeats itself on the crystalline surface, any translation along the x- or y-direction by a distance equal to the lattice vector of the cell will result in the same interaction energy as given by (1.3). The potential function $w(z)\big]_{x,y}$ must be periodic with respect to the geometry of the cell.

Let us consider some limiting cases of adsorption. If the gas molecules are small in comparison to the atoms of the solid, the adsorption will take place mostly at the S sites. The gas molecule will vibrate about the equilibrium position and will only rarely wander to neighboring S sites. In other words, once the gas molecule is captured on the surface, it cannot freely execute lateral translations. This type of adsorption is *localized*. On the other hand, if the gas molecule is large in comparison to the solid atoms, the differences among the S, SP, and A sites have negligible influences on the adsorbed gas. The molecule is free to travel from one spot to the other, actively engaged in surface diffusion. This type of adsorption is called *mobile* adsorption. In the limit of small solid atoms, the surface appears *smooth* to the gas molecules. We can replace the sum over the layers of atoms parallel, for example, to the (100) plane by an integral. Thus (1.3) becomes

$$w^{10\text{-}4}(z) = \sum_{k \text{ layers}} \int_{-\infty}^{+\infty} dx \int_{-\infty}^{+\infty} dy \, 4\varepsilon_{gs} \left[ \left( \frac{\sigma_{gs}^{12}}{(x^2+y^2+z_k^2)^6} \right) - \left( \frac{\sigma_{gs}^{6}}{(x^2+y^2+z_k^2)^3} \right) \right] \tag{1.4}$$

$$= 2\pi\varepsilon_{gs} \sum_{k \text{ layers}} \left[ \frac{2}{5} \left( \frac{\sigma_{gs}}{\sigma_{ss}} \right)^{12} \left( \frac{1}{z^*+k} \right)^{10} - \left( \frac{\sigma_{gs}}{\sigma_{ss}} \right)^{6} \left( \frac{1}{z^*+k} \right)^{4} \right]$$

The resulting interaction is given by a Lennard-Jones 10-4 potential. k is the depth index for layers beneath the surface, $z_k$ being the vertical distance from the kth layer of

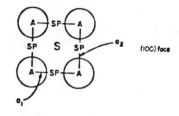

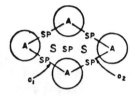

FIGURE XV.1. *The (100) and (111) faces of a cubic crystal. A: atoms of the solid. S: sites of adsorption. SP: saddle points posing as barriers against surface migration.* $a_1$ *and* $a_2$ *are lattice vectors.*

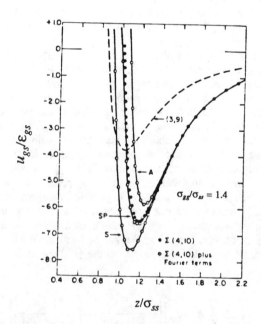

FIGURE XV.2. *The energies of interaction* $u_{gs}$ *between a gas molecule and the smooth solid surface. A: atom. S: site. SP: saddle point. •: LJ 10:4 potential. O: LJ 10:4 potential with Fourier terms. Dashed line: LJ 9:3 potential. (Steele, Pergamon, 1974)*

solid atoms to the gas molecule. $\sigma_{ss}$ is the Lennard-Jones size parameter for the solid atom–solid atom interaction. $z^* = z/\sigma_{ss}$ is the reduced vertical distance of the gas molecule above the surface. The integration is standard and is left as an exercise. We note that (1.4) differs from (1.3) in that (1.4) is a function of $z$ only while (1.3) depends on both the $x$- and $y$-coordinates. Note that (1.4) makes no distinction of the S, SP, and A sites. The black circles in Figure XV.2 show the energies of (1.4). The curve is very close to the SP curve. Furthermore, if the gas molecule is far from the surface, the layers of solid atoms beneath the surface would appear to be compacted. The summation over the $k$ layers can be approximated by an integration

$$
w^{9-3}(z) = 2\pi\varepsilon_{gs} \int_0^\infty dk \left[ \frac{2}{5} \left[ \frac{\sigma_{gs}}{\sigma_{ss}} \right]^{12} \left[ \frac{1}{z^*+k} \right]^{10} - \left[ \frac{\sigma_{gs}}{\sigma_{ss}} \right]^6 \left[ \frac{1}{z^*+k} \right]^4 \right]
$$

$$
= \frac{2\pi}{3}\varepsilon_{gs} \left[ \frac{2}{15} \left[ \frac{\sigma_{gs}}{\sigma_{ss}} \right]^{12} \left[ \frac{1}{z^*} \right]^9 - \left[ \frac{\sigma_{gs}}{\sigma_{ss}} \right]^6 \left[ \frac{1}{z^*} \right]^3 \right]
$$

(1.5)

In Figure XV.2, this function is shown as the dashed curve. It is seen that (1.5) is very different from any of the A, SP, and S curves. It is valid only at large $z$ (for $z > 4\ \sigma_{ss}$).

Let us return to the question of periodicity of the potential (1.3). If we choose the position of one of the A sites as origin, for the (100) surface of the cubic lattice the potential (1.3) can rewritten as

$$
\left[ \frac{w(r)}{\varepsilon_{gs}} \right] = 2\pi \left[ \sum_k \left\{ \frac{2}{5} \left[ \frac{\sigma_{gs}}{\sigma_{ss}} \right]^{12} \left[ \frac{1}{z_k^*} \right]^{10} - \left[ \frac{\sigma_{gs}}{\sigma_{ss}} \right]^6 \left[ \frac{1}{z_k^*} \right]^4 \right\} \right.
$$

(1.6)

$$
+ [\cos 2\pi x^* + \cos 2\pi y^*] \left\{ \frac{1}{15} \left[ \frac{\sigma_{gs}}{\sigma_{ss}} \right]^{12} \left[ \frac{\pi}{z_k^*} \right]^5 K_5(2\pi z_k^*) - 4 \left[ \frac{\sigma_{gs}}{\sigma_{ss}} \right]^6 \left[ \frac{\pi}{z_k^*} \right]^2 K_2(2\pi z_k^*) \right\}
$$

$$
\left. + \cos(2\pi x^*) \cos(2\pi y^*) \left\{ \frac{2}{15} \left[ \frac{\sigma_{gs}}{\sigma_{ss}} \right]^{12} \left[ \frac{\sqrt{2\pi}}{z_k^*} \right]^5 K_5(2\sqrt{2}\pi z_k^*) - 8 \left[ \frac{\sigma_{gs}}{\sigma_{ss}} \right]^6 \left[ \frac{\sqrt{2\pi}}{z_k^*} \right]^2 K_2(2\sqrt{2}\pi z_k^*) \right\} \right]
$$

where $x^* = x/\sigma_{ss}$, $y^* = y/\sigma_{ss}$, and $z_k^* = z_k/\sigma_{ss}$ is the vertical distance from the $k$th layer of solid atoms (the layers being parallel to the (100) surface) to the gas molecule (reduced by the $\sigma_{ss}$ parameter). $K_5$ and $K_2$ are the modified Bessel functions, $K_n(x)$, of the second kind of orders 5 and 2, respectively. This accounts for the damping of the potential. Also the first term in the summation is the Lennard-Jones 10-4 potential similar to (1.4). The second and third terms give the periodic effects (via the cosine functions). The white circles in Figure XV.2 give the potential curves for the S, SP, and A sites. They reproduce very well the discrete sums of (1.3). The derivation of (1.6) is rather lengthy (through a Fourier expansion) and is not given here (see, e.g., Steele [11]). The above discussions are valid for Lennard-Jones type interactions. They are applicable to the noble gases, such as adsorption of He, Ne, and Ar on the (100) and (111) surfaces of a xenon crystal [11].

For the adsorption of rare gases on the graphite basal plane (Figure XV.3), the following partially integrated Lennard-Jones 10-4 potential can be used:

$$w(z) = \pi \rho_a C_6 \left\{ \frac{\sigma_{gs}^6}{5z^{10}} - \frac{1}{2} \sum_i \frac{1}{z_i^4} \right\} \tag{1.7}$$

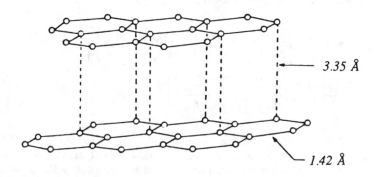

3.35 Å

1.42 Å

*FIGURE XV.3. Schematics of a graphite lattice. The bond length is 1.42 Å. The interplanar distance is 3.35 Å. (Steele, Pergamon, 1974)*

where $C_6$ is the constant in the expression for the pairwise attractive energy $C_6/r^6$, and is equal to $4\varepsilon_{gs}\sigma_{gs}^6$. $\rho_a$ is the solid number density. For graphite, $\rho_a = 0.382$ atoms/Å². The summation is on the $i$th layer of carbon atoms parallel to the top basal plane, and $z_i$ is the vertical distance from the $i$th layer to the gas molecule. Table XV.2 gives the interaction energies of helium, neon, argon, krypton, and xenon on the graphite surface [12]. It is seen that the direct sums for SP points have a well depth of similar magnitude to those of the A and S points. This indicates that there is little barrier to lateral motions for the adsorbed molecules. The adsorption can be treated as mobile.

For the adsorption of simple gas on ionic solids (e.g., argon on NaCl crystals), the potential energy due to the ion-induced dipole, ion-quadrupole, and ion-induced quadrupole interactions should be added to the existing dispersion and electron repulsion energies discussed earlier. The electric potential energy per unit charge at a distance $r_{ci}$ from a solid ion of type $c$ (e.g., Cl⁻ or Na⁺) at the $i$th position inside the solid can be

*Table XV.2. Interaction Energies of Noble Gases on Graphite Surface*

| Site | Gas-Solid Interaction, $\varepsilon_{gs}/k$ (Kelvin) | | | | |
|------|------|------|------|------|------|
|      | *He* | *Ne* | *Ar* | *Kr* | *Xe* |
| S    | 252  | 514  | 1140 | 1373 | 1838 |
| SP   | 235  | 485  | 1107 | 1335 | 1805 |
| A    | 232  | 481  | 1103 | 1332 | 1801 |

written as

$$w_{elec}(r) = \sum_c \sum_i \frac{q_c}{r_{ci}} \tag{1.8}$$

where $q_c$ is the charge on an ion of type $c$ (for $Na^+$, $q_c = |e|$, $e$ being the electron charge; for $S^{-2}$, $q_c = -2|e|$). Expression (1.8) can be approximated (to within 1%) by the equation

$$w_{elec}(z) = -\frac{16e}{\sqrt{2}\,a} \left[ \frac{e^{-2\sqrt{2}\pi(z/a)}}{1 + e^{-\pi\sqrt{2}}} \right] \cos\left[2\pi\frac{x}{a}\right] \cos\left[2\pi\frac{y}{a}\right] \tag{1.9}$$

for adsorption on the (100) face of NaCl type lattices, $a$ being the lattice vector length (or length of the sides of the unit cell). The $x$- and $y$-coordinates have as origin the position of an anion. If the dispersion and electronic repulsion forces can be represented by (1.6), the complete energy function should be the sum of all contributions

$$w_{total}(z) = w_{LJ\,(100)}(z) + w_{elec}(z) + \text{etc.} \tag{1.10}$$

Some of the interaction energies between the simple gases (e.g., argon) and the ionic crystals of NaCl, KCl, and KBr are given in Table XV.3 [13].

Table XV.3. *Interaction Energies of Gases over Ionic Crystals*

| System | $\varepsilon_{gs}$, cal/mol |
|---|---|
| Ar-NaCl (100) | 1876 |
| $N_2$-NaCl (100) | 2308 |
| Ar-KCl (100) | 1792 |
| $N_2$-KCl (100) | 2021 |
| Ar-KBr (100) | 2228 |
| $N_2$-KBr (100) | 2389 |
| Ar-KCl (111) | 2030 ($Cl^-$) |
| | 1307 ($K^+$) |

For rare gases adsorbed on metallic surfaces, the interaction cannot be assumed pairwise additive. The expression for the potential function is much more complicated. It involves contributions from the repulsive energy, the image-charge attraction, and polarization energies.

We have shown some simple examples of gas-solid interactions. Other adsorptive systems, such as surfactant molecules on substrates, enzyme molecules on membranes, colloidal particles, and electrical double layers abound. Specialized texts should be consulted.

## XV.2. *Interfacial Thermodynamics*

Let us consider a system of $N$ fluid molecules occupying the semi-infinite region $z > 0$ with Cartesian coordinates $x$, $y$, and $z$. The solid wall occupies the region $z \leq 0$. In the thermodynamic limit, $N \rightarrow \infty$ and $V \rightarrow \infty$, with constant ratio $N/V$. The total energy of this system consists of the kinetic energy (the translational, rotational, and vibrational energies) and the potential energy due to intermolecular interactions and interactions with the solid wall. Thus the Hamiltonian of the system is

$$H_N = KE + PE \tag{2.1}$$

where

$$KE = E_{\text{trans}} + E_{\text{rot}} + E_{\text{vib}} \tag{2.2}$$

$$PE = \sum_{i<j} u(r_{ij}) + \sum_k w(z_k) \tag{2.3}$$

where $E_{\text{trans}}$, $E_{\text{rot}}$, and $E_{\text{vib}}$ are the translational, rotational, and vibrational energies; $r_{ij}$ is the distance between molecular centers $i$ and $j$; and $z_k$ is the vertical distance of the $k$th molecule from the wall. $u(r)$ is the bath pair potential, and $w(z)$ is the potential of the wall. For anisotropic molecules, the angle dependence should be included in the arguments. Note that we have assumed pairwise additivity for the bath potential energy. If three-body forces are present, they should also be included. The canonical ensemble partition function is

$$Z_N = \frac{q_r q_v}{N! h^{3N}} \int d\mathbf{p}^N \, dr^N \exp\left[-\beta(E_{\text{trans}}) - \beta[\sum_{i,j}^N u(r_{ij}) + \sum_k^N w(z_k)]\right] \tag{2.4}$$

where $q_r$ and $q_v$ are the rotational and vibrational partition functions (separable in this case) of the $N$ gas molecules, respectively. Integration over the momenta gives the de Broglie thermal wavelength $\Lambda$:

$$Z_N = \frac{q_r q_v}{N! \Lambda^{3N}} \int dr^N \exp\left[-\beta[\sum_{i<j}^N u(r_{ij}) + \sum_k^N w(z_k)]\right] \tag{2.5}$$

If the wall forces are short-ranged, say a distance of a few molecular diameters, the bulk fluid would not experience the effects of the wall. Thus the bulk fluid would behave as if the wall did not exist. We denote this bulk fluid behavior by a superscript $g$ (for gas). Its partition function is given by (2.5) with all $w(\cdot) = 0$

$$Z_N^g = \frac{q_r q_v}{N! \Lambda^{3N}} \int dr^N \exp[-\beta \sum_{i<j}^N u(r_{ij})] \tag{2.6}$$

For molecules near the wall, the partition function $Z_N^s$ is given by (2.5). The superscript $s$ denotes the solid. The division of the gas phase into the surface region and the bulk region is arbitrary, since the transition is gradual. The wall forces diminish slowly. Thus the surface region extends into the bulk fluid until the surface effects disappear

eventually. It is safe to take the bulk properties at $z \to \infty$. The thermodynamic properties are related, as usual, to the partition function by

$$A^s = -kT \ln Z_N^s \tag{2.7}$$

$$U^s = -k \left[ \frac{\partial \ln Z_N^s}{\partial (1/T)} \right]_{V,N,\Sigma}$$

$$P^s = kT \left[ \frac{\partial \ln Z_N^s}{\partial V^s} \right]_{N,T,\Sigma}$$

where $\Sigma$ is the interfacial surface area. In the surface region, the surface forces modify
the thermodynamic pressure of the bulk fluid. If we denote the diagonal elements of the
pressure tensor by $P_{xx}$, $P_{yy}$, and $P_{zz}$, the anisotropy of the medium causes $P_{zz}$ to differ
from the values of $P_{xx}$ and $P_{yy}$. We define a spreading pressure $\phi^s$ by

$$\phi^s \equiv \frac{1}{\Sigma} \int d\mathbf{r} \left[ \frac{P_{xx}(r) + P_{yy}(r)}{2} - P_{zz}(r) \right] \tag{2.8}$$

For homogeneous surfaces, i.e., surfaces whose potential $w(\cdot)$ is independent of the coordinates $x$ and $y$, $P_{xx} = P_{yy}$. Thus

$$\phi^s = \frac{1}{\Sigma} \int d\mathbf{r} \, [P_{xx}(\mathbf{r}) - P_{zz}(\mathbf{r})] \tag{2.9}$$

In the bulk fluid, $P_{xx} = P_{zz}$; therefore $\phi^s = 0$. $\phi^s$ has the same units (force per unit length)
as the surface tension $\gamma$: in fact, $\phi^s = -\gamma$. $\phi^s$ is related to the partition function $Z_N^s$ by

$$\phi^s = kT \left[ \frac{\partial \ln Z_N^s}{\partial \Sigma} \right]_{N,T,V} \tag{2.10}$$

When the adsorbate phase is in equilibrium with the bulk fluid, the chemical potentials
are equal: $\mu^s = \mu^g$ and

$$\mu^s = \left[ \frac{\partial A^s}{\partial N} \right]_{T,\Sigma} = -kT \left[ \frac{\partial \ln Z_N^s}{\partial N} \right]_{T,\Sigma} \approx -kT \left[ \ln \frac{Z_{N+1}^s}{Z_N^s} \right] \tag{2.11}$$

where we have assumed $N$ large. A similar set of equations hold for $A^g$, $U^g$, $P^g$, and $\mu^g$.

As gas is adsorbed on the surface, heat is evolved and there is enthalpy change
per molecule adsorbed. To express this quantity, we define the *isosteric heat* of adsorption $q_{st}$

$$q_{st} \equiv \left( U^g + \frac{P^g}{\rho^g} \right) - \left( U^s + \frac{P^s}{\rho^s} \right) \tag{2.12}$$

where

$$U^s = -k \frac{\partial^2 \ln Z_N^s}{\partial N^s \partial(1/T)}, \qquad U^g = -k \frac{\partial^2 \ln Z_N^g}{\partial N^g \partial(1/T)} \qquad (2.13)$$

When the bulk gas is an ideal gas, $P^g/\rho^g = kT$, $Z_N^g = (V^g)^{N^g}/N^g!\Lambda^{3N}$, we have simply

$$q_{st} = kT + k \frac{\partial^2 N \ln(Q_N^s/V^N)}{\partial N^s \partial(1/T)} - \frac{P^s}{\rho^s} \qquad (2.14)$$

where $Q_N^s$ is the configurational integral of the adsorbate phase. $q_{st}$ is the molar enthalpy difference between that of the bulk gas and that of the adsorbed gas, i.e., the amount of heat evolved per molecule as the molecule is transferred from the bulk region to the surface region.

Surface quantities can be expressed in terms of the distribution functions, such as the spreading pressure, the energy, and the excess number. The derivation is similar to the case of bulk properties discussed earlier. The spreading pressure $\phi^s$ is defined by (2.10). Differentiation with respect to the surface area $\Sigma$ can be carried out using the method of Green (scaling the $x$–, $y$– and $z$-coordinates by the factors: $x/\sqrt{\Sigma}$, $y/\sqrt{\Sigma}$, and $z\Sigma/V$). The result is [14]

$$\phi^s \Sigma = \frac{1}{2} \int d\mathbf{r}_1\, d\mathbf{r}_2\, \rho^{(2)}(\mathbf{r}_1,\mathbf{r}_2) \left[ \frac{du(r_{12})}{dr_{12}} \right] \left[ \frac{z_{12}^2 - x_{12}^2}{r_{12}} \right] \qquad (2.15)$$

$$- \int d\mathbf{r}_1\, \rho^{(1)}(\mathbf{r}_1)[\nabla_1 w(\mathbf{r}_1) \cdot \mathbf{r}^{*1}]$$

where $\mathbf{r}^{*1} = (x_1/2, y_1/2, -z_1)$, $r_{12} = |\mathbf{r}_{12}| = |\mathbf{r}_2 - \mathbf{r}_1|$. $\rho^{(1)}$ and $\rho^{(2)}$ are the singlet and pair correlation functions subject to the influence of the surface $w$. Formula (2.15) is valid for pairwise additive potentials. For solids with homogeneous planar surfaces, $w$ is a function of $z$ only. Equation (2.15) is simplified to

$$\phi^s = \frac{1}{2} \int_{-\infty}^{+\infty} dz_1 \int d\mathbf{r}_{12}\, \rho^{(2)}(z_1, r_{12}) \frac{du(r_{12})}{dr_{12}} \frac{z_{12}^2 - x_{12}^2}{r_{12}} \qquad (2.16)$$

$$+ \int_{-\infty}^{+\infty} dz_1\, z_1 \rho^{(1)}(z_1) \frac{dw(z_1)}{dz_1}$$

Note that $x_{12} = x_2 - x_1$, $z_{12} = z_2 - z_1$, $\nabla_1 = \partial(\cdot)/\partial \mathbf{r}_1$, $\rho^{(2)}(\mathbf{r}_1,\mathbf{r}_2) = \rho^{(2)}(z_1, r_{12})$, and $\rho^{(1)}(\mathbf{r}_1) = \rho^{(1)}(z_1)$. The local internal energy $U^s(\mathbf{r})$ near the wall is

$$U^s(\mathbf{r}) = w(\mathbf{r})\rho^{(1)}(\mathbf{r}) + \frac{1}{2} \int d\mathbf{r}'\, u(\mathbf{r},\mathbf{r}')\rho^{(2)}(\mathbf{r},\mathbf{r}') \qquad (2.17)$$

When averaged over the surface region, the average adsorbate surface energy is

$$<U^s> = \int dr \ w(\mathbf{r})\rho^{(1)}(\mathbf{r}) + \frac{1}{2}\int dr \ dr' \ u(\mathbf{r},\mathbf{r}') \ \rho^{(2)}(\mathbf{r},\mathbf{r}') \tag{2.18}$$

where $<U^s>$ is the average configurational energy, in excess of the kinetic (ideal gas) contribution. Another quantity of interest in adsorption is the excess number of adsorption (or excess surface density) $N^e$ defined as

$$N^e \equiv \int_0^\infty dz \ [\rho^{(1)}(z) - \rho^g] \tag{2.19}$$

where $\rho^{(1)}$ is the singlet density near the wall and $\rho^g$ is the bulk number density. This quantity is a measure of the net amount of gas molecules adsorbed onto the surface in excess of the same number given by the bulk density. Thus $N^e$ indicates *wetting* (or adhesion) on the surface: if $N^e > 0$, the wall is adhesive; if $N^e = 0$, the wall does not affect the surface density profiles; and if $N^e < 0$, the wall repels the gas molecules. Our formulation here is valid on the molecular scale $\sim 3$ Å. This should be distinguished from the discussion of wetting on the *mesoscale* (from 30 Å to 300 Å).

Due to the existence of one additional external deformation coordinate $\Sigma$, the energy levels $\varepsilon_i$ now depend explicitly on the surface area. From results in Chapter II, the change in the average energy is

$$d<U>^s = \sum_i \varepsilon_i dp_i + \sum_i p_i d\varepsilon_i = \sum_i \varepsilon_i dp_i - <P^s> dV - <\phi^s> d\Sigma \tag{2.20}$$

where

$$d\varepsilon_i = - P_i^s dV - \phi_i^s d\Sigma \tag{2.21}$$

where $\varepsilon_i$, $P_i^s$, and $\phi_i^s$ are the microscopic energy level, the microscopic pressure, and the spreading pressure of the $i$th state, respectively, and $\sum_i \varepsilon_i dp_i$ is the heat $dQ$. The first law for energy conservation reads

$$dU^s = dQ - P^s dV - \phi^s d\Sigma \tag{2.22}$$

The term $\phi_s d\Sigma$ accounts for variations in the surface area.

## XV.3. *The Lattice Gas Model*

We have discussed two limiting cases of adsorption: (1) localized adsorption and (2) mobile adsorption. The former is valid in the limit of large solid atoms and small gas molecules, or strong attraction at the S sites and high barriers at the SP sites. In this section, we consider localized adsorption. When the gas molecule is adsorbed on the S site, it vibrates around the S point and occasionally "desorbs" from the surface. These sites could be considered as lattice points. This model of adsorption is called the *lattice gas model*. For a review of such models see Pandit et al. [15]. The Hamiltonian of $N$ such molecules is

$$H_N = KE + PE \tag{3.1}$$

The kinetic energy consists of the vibrational energies in the $x$-, $y$-, and $z$-directions (i.e., the occasional desorption dynamics is neglected). The PE term is set to zero here. We further distinguish *monolayer* and *multilayer* adsorption. For monolayer adsorption, the thickness of the adsorbed layer is about one molecular diameter. The attractive forces of the wall decay rapidly with distance, and only those gas molecules near the wall are affected. (This case is more likely in chemisorption, since chemical bonding occurs between adjacent molecules.) Thus only one layer of molecules is formed on the surface. Let there be $B$ adsorption sites on the surface, and let $N$ of them be occupied by the adsorbate molecules. This leaves $(B - N)$ sites unoccupied. The maximum value for $N$ is $N = B$. The situation should be viewed dynamically. At a given moment, there are $n$ gas molecules moving onto the surface and adsorbed at the active sites. At the same time, $n'$ adsorbed molecules on the surface desorb from some other sites and return to the gas phase. At equilibrium, the rate of adsorption is equal to the rate of desorption $n = n'$. The average number of sites occupied, $<N>$, is a constant. It can be expected that as the gas pressure is increased, more molecules tend to adsorb on the surface. When the pressure is lowered, molecules already adsorbed will desorb from the surface. In physisorption, the potentials are longer-ranged than the chemical forces. More gas molecules come under the influence of the surface forces, and multilayers build up. The total number of molecules adsorbed, $N$, in all the layers may very well exceed the number of active sites, $B$. In fact, as the bulk pressure approaches the vapor pressure of the gas, $N$ approaches infinity and *condensation* occurs (see Pandit [16]; there are several transitions taking place). We discuss two simple theories for lattice gas below: the *Langmuir theory* and the *Brunauer-Emmett-Teller (BET) theory*.

## LANGMUIR ADSORPTION ISOTHERM

The Langmuir theory treats the localized monolayer adsorption. The bulk gas is assumed to be an ideal gas. As the gas molecules are attached to the S sites on the surface, no lateral interaction between neighboring sites is allowed. The molecules vibrate in the $x$-, $y$-, and $z$-directions with frequencies $v_x$, $v_y$, and $v_z$ on the order of $10^{12}$ s $^{-1}$. This situation occurs at temperatures $kT$ lower than the energy barrier $\Delta U_0$ across the saddle points.

The partition function for adsorbate molecules factors into the partition functions of the cells. Thus

$$Z_N^s = q^N \tag{3.2}$$

where $q$ is the partition function of one single ideal-gas molecule adsorbed on the site and performing harmonic oscillations

$$q = q_x q_y q_z e^{-\beta U_0} \tag{3.3}$$

where $q_x$, $q_y$, and $q_z$ are the vibrational partition functions in the $x$-, $y$-, and $z$-directions, respectively, and

$$q_x = \frac{e^{-hv_x/2kT}}{1 - e^{-hv_x/kT}} \tag{3.4}$$

(similarly for $q_y$ and $q_z$). $U_0$ is the minimum energy of the potential at the S site. The factor $\exp(-\beta U_0)$ is added to ensure that the energies of interaction of the adsorbate molecules and the energies of interaction of the bulk molecules share the same reference energy level.

For $N$ identical molecules adsorbed on $B$ sites ($N$ being less than or at most equal to $B$), there are $B!/N!(B-N)!$ combinations. Thus the degeneracies are of the same magnitude. The partition function $Z_N^s$ takes this into account:

$$Z_N^s = \frac{B!}{N!(B-N)!} q^N \tag{3.5}$$

The Helmholtz free energy is

$$A^s = -kT \ln Z_N^s = -kT [B \ln B - N \ln N - (B-N) \ln (B-N) + N \ln q] \tag{3.6}$$

We have used the Stirling approximation for the logarithms. Let the ratio $N/B = \theta$; i.e., $\theta$ is the fraction of sites occupied. The spreading pressure $\phi^s$ is

$$\phi^s = -\left.\frac{\partial A^s}{\partial \Sigma}\right|_{T,N,V} \tag{3.7}$$

We denote specific area (i.e., surface area per site) by $a$; then $\Sigma = Ba$, and

$$\frac{\phi^s}{kT} = -\frac{1}{a}\frac{\partial \ln Z_N^s}{\partial B} = \frac{1}{a} \ln(1-\theta) \tag{3.8}$$

On the other hand, the chemical potential $\mu^s$ is given by

$$\frac{\mu^s}{kT} = -\left.\frac{\partial \ln Z_N}{\partial N}\right|_{B,T} = \ln\frac{\theta}{1-\theta} - \ln q \tag{3.9}$$

The adsorbate phase is in equilibrium with the bulk gas, thus $\mu^s = \mu^g$. For an ideal gas

$$\frac{\mu^g}{kT} = \ln P + \ln\frac{\Lambda^3}{kT} \tag{3.10}$$

Equating (3.9 and 10) gives

$$\theta = \frac{cP}{1+cP} \tag{3.11}$$

where

$$c \equiv q\left[\frac{\Lambda^3}{kT}\right] \tag{3.12}$$

Equation (3.11) is called the *Langmuir isotherm*, giving the amount of gas adsorbed as a function of gas pressure at a fixed temperature. At low $P$, the adsorption is linear with pressure; i.e., $\theta \approx cP$, as $P \to 0$. When $P \to \infty$, the surface goes to complete coverage $\theta = 1$. The real gas will "condense" for $P$ equal to the vapor pressure before $P$ reaches infinity. In addition, real gases form multilayers on the surface. This behavior is approximated by the BET theory.

## THE BET ADSORPTION ISOTHERM

In BET theory we allow multilayers to build up on the surface. Let there be $B$ sites on the surface and $N$ molecules adsorbed on the surface ($N$ may be greater than $B$ in this case). Again we assume that the molecules are ideal gas and that there are no lateral interactions between neighboring molecules. Since multilayers are allowed, each site S on the surface may accumulate a stack of molecules numbering from 0 at the bottom to a maximum of, say, $m$ molecules. Let there be $b_0$ unoccupied sites, $b_1$ sites with only one adsorbed molecule, $b_2$ sites with a stack of two molecules, etc., up to $b_m$ sites with stacks of $m$ molecules. Material balances give

$$b_0 + b_1 + \cdots + b_m = B \tag{3.13}$$

and

$$\sum_{i=0}^{m} i b_i = N$$

Let the partition function of a single stack of $i$ molecules be $q(i)$, the nature of which remains to be specified. The canonical partition function of $N$ adsorbate molecules is

$$Z_N^s = \sum_{(b)} \frac{B! q(0)^{b_0} q(1)^{b_1} q(2)^{b_2} \cdots q(m)^{b_m}}{b_0! b_1! \cdots b_m!} \tag{3.14}$$

where the summation is over the set of all possible partitions $(b_0, b_1, \ldots, b_m)$ of $N$, such that (3.13) is satisfied. For BET, it is more convenient to consider a grand canonical partition function

$$\Xi = \sum_{N=0}^{mB} Z_N^s z^N \tag{3.15}$$

where $z$ is the activity $\exp(\mu/kT)$. Substitution of (3.14) into (3.15) and use of the binomial expansion theorem yield

$$\Xi = \left[ q(0) + z q(1) + z^2 q(2) + \cdots + z^m q(m) \right]^B \equiv \xi^B \tag{3.16}$$

where

$$\xi \equiv \sum_{i=0}^{m} z^i q(i) \tag{3.17}$$

Since for a grand canonical ensemble $\phi^s \Sigma / kT = \ln \Xi$, we have

$$\frac{\phi^s \Sigma}{kT} = B \ln \xi \tag{3.18}$$

Also, the average number of molecules per site, $N_s$, is the average number $<N>$ of the grand canonical ensemble divided by the number of adsorption sites $B$:

$$N_s = \frac{<N>}{B} = \frac{1}{B} z \left[ \frac{\partial \ln \Xi}{\partial z} \right]_{B,T} = z \left[ \frac{\partial \ln \xi}{\partial z} \right]_T \tag{3.19}$$

The BET theory makes further assumptions regarding the nature of the *stack partition functions* $q(i)$. For empty sites, there is no energy, $q(0) = 1$. The first layer of adsorbate molecules is attached to the surface with the partition function $q(1)=q_1$. As the second layer of molecules is adsorbed, it is attached to the previous layer (i.e., on top of the first layer). This contributes to the partition function by a factor of $q_2$; i.e., $q(2) = q_1 q_2$. That $q(2)$ factors into $q_1$ and $q_2$ shows that the two layers of molecules are uncorrelated, a restrictive assumption for BET. Higher layers of molecules are similarly attached to the previous layers already in existence. BET also makes the approximation that the contributions from these layers to the partition function stay the same as that from the second layer, i.e., $q_2$.

$$q(0) = 1, \qquad\qquad q(1) = q_1 \tag{3.20}$$

$$q(2) = q_1 q_2, \qquad\qquad q(3) = q_1 q_2^2$$

$$q(n) = q_1 q_2^{(n-1)}, \qquad n = 2,3,4,...$$

The quantity $\xi$ can now be expressed as

$$\xi = 1 + \frac{q_1 z}{1 - q_2 z} \tag{3.21}$$

Carrying out the differentiation in (3.19) gives

$$N_s = \frac{q_1 z}{(1 - q_2 z + q_1 z)(1 - q_2 z)} \tag{3.22}$$

Since the adsorbate phase is in equilibrium with the bulk gas, $z^s = z^g$, and $z^g = \exp(\mu^g/kT) = P\Lambda^3 / kT$ (for ideal gas), we have

$$N_s = \frac{c_1 P}{(1 - c_2 P + c_1 P)(1 - c_2 P)} \quad \text{or} \quad \frac{cx}{(1 - x + cx)(1 - x)} \tag{3.23}$$

where

$$c_1 \equiv q_1 \frac{\Lambda^3}{kT}, \qquad c_2 \equiv q_2 \frac{\Lambda^3}{kT}, \qquad c \equiv \frac{c_1}{c_2}, \qquad x \equiv c_2 P \qquad (3.24)$$

Equation (3.23) is the *BET isotherm*. Its behavior is shown in Figure XV.4. For the chosen value $c_1/c_2 = 157$, it is shown that monolayer coverage is exceeded at a reduced pressure $c_2 P = 0.2$. As $c_2 P$ approaches 0.8, multilayers are formed ($N_s > 4$). (b L F

In practice, the Langmuir isotherm (3.11) and the BET isotherm (3.23) are used with the constants $c$, $c_1$, and $c_2$ treated as adjustable parameters in order to match the experimental data. They are functions of temperature (see definitions). Despite the crudeness of these models, they are useful for summarizing experimental information in a compact form. However, one should not read too much significance into the constants, since real molecules interact not only with the solid surfaces but also with neighboring molecules. The mechanisms of adsorption do not follow the Langmuir, or the BET theories.

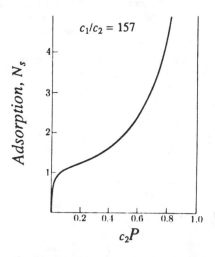

FIGURE XV.4. *The BET adsorption isotherm at the parameter ratio $c_1/c_2$ = 157. The rapid increase of adsorption at $c_2 P \sim 0.85$ indicates onset of vapor condensation. (Hill, Addison, 1960)*

Recent developments in statistical mechanics have enabled us to study more realistic interaction forces: first excluded volumes are taken into account (e.g., adsorption of hard spheres on a hard wall); then attractive and repulsive forces come into play, such as adsorption of Lennard-Jones molecules to Lennard-Jones 10-4 walls. More recently, topics such as the adsorption of ions and dipoles to a charged wall (electrical double layers) and the coexistence of solid and liquid phases at the melting line have been explored. Several studies have considered the adsorption of long-chain molecules onto a solid surface, and adsorbates inside a micropore (capillary or membrane networks). These systems will not be treated here. We shall consider a few simple cases below.

## XV.4.  *Adsorption of Hard Spheres on a Hard Wall*

The first step in modification of the simple models discussed so far is to endow the bulk molecules with an excluded volume, as in hard spheres. The substrate is given as a continuous solid with a hard and structureless surface. The smooth hard wall lacks any particular sites of registry. The adsorption is therefore mobile. The excess number of adsorption $N^e$ is determined by the density and structure of hard spheres near the surface. Such a model for adsorption is clearly simplistic. However, it is a useful one toward real molecules, by allowing the effects of the *excluded volume* to emerge and serving at the same time as an example of the statistical mechanics methods.

The Hamiltonian for this model is given by

$$H_N(\mathbf{r}^N,\mathbf{p}^N) = \sum_{i=1}^{N} \frac{p_i^2}{2m} + \sum_{i<j}^{N} u(r_{ij}) + \sum_{k=1}^{N} w(\mathbf{r}_k) \tag{4.1}$$

where $u(\cdot)$ is the hard-sphere pair potential, and $w(\cdot)$ is the potential of the hard wall:

$$u(r) = +\infty, \qquad\qquad r \le d \tag{4.2}$$

$$= 0, \qquad\qquad r > d$$

where $d$ is the hard-sphere diameter; and

$$w(z) = +\infty, \qquad\qquad z \le d/2 \tag{4.3}$$

$$= 0, \qquad\qquad z > d/2$$

where $z$ is the vertical distance measured from the surface. Snook and Henderson [17] have made a Monte Carlo simulation of this system and reported results on the singlet density $\rho^{(1)}(z)$ and the pair density $\rho^{(2)}(\mathbf{r}_1,\mathbf{r}_2, w) = \rho^{(2)}(z_1,z_2;r_{12})$, where $z_1$ and $z_2$ are vertical distances of molecules 1 and 2 from the solid surface and $r_{12}$ is the linear distance between the two molecules. The density states studied are $\rho d^2 = 0.57, 0.755, 0.81$, and $0.91$. Figure XV.5 shows the singlet density profile $\rho^{(1)}(z)$ near the wall. It is seen that sharper profiles are formed at higher densities. Due to the cutoff distances in simulation, only two peaks are shown here (i.e. bilayers). Adsorbed hard spheres form multilayers on a hard wall; neither is the adsorption localized. The pair correlation functions $g(z_1,z_2,r_{12})$ at different values of $z$ and $r_{12}$ are compared in Figure XV.6. The obvious question is what effects, if any, does the wall have on the liquid structure? At large vertical distances (i.e., for bath molecules far removed from the surfaces), the pcf $g(z_2,z_2,r_{12})$ obtained from MC simulation approaches that of a uniform hard-sphere fluid. Figure XV.6 gives the variations of the pcf with $r$ at different values of $z_1 = z_2$ (i.e., we look at the radial distributions of molecules on planes parallel to the surface). The curves are marked with values $z_1/d$ ($= z_2/d$.) The solid curves are the pcf for uniform hard spheres. At $\rho d^3 = 0.81$ and $z/d = 0.03$ (a plane adjacent to the surface), $g(z,z,r_{12})$ shows a higher contact value than the uniform pcf. Such enhancement is due to the quasi-two-dimensional nature of the hard spheres near the wall. The movements of the adsorbed spheres are constrained by the wall and must remain in the vicinity thereof. Freedom of movement is greatly restricted in the half-plane. This corresponds effectively to an *apparent* higher density, and thus a higher contact value.

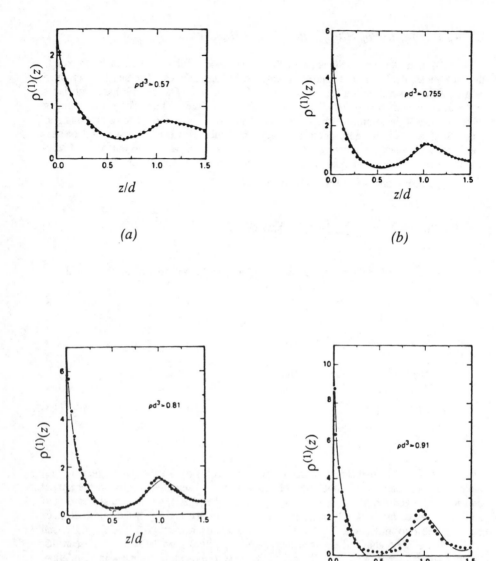

*FIGURE XV.5. Density profiles for hard spheres near a hard wall.* •: MC *data. Line: GMSA (generalized mean spherical approximation). (a) At reduced density* $\rho^* = \rho d^3 = 0.57$. *(b)* $\rho^* = 0.755$. *(c)* $\rho^* = 0.81$. *(d)* $\rho^* = 0.91$. *(Snook and Henderson, J. Chem. Phys., 1978)*

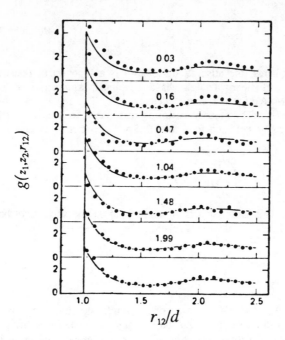

FIGURE XV.6. *Pair correlation functions for the hard-sphere/hard-wall system in parallel layers* $z_1 = z_2$ *and* $\rho d^3 = 0.81$. ●: *MC data. Line: the pcf* $g(r_{12})$ *of bulk fluid. The numbers give the values of* $z_1 = z_2$ *(in units of d). (Snook and Henderson, J. Chem. Phys., 1978)*

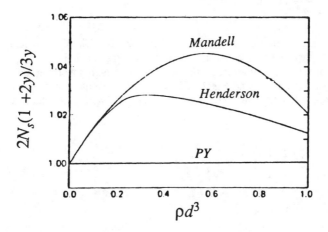

FIGURE XV.7. *The excess number* $N^e$ *of hard spheres near a hard wall. Comparison of PY, Mandell [20], and Henderson [21] theories. Note that the PY (scaled particle theory) adsorption is used as the reference. (Henderson, J. Chem. Phys., 1978)*

For hard spheres, the adsorption isotherm is a function of density only. Waisman, Henderson, and Lebowitz [18] have developed a generalized mean spherical approximation (GMSA) for the singlet density $\rho^{(1)}$. The results are shown in Figure XV.6 as solid curves. The theory performs reasonably well except at high densities. The excess number $N^e$ calculated from the GMSA $\rho^{(1)}(z)$ can be expressed as

$$N^e = \frac{9y(1+2y) + y^2(4-y)(1-4y)}{6[(1+2y)^2 - y^3(4-y)]} \tag{4.4}$$

where $y = \pi\rho d^3/6$. This expression is an approximation and should not be used for high densities. For hard spheres, the exact relation [19]

$$\rho^{(1)}(0) = \frac{P}{kT} \tag{4.5}$$

holds, where $\rho^{(1)}(0)$ is the contact value of the singlet density; and $P$ is the pressure of bulk hard spheres. Simulation results on $N^e$ have poor statistics because the integral $\int dz\,[\rho^{(1)}(z)-\rho]$ contains a difference term $\rho^{(1)}(z)-\rho$ that is close to zero. One could also use the Percus-Yevick theory to find $N^e$ (see section XV.6.). The result is

$$N^e = \frac{3y}{2(1+2y)} \tag{4.6}$$

Mandell [20] has extended the scaled particle theory to obtain an equation for $N^e$

$$N^e = \frac{3y}{2(1+2y)} (1 + 0.3y - 0.5y^2) \tag{4.7}$$

These three expressions are compared in Figure XV.7. The GMSA and the extended scaled particle theory are more accurate than the PY results [21].

It is of interest to investigate the adsorption of hard spheres to soft repulsive and attractive walls. Abraham and Singh [22] have made Monte Carlo simulations of hard spheres near a WCA-type repulsive Lennard-Jones 9:3 wall, characterized by the potential

$$w(z) = \pi\rho_s\varepsilon\sqrt{\frac{10}{3}}\left[\left[\frac{\sigma_s}{z}\right]^9 - \left[\frac{\sigma_s}{z}\right]^3\right] + \varepsilon_{min}, \qquad \text{for } z < z_0 \tag{4.8}$$

$$= 0, \quad \text{for } z \geq z_0$$

where $\varepsilon_{min}$ is the absolute value of the minimum energy for the LJ 9:3 potential; $\rho_s$ is the solid number density, taken to be 0.85; $\sigma_s^6 = 2/15$; and $z_0^6 = 2/5$. All quantities are measured in units of the hard-sphere diameter $d$. Simulations for both the LJ 9:3 wall (eq. (1.5)) and the repulsive LJ 9:3 wall (eq. (4.8)) were performed. The results are presented in Figure XV.8. The reduced temperature is $T^*=kT/\varepsilon=1.0$ and the bulk fluid densities are $\rho d^3 = 0.63$ and $\rho\sigma^3 = 0.65$. An interesting observation is that the adsorption of the hard spheres to a repulsive wall gives a peak (about 2.0) much lower than the peak for an attractive wall (2.78 for LJ 9:3 attractive potential). The attractive wall has a greater tendency to adsorb the hard spheres. The bulk fluid, which is composed of hard spheres, is purely repulsive and cannot pull the adsorbate away from the LJ 9:3 wall. Abraham and Singh [23] presented another simulation on LJ 12:6 fluid with the LJ 9:3 wall. The peak height (=1.4) is even lower than the hard-spheres/LJ 9:3 wall case. Since the wall is the same in both cases, the difference must lie in the bulk fluid: the hard spheres, being purely repulsive, lose their molecules to an attractive wall; the LJ 12:6 fluid, being attractive, is able to retain some of the molecules in the bulk. This point is abundantly illustrated in Figure XV.9, where the wall is hard and the fluid is of the LJ 12:6 type. It is seen that the LJ fluid does not even "wet" the wall. The cohesion within the LJ 12:6 fluid holds the adsorbate back from the wall. This led Abraham and Singh [24] to claim that a free surface is formed in the vicinity of the wall.

## XV.5. *Adsorption of Lennard-Jones Molecules*

In the last section, we discussed the excluded volume effects on adsorption. A more realistic model, the Lennard-Jones fluid, with attractive interaction is presented here. Two types of Lennard-Jones walls: the f.c.c. (100) and smooth 10-4 solids have been studied by computer simulation [25]. The wall potential for the (100) surface depends on the $x$-$y$ coordinates (see eq. (1.6)) as well as on $z$. Two parallel planes composed of solid LJ atoms were used in the simulation. The solid atoms were arranged at the f.c.c. sites with density $\rho_s\sigma^3 = 1.01$. The first basal plane was positioned at $z = 0$, the second at $z = -2.69$ Å. The potential curves at the S, SP, and A sites are displayed in Figure XV.10. On the other hand, the LJ 10-4 wall potential is given by

$$w^{10-4}(z) = 2\pi\rho\varepsilon \sum_{i=1}^{2} \frac{2\sigma^2}{5} \left[ \left[\frac{\sigma}{z_i}\right]^{10} - \left[\frac{\sigma}{z_i}\right]^{4} \right] \tag{5.1}$$

where $\varepsilon/k = 119.8$ K, $\sigma = 3.405$ Å, and $\rho$, the surface number density, is such that $\rho\sigma^2 = 0.8$ (this corresponds to a solid density $\rho_s\sigma^3 = 1.01$). The parameters chosen correspond to argon. For the first layer $z_1 = z$, and for the second layer $z_2 = z + 2.69$ Å. It was further proposed that an effective wall potential be given by averaging the Boltzmann factor, i.e.,

$$\exp[-\beta w^{\text{eff}}(z)] \equiv \frac{1}{A_u} \int_{A_u} dx \, dy \, \exp[-\beta w(x,y,z)] \tag{5.2}$$

where $w^{\text{eff}}$ is the effective wall potential, and $A_u$ is the area of the (100) unit cell. The resulting $w^{\text{eff}}$ is a softer potential than the SP interaction and is located between the SP curve and the S curve (with a potential minimum close to that of the SP curve; see Figure XV.10). We note that due to Boltzmann averaging, $w^{\text{eff}}$ is temperature dependent.

At $T^* = kT/\varepsilon = 1.0$ and $\rho\sigma^3 = 0.63$, the singlet density $\rho^{(1)}(z)$ is shown for the three wall systems, the (100) wall, the 10-4 wall, and the Boltzmann weighted wall (Figures XV.11 and 12). Figure XV.11 shows that the density profile of the LJ 10-4 wall does not correctly represent the (100) surface. The peak of the singlet density for the 10-4 wall is too high and occurs at larger $z$. The oscillations of the LJ 10-4 curve are out of phase with respect to the (100) curve. On the other hand, the density profile of the Boltzmann weighted potential is almost in phase with the (100) density curve. It slightly underestimates the first and second peaks.

Broughton et al. [26] have studied the melting of the f.c.c. (111) and (100) solids

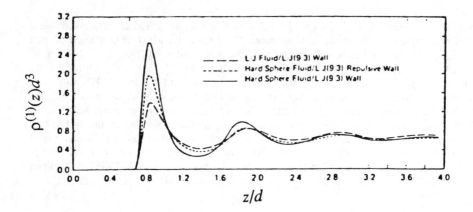

FIGURE XV.8. The density profile $\rho^{(1)}(z)d^3$ for the hard-sphere/LJ 9:3 repulsive wall system (dotted line); the hard-sphere/LJ 9:3 wall system (solid line); and the LJ 12:6 fluid/LJ 9:3 wall system (dashed line). The temperature $T^* = 1$, and $\rho^* = 0.65$.. (Abraham and Singh, J. Chem. Phys., 1978)

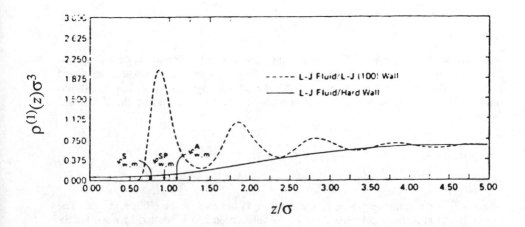

FIGURE XV.9. Comparison of the density profiles $\rho^{(1)}(z)\sigma^3$ for the LJ 12:6 fluid/LJ (100) wall system (dotted line); and the LJ 12:6 fluid/hard-wall system (solid line). $\phi$ gives the locations of the minimum energies for the A, S, and SP interactions of the (100) wall. The temperature $T^* = 1$, and $\rho^* = 0.63$. The hard wall is not "wetted" by the LJ fluid. (Abraham, J. Chem. Phys., 1978)

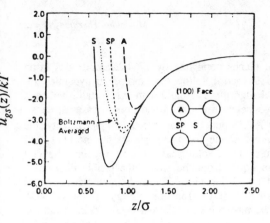

FIGURE XV.10. The interaction energy of the (100) surface of an fcc LJ solid at sites A, S, and SP. The Boltzmann averaged wall potential is also shown. The temperature $T^* = 1$. (Abraham, J. Chem. Phys., 1978)

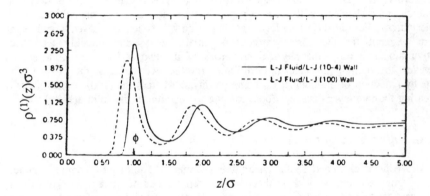

FIGURE XV.11. Comparison of the density profiles of the LJ 12:6 fluid/LJ (100) wall system (dotted line), and the LJ 12:6 fluid/LJ 10:4 wall system (solid line). $\phi$ denotes the potential minimum of the 10:4 wall. (Abraham, J. Chem. Phys., 1978)

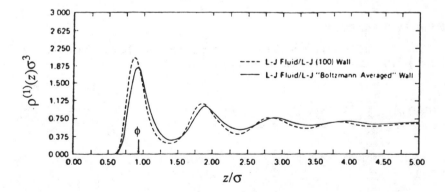

FIGURE XV.12. Comparison of the density profiles of the LJ 12:6 fluid/LJ (100) wall system (dotted line), and the LJ 12:6 fluid/Boltzmann averaged wall system (solid line). $\phi$ denotes the well minimum of the Boltzmann averaged potential. (Abraham, J. Chem. Phys., 1978)

into liquid. They used 14 to 16 crystal planes populated by 1764 LJ molecules in an f.c.c. configuration. The molecules were allowed to follow their molecular dynamics trajectories by a Nordsieck-Gear numerical algorithm. Equilibrium was established at a reduced temperature of $T^* = 0.67 \pm 0.005$ between the solid and liquid phases at densities $\rho^{*s} = 0.965$ and $\rho^{*L} = 0.83$. Figures XV.13 and 14 show the the singlet density, the energy density, and the average temperature of the coexisting phases for the (111) and (100) surfaces, respectively. The structure of the solid phase is quite regular. Beyond the Gibbs dividing surface (dividing the liquid from the solid), the profile dies down after four to five oscillations, typical of liquid behavior. The melting of the crystal is easily seen from the parallel layer radial distribution function $g(z,z,r)$. Figures XV.15 and 16 show the radial distributions of the molecules in different layers, rising from the crystal planes into the molten region. For the (111) surface, this transition occurs between the sixth and the seventh layers (numbers were enumerations used in the computer program). The solidlike radial distribution is characterized by a deep first minimum and a shoulder on the second peak. The shoulder is finally melted away in the liquid phase. For the (100) surface, melting occurs between the seventh and eighth layers. Except for the surface excess potential energy, and possibly the excess configurational entropy, the physical properties of the two surfaces are quite similar. However, the melting mechanism is different for the two surfaces: the (111) surface melts by reduction of the number of atoms per layer, whereas the (100) surface melts by increasing the interlayer distances (both causing a reduction of the solid density).

## XV.6. *Integral Equation Theories*

We have examined so far adsorptive systems exemplified by computer simulations. In this section, we look at theoretical interpretations of adsorption. Historically, for physisorption at low densities, virial expansions of $\rho^{(1)}$ in terms of the bulk density were used [27]. Following advances in liquid state theories, other approaches have been devised, notably integral equations, BBGKY hierarchy for nonuniform systems, density functional methods, and related mean-field theories (such as the van der Waals theory). We shall not discuss all these methods, but will refer the reader to a number of excellent reviews [28]. In view of our *local* approach, we shall discuss only the integral equations, the BBGKY, and the density functional approaches.

### THE BBGKY HIERARCHY

The wall-bath system is an inhomogeneous system due to the influence of wall forces $w(\cdot)$. The Hamiltonian of the fluid molecules is

$$H_N(\mathbf{r}^N,\mathbf{p}^N) = \sum_{i=1}^{\infty} \frac{p_i^2}{2m} + \sum_{1=i<j}^{\infty}\sum u(r_{ij}) + \sum_{k=1}^{\infty} w(\mathbf{r}_k) \tag{6.1}$$

The canonical partition function has been given earlier in (2.5). The singlet density $\rho^{(1)}(1; w)$ is

$$\rho^{(1)}(1; w) = \frac{N}{Q_N}\int d\{N-1\}\ \exp[-\beta\textstyle\sum u(r_{ij}) - \beta\textstyle\sum w(\mathbf{r}_k)] \tag{6.2}$$

where $Q_N$ is the configurational integral. If we apply the method of BBGKY to (6.2), we obtain the form

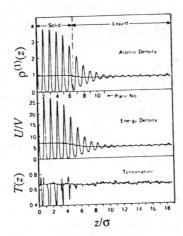

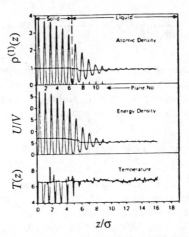

FIGURE XV.13. Melting of the LJ fcc (111) interfaces. Shown are the density profile, energy density, and temperature fluctuations from a molecular dynamics simulation. The vertical dashed line locates the Gibbs dividing surface. $T^* = 0.678$, density of solid $\rho^* = 0.964$, density of liquid $\rho^* = 0.830$. (Broughton, Bonissent, and Abraham, J. Chem. Phys., 1981)

FIGURE XV.14. Melting of the LJ fcc (100) interfaces. Shown are the density profile, energy density, and temperature fluctuations from a molecular dynamics simulation. The vertical dashed line locates the Gibbs dividing surface. $T^* = 0.665$, density of solid $\rho^* = 0.965$, density of liquid $\rho^* = 0.834$. (Broughton, Bonissent, and Abraham, J. Chem. Phys., 1981)

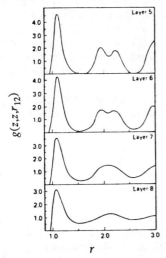

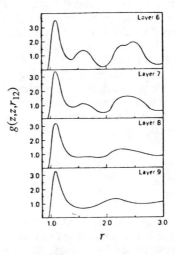

FIGURE XV.15. Pair correlation functions in parallel planes from the melting (111) surface. Refer to Figure XV.13 for the numbering of the layers. Layer 5 solid is inside the Gibbs dividing surface, whereas layer 8 is in the liquid region. (Broughton, Bonissent, and Abraham, J. Chem. Phys., 1981)

FIGURE XV.16. Pair correlation functions in parallel planes from the melting (100) surface. Refer to Figure XV.14 for the numbering of the layers. Layer 6 solid is inside the Gibbs dividing surface, whereas layer 9 is in the liquid region. (Broughton, Bonissent, and Abraham, J. Chem. Phys., 1981)

$$\frac{\partial}{\partial \mathbf{r}_1} \ln \rho^{(1)}(\mathbf{r}_1) = \frac{1}{kT}\frac{\partial}{\partial \mathbf{r}_1} w(\mathbf{r}_1) - \int d\mathbf{r}_2\, \rho^{(1)}(\mathbf{r}_2) g^{(2)}(\mathbf{r}_1,\mathbf{r}_2)\frac{1}{kT}\frac{\partial}{\partial \mathbf{r}_1} u(r_{12}) \qquad (6.3)$$

where we have omitted the sign of inhomogeneity $w$. This expression is exact and is the first equation in the BBGKY hierarchy for inhomogeneous systems. However, this equation cannot be solved without a closure. Furthermore, the Kirkwood superposition approximation for the nonuniform pcf $g^{(2)}(\mathbf{r}_1,\mathbf{r}_2)$ could be introduced:

$$g^{(2)}(\mathbf{r}_1,\mathbf{r}_2;w) = g^{(2)}(z_1,z_2;r_{12};w) \approx g^{(1)}(z_1;w)g^{(1)}(z_2;w)g^{(2)}(r_{12},w=0) \qquad (6.4)$$

The superposition principle says that the pcf under the influence of the wall potential $w$ is approximated by the product of the nonuniform singlet correlation functions $g^{(1)}$ at the vertical distances $z_1$ and $z_2$ and the bulk pcf $g^{(2)}$ in the absence of the wall potential. Thus if the bulk $g^{(2)}$ is known, eq. (6.3) is closed by (6.4). An iterative procedure can be employed to find $\rho^{(1)}$ for the given density, temperature, and wall potential. Fischer and Methfessel [29] have solved eq. (6.3) without using the superposition by substituting for $g^{(2)}(\mathbf{r}_1,\mathbf{r}_2;w)$ the bulk pcf at the average (mean-field) density:

$$\overline{\rho} \equiv \frac{1}{V}\int_V d\mathbf{r}\, \rho^{(1)}(\mathbf{r}+\mathbf{r}_c) \qquad (6.5)$$

where $\mathbf{r}_c = (\mathbf{r}_1+\mathbf{r}_2)/2$ and $V = 4\pi d^3/3$, $d$ being the hard-sphere diameter. This approximation has recently been shown to yield exact results in one dimension [30]. Results were presented for the systems (1) hard-spheres fluid/LJ 9:3 wall, and (2) LJ 12:6 fluid/LJ 9:3 wall. The density was taken to be $\rho^* = 0.65$ (for both (1) and (2)), and the temperature was $T^* = 1.0$ (for (2)). (Note that, for equivalence, the hard-sphere diameter $d$ is taken to be $\sigma$.) The results are shown in Figure XV.17. Good agreement with simulation data of Abraham and Singh [31] was obtained.

For hard spheres near a hard wall (eq. (4.2)), Navascués and Tarazona [32] gave the following integral equation for the singlet density (after substitution of the superposition relation (6.4)):

$$\rho^{(1)}(z) = \rho\, \exp\left\{2\pi y(d)\int_{-d}^{+d} dz_{12}\, [\rho-\rho^{(1)}(z+z_{12})]\frac{d^2-z_{12}^2}{2}\right\} \qquad (6.6)$$

where $y(d)$ is the background function of the bulk fluid evaluated at $r=d$. The wall potential $w$ has been shifted to $w(z) = \infty$ for $z < 0$ and $w(z) = 0$, for $z \geq 0$. For packing fractions, $y = \pi\rho d^3/6 = 0.105, 0.209, 0.314,$ and $0.419$, the density profiles are presented in Figure XV.18. We see again that the adsorption of hard spheres is multilayer.

## GROWING ADSORBENT MOLECULE

In this approach a mixture of molecules of species $A$ (the adsorbent) and species $B$ (the adsorbate) vastly different in size is used. The molecules of species $A$ are identified as the *adsorbent* molecules and are allowed to grow indefinitely. Eventually, an adsorbing wall is created as $\sigma_A \to \infty$ and $\rho_A \to 0$. This wall is surrounded by many small $B$ molecules. The method was proposed by Henderson, Abraham, and Barker [33] and is also called the HAB approach. The mathematical conditions have been further examined by Gruber et al. [34] The theoretical basis is the Ornstein-Zernike equation written for a mixturer:

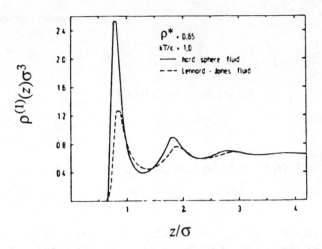

FIGURE XV.17. *Density profiles of the hard-sphere/LJ 9:3 wall system (solid line), and the LJ 12:6 fluid/LJ 9:3 wall system (dashed line).* $\rho^* = 0.65$; $T^* = 1$. *(Fischer and Methfessel, Phys. Rev., 1980)*

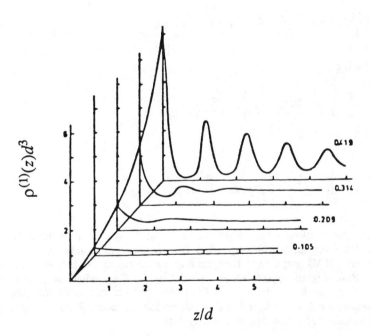

FIGURE XV.18. *Density profiles of hard spheres near a hard wall at different bulk densities using the superposition in the BBGKY equations. (Navascués and Tarazona, Mol. Phys., 1979)*

$$h_{AB}(\mathbf{r}_1,\mathbf{r}_2) - C_{AB}(\mathbf{r}_1,\mathbf{r}_2) = \rho_A \int d\mathbf{r}_3\ h_{AA}(\mathbf{r}_1,\mathbf{r}_3)C_{AB}(\mathbf{r}_3,\mathbf{r}_2) \tag{6.7}$$

$$+ \rho_B \int d\mathbf{r}_3\ h_{AB}(\mathbf{r}_1,\mathbf{r}_3)C_{BB}(\mathbf{r}_3,\mathbf{r}_2)$$

Equivalently,

$$h_{BA}(\mathbf{r}_1,\mathbf{r}_2) - C_{BA}(\mathbf{r}_1,\mathbf{r}_2) = \rho_A \int d\mathbf{r}_3\ h_{BA}(\mathbf{r}_1,\mathbf{r}_3)C_{AA}(\mathbf{r}_3,\mathbf{r}_2) \tag{6.8}$$

$$+ \rho_B \int d\mathbf{r}_3\ h_{BB}(\mathbf{r}_1,\mathbf{r}_3)C_{BA}(\mathbf{r}_3,\mathbf{r}_2)$$

We know from symmetry, $h_{AB} = h_{BA}$ and $C_{AB} = C_{BA}$. Recall that in order to make the adsorbent molecules into a wall, we let the size of the $A$ molecules grow and let the density $\rho_A$ go to zero. We first take the limit $\rho_A \rightarrow 0$,

$$h_{AB}(r) - C_{AB}(r) = \frac{2\pi\rho_B}{r} \int_0^{\infty} dt\ th_{BB}(t) \int_{|r-t|}^{r+t} ds\ sC_{AB}(s) \tag{6.9}$$

or, equivalently,

$$h_{AB}(r) - C_{AB}(r) = \frac{2\pi\rho_B}{r} \int_0^{\infty} dt\ tC_{AB}(t) \int_{|r-t|}^{r+t} ds\ sh_{BB}(s) \tag{6.10}$$

Similarly, from (6.7)

$$h_{AB}(r) - C_{AB}(r) = \frac{2\pi\rho_B}{r} \int_0^{\infty} dt\ th_{AB}(t) \int_{|r-t|}^{r+t} ds\ sC_{BB}(s) \tag{6.11}$$

and

$$h_{AB}(r) - C_{AB}(r) = \frac{2\pi\rho_B}{r} \int_0^{\infty} dt\ tC_{BB}(t) \int_{|r-t|}^{r+t} ds\ sh_{AB}(s) \tag{6.12}$$

We now have four equivalent equations. Next we let the size of the $A$ molecules grow to infinity. However, numerically eq. (6.9) to (6.12) become unstable as the size of $A$ goes to infinity. HAB proposed shifting the origin from the molecular center to the surface of $A$. Referring to Figure XV.19, we define $r = R + z$, where $R$ is the size of molecule $A$ (e.g., for hard spheres, $R$ is the radius; for Lennard-Jones molecules, $R$ is $\sigma/2$). Similarly, $s = R + s'$, where $z$ and $s'$ are vertical distances from the surface. We denote the functions after the shift also with a prime:

$$h'_{AB}(z) \equiv h_{AB}(R+z) \qquad \text{and} \qquad C'_{AB}(z) \equiv C_{AB}(R+z) \tag{6.13}$$

As $R \rightarrow \infty$, we observe from the geometry that $s/r = (s'+R)/(z+R) = 1$. Thus the four equations (6.9 to 12) become

$$h'_{AB}(z) - C'_{AB}(z) = 2\pi\rho_B \int\limits_0^\infty dt \ t h_{BB}(t) \int\limits_{z-t}^{z+t} ds' \ C'_{AB}(s') \tag{6.14}$$

$$h'_{AB}(z) - C'_{AB}(z) = 2\pi\rho_B \int\limits_{-\infty}^{+\infty} ds' \ C'_{AB}(s') \int\limits_{|z-s'|}^\infty dt \ t h_{BB}(t)$$

$$h'_{AB}(z) - C'_{AB}(z) = 2\pi\rho_B \int\limits_{-\infty}^{+\infty} ds' \ h'_{AB}(s') \int\limits_{|z-s'|}^\infty dt \ t C_{BB}(t)$$

$$h'_{AB}(z) - C'_{AB}(z) = 2\pi\rho_B \int\limits_0^\infty dt \ t C_{BB(t)} \int\limits_{z-t}^{z+t} ds' \ h'_{AB}(s')$$

The limits of integration have been adjusted according to the growing molecule. Any one of the four equations is correct to use. However, given a particular case, one form may be more convenient than the others. This growing adsorbent model has recently been applied to the hard spheres/hard wall and Lennard-Jones fluid/Lennard Jones wall adsorptives. Also the adsorption of hard diatomic dumbbells to a hard wall has also been studied by this method. We note that to solve the above OZ equations, a closure relation is needed. Henderson et al. [35], Fischer [36], and Navascués et al. [37], have solved the above equations with the PY approximation for a hard-spheres/hard-wall system; i.e.,

$$C'_{AB}(z) = h'_{AB}(z) - y'_{AB}(z) + 1, \qquad \text{(PY)} \tag{6.15}$$

where $y'_{AB}(z)$ is the background correlation function for wall-particle interaction, and $y_{AB} = g_{AB} \exp(+\beta w)$. At $\rho d_{BB}^3 = 0.8$, the density profile, $g'_{AB}(z)$, is shown in Figure XV.20 (Fischer [38]). Comparison with the superposition (SA) results based on BBGKY is made. The SA results are strongly oscillatory and exhibit peaks with longer periodicity. Figure XV.21 compares the contact value of $\rho^{(1)}(z)$. PY results are less accurate than the SA method.

## LENNARD-JONES MOLECULES ON LENNARD-JONES WALL

Smith and Lee [39] have solved the PY and HNC closures for a Lennard-Jones 12:6 fluid adsorbed on a Lennard-Jones 9:3 wall. The Lennard-Jones 9:3 potential is

$$w(z) = \pi\rho_s^* \varepsilon \sqrt{\frac{10}{3}} \left[ \left[ \frac{\alpha\sigma}{z} \right]^9 - \left[ \frac{\alpha\sigma}{z} \right]^3 \right] \tag{6.16}$$

where $\rho_s^* = 0.85$ is the average solid density. $\alpha = \sqrt[6]{2/15} = 0.71475$ for a Lennard-Jones 12:6 solid, and $\alpha = 0.5621$ for the argon-graphite system. The PY closure is given by (6.15); and the HNC closure is given by

$$\text{(HNC)} \qquad C'_{AB}(z) = h'_{AB}(z) - \ln y'_{AB}(z) \tag{6.17}$$

These closures combined with the OZ relation (6.14) were used by Smith et al. (*Ibid.*) to solve for the LJ 12:6/LJ 9:3 system at $T^* = kT/\varepsilon = 1.0$ and $\rho\sigma^3 = 0.65$. The bulk structures, $h_{BB}(r)$ and $C_{BB}(r)$, needed were obtained from the MD tabulations of Verlet [40]

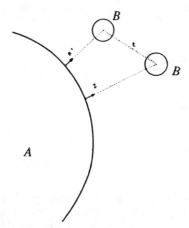

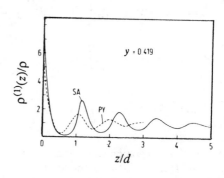

FIGURE XV.19. Geometry of the growing adsorbent molecule A, and the surrounding adsorbate
molecules B. The interparticle distances are
measured from the surface of the adsorbent
molecule. (Smith and Lee, J. Chem. Phys., 1979)

FIGURE XV.20. Density profile of hard spheres
near a hard wall. PY: results from PY closure to
the growing adsorbent molecule. SA: results from
the superposition approximation to the BBGKY
equation. (Fischer, Mol. Phys., 1977)

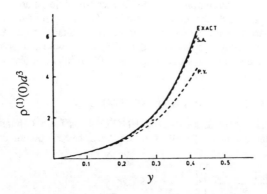

FIGURE XV.21. Comparison of the contact values of the singlet density
of hard spheres on a hard wall. Exact: from Fisher's formula using the
Carnahan-Starling equation. PY: from PY closure. SA: from superposition approximation. SA gives better agreement with exact results.
(Navascués and Tarazona, Mol. Phys., 1979)

for the LJ 12:6 fluid. The results are given in Figure XV.22. The HNC results are shown as broken lines, the PY results as symbols "+", and the MC results (from Abraham and Singh [41]) by the dotted lines. The HNC curve is closer to the MC data than the PY results. However, HNC is still inadequate. It overestimates the structure.

Other approaches have been proposed, such as the blip function method of Abraham and Singh [42], the perturbation formulations by Lee [43], the MSA methods of Waisman, Henderson, and Lebowitz [44], and Sullivan and Stell [45], etc. Most of these theories were applied to simple adsorptive systems, such as hard spheres near a soft repulsive wall, or LJ molecules near a soft or LJ wall. None of these approaches is, however, completely satisfactory. The difficulties in adsorption studies arise from the inhomogeneities due to the wall; i.e., additional disturbances are introduced into the bulk fluid by the external wall. The theories that were designed for homogeneous fluids are subject to a very severe test.

## HARD DIATOMIC MOLECULES ON A HARD WALL

Dumbbell molecules adsorbed to a wall exhibit not only the excluded volume behavior of hard bodies but also orientational order; i.e., for diatomics the molecules could align themselves either parallel or perpendicular to the surface. It is of interest to examine the factors that influence the orientational order. Sullivan et al. [46] have used the RISM model, described previously, to calculate the density profiles of such a system. The correlation functions obtained are of the site-site type. The growing adsorbent approach is used here. To start, a mixture of diatomic molecules and hard spheres is considered. Next, the size of the hard spheres is later allowed to grow to infinity as the density of the hard spheres goes to zero. The procedure of Stell, Patey, and Høye [47] is adopted, treating the hard dumbbells as an equimolar mixture of the constituent (fused) hard spheres. Thus, to begin with, there are three types of hard spheres in the mixture: the adsorbent hard spheres and spheres constituting the atoms of the dumbbells. An OZ-like equation is generated by identifying the intramolecular and intermolecular correlations appearing in the atomic OZ equation. In general, there are $m$ sites in a structured molecule (for dumbbells, $m=2$). Let subscript $A$ denote the adsorbent hard spheres, and $B$ the bath particles (the hard spheres that eventually fuse into polyatomics). The RISM site-site OZ equations could be written as (with * denoting matrix multiplication and/or convolution)

$$\mathbf{h}_B = \omega * C_B * \omega + \rho\omega * C_B * \mathbf{h}_B \qquad (6.18)$$

and

$$\mathbf{h}_A = \omega * C_A + \rho\omega * C_B * \mathbf{h}_A$$

where $\mathbf{h}_B$ is an $m \times m$ matrix with elements $h_{B,\alpha\gamma}(r)$, and $1 \leq \alpha, \gamma \leq m$ are labels for the sites in different molecules; $\mathbf{h}_A$ is a column vector, $h_{A,\alpha}$, $1 \leq \alpha \leq m$, denoting total correlations between site $\alpha$ and the wall. $\omega$ is the intramolecular correlation function.

$$\omega_{\alpha\gamma}(r) = \delta_{\alpha\gamma}\delta_D(r) + (1-\delta_{\alpha\gamma})s_{\alpha\gamma}(r) \qquad (6.19)$$

which for rigid sites becomes

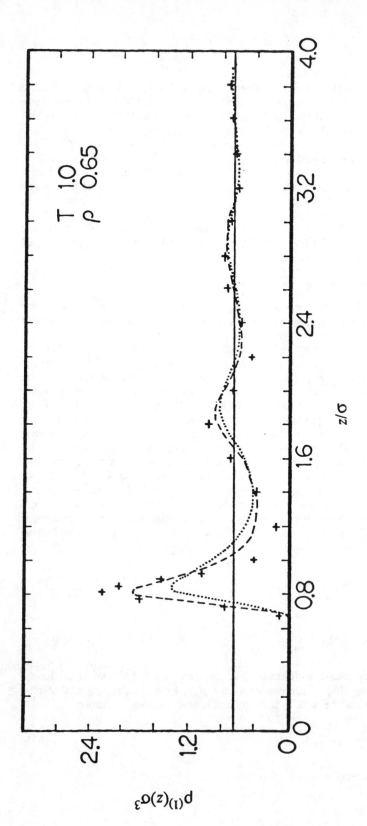

FIGURE XV.22. Density profile of the LJ 12:6 fluid on an LJ 9:3 wall. +++: HNC results. ---: PY results. ...: MC data of Abraham and Singh [41]. T* = 1.0. ρ* = 0.65. (Smith and Lee, J. Chem. Phys., 1979)

$$s_{\alpha\gamma}(r) = \frac{1}{4\pi L_{\alpha\gamma}^2} \delta_D(r, L_{\alpha\gamma}) \qquad (6.20)$$

where $L_{\alpha\gamma}$ is the $\alpha\gamma$ bond length, and $\delta_D$ is the (symmetric) Dirac delta function. For homonuclear diatomics (see Figure XV.23), the sites are two fused hard spheres of equal size with bond length $L$. Since both sites are equivalent, we set

$$h_{A,\alpha}(r) \equiv h_A(r), \qquad C_{A,\alpha} \equiv C_A(r), \qquad C_{B,\alpha\gamma}(r) \equiv C_B(r), \qquad \forall\,\alpha,\gamma \qquad (6.21)$$

We write

$$\sum_{\gamma=1}^{m} \omega_{\alpha\gamma}(r) \equiv \omega_T(r) \qquad (6.22)$$

which is independent of $\alpha$. Equation (6.18) reduces to the scalar form

$$h_A = \omega_T * C_A + m\rho\omega_T * C_B * h_A \qquad (6.23)$$

Here * denotes convolution only. As before, we set $r = R+z$ and let $R$ go to $\infty$. Equation (6.23) can be solved with a closure. According to RISA (see Chapter XIV),

$$g_{A,\alpha}(z) = 1 + h_{A,\alpha}(z) = 0, \qquad \text{for } z < \sigma_{min}/2 \qquad (6.24)$$

and

$$C_{A,\alpha}(z) = 0, \qquad \text{for } z > \sigma_{min}/2 \qquad (6.25)$$

where $\sigma_{min}/2$ is the minimum distance of approach between the diatomic molecule and the wall. In general, $\sigma_{min}$ is dependent on the overall geometry of the polyatomic molecules, not just on the diameters themselves. For homonuclear dumbbells, $\sigma_{min} = \sigma$, the diameter of the atom. Equations (6.23 to 25) were solved by Sullivan et al. [46] using a Wiener-Hopf factorization method. The results are presented in Figures XV.24 and 25, together with the MC results. Four conditions are displayed: $\rho\sigma^3 = 0.1, 0.3, 0.4$, and 0.5, and bond length $L/\sigma = 0.3$. Several peaks are displayed: the first at $z = \sigma/2$, representing the closest approach of the dumbbells parallel to the wall. The second peak located at $z = \sigma/2 + L$ represents the adsorption of a dumbbell vertical to the surface. The increase of the first peak relative to the second peak as density is increased or the elongation ($L/\sigma$) is increased is a clear indication that more and more dumbbells are adsorbed parallel to the surface. This trend shows up in the MC data as well. The RISA site-site correlation functions are not accurate at the first peak (being too small) and underestimate the second peak. However, RISA results are qualitatively correct in the orientational structure at the surface. At high densities, the parallel orientation is preferred over the vertical orientation. The preference could be understood by considering the geometry of packing: parallel packing is more compact (i.e., more dumbbells per volume) than vertical packing. However, the qualitative agreement between RISA and simulation in the hard-dumbbells/hard-wall case disappears when the wall becomes attractive (see [48]).

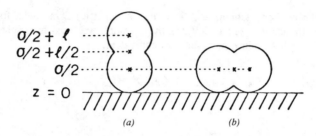

FIGURE XV.23. Adsorption of diatomics on a structureless wall. (a) perpendicular, (b) parallel. (Sullivan at al., Mol. Phys., 1981)

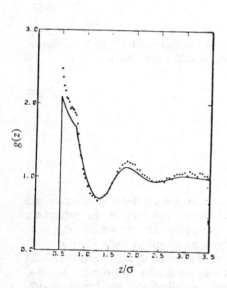

FIGURE XV.24. Comparison of RISA and MC density profiles g(z) of adsorbed hard dumbbells on hard wall. Line: RISA results. · · · : MC data. $\rho\sigma^3 = 0.40$. Bond length of the diatomic $L/\sigma = 0.3$. (Sullivan at al., Mol. Phys., 1981)

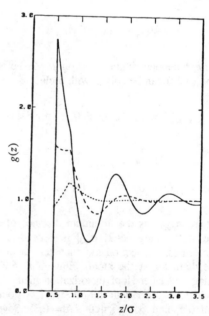

FIGURE XV.25. Comparison of RISA density profiles g(z) of adsorbed hard dumbbells on hard wall at three densities. —: $\rho\sigma^3 = 0.5$. ---: $\rho\sigma^3 = 0.3$. · · · : $\rho\sigma^3 = 0.1$. Higher density favors parallel orientation on the surface. Bond length of the diatomic $L/\sigma = 0.3$. (Sullivan at al., Mol. Phys., 1981)

## XV.7. *Density Functional Approach*

The density functional approach to adsorption was proposed by Ebner, Saam, and Stroud (ESS) [49] in 1976 by minimization of the grand potential $\Omega$. $\Omega$ is treated as a functional of the singlet density $\rho^{(1)}$:

$$\Omega \equiv - PV = - kT \ln \Xi \tag{7.1}$$

Mermin [50] proposed that there exists such a functional $\Omega[\rho^{(1)}]$ such that the minimum value of

$$\Omega = A[\rho^{(1)}] + \int d\mathbf{r} \, \rho^{(1)}(\mathbf{r})[w(\mathbf{r}) - \mu] \tag{7.2}$$

with respect to variations in $\rho^{(1)}(\mathbf{r})$ at constant $\mu$, $T$, $V$, and $w(\mathbf{r})$ gives the equilibrium grand free energy of the system. The procedure is equivalent to the minimization of the free energy. $\rho^{(1)}(r)$ producing the minimum is also the equilibrium density. In eq. (7.2), $A[\rho^{(1)}]$ is the Helmholtz free energy expressed as a functional of the singlet density. Its variational derivative is defined as

$$v(\mathbf{r}) \equiv - \left. \frac{\delta A[\rho^{(1)}]}{\delta\rho^{(1)}(\mathbf{r})} \right|_{T,\mu,V} \tag{7.3}$$

We remark that, for any given density $\rho^{(1)}(\mathbf{r})$, $v(\mathbf{r})$ is simply the corresponding external potential that would produce this nonuniform density. The stationary condition for (7.2) requires

$$\left. \frac{\delta\Omega}{\delta\rho^{(1)}(\mathbf{r})} \right|_{T,\mu,V} = - v(\mathbf{r}) + w(\mathbf{r}) = 0 \tag{7.4}$$

The null value is to guarantee the minimization of $\Omega$. From the functional calculus of Lebowitz and Percus [51], we know that the derivative of $v(r)$ is related to the direct correlation function by

$$\frac{\delta[-\beta v(\mathbf{r})]}{\delta\rho^{(1)}(\mathbf{r}')} = \frac{\delta_D(\mathbf{r},\mathbf{r}')}{\rho^{(1)}(\mathbf{r})} - C(\mathbf{r},\mathbf{r}') \tag{7.5}$$

To integrate, we choose a path in the function space characterized by the initial and final densities, $\rho_0^{(1)}(\mathbf{r})$ and $\rho^{(1)}(\mathbf{r})$, respectively, with a homotopic parameter $\alpha$ ($0 \le \alpha \le 1$):

$$\rho_\alpha^{(1)}(\mathbf{r}) \equiv \rho_0^{(1)}(\mathbf{r}) + \alpha\left[\rho^{(1)}(\mathbf{r}) - \rho_0^{(1)}(\mathbf{r})\right] \tag{7.6}$$

The integral is

$$\beta v(\mathbf{r}) - \beta v_0(\mathbf{r}) = -\ln\left[\frac{\rho^{(1)}(\mathbf{r})}{\rho_0^{(1)}(\mathbf{r})}\right] + \int_0^1 d\alpha \int d\mathbf{r}' \, C(\mathbf{r},\mathbf{r}'; \alpha)\left[\rho^{(1)}(\mathbf{r}') - \rho_0^{(1)}(\mathbf{r}')\right] \qquad (7.7)$$

where $C(\cdot)$ is the nonuniform direct correlation function in (7.5). If the reference state is a homogeneous fluid, i.e., $v_0(\mathbf{r}) = 0$ and $\rho_0^{(1)}(\mathbf{r}) = \rho$, (7.7) could be further integrated (taking into account the stationary condition (7.4))

$$\beta(\Omega - \Omega_0) = \int d\mathbf{r} \, \beta w(\mathbf{r})[\rho^{(1)}(\mathbf{r}) - \rho] + \int d\mathbf{r} \, \rho^{(1)}(\mathbf{r}) \ln[\rho^{(1)}(\mathbf{r})/\rho] \qquad (7.8)$$

$$- \int d\mathbf{r} \, [\rho^{(1)}(\mathbf{r}) - \rho] - \int_0^1 d\alpha \int_0^\alpha d\alpha' \int d\mathbf{r} \int d\mathbf{r}' \, C(\mathbf{r},\mathbf{r}'; \alpha')[\rho^{(1)}(\mathbf{r}) - \rho][\rho^{(1)}(\mathbf{r}') - \rho]$$

$\Omega_0$ is the grand free energy of the uniform system. The theorem of Mermin guarantees the uniqueness of $\Omega$. Saam and Ebner [52] have applied (7.8) to the calculation of adsorbed Lennard-Jones molecules. They made several approximations regarding the nonuniform dcf. First, the Percus-Yevick approximation was used to calculate the uniform dcf $C(r;\bar{\rho})$. Second, a mean density $\bar{\rho}$ was chosen as

$$\bar{\rho} = \frac{\rho^{(1)}(\mathbf{r}) + \rho^{(1)}(\mathbf{r}')}{2} \qquad (7.9)$$

and the nonuniform dcf was approximated by a uniform dcf at this mean density. Third, instead of evaluating the dcf at a given coupling strength $\alpha$, a new dcf was defined

$$\hat{C}(\mathbf{r},\mathbf{r}') \equiv 2 \int_0^1 d\alpha \int_0^\alpha d\alpha' \, C(\mathbf{r},\mathbf{r}'; \alpha') \qquad (7.10)$$

(This new dcf was solved by the PY scheme.) For an LJ 9:3 wall (simulating solid $CO_2$), they calculated the density profiles for adsorbed argon molecules (Figure XV.26). The oscillations intensify at higher densities, indicating a *thin-to-thick film transition*. This behavior has been discussed recently in a study on *wetting* phenomena [53].

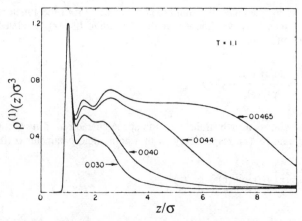

FIGURE XV.26. *The density profiles of argon adsorbed on $CO_2$ surface according to the density functional theory of Ebner-Saam-Stroud [52]. Four density values are displayed. $T^* = kT/\varepsilon = 1.1$. Reduced density $\rho^* = \rho\sigma^3$. The growth of the film thicknesses is clearly visible.*

# References

[1] A.W. Adamson, *Physical Chemistry of Surfaces* (J. Wiley, New York, 1972).

[2] P.G. de Gennes, Rev. Mod. Phys. **57**, 827 (1985).

[3] R. Evans, Adv. in Phys. **28**, 143 (1979).

[4] R. Pandit, M. Schick, and M. Wortis, Phys. Rev. B **26**, 5112 (1982).

[5] P.G. de Gennes, Rev. Mod. Phys. **57**, 827 (1985).

[6] D.E. Sullivan and M.M. Telo da Gama, in *Fluid Interfacial Phenomena*, edited by C.A. Croxton (Wiley, New York, 1986).

[7] H.T. Davis and L.E. Scriven, Adv. Chem. Phys. **49**, 357 (1982).

[8] E. Dickson and M. Lal, Advances in Molecular Relaxation and Interaction Processes **17**, 1 (1980).

[9] D. Nicholson and N.G. Parsonage, *Computer Simulation and the Statistical Mechanics of Adsorption* (Academic Press, New York, 1982).

[10] J.K. Percus, "Non-Uniform Fluids," in *The Liquid State of Matter: Fluids, Simple and Complex*, edited by E.W. Montroll and J.L. Lebowitz (North-Holland, Amsterdam, 1982), pp. 31ff.

[11] W.A. Steele, *The Interaction of Gases with Solid Surfaces* (Pergamon, Oxford, 1974).

[12] W.A. Steele, *The Interaction of Gases with Solid Surfaces* (Pergamon, Oxford, 1974).

[13] W.A. Steele, *The Interaction of Gases with Solid Surfaces* (Pergamon, Oxford, 1974).

[14] See, e.g., G. Navascué and M.V. Berry, Mol. Phys. **34**, 649 (1977).

[15] R. Pandit, M. Schick, and M. Wortis, *Ibid.* (1982).

[16] R. Pandit, M. Schick, and M. Wortis, *Ibid.* (1982).

[17] I.K. Snook and D. Henderson, J. Chem. Phys. **68**, 2134 (1978).

[18] E. Waisman, D. Henderson, and J.L. Lebowitz, Mol. Phys. **32**, 1373 (1976).

[19] I.Z. Fisher, *Statistical Theory of Liquids* (University of Chicago Press, Chicago, 1964).

[20] M.J. Mandell, J. Chem. Phys. **65**, 813 (1976).

[21] D. Henderson, J. Chem. Phys. **68**, 780 (1978).

[22] F.F. Abraham and Y. Singh, J. Chem. Phys. **68**, 4767 (1978).

[23] F.F. Abraham and Y. Singh, *Ibid.* (1978).

[24] F.F. Abraham and Y. Singh, *Ibid.* (1978).

[25] F.F. Abraham, J. Chem. Phys. **68**, 3713 (1978).

[26]  J.Q. Broughton, A. Bonissent, and F.F. Abraham, J. Chem. Phys. **74**, 4029 (1981).

[27]  See, e.g., A. Bellemans, Physica **28**, 493 and 617 (1962); J. Fischer, Mol. Phys. **35**, 897 (1978); and D.A. McQuarrie and J.S. Rowlinson, Mol. Phys. **60**, 977 (1987).

[28]  See Pandit et al., Sullivan et al., and de Gennes cited above.

[29]  J. Fischer and M. Methfessel, Phys. Rev. A **22**, 2836 (1980).

[30]  H.T. Davis, J. Chem. Phys. **85**, 6808 (1986).

[31]  F.F. Abraham and Y. Singh, *Ibid.* (1978).

[32]  G. Navascués and P. Tarazona, Mol. Phys. **37**, 1077 (1979).

[33]  D. Henderson, F.F. Abraham, and J.A. Barker, Mol. Phys. **31**, 1291 (1976).

[34]  C. Gruber, J.L. Lebowitz, and P.A. Martin, J. Chem. Phys. **75**, 944 (1981).

[35]  D. Henderson, F.F. Abraham, and J.A. Barker, *Ibid.* (1976).

[36]  J. Fischer, Mol. Phys. **33**, 75 (1977).

[37]  G. Navascué and P. Tarazona, Mol. Phys. **37**, 1077 (1979).

[38]  J. Fischer, *Ibid.* (1977).

[39]  L.S. Smith and L.L. Lee, J. Chem. Phys. **71**, 4085 (1979).

[40]  L. Verlet, Phys. Rev. **165**, 201 (1968).

[41]  F.F. Abraham and Y. Singh, J. Chem. Phys. **68**, 4767 (1978).

[42]  F.F. Abraham and Y. Singh, J. Chem. Phys. **67**, 2384 (1977).

[43]  L.L. Lee, J. Chem. Phys. **73**, 4050 (1980).

[44]  E. Waisman, D. Henderson, and J.L. Lebowitz, Mol. Phys. **32**, 1373 (1976).

[45]  D.E. Sullivan and G. Stell, J. Chem. Phys. **69**, 5450 (1978).

[46]  D.E. Sullivan, R. Barker, C.G. Gray, W.B. Streett, and K.E. Gubbins, Mol. Phys. **44**, 597 (1981); S.M. Thompson, K.E. Gubbins, D.E. Sullivan, and C.G. Gray, Mol. Phys. **51**, 21 (1984).

[47]  G. Stell, G.N. Patey, and J.S. Høye, in Adv. Chem. Phys. **48**, edited by S. Rice (Wiley, New York, 1981).

[48]  S.M. Thompson, K.E. Gubbins, D.E. Sullivan, and C.G. Gray, Mol. Phys. **51**, 21 (1984).

[49]  C. Ebner, W.F. Saam, and D. Stroud, Phys. Rev. A **14**, 2264 (1976).

[50]  N.D. Mermin, Phys. Rev. **137**, A1441 (1964).

[51]  J.L. Lebowitz and J.K. Percus, J. Math. Phys. **4**, 116 (1963).

[52]  C. Ebner and W.F. Saam, Phys. Rev. Lett. **38**, 1486 (1977); W.F. Saam and C. Ebner, Phys. Rev. A **17**, 1768 (1978).

[53] D.E. Sullivan and M. M. Telo da Gama, in *Fluid Interfacial Phenomena*, edited by C. A. Croxton (Wiley, New York, 1986) pp. 45ff.

## Exercises

1. Derive the Lennard-Jones 10-4 and 9-3 potentials, (1.4) and (1.5), by carrying out the indicated integration, *smoothing* over the solid atoms.

2. Numerically evaluate the $w^{eff}$ of (5.2) by carrying out the averaging for a LJ 12:6 potential $w$ with $\sigma = 3.405$ Å and $\varepsilon/k = 119.8$ K for the graphite surface shown in Figure XV.3.

3. Derive the Navascués-Tarazona equation (6.6) from the BBGKY hierarchy and superposition approximation.

4. Use the density functional method and the Fischer-Methfessel formulation to solve the same problem listed for equation (6.16). Compare the answers with the PY and HNC results.

5. Discuss if the BBGKY method developed in Chapter XIV for polyatomics could be used to recalculate the examples for diatomic adsorption as listed in Figures XV.24 and 25. What kind of mathematical tools are needed to solve these equations? What appropriate assumptions would have to be made? Could we solve the problem for low densities at first?

# APPENDIX A

# INTERMOLECULAR POTENTIALS

Molecules interact with forces derived from physical origins. These intermolecular forces are sources of the nonideal behavior. The total interaction energy of the $N$ molecules is the sum of interactions coming from pairs, triplets, quadruplets,... of molecules. In the majority of cases, the principal contribution comes from the pair interactions. The range of interaction between a pair of molecules can be conveniently divided into the short-range ($r < \sigma$), close-range ($\sigma < r < r^*$), and long-range ($r > > r^*$) parts (see Figure IX.1). The demarcation between these regions is somewhat arbitrary, since different regions may overlap. The origins of the intermolecular forces are explored below.

1. *Short Range Forces:* In this region, the forces of interaction are repulsive. They derive from the valence energies (or chemical forces). The calculation must be done in accordance with quantum mechanics.

2. *Close Range Forces: Residual* valence forces, such as hydrogen bonds, operate in this region. Charge-transfer complexes, also known as *electron donor-acceptor* complexes, are also formed within this region. These forces must be treated quantum mechanically.

3. *Long Range Forces*: At least three different types of interaction take part in this region: (i) *dispersion forces*, (ii) *induction forces*, and (iii) *electrostatic forces*.

## A.1. *Origins of Intermolecular Forces*

### SHORT-RANGE FORCES

The short-range forces can be divided into *electrostatic* and *exchange* contributions. They are evaluated from quantum Schrödinger equation. The electrostatic part is from the Coulomb interactions of pairs— electron-electron, electron-nucleus and nucleus-nucleus. The exchange part comes from the Pauli exclusion in quantum mechanics. These forces are repulsive in nature and can usually be fitted with a simple exponential (or power) law:

$$u_{rep}(r) = me^{-nr} \tag{A.1}$$

### DISPERSION FORCES

For a nonpolar molecule, the center of positive charges coincides with the center of negative charges. However, instantaneous fluctuations of the electron charges cause a temporary separation of the charge centers. This transient dipole moment induces a dipole moment in the neighboring molecules. The interaction is called the dispersion force (a term derived from the optical refractive index that depends on the same characteristic electron frequencies $v_0$ of the ionization potential). London [1] has used the quantum Schrödinger equation to evaluate such interactions. The energy is expressed as

$$u_{disp}(r) = -\frac{C_6}{r^6} - \frac{C_8}{r^8} - \frac{C_{10}}{r^{10}} - \cdots \tag{A.2}$$

The constant $C_6$, for example, is related to the properties of the molecule

$$C_6 = \frac{3}{4}\,\alpha^2 I \tag{A.3}$$

where $\alpha$ is the polarizability and $I$ is the first ionization potential ($\sim h v_0$). The magnitude of dispersion energies is small compared to that of the permanent dipoles. Dispersion forces are attractive for all orientations.

The dispersion interaction also gives rise to three-body forces. Quantum chemistry gives the following contribution

$$u_{disp}^{(3)} = \frac{A(1 + 3c_1 c_2 c_3)}{(r_{12} r_{23} r_{31})^3} \tag{A.4}$$

where $c_i = \cos \gamma_i$ is the cosine of the internal angle $\gamma_i$ formed by sides $r_{ij}$ and $r_{ik}$. This formula gives the Axilrod-Teller [2] energy, and is important in dense fluids where molecules are close together.

**INDUCTION FORCES**

Induction energies result from interaction of the polarization in one molecule (i.e., distortion of the electron charge distribution in the presence of an external electric field) with the multipole moments of another molecule. This interaction could be treated classically. The induced dipole in a molecule experiencing an electric field $E$ is

$$\mu_{ind} = \alpha E \tag{A.5}$$

$\alpha$ is the polarizability. It is a scalar quantity for homogeneous media, and a tensor in anisotropic polarization. $\alpha$ measures the ease that the arrangements of electrons in a molecule are deformed by an external electric field. For example, the interaction between an induced dipole and a permanent dipole is

$$u_{\mu\text{-ind }\mu}(r_{12}) = -\frac{\alpha_1 \mu_2^2}{r_{12}^6} \tag{A.6}$$

Higher induction forces are given in terms of higher inverse powers of $r$.

**ELECTROSTATIC FORCES**

Electrostatic forces are the classical Coulomb interactions between charge distributions. Since each molecule is a charge distribution, when a pair of molecules approach each other Coulomb forces are established between the two. Individual charge distribution could be resolved into its components: i.e., dipoles, quadrupoles, octopoles, hexadecapoles, and higher multipoles. The total energy due to electrostatic interactions is the sum of interactions between these polar moments. This representation is called the *multipole expansion*. Classical electrostatics gives the electric potential $\phi$ at a point $\mathbf{r}$ due to a charge distribution (a collection of charges $q_i$ fixed at $\mathbf{r}_i$ in space) as [3]

$$\phi(\mathbf{r}) = \sum_i \frac{q_i}{|\mathbf{r} - \mathbf{r}_i|} \tag{A.7}$$

The multipole expansion in Cartesian tensors is derived from the Taylor expansion of $|\mathbf{r} - \mathbf{r}_i|^{-1}$ in terms of $\mathbf{r}_i$

$$\phi(\mathbf{r}) = \sum_i q_i \left\{ \frac{1}{r} + (-\mathbf{r}_i)\cdot\nabla(\frac{1}{r}) + \frac{1}{2!}(-\mathbf{r}_i)(-\mathbf{r}_i):\nabla\nabla(\frac{1}{r}) \right.$$
$$\left. + \frac{1}{3!}(-\mathbf{r}_i)(-\mathbf{r}_i)(-\mathbf{r}_i)\cdot\nabla\nabla\nabla(\frac{1}{r}) + \cdots \right\} \tag{A.8}$$

The polar moments are defined to be the coefficients of the dyadic gradient tensors:

$$Q \equiv \sum_i q_i, \qquad \mu \equiv \sum_i q_i \mathbf{r}_i, \tag{A.9}$$

$$\Theta \equiv \sum_i q_i \mathbf{r}_i \mathbf{r}_i, \qquad \Omega \equiv \sum_i q_i \mathbf{r}_i \mathbf{r}_i \mathbf{r}_i$$

etc. Here $Q$ is the total charge of the distribution, $\mu$ the dipole moment, $\Theta$ the quadrupole moment, and $\Omega$ the octopole moment. This expansion gives an interpretation of the dipolar/multipolar moments. In practice, one could also employ a different mathematical representation, the *spherical harmonic expansion* of the potential function. Spherical harmonics are based on the Legendre polynomials and their orthogonal properties. Given $\phi(\mathbf{r})$ in (A.7), we could make the following expansion

$$\phi(\mathbf{r}) = \sum_i \sum_l \sum_m \left[ \frac{4\pi}{2l+1} \right] q_i r_i^l \, Y_{lm}(\omega_i) Y_{lm}^*(\omega) \, \frac{1}{r^{l+1}} \tag{A.10}$$

In the sums, $l$ goes from 0 to $\infty$, and m goes from $-l$ to $+l$. We define the spherical multipole moment $Q_{lm}$ (the $m$th component of $l$th order) as

$$Q_{lm} \equiv \sum_i q_i r_i^l \, Y_{lm}(\omega_i) \tag{A.11}$$

These moments are related to the Cartesian moments by well defined relations. For example,

$$Q_{00} = (4\pi)^{-1/2}Q, \qquad Q_{10} = \left[\frac{3}{4\pi}\right]^{1/2}\mu_z \qquad\qquad (A.12)$$

$$Q_{20} = \left[\frac{5}{4\pi}\right]^{1/2}Q_{zz}$$

etc. (See Gray and Gubbins [4]).

## HYDROGEN BONDING FORCES

Hydrogen bonds are considered to be *residual* chemical forces due to their weak energies as compared to chemical bonds. The bond strength of most hydrogen bonds lies within 8 ~ 40 kJ/mol; whereas covalent bond strength is about 200 ~ 400 kJ/mol. Hydrogen bonds could be formed between molecules of the same species, such as water, hydrogen fluoride, ammonia, alcohols (methanol and ethanol), amines, and acids (e.g., acetic acid). The hydrogen atom in one molecule (e.g., H in HF) is "loosely" attached to the electro-negative atom (F) in the other molecule. This attachment accompanied by energy release is called a hydrogen bond. The consequence is the formation of dimers, trimers, etc., for $(HF)_n$, up to hexamers. This self-aggregation is called *association*. On the other hand, hydrogen bonds can form between molecules of different species, such as chloroform (the H in $HCCl_3$) and acetone (the O in $O=C(CH_3)_2$). The cross-bonding is called *solvation*. Hydrogen bonds are stronger than long-range forces (e.g., water-water interaction is 17.7 times stronger than argon-argon dispersion interaction) and highly directional. Recently, *ab initio* potential models have been proposed for the water-water interaction (e.g., MCY [5]).

Other intermolecular interactions, such as charge transfer complexes, could be found in standard texts [6]. In the following, we shall discuss potential models that are essentially empirical and are used primarily to represent, sometimes crudely, real interaction energies.

## A.2. *Potential Models*

The quantum mechanical (or *ab initio*) calculations of the intermolecular potentials for structured molecules are very involved and the results in many cases are not readily usable in statistical mechanical calculations. Thus empirical or semi-empirical formulas have been proposed to approximate the actual potential. The loss of rigor is made up by gains in computational ease. Some potential models are *extreme* simplifications of reality, such as the hard sphere model. Others are more sophisticated approximations to real fluid interactions. We have already discussed some of the models, such as the Lennard-Jones potential, the exp-6 (Williams) potential, the square-well potential, the Coulomb potential, the dipole-dipole, and multipole-multipole interactions. For structured molecules, we have introduced the interaction site model. For elongated molecules, we have discussed the dumbbell model and the convex body models. In the following, we shall discuss some accurate potentials that are used for real fluids.

### EXP-6 POTENTIAL

The exp-6 potential is based on an $r$-dependence of exponential repulsion and inverse 6th power dispersion:

$$u(r) = \varepsilon \left[ \frac{h}{h-6} \right] \left\{ \left[ \frac{6}{h} \right] e^{h(1 - r/r_0)} - \left[ \frac{r_0}{r} \right]^6 \right\} \tag{A.13}$$

The coefficients $\varepsilon$, $h$, and $r_0$ for some "spheicalized" molecules are given in Table A.1.

*Table A.1. Parameters of Exp-6 Potnetial for Some Simple Molecules*

| Molecule | $\varepsilon/k$ (K) | $r_0$ (Å) | $h$ |
|----------|-----------|-----------|-----|
| Ar | 152.0 | 3.644 | 18 |
| Kr | 213.9 | 3.919 | 18 |
| Xe | 326.0 | 4.167 | 22 |
| $CH_4$ | 225.5 | 3.868 | 24 |

## KIHARA POTENTIAL

In the Kihara potential, each molecule is assumed to have an impenetrable hard convex core. For example, for benzene the convex core is a hexagon, for $CF_4$, a tetrahedron. The attractive forces depend on the shortest distance between the two bodies. In the spherical case, the core is a sphere with diameter $d$. Thus the potential is

$$u(r) = 4\varepsilon \left\{ \left[ \frac{\sigma - d}{r - d} \right]^{12} - \left[ \frac{\sigma - d}{r - d} \right]^6 \right\} \tag{A.14}$$

The parameters for some common gases are listed in Table A.2.

*Table A.2. The Parameters for the Spherical Kihara Potential*

| Molecule | $d$ (Å) | $\sigma$ (Å) | $\varepsilon/k$ (K) |
|----------|---------|--------------|-----------|
| Ar | 0.121 | 3.314 | 146.52 |
| Kr | 0.144 | 3.533 | 213.73 |
| Xe | 0.173 | 3.880 | 298.15 |
| $N_2$ | 0.250 | 3.526 | 139.2 |
| $CO_2$ | 0.615 | 3.760 | 424.16 |

## BARKER-FISHER-WATTS POTENTIAL

Barker, Fisher, and Watts [7] have proposed an accurate potential for argon-argon interaction. It has the form

$$u(r) = \varepsilon \left\{ e^{\alpha(1 - r^*)} \sum_{i=0}^{5} A_i (r^* - 1)^i - \sum_{j=0}^{2} \frac{C_j}{r^{*2j+6} + \delta} \right\} \tag{A.15}$$

where $r^* = r/r_{min}$. The coefficients $A_i$, $\varepsilon$, $\alpha$, $r_{min}$, and $\delta$ have also been determined for argon, krypton and xenon. They are listed below

Table A.3.  *Parameters of the BFW Potential for Argon, Krypton, and Xenon*

|                      | Ar-Ar   | Kr-Kr   | Xe-Xe   |
|----------------------|---------|---------|---------|
| $\varepsilon/k$ $(K)$ | 142.2   | 201.9   | 282.4   |
| $r_{min}$ $(\text{Å})$ | 3.761   | 4.007   | 4.362   |
| $\sigma$ $(\text{Å})$ | 3.361   | 3.573   | 3.890   |
| $A_0$                | 0.278   | 0.235   | 0.240   |
| $A_1$                | −4.504  | −4.787  | −4.817  |
| $A_2$                | −8.331  | −9.200  | −10.90  |
| $A_3$                | −25.27  | −8.000  | −25.00  |
| $A_4$                | −102.0  | −30.00  | −50.70  |
| $A_5$                | −113.3  | −205.8  | −200.0  |
| $C_0$                | 1.107   | 1.063   | 1.054   |
| $C_2$                | 0.170   | 0.170   | 0.166   |
| $C_4$                | 0.013   | 0.014   | 0.032   |
| $\alpha$             | 12.5    | 12.5    | 12.5    |
| $\delta$             | 0.01    | 0.01    | 0.01    |

## References

[1]   F. London, Z. Phys. Chem. **B11**, 222 (1930); Trans. Faraday Soc. **33**, 8 (1937).

[2]   B.M. Axilrod and E. Teller, J. Chem. Phys. **11**, 299 (1943); J. Chem. Phys. **17**, 1349 (1949); J. Chem. Phys. **19**, 71 (1951).

[3]   C.J.F. Böttcher, *Theory of Electric Polarization*, 2nd ed. vol. 1 (Elsevier, Amsterdam, 1973).

[4]   C.G. Gray and K.E. Gubbins, *Theory of Molecular Fluids*, Vol. 1 (Clarendon Press, Oxford, 1984), p. 58.

[5]   O. Matsuoka, E. Clementi, and M. Yoshimine, J. Chem. Phys. **64**, 1351 (1976); E. Clementi, G. Corongiu, Int. J. Quantum Chem. **10**, 31 (1983).

[6]   See, e.g., A.D. Buckingham, in *Physical Chemistry*, vol 4, edited by H. Eyring, D. Henderson, and W. Jost (Academic Press, New York, 1970), pp. 349-386.

[7]   J.A. Barker, R.A. Fisher, and R.O. Watts, Mol. Phys. **21**, 657 (1972).

# APPENDIX B

```
C_____
C
C
C              GILLAN'S METHOD OF SOLUTION
C                FOR INTEGRAL EQUATIONS
C
C                 Mol. Phys. 38, 79 (1979)
C
C   NOTE:  VARIABLE NAMES ARE KEPT CONSISTENT
C          WITH THOSE IN GILLAN'S PAPER.
C          THE VARIABLE NAMES GAMMA(), GAMMAP(), DGAMMA(), ETC.
C          REFER TO GILLAN'S GAMMA CORRELATION FUNCTIONS.
C
C          THE RADIUS,TEMP,AND DENSITY (R,TSTAR,RHO)
C          ARE TAKEN TO BE REDUCED (i.e. R=R/SIGMA,
C          TSTAR=kT/EPSILON, AND RHO=RHO*(SIGMA)**3).
C
C
C_____

       REAL R(0:511),F(0:511),K(0:511),P(7,0:511)
       REAL Q(7,0:511),GAMMA(0:511),A(7)
       REAL DGAMMA(0:511),DGAMAP(0:511),FTC(0:511),D(7),GAMMAP(0:511)
       REAL AP(7),JINV(7,7),C(0:511)
       REAL PI,SUM,RHO,DR,TSTAR,DK,MU(100)
       INTEGER I,J,ALPHA,BETA,N,NUMNOD,RC,NRC,LIMIT,NRCYCL(100)
       LOGICAL CONVRG
       TSTAR=1.6694
       RHO=0.6661688
       N=511
       NUMNOD=7
       DR=.05
       PI=3.1415927
       CALL PQGEN (P,Q,N,NUMNOD,LIMIT)
       CALL INIT (R,F,K,GAMMA,DR,N,DK,TSTAR,DGAMMA)
       CALL DECOMP (Q,P,GAMMA,A,DGAMMA,NUMNOD,LIMIT,N)
C      PRINT 822
C 822  FORMAT (/,1X,'**********  BEGIN MAIN LOOP  **********')
       DO 3000 RC=1,100
       DO 1000 NRC=1,100
C  RC ==> REFINEMENT CYCLE
C  NRC ==> NEWTON-RAPHSON CYCLE
       NRCYCL(RC)=NRC
C      PRINT 726,RC
C      PRINT 727,NRC
  726  FORMAT (/,1X,'REFINEMENT  CYCLE   = ',I3)
  727  FORMAT (1X,  'NEWTON-RAPHSON CYCLE = ',I3)
       CALL GCALC (P,A,DGAMMA,GAMMA,N,NUMNOD)
       CALL GPCALC (F,GAMMA,R,K,FTC,GAMMAP,N,PI,DR,DK,RHO)
       CALL DECOMP (Q,P,GAMMAP,AP,DGAMAP,NUMNOD,LIMIT,N)
C
C  TEST FOR NRC CONVERGENCE     BY EQN 31
       CALL NRTEST (A,AP,D,NUMNOD,CONVRG)
       IF (CONVRG) GOTO 2000
```

```
      CALL JACOB (R,Q,P,FTC,K,F,JINV,DR,DK,N,NUMNOD,RHO,LIMIT)
      CALL NEWA (JINV,D,A,NUMNOD)
1000  CONTINUE
C             NRC LOOP COMPLETE
2000  CONTINUE
C          RC LOOP
C
C TEST FOR RC CONVERGENCE - CALC OF MU - EQN 32
      CALL RCTEST (GAMMA,GAMMAP,MU,N,DR,RC,CONVRG)
      IF (CONVRG) GO TO 5000
C     PRINT 989
C 989 FORMAT (1X,70('*'))
      DO 45 I = 0,N
   45 DGAMMA(I)=DGAMAP(I)
3000  CONTINUE
C             RC LOOP COMPLETE
5000  CONTINUE
C     PRINT 909
C 909 FORMAT (/,1X,'************ CONVERGENCE ************')
      CALL OUTPUT(NRCYCL,MU,RHO,TSTAR,F,GAMMA,R,RC,N)
      END
C ******************************************************
      SUBROUTINE PQGEN (P,Q,N,NUMNOD,LIMIT)
C
C THIS ROUTINE GENERATES THE P AND Q BASIS FUNCTIONS
C
C NOTE THAT EQN 17B IS INCORRECT IF I0 IS DEFINED AS 1
C       (AS GILLAN DOES IN THE PAPER)
C HOWEVER, EQNS 17A & 17B CAN BOTH BE USED AS PRESENTED IF
C I0 IS DEFINED TO BE 0 (THUS THE LOWER LIMIT OF THE INDEX MUST
C BE CHANGED FROM 1 TO 0 FOR ALL P FUNCTIONS IN 17A & 17B
C
C THEREFORE, FOR ADDED EASE IN PROGRAMMING AND FOR INCREASED
C UNDERSTANDABILITY, APPROPRIATE ARRAYS WILL BEGIN WITH
C INDEX=0, I.E., P(ALPHA,0) Q(ALPHA,0) GAMMA(0) WHERE R(0)=0
C
      REAL P(7,0:511),Q(7,0:511),RR(7,7),BB(7,7),WKAREA(10)
      REAL IA(0:8),SUM
      INTEGER ALPHA,LIMIT,BETA,NUMNOD,I
C STORE INDECES OF NODES IN IA()
C AS IN GILLAN'S PAPER FOR THE L-J POTENTIAL
C SET BASE FUNCTION NODES AT  R/SIGMA=.2,.4,.65,.95,1.1,1.2,1.5
      DATA IA(0),IA(1),IA(2),IA(3),IA(4),IA(5),IA(6),IA(7)
     1 /0.,4.,8.,13.,19.,21.,24.,30./
C     PRINT 11
C 11  FORMAT (3X,'SUBROUTINE PQGEN')
C INITIALIZE BASE FUNCTIONS AND CONJUGATES TO ZERO
      DO 4 ALPHA=1,NUMNOD
      DO 40 I=0,N
      Q(ALPHA,I)=0.
   40 P(ALPHA,I)=0.
    4 CONTINUE
C INITIALIZE R & B MATRICES TO ZERO
      DO 99 ALPHA=1,NUMNOD
      DO 98 BETA=1,NUMNOD
      RR(ALPHA,BETA)=0.
   98 BB(ALPHA,BETA)=0.
   99 CONTINUE
C CALC P(1,I)                      ACCORDING TO EQN 17B
      ALPHA=1
```

```
      DO 5 I = 0,10
    5 IF (I.LE.IA(1)) P(ALPHA,I)=(IA(1)-I)/IA(1)
C CALC P(2,I) - P(7,I)                  USING EQN 17A
      DO 10 ALPHA=2,NUMNOD
      CALL PGEN (P,ALPHA,IA(ALPHA-2),IA(ALPHA-1),IA(ALPHA),LIMIT)
   10 CONTINUE
C CALC BASE FUNCTION CONJUGATES Q(ALPHA,I)
C FIRST CALC R MATRIX                  BY EQN 20
      DO 15 ALPHA=1,NUMNOD
      DO 50 BETA=1,NUMNOD
      SUM=0.
      DO 20 I=0,LIMIT
   20 SUM=SUM+P(ALPHA,I)*P(BETA,I)
   50 RR(ALPHA,BETA)=SUM
   15 CONTINUE
C LINV1F STORES THE INVERSE OF RR(ALPHA,BETA) IN BB(ALPHA,BETA)
      CALL LINV1F (RR,NUMNOD,NUMNOD,BB,0,WKAREA,IER)
C
C  CALC Q(ALPHA,I)                     FROM EQN 18
      DO 100 ALPHA=1,NUMNOD
      DO  90 I=0,LIMIT
      SUM=0.
      DO 80 BETA=1,NUMNOD
   80 SUM=SUM+BB(ALPHA,BETA)*P(BETA,I)
   90 Q(ALPHA,I)=SUM
  100 CONTINUE
C     PRINT 105
C     DO 150 ALPHA=1,NUMNOD
C 150 PRINT 101,(RR(ALPHA,BETA),BETA=1,NUMNOD)
C 101 FORMAT (1X,'R',7(2X,F12.6))
C     DO 160 ALPHA=1,NUMNOD
C 160 PRINT 102,(BB(ALPHA,BETA),BETA=1,NUMNOD)
C 102 FORMAT (1X,'B',7(2X,F12.6))
C 105 FORMAT (//)
C     PRINT 501
C     DO 511 I=0,35
C 511 PRINT 500, I,P(1,I),P(2,I),P(3,I),P(4,I),P(5,I),P(6,I),P(7,I)
C     PRINT 601
C     DO 555 I=0,35
C 555 PRINT 500,I,Q(1,I),Q(2,I),Q(3,I),Q(4,I),Q(5,I),Q(6,I),Q(7,I)
C 500 FORMAT (1X,I3,7(2X,F10.7))
C 501 FORMAT (/,3X,'I',5X,'P(1,I)',6X,'P(2,I)',6X,'P(3,I)',6X,
C    1 'P(4,I)',6X,'P(5,I)',6X,'P(6,I)',6X,'P(7,I)')
C 601 FORMAT (//,3X,'I',5X,'Q(1,I)',6X,'Q(2,I)',6X,'Q(3,I)',6X,
C    1 'Q(4,I)',6X,'Q(5,I)',6X,'Q(6,I)',6X,'Q(7,I)')
      RETURN
      END
C ****************************************************
      SUBROUTINE PGEN (P,ALPHA,LO,MED,HI,LIMIT)
      REAL P(7,0:511)
      REAL LO,MED,HI
      INTEGER ALPHA,I,LIMIT
C CALC P(ALPHA,I) FOR ALPHA > 2        BY EQN 17A
C     PRINT 11
C  11 FORMAT (6X,'SUBROUTINE PGEN')
C MANY LOOPS ARE LIMITED TO I=LIMIT=30 SINCE ALL P(ALPHA,I>30)=0
C P(ALPHA,I) AND Q(ALPHA,I) = 0 FOR I > LIMIT
      LIMIT=30
      DO 50 I=0,LIMIT
      IF ((I.LE.LO).OR.(I.GT.HI)) P(ALPHA,I)=0.
```

```
         IF ((I.GT.LO).AND.(I.LE.MED))  P(ALPHA,I)=(I-LO)/(MED-LO)
         IF ((I.GT.MED).AND.(I.LE.HI))  P(ALPHA,I)=(HI-I)/(HI-MED)
  50   CONTINUE
       RETURN
       END
C ******************************************************
       SUBROUTINE INIT (R,F,K,GAMMA,DR,N,DK,TSTAR,DGAMMA)
C
C  THIS ROUTINE INITIALIZES VALUES FOR SEVERAL ARRAYS
C
       REAL R(0:511),F(0:511),K(0:511),GAMMA(0:511),DGAMMA(0:511)
       REAL DK,DR,TSTAR,S,S6,PI
       DOUBLE PRECISION U
       INTEGER N,I,J
C      PRINT 12
C  12  FORMAT (3X,'SUBROUTINE INIT')
       PI=3.1415927
       DO 11 I=0,40
  11   F(I)=-1.
       DO 50 I=0,N
  50   DGAMMA(I)=0.
C CALC REDUCED R(I) VALUES
       DO 5 I=0,N
       R(I)=REAL(I)*DR
C      PRINT 2323,R(I),I,DR
   5   CONTINUE
C2323  FORMAT (1X,'R=',F6.3,2X,I3,2X,F7.3)
C CALC MEYER FACTORS F(I) USING A L-J POTENTIAL
       DO 10 I=15,N
       S=1./R(I)
       S6=S**6
       U=4./TSTAR*S6*(S6-1.)
       F(I)=EXP(-U) - 1.
  10   CONTINUE
C CALC THE FOURIER TRANSM FACTORS K(J)          BY EQN 17
       DK=2.*PI/(REAL(N+1)*DR)
C      PRINT 501,DK
C 501  FORMAT (1X,'DK = ',F10.7)
       DO 15 J=0,N
  15   K(J)= REAL(J)*DK
       CALL GETG (R,GAMMA,N)
C      PRINT 400
C 400  FORMAT (//,2X,'INITIAL VALUES FROM SUBROUTINE    INIT')
C      PRINT 600
C      PRINT 605
C      DO 20 I=0,500,10
C 20    PRINT 500,I,R(I),F(I),K(I),GAMMA(I)
C 600  FORMAT (/,2X,'I',9X,'R',13X,'F',14X,'K',13X,'G')
C 605  FORMAT (1X,90('-'))
C 500  FORMAT (1X,I3,2(2X,F10.5,2X,G15.7))
       RETURN
       END
C ******************************************************
       SUBROUTINE GETG (R,GAMMA,N)
C
C  THE FOLLOWING GENERATES THE INITIAL GUESS FOR GAMMA(I) VALUES
C  THE FUNCTION USED WAS OBTAINED BY A NLLSQ FIT OF DATA
C     (USING A TWO-BAND LORENTZ FUNCTION)
C  FOR THE CASE OF RHOSTAR=0.01 AND TSTAR=1.0 SOLVED BY THE
C  PY.CONV PROGRAM
```

```fortran
C
      REAL GAMMA(0:511),R(0:511),A1,A2,A3,A4,A5,A6,A8,A9
      INTEGER I,N
C     PRINT 12
C 12  FORMAT (6X,'SUBROUTINE GETG')
      A1=.01577
      A2=.05409
      A3=.3298
      A4=3.0931E-4
      A5=2.2865
      A6=.177106
      DO 50 I=0,N
      A8=R(I)-A2
      A9=R(I)-A5
C     GAMMA(I)=-1.0
      GAMMA(I)=A1/(A8*A8+A3*A3) + A4/(A9*A9+A6*A6)
C     IF ((R(I).GE.1.).AND.(R(I).LE.1.4)) GAMMA(I)=-1.0
   50 CONTINUE
      RETURN
      END
C ******************************************************
      SUBROUTINE DECOMP (Q,P,GAMMA,AA,DGAMMA,NUMNOD,LIMIT,N)
C
C  ROUTINE TO DECOMPOSE GAMMA INTO A(ALPHA) AND DELTAGAMMA(I)
C     (OR GAMMAPRIME INTO APRIME AND DELTAGAMMAPRIME)
      REAL Q(7,0:511),P(7,0:511),GAMMA(0:511),AA(7),DGAMMA(0:511)
      REAL SUM
      INTEGER ALPHA,I,NUMNOD,LIMIT,N
C     PRINT 101
C 101 FORMAT (3X,'SUBROUTINE DECOMP')
C INITIALIZE A(ALPHA) TO ZERO
      DO 4 ALPHA=1,NUMNOD
    4 AA(ALPHA)=0.
C     PRINT 19
C CALC APRIME(ALPHA)                 FROM EQN 22
      DO 40 ALPHA = 1,NUMNOD
      DO 10 I=0,LIMIT
   10 AA(ALPHA)= AA(ALPHA)+Q(ALPHA,I)*GAMMA(I)
C     PRINT 25,ALPHA,AA(ALPHA)
   40 CONTINUE
C CALC DELTAGAMMA                    FROM EQN 10
      DO 5 I = 0,N
      SUM=0.
      DO 50 ALPHA=1,NUMNOD
   50 SUM=SUM+AA(ALPHA)*P(ALPHA,I)
      DGAMMA(I)=GAMMA(I)-SUM
    5 CONTINUE
C     PRINT 21
C     PRINT 22, (DGAMMA(I),I=0,99)
C 19  FORMAT (/,7X,'VALUES OF A(ALPHA)')
C 20  FORMAT (1X,7(1X,F15.7))
C 21  FORMAT (/,7X,'VALUES OF DGAMMA(I) ,I=0,99')
C 22  FORMAT (1X, 7(1X,F15.7))
C 25  FORMAT (7X,'A(',I1,')= ',F15.7)
      RETURN
      END
C ********************************************************
      SUBROUTINE NRTEST (A,AP,D,NUMNOD,CONVRG)
C
C ROUTINE TO TEST FOR CONVERGENCE OF A NEWTON-RAPHSON CYCLE
```

```
C
      REAL A(7),AP(7),D(7),DTEST,SUM
      INTEGER NUMNOD,ALPHA
      LOGICAL CONVRG
C   TEST FOR NRC CONVERGENCE      BY EQN 31
      DTEST=5.
      CONVRG=.FALSE.
      SUM=0.
      DO 5 ALPHA=1,NUMNOD
      D(ALPHA)=A(ALPHA)-AP(ALPHA)
  5   SUM=SUM+D(ALPHA)*D(ALPHA)
C TEST FOR EXCESSIVE CHANGE IN ALPHA AS PER          EQNS 33 & 34
      IF (SUM.LT.DTEST) GOTO 522
      PRINT 523
 523  FORMAT (1X,'D(ALPHA) LIMIT SURPASSED')
      PRINT 524, (D(ALPHA),ALPHA=1,NUMNOD)
 524  FORMAT (1X,'INITIAL D(ALPHA)S : ',7G12.6)
      DO 29 ALPHA=1,NUMNOD
  29  D(ALPHA)=SQRT(DTEST)*D(ALPHA)/SQRT(SUM)
      PRINT 525, (D(ALPHA),ALPHA=1,NUMNOD)
 525  FORMAT (1X,'ADJUSTED D(ALPHA)S : ',7G12.6)
 522  CONTINUE
C     PRINT 999,SQRT(SUM)
C 999 FORMAT (/,1X,'RMS D(ALPHA)= ',G15.7)
      IF (SQRT(SUM).LT.1.0E-5) CONVRG=.TRUE.
      RETURN
      END
C ********************************************************
      SUBROUTINE GCALC (P,A,DGAMMA,GAMMA,N,NUMNOD)
C
C ROUTINE TO CALCULATE NEW GAMMA FROM A(ALPHA) AND DELTAGAMMA
C
      REAL A(7),P(7,0:511),GAMMA(0:511),DGAMMA(0:511),SUM
      INTEGER I,N,NUMNOD,ALPHA
C     PRINT 25
C CALC GAMMAPRIME(I)              FROM EQN 10
      DO 50 I=0,N
      SUM=0.
      DO 40 ALPHA=1,NUMNOD
  40  SUM=SUM+A(ALPHA)*P(ALPHA,I)
  50  GAMMA(I)=SUM+DGAMMA(I)
C     DO 20 I=0,100,2
C 20  PRINT 22,I,GAMMA(I)
C 22  FORMAT (7X,'GAMMA(',I2,') = ',G15.7)
C     PRINT 23
C 23  FORMAT (1X,'GCALC DONE')
C 25  FORMAT (3X,'SUBROUTINE GCALC')
      RETURN
      END
C ********************************************************
      SUBROUTINE GPCALC (F,GAMMA,R,K,FTC,GAMMAP,N,PI,DR,DK,RHO)
C
C ROUTINE TO CALCULATE GAMMAPRIME FROM GAMMA
C
      REAL F(0:511),GAMMA(0:511),R(0:511),K(0:511),FTC(0:511)
      REAL C(0:511),FTG(0:511),GAMMAP(0:511),COEFF
      REAL PI,DR,DK,RHO
      INTEGER I,N
C     PRINT 22
C 22  FORMAT (3X,'SUBROUTINE GPCALC')
```

```
C CALC NEW C(I)                          USING EQN 16A
      DO 5 I=0,N
    5 C(I)=(1.+GAMMA(I))*F(I)
C CALC FOURIER TRNSFM OF C(I)                 USING EQN 16B
      COEFF=4.*PI
      CALL FFT (COEFF,DR,R,C,FTC,K,N)
C CALC FOURIER TRNSM OF GAMMAPRIME(I)            BY EQN 16C
      DO 15 I=0,N
   15 FTG(I)=RHO*FTC(I)*FTC(I)/(1.-RHO*FTC(I))
C INVERSE FOURIER TRNSFM TO GET GAMMAPRIME        BY EQN 16D
      COEFF=1./(2.*PI*PI)
      CALL FFT (COEFF,DK,K,FTG,GAMMAP,R,N)
C     PRINT 100
C     PRINT 605
C 100 FORMAT (/,2X,'I',10X,'C',14X,'FTC',17X,'FTG',12X,'GP')
C     DO 150 I=0,30
C 150 PRINT 101,I,C(I),FTC(I),FTG(I),GAMMAP(I)
C 101 FORMAT (1X,I3,4(2X,G15.7))
C 605 FORMAT (1X,90('-'))
      RETURN
      END
C ********************************************************
      SUBROUTINE FFT (COEFF,DX,X,C,FTC,K,N)
C
C ROUTINE FOR CALCULATING FOURIER AND INVERSE FOURIER TRSFMS
C
      REAL X(0:511),C(0:511),FTC(0:511),K(0:511)
      REAL SUM,COEFF,DX,Z,PI
      INTEGER I,J,N,FN
C     PRINT 99
      PI=3.1415927
C***CALC ELEMENT FTC(I) AT I=0
C      AS THE LIMIT AS K(J) GOES TO ZERO
      FN=(N+1)/2-1
      SUM=0.
      DO 2 I=1,FN
    2 SUM=SUM+X(I)*X(I)*C(I)
      FTC(0)=COEFF*DX*SUM
C***CALC REMAINING ELEMENTS IN FTC(I)
      DO 20 J=1,N
      SUM=0.
      DO 15 I=1,FN
   15 SUM=SUM+X(I)*SIN(K(J)*X(I))*C(I)
   20 FTC(J)=COEFF*DX*SUM/K(J)
C 99 FORMAT (6X,'FFT ENTERED')
      RETURN
      END
C ********************************************************
      SUBROUTINE JACOB (R,Q,P,FTC,K,F,JINV,DR,DK,N,NUMNOD,RHO
     1 ,LIMIT)
C
C ROUTINE TO CALCULATE AND INVERT THE JACOBIAN MATRIX
C
      REAL R(0:511),F(0:511),K(0:511),P(7,0:511),Q(7,0:511)
      REAL DGPDG(0:31,0:31),FTC(0:511),WKAREA(10),JJ(7,7),Y
      REAL ZZ(0:250),JINV(7,7),DD(0:70),DR,DK,RHO,Z,SUM,PI,DELTA
      INTEGER I,J,M,L,LIMIT,ALPHA,BETA,NUMNOD
C     PRINT 11
C 11 FORMAT (3X,'SUBROUTINE JACOB')
      PI=3.1415927
```

```
          DO 10 I = 0,(N+1)/2-1
          Z=RHO*FTC(I)
          Z=Z/(1.-Z)
    10    ZZ(I)=2.*Z + Z*Z
C EVALUATE DGPDG(I,J) FOR CASE I=0         USING EQNS 27 & 28
          I=0
          DO 900 J=0,LIMIT
          SUM=0.
          DO 950 M=0,(N+1)/2-1
   950    SUM=SUM+K(M)*ZZ(M)*SIN(K(M)*R(J))
   900    DGPDG(I,J)=(2.*DR*R(J)/PI)*F(J)*DK*SUM
C EVALUATE REMAINING DGPDG(I,J)            USING EQNS 25 & 26
          DO 20 L=0,2*LIMIT+1
          SUM=0.
          DO 30 M=0,(N+1)/2-1
    30    SUM=SUM+ZZ(M)*COS(K(M)*R(L))
    20    DD(L)=DK*SUM
C NOTE THAT ONLY DGPDG(0-LIMIT,0-LIMIT) WILL BE NEEDED LATER
C AND THAT WE TAKE D(L)=D(-L)
          DO 40 I = 1,LIMIT
          DO 50 J = 0,LIMIT
          L=ABS(I-J)
          M=I+J
          DGPDG(I,J)=DR*R(J)/(PI*R(I))*F(J)*(DD(L)-DD(M))
    50    CONTINUE
    40    CONTINUE
C WE NOW FILL THE JACOBIAN MATRIX                    BY EQN 23
          DO 60 ALPHA=1,NUMNOD
          DO 70 BETA=1,NUMNOD
          DELTA=0.
          IF (ALPHA.EQ.BETA) DELTA=1.
          SUM=0.
          DO 100 I = 1,LIMIT
          DO 110 J = 1,LIMIT
   110    SUM=SUM+Q(ALPHA,I)*DGPDG(I,J)*P(BETA,J)
   100    CONTINUE
    70    JJ(ALPHA,BETA)=DELTA-SUM
    60    CONTINUE
C USE THE IMSL ROUTINE TO INVERT THE JACOBIAN
          CALL LINV1F (JJ,NUMNOD,NUMNOD,JINV,0,WKAREA,IER)
          RETURN
          END
C ***********************************************************
          SUBROUTINE NEWA (JINV,D,A,NUMNOD)
C
C ROUTINE TO CALCULATE THE NEW VALUES OF A(ALPHA)
C
          REAL JINV(7,7),D(7),A(7),SUM
          INTEGER NUMNOD,ALPHA,BETA
C     PRINT 11
C 11   FORMAT (3X,'SUBROUTINE NEWA')
C EVALUATE THE NEW A(ALPHA)                          BY EQN 13
          DO 150 ALPHA=1,NUMNOD
          SUM=0.
          DO 140 BETA=1,NUMNOD
   140    SUM=SUM+JINV(ALPHA,BETA)*D(BETA)
          A(ALPHA)=A(ALPHA)-SUM
   150    CONTINUE
          RETURN
          END
```

```
C ********************************************************
      SUBROUTINE RCTEST (GAMMA,GAMMAP,MU,N,DR,RC,CONVRG)
C
C ROUTINE TO TEST FOR CONVERGENCE OF A REFINEMENT CYCLE
C
      REAL GAMMA(0:511),GAMMAP(0:511),MU(100),DR,SUM,DIFF
      INTEGER N,RC,I
      LOGICAL CONVRG
C TEST FOR RC CONVERGENCE  - CALC OF MU -  EQN 32
C      PRINT 797
C 797  FORMAT ( 3X,'I',8X,'GAMMAP(I)',8X,'GAMMA(I)',9X,'DIFF')
      CONVRG=.FALSE.
      SUM=0.
      DO 50 I = 0,N
      DIFF=GAMMAP(I)-GAMMA(I)
C      PRINT 737,I,GAMMAP(I),GAMMA(I),DIFF
  50  SUM=SUM+DIFF*DIFF
C 737  FORMAT (2X,I3,4X,3(G15.6,3X))
      MU(RC)=SQRT(DR*SUM)
      PRINT 888,MU(RC)
  888 FORMAT (/,21X,'MU = ',F12.7)
      IF (MU(RC).LT.1.E-5) CONVRG=.TRUE.
      RETURN
      END
C ********************************************************
      SUBROUTINE OUTPUT(NRCYCL,MU,RHO,TSTAR,F,GAMMA,R,RC,N)
C
C ROUTINE TO OUTPUT RESULTS
C
      REAL MU(100),F(0:511),GAMMA(0:511),R(0:511),RHO,TSTAR
      INTEGER NRCYCL(100),RC,I,N
      REAL C(0:511),H(0:511),G(0:511)
      PRINT 7777
 7777 FORMAT (//,3X,'**********  SUMMARY OF PROCESS  **********')
      PRINT 7779
 7779 FORMAT (/,7X,'REFINEMENT',2X,'NEWT-RAPHN',
     1 /,7X,' CYCLE ',2X,' CYCLE ',6X,'MU')
      DO 20 I=1,RC
  20  PRINT 7778,I,NRCYCL(I),MU(I)
 7778 FORMAT (10X,I3,9X,I3,5X,G15.6)
      PRINT 7776
 7776 FORMAT (/,5X,'**********  END OF SUMMARY ************')
      PRINT 1115
 1115 FORMAT (//,1X,'VALUES OBTAINED FOR THE CORRELATION FUNCTIONS')
      PRINT 1114,RHO,TSTAR
 1114 FORMAT (/,5X,'FOR: RHOSTAR = ',F6.3,'  AT TSTAR = ',F6.3)
      PRINT 1111
      DO 1112 I=0,48
      C(I)=(1.+GAMMA(I))*F(I)
      H(I)=GAMMA(I) + C(I)
      G(I)=H(I) + 1.
 1112 PRINT 1113,I,R(I),GAMMA(I),C(I),H(I),G(I)
 1111 FORMAT (/,3X,'I', 5X,'R(I)', 9X,'GAMMA(I)',12X,'C(I)',
     1 15X,'H(I)',15X,'G(I)')
 1113 FORMAT (1X,I3,4X,F5.2,4(4X,G15.6))
      DO 1116 I=52,100,4
      C(I)=(1.+GAMMA(I))*F(I)
      H(I)=GAMMA(I) + C(I)
      G(I)=H(I) + 1.
 1116 PRINT 1117,I,R(I),GAMMA(I),C(I),H(I),G(I)
```

```
1117  FORMAT (1X,I3,4X,F5.2,4(4X,G15.6))
      RETURN
      END
```

# APPENDIX C

```
C_____
C
C
C       MOLECULAR DYNAMICS PROGRAM IN THE N-V-E ENSEMBLE
C       USING A FIFTH-ORDER PREDICTOR-CORRECTOR METHOD
C          TO SOLVE THE EQUATIONS OF MOTION
C
C
C
C_____



      DOUBLE PRECISION D4AX,D4AY,D4AZ,D3AX,D3AY,D3AZ,D2AX,D2AY,D2AZ,
     1    DACX,DACY,DACZ,SUMD4,SUMD3,SUMD2,SUMDAC,SUMACC,SUMD4A,SUMD3A,
     2    SUMD2A,SUMDA,SUMA,RK12,RK3,RK123,RK4,RK14,RK1234,RKS4,RK5
     3    ,X5(256),Y5(256),Z5(256),X4(256),Y4(256),Z4(256),X3(256),
     4    Y3(256),Z3(256),X2(256),Y2(256),Z2(256),X1(256),Y1(256),
     5    Z1(256),DEL5,DEL4,DEL3,DELSQ,DELTA
      COMMON /VELO/ VX(32,2000),VY(32,2000),VZ(32,2000)
      COMMON /POS/ X0(256),Y0(256),Z0(256)
      COMMON /VEL/ X1,Y1,Z1
      COMMON /DER/ X2,Y2,Z2,X3,Y3,Z3,X4,Y4,Z4,X5,Y5,Z5
      COMMON /FOR/ FX(8,256),FY(8,256),FZ(8,256)
      COMMON /D/ DAX(256),DAY(256),DAZ(256),X0L(256),Y0L(256),Z0L(256)
      COMMON /NABLST/ LIST(11000),NABORS(256)
      COMMON /PROP/ IDIST(300),ISUM
      COMMON /SUMD/ SUMACC,SUMDAC,SUMD2,SUMD3,SUMD4
      COMMON /TIME/ DEL5,DEL4,DEL3,DELSQ,DELTA
      COMMON /SS/ SUMD4A,SUMD3A,SUMD2A,SUMDA,SUMA,XSUM,SUMV,SUME
      COMMON /KAB/ MAXKA,MAXKB
      COMMON /PROPA/ RDEL,RMAX,RDMAX,RLIST,FSHFT,ESHFT,ESHFTA,NP1,NP2,
     A              KSORT
C
C
C === SET NUMBER OF PARTICLES IN PRIMARY CELL
      NP=256
      PART=NP
      NP1=NP-1
      NP2=0.5*PART+0.01
      NP22=NP-2
C
C === SET VALUES OF PHYSICAL CONSTANTS
      SIGMA=3.405
      EPSI=120.
      AVO=6.0225E+23
      BOLTZ=1.38054
      WTMOL=39.994
      THIRD=1./3.
      PI=3.14159265
C
C === SET DESIRED FLUID STATE CONDITIONS
      DR=0.931
      TR=0.765
```

```
      T=TR*EPSI
      VOL=PART/DR
      VSCALE=THIRD/TR
      VOLCO=AVO*(SIGMA*1.E-8) **3/DR
C
C === SET RUN FLAGS AND PARAMETERS
      IFLG=0
      KB=0
      KSAVE=10
      KSORT=10
      KWRITE=50
      NVELO=8
      MAXKB=2600
      MAXKA=0
      TDIST=0.
      XDIST=0.2
C
C === SET TIME-STEP AND ITS MULTIPLES
      DELTA=0.0025
      DELSQ=DELTA*DELTA
      DELTSQ=0.5*DELSQ
      DEL3=DELTA*DELSQ
      DEL4=DELSQ*DELSQ
      DEL5=DEL4*DELTA
      DEL6=DEL5*DELTA
      TSTEP=SQRT(WTMOL*SIGMA**2/AVO/EPSI/BOLTZ)*1.E+12
      TT1=TSTEP*DELTA*1.E-12
C
C === SET PARAMETERS IN PREDICTOR-CORRECTOR METHOD
C
C === SET DISTANCES FOR POTENTIAL CUT-OFF, VERLET LIST, ETC.
      CUBE=VOL**THIRD
      CUBE2=0.5*CUBE
      RC=2.5
      RMAX=RC*RC
      RLIST=(RC+0.3)**2
      RDMAX=CUBE2*CUBE2
      IF(RLIST.GT.RDMAX) RLIST=RDMAX
      IF(RMAX.GT.RDMAX) RMAX=(CUBE2-0.3)**2
C
C === SCALE FACTORS FOR VELOCITIES DURING EQUILIBRIUM
      AHEAT=DELSQ*PART*3.*TR
      RTR=DELTA*SQRT(3.*TR)
C
C === INCREMENT FOR SAMPLING FOR G(R)
      RDEL=0.025
      NY=CUBE2/RDEL-1
C
C === SHIFTED-FORCE CONSTANTS
      RRMAX=1./SQRT(RMAX)
      RRMAX6=RRMAX**6
      ESHFT=RRMAX6*(28.-52.*RRMAX6)
      ESHFTA=48.*RRMAX*RRMAX6*(RRMAX6-0.5)
      FSHFT=ESHFTA
C
C === CORRECTIONS FOR LONG-RANGE INTERACTIONS
      RC3=RC**3
      RC9=RC3**3
      CORE=8.*PI*DR*(1./9./RC9-THIRD/RC3)
      CORV=96.*PI*DR*(0.5*THIRD/RC3-1./9./RC9)
```

```
      DE=CORE
      DP=-VSCALE*CORV
C
C === INITIALIZE SUM ACCUMULATORS
      ISUM=0
      XSUM=0.
      SUME=0.
      SUMV=0.
      SUMA=0.
      SUMDA=0.
      SUMD2A=0.
      SUMD3A=0.
      SUMD4A=0.
      DO 510 K=1,NY
  510 IDIST(K)=0
C
C === PRINT PARAMETERS
      WRITE(6,900)
  900 FORMAT(1H1///)
      WRITE(6,902)
  902 FORMAT(7X,49('*'))
      WRITE(6,904)
  904 FORMAT(7X,'*',T56,'*')
      WRITE(6,906) NP
  906 FORMAT(7X,'*',2X,'MOLECULAR DYNAMICS FOR',I4,' L-J PARTICLES',
     A    T56,'*')
      WRITE(6,904)
      WRITE(6,902)
      WRITE(6,904)
      WRITE(6,908) EPSI,SIGMA
  908 FORMAT(7X,'*',2X,'EPSI/K = ',F7.3,T36,'SIGMA = ',F7.3,T56,'*')
      WRITE(6,910) TR,T
  910 FORMAT(7X,'*',2X,'TR    = ',F7.3,T36,'  T = ',F7.3,T56,'*')
      WRITE(6,912) DR,VOL
  912 FORMAT(7X,'*',2X,'D     = ',F7.3,T36,'VOL  = ',F7.3,T56,'*')
      WRITE(6,914) CUBE,RMAX
  914 FORMAT(7X,'*',2X,'CUBE  = ',F7.3,T36,'RMAX = ',F7.3,T56,'*')
      WRITE(6,916) RC,RDEL
  916 FORMAT(7X,'*',2X,'RC    = ',F7.3,T36,'DELR = ',F7.3,T56,'*')
      WRITE(6,918) RLIST,RDMAX
  918 FORMAT(7X,'*',2X,'RLIST = ',F7.3,T36,'RDMAX = ',F7.3,T56,'*')
      WRITE(6,920) DELTA,NY
  920 FORMAT(7X,'*',2X,'DELTA = ',F7.3,T36,'NY    = ',I3,T56,'*')
      WRITE(6,904)
      WRITE(6,922) TSTEP
  922 FORMAT(7X,'*',2X,'TIME UNIT = ',F6.3,'E-12 SEC',T56,'*')
      WRITE(6,924) TT1
  924 FORMAT(7X,'*',2X,'TIME STEP = ',E10.3,' SEC',T56,'*')
      WRITE(6,904)
      WRITE(6,926) DE
  926 FORMAT(7X,'*',2X,'ENERGY   CORRECTION = ',F7.3,T56,'*')
      WRITE(6,928) DP
  928 FORMAT(7X,'*',2X,'PRESSURE CORRELATION = ',F7.3,T56,'*')
      WRITE(6,904)
      WRITE(6,902)
  625 FORMAT(3X,'DA=',G12.5,6X,'REDUCED K2= ',G12.5)
  666 FORMAT(6G12.5)
  667 FORMAT(3F10.4,I5)
C
C
```

```
C -----------------------------------------------------------------
C
C === LOAD INITIAL POSITIONS OF ATOMS
      CALL FCC(CUBE,NP)
C
C === LOAD INITIAL VELOCITIES OF ATOMS
      CALL INTVEL(AHEAT,RTR,PART,NP)
C
C === ASSIGN INITIAL ACCELERATIONS BASED ON INITIAL POSITIONS
      CALL EVAL(TOTV,TOTE,CUBE,CUBE2,KB,NP,NABTOT)
C
C === SCALE ACCELERATIONS AND STORE STARTING POSITIONS
      DO 530 I=1,NP
      X2(I)=FX(1,I)*DELTSQ
      Y2(I)=FY(1,I)*DELTSQ
      Z2(I)=FZ(1,I)*DELTSQ
      X0L(I)=X0(I)
      Y0L(I)=Y0(I)
      Z0L(I)=Z0(I)
  530 CONTINUE
C
C
C ********************************************************************
C
C === ENTER MAIN LOOP OF SIMULATION
      DO 599 NTIMES=1,MAXKB
      KB=KB+1
      CALL PREDCT(NP)
      CALL EVAL(TOTV,TOTE,CUBE,CUBE2,KB,NP,NABTOT)
      CALL CORR(DELTSQ,CUBE,NP)
C
C === CALCULATE MEAN SQUARE DISPLACEMENT & KINETIC ENERGY
      TDIST=0.
      SUMVEL=0.
      DO 541 I=1,NP
      TDIST=TDIST + DAX(I)**2+DAY(I)**2+DAZ(I)**2
      SUMVEL=SUMVEL + X1(I)**2+Y1(I)**2+Z1(I)**2
  541 CONTINUE
      TDIST=TDIST/PART
      EK=SUMVEL/(2.*PART*DELSQ)
C
      IF(IFLG.LT.1) GO TO 410
C === STORE VELOCITIES FOR VELOCITY AUTOCORRECTION FUNCTION
      DO 400 I=1,NP,NVELO
      J=I/NVELO+1
      VX(J,KB)=X1(I)/DELTA
      VY(J,KB)=Y1(I)/DELTA
      VZ(J,KB)=Z1(I)/DELTA
  400 CONTINUE
C
C
C === ACCUMMULATE SUMS FOR PROPERTY AVERAGES
      XSUM=XSUM+SUMVEL
      SUME=SUME+TOTE
      SUMV=SUMV+TOTV
      SUMA=SUMA+SUMACC
      SUMDA=SUMDA+SUMDAC
      SUMD2A=SUMD2A+SUMD2
      SUMD3A=SUMD3A+SUMD3
      SUMD4A=SUMD4A+SUMD4
```

```
C
C === PROPERTY CALCULATION & PRINT-OUT AT INTERVALS
      IF(MOD(KB,KWRITE).NE.0) GO TO 550
      FKB=FLOAT(KB)*PART
      TMP=XSUM/(3.*DELSQ*FKB)
      ENR=(SUME/FKB+CORE)
      VIR=VSCALE*(SUMV/FKB+CORV)
      PRES=TMP*DR*(1.-VIR)
      E1=TOTE/PART+CORE
      ETOT=E1+EK
      P1=TMP*DR*(1.-VSCALE*(TOTV/PART+CORV))
      RLTIM=DELTA*FLOAT(KB)*TSTEP
      RK0=TMP*3.
      RK1=SUMA/(FKB*RK0)
      RK01=RK0*RK1
      RK2=SUMDA/(FKB*RK01)-RK1
      RK12=RK01*RK2
      RK3=SUMD2A/(FKB*RK12)-SUMDA*(RK1+RK2)/(FKB*RK12)
      RK123=RK12*RK3
      RK4=SUMD3A/(FKB*RK123)-2.*SUMD2A*(RK1+RK2+RK3)/(FKB*RK123)+
     1    SUMDA*(RK1+RK2+RK3)**2/(FKB*RK123)-RK1*RK3/RK2
      RK14=RK1*RK3+RK1*RK4+RK2*RK4
      RK1234=RK123*RK4
      RKS4=RK1+RK2+RK3+RK4
      RK5=SUMD4A/(FKB*RK1234)-2.0*SUMD3A*RKS4/(FKB*RK1234)+SUMD2A*
     U    (RKS4**2+2.0*RK14)/(FKB*RK1234)-2.0*SUMDA*RKS4*RK14/
     2    (FKB*RK1234)+RK14*RK14/(RK2*RK3*RK4)
C === PRINT RUN-TABLE HEADING
      WRITE(6,930)
 930  FORMAT(////2X,'KB',5X,'TIME',4X,'PRES',6X,'P1',4X,'ENRG',
     A    5X,'E1',6X,'DIST',4X,'TEMP',5X,'NAB',4X,'TOT ENR'/)
      WRITE(6,940) KB,RLTIM,PRES,P1,ENR,E1,TDIST,TMP,NABTOT,ETOT
 940  FORMAT(1H ,I4,6F8.3,F9.4,I8,2X,F8.4,/)
      WRITE(6,931) RK0,RK1,RK2,RK3,RK4,RK5
 931  FORMAT(5X,'K0= ',G12.4,'   K1= ',G12.4,'   K2= ',G12.4,'   K3= ',
     1    G12.4,'   K4= ',G12.4,'   K5= ',G12.4,/)
C
 410  CONTINUE
C === DURING FIRST OF RUN, SCALE VELOCITIES FOR TEMPERATURE
 550  IF(IFLG.LT.1) CALL EQBRAT(SUMVEL,AHEAT,TDIST,XDIST,NP,IFLG,NY,KB)
 599  CONTINUE
C === PRINT G(R) AT INTERVALS
      CALL RDF(DR,RDEL,NY)
C
      CALL VELOCO(NP,NVELO,DELTA,MAXKA,TMP)
C
      STOP
      END
      SUBROUTINE FCC(CUBE,NP)
C
C === CALCULATE POSITIONS OF SITES ON FACE-CENTERED CUBIC LATTICE
C       BASED ON A BOX OF SIDE = 100
C
      COMMON /POS/ X0(256),Y0(256),Z0(256)
      NC=(FLOAT(NP)/4.)**(1./3.)+0.1
      XL=100./FLOAT(NC)
      Y=0.5*XL
      X0(1)=0.
      Y0(1)=0.
      Z0(1)=0.
```

```
      X0(2)=0.
      Y0(2)=Y
      Z0(2)=Y
      X0(3)=Y
      Y0(3)=0.
      Z0(3)=Y
      X0(4)=Y
      Y0(4)=Y
      Z0(4)=0.
      M=0
      DO 10 I=1,NC
      DO 10 J=1,NC
      DO 10 K=1,NC
      DO 11 IJ=1,4
      X0(IJ+M)=X0(IJ)+XL*(K-1)
      Y0(IJ+M)=Y0(IJ)+XL*(J-1)
      Z0(IJ+M)=Z0(IJ)+XL*(I-1)
   11 CONTINUE
      M=M+4
   10 CONTINUE
C
C === SCALE POSITIONS TO BOX OF SIDE = CUBE
      DO 12 I=1,NP
      X0(I)=X0(I)*CUBE*0.01
      Y0(I)=Y0(I)*CUBE*0.01
      Z0(I)=Z0(I)*CUBE*0.01
   12 CONTINUE
      RETURN
      END
      SUBROUTINE INTVEL(AHEAT,RTR,PART,NP)
C
C === ASSIGN INITIAL VELOCITIES TO ATOMS
C
      DOUBLE PRECISION D4AX,D4AY,D4AZ,D3AX,D3AY,D3AZ,D2AX,D2AY,D2AZ,
     1    DACX,DACY,DACZ,SUMD4,SUMD3,SUMD2,SUMDAC,SUMACC,SUMD4A,SUMD3A,
     2    SUMD2A,SUMDA,SUMA,RK12,RK3,RK123,RK4,RK14,RK1234,RKS4,RK5
     3    ,X5(256),Y5(256),Z5(256),X4(256),Y4(256),Z4(256),X3(256),
     4    Y3(256),Z3(256),X2(256),Y2(256),Z2(256),X1(256),Y1(256),
     5    Z1(256)
      COMMON /VEL/ X1,Y1,Z1
      IRR=34567
      SUMX=0.
      SUMY=0.
      SUMZ=0.
      DO 200 I=1,NP
      IRR=IRR*65539
      XX=FLOAT(IRR)*0.4656613E-9
      IRR=IRR*65539
      YY=FLOAT(IRR)*0.4656613E-9
      IRR=IRR*65539
      ZZ=FLOAT(IRR)*0.4656613E-9
      XYZ=1./SQRT(XX*XX+YY*YY+ZZ*ZZ)
      X1(I)=XX*XYZ*RTR
      Y1(I)=YY*XYZ*RTR
      Z1(I)=ZZ*XYZ*RTR
      SUMX=SUMX+X1(I)
      SUMY=SUMY+Y1(I)
      SUMZ=SUMZ+Z1(I)
  200 CONTINUE
C
```

```
C === SCALE VELOCITIES SO THAT TOTAL MOMENTUM = ZERO
      X=0.
      DO 210 I=1,NP
      X1(I)=X1(I)-SUMX/PART
      Y1(I)=Y1(I)-SUMY/PART
      Z1(I)=Z1(I)-SUMZ/PART
      X=X+X1(I)**2+Y1(I)**2+Z1(I)**2
  210 CONTINUE
C
C === SCALE VELOCITIES TO DESIRED TEMPERATURE
      HEAT=SQRT(AHEAT/X)
      DO 220 I=1,NP
      X1(I)=X1(I)*HEAT
      Y1(I)=Y1(I)*HEAT
      Z1(I)=Z1(I)*HEAT
  220 CONTINUE
      RETURN
      END
      SUBROUTINE PREDCT(NP)
C
C === USE TAYLOR SERIES TO PREDICT POSITIONS & THEIR DERIVATIVES
C          AT NEXT TIME-STEP
C
      DOUBLE PRECISION D4AX,D4AY,D4AZ,D3AX,D3AY,D3AZ,D2AX,D2AY,D2AZ,
     1    DACX,DACY,DACZ,SUMD4,SUMD3,SUMD2,SUMDAC,SUMACC,SUMD4A,SUMD3A,
     2    SUMD2A,SUMDA,SUMA,RK12,RK3,RK123,RK4,RK14,RK1234,RKS4,RK5
     3    ,X5(256),Y5(256),Z5(256),X4(256),Y4(256),Z4(256),X3(256),
     4    Y3(256),Z3(256),X2(256),Y2(256),Z2(256),X1(256),Y1(256),
     5    Z1(256),DEL5,DEL4,DEL3,DELSQ,DELTA
      COMMON /POS/ X0(256),Y0(256),Z0(256)
      COMMON /VEL/ X1,Y1,Z1
      COMMON /DER/ X2,Y2,Z2,X3,Y3,Z3,X4,Y4,Z4,X5,Y5,Z5
      COMMON /FOR/ FX(8,256),FY(8,256),FZ(8,256)
      DO 300 I=1,NP
      X0(I)=X0(I)+X1(I)+X2(I)+X3(I)+X4(I)+X5(I)
      Y0(I)=Y0(I)+Y1(I)+Y2(I)+Y3(I)+Y4(I)+Y5(I)
      Z0(I)=Z0(I)+Z1(I)+Z2(I)+Z3(I)+Z4(I)+Z5(I)
      FX(1,I)=0.
      FY(1,I)=0.
      FZ(1,I)=0.
  300 CONTINUE
      RETURN
      END
      SUBROUTINE EVAL(TOTV,TOTE,CUBE,CUBE2,KB,NP,NABTOT)
C
C === EVALUATE FORCES ON ATOMS USING PAIRWISE ADDITIVE LENNARD-
C        JONES (6-12) INTERMOLECULAR POTENTIAL
C
      DOUBLE PRECISION D4AX,D4AY,D4AZ,D3AX,D3AY,D3AZ,D2AX,D2AY,D2AZ,
     1    DACX,DACY,DACZ,SUMD4,SUMD3,SUMD2,SUMDAC,SUMACC,SUMD4A,SUMD3A,
     2    SUMD2A,SUMDA,SUMA,RK12,RK3,RK123,RK4,RK14,RK1234,RKS4,RK5
      COMMON /POS/ X0(256),Y0(256),Z0(256)
      COMMON /FOR/ FX(8,256),FY(8,256),FZ(8,256)
      COMMON /NABLST/ LIST(11000),NABORS(256)
      COMMON /PROP/ IDIST(300),ISUM
      COMMON /PROPA/ RDEL,RMAX,RDMAX,RLIST,FSHFT,ESHFT,ESHFTA,NP1,NP2,
     A          KSORT
      K=0
      TOTE=0.
      TOTV=0.
```

```
C
C === SET FLAG FOR LIST UP-DATE
      LOHK=2
      IF(MOD(KB,KSORT).EQ.0) LOHK=1
C
C === OUTER LOOP OVER ATOMS
      DO 490 I=1,NP1
      IF(LOHK.EQ.1) GO TO 410
      JBEGIN=NABORS(I)
      JEND  =NABORS(I+1)-1
      IF(JBEGIN.GT.JEND) GO TO 490
      GO TO 415
 410  NABORS(I)=K+1
      JBEGIN=I+1
      JEND=NP
C
C === STORE POSITION OF ATOM I
 415  XI=X0(I)
      YI=Y0(I)
      ZI=Z0(I)
C
C === INNER LOOP OVER ATOMS
      DO 495 JX=JBEGIN,JEND
      J=JX
      IF(LOHK.EQ.2) J=LIST(JX)
C
C === DISTANCE BETWEEN ATOM I AND ATOM J
      X=XI-X0(J)
      Y=YI-Y0(J)
      Z=ZI-Z0(J)
C
C === FIND IMAGE OF ATOM J CLOSEST TO ATOM I
      XR=ABS(X)
      YR=ABS(Y)
      ZR=ABS(Z)
      IF(XR.GT.CUBE2) X=(XR-CUBE)*SIGN(1.,X)
      IF(YR.GT.CUBE2) Y=(YR-CUBE)*SIGN(1.,Y)
      IF(ZR.GT.CUBE2) Z=(ZR-CUBE)*SIGN(1.,Z)
      RSQ=X*X+Y*Y+Z*Z
      RCTR=SQRT(RSQ)
C
C === INCREMENT COUNTER FOR G(U) & UPDATE LIST AT INTERVALS
      IF(LOHK.NE.1) GO TO 419
      IF(RSQ.GT.RDMAX) GO TO 495
      IJ=RCTR/RDEL+.5
      IDIST(IJ)=IDIST(IJ)+1
      IF(RSQ.GT.RLIST) GO TO 495
      K=K+1
      LIST(K)=J
 419  IF(RSQ.GT.RMAX) GO TO 495
C
C === CHECK IF TWO PARTICAL GET TOGETHER
      MORT=1./RCTR-1.4
  69  FORMAT(1X,'ACCEL--MORT=',I2,', A MOLECULE I= ',I3,', J=',I3)
      IF(MORT.GT.0) WRITE(6,69) MORT,I,J
C === EVALUATE FORCES ON ATOMS
      RSI=1./RSQ
      R6=RSI**3
      RPL=48.*R6*(R6-0.5)-RCTR*FSHFT
      RP=RPL*RSI
```

```
      REPX=RP*X
      FX(1,I)=FX(1,I)+REPX
      FX(1,J)=FX(1,J)-REPX
      REPY=RP*Y
      FY(1,I)=FY(1,I)+REPY
      FY(1,J)=FY(1,J)-REPY
      REPZ=RP*Z
      FZ(1,I)=FZ(1,I)+REPZ
      FZ(1,J)=FZ(1,J)-REPZ
C
C === ACCUMMULATE ENERGY AND VIRIAL
      TOTE=TOTE+4.*R6*(R6-1.)+ESHFT+RCTR*ESHFTA
      TOTV=TOTV-RPL
  495 CONTINUE
  490 CONTINUE
      IF(LOHK.NE.1) RETURN
      NABORS(NP)=K+1
      ISUM=ISUM+NP2
      NABTOT=K
      RETURN
      END
      SUBROUTINE CORR(DELTSQ,CUBE,NP)
C
C === CORRECT PREDICTED POSITIONS AND THEIR DERIVATIVES
C
      DOUBLE PRECISION D4AX,D4AY,D4AZ,D3AX,D3AY,D3AZ,D2AX,D2AY,D2AZ,
     1    DACX,DACY,DACZ,SUMD4,SUMD3,SUMD2,SUMDAC,SUMACC,SUMD4A,SUMD3A,
     2    SUMD2A,SUMDA,SUMA,RK12,RK3,RK123,RK4,RK14,RK1234,RKS4,RK5
     3    ,X5(256),Y5(256),Z5(256),X4(256),Y4(256),Z4(256),X3(256),
     4    Y3(256),Z3(256),X2(256),Y2(256),Z2(256),X1(256),Y1(256),
     5    Z1(256),DEL5,DEL4,DEL3,DELSQ,DELTA
      COMMON /POS/ X0(256),Y0(256),Z0(256)
      COMMON /VEL/ X1,Y1,Z1
      COMMON /DER/ X2,Y2,Z2,X3,Y3,Z3,X4,Y4,Z4,X5,Y5,Z5
      COMMON /SUMD/ SUMACC,SUMDAC,SUMD2,SUMD3,SUMD4
      COMMON /TIME/ DEL5,DEL4,DEL3,DELSQ,DELTA
      COMMON /FOR/ FX(8,256),FY(8,256),FZ(8,256)
      COMMON /D/ DAX(256),DAY(256),DAZ(256),X0L(256),Y0L(256),Z0L(256)
      SUMACC=0.
      SUMDAC=0.
      SUMD2=0.
      SUMD3=0.
      SUMD4=0.
      DO 690 I=1,NP
      X2(I)=DELTSQ*FX(1,I)
      Y2(I)=DELTSQ*FY(1,I)
      Z2(I)=DELTSQ*FZ(1,I)
      X1(I)=X1(I)+DELSQ*(-19.0*FX(5,I)+106.0*FX(4,I)-264.0*FX(3,I)+
     1    646.0*FX(2,I)+251.0*FX(1,I))/720.0
      Y1(I)=Y1(I)+DELSQ*(-19.0*FY(5,I)+106.0*FY(4,I)-264.0*FY(3,I)+
     1    646.0*FY(2,I)+251.0*FY(1,I))/720.0
      Z1(I)=Z1(I)+DELSQ*(-19.0*FZ(5,I)+106.0*FZ(4,I)-264.0*FZ(3,I)+
     1    646.0*FZ(2,I)+251.0*FZ(1,I))/720.0
      ACX=FX(1,I)
      ACY=FY(1,I)
      ACZ=FZ(1,I)
      DACX=(-60.0*FX(8,I)+490.00*FX(7,I)-1764.0*FX(6,I)+3675.0*FX(5,I)
     1    -4900.0*FX(4,I)+4410.0*FX(3,I)-2940.0*FX(2,I)+1089.0*FX(1,I)
     2    )/(420.0*DELTA)
      DACY=(-60.0*FY(8,I)+490.00*FY(7,I)-1764.0*FY(6,I)+3675.0*FY(5,I)
```

```
1       -4900.0*FY(4,I)+4410.0*FY(3,I)-2940.0*FY(2,I)+1089.0*FY(1,I)
2       )/(420.0*DELTA)
   DACZ=(-60.0*FZ(8,I)+490.00*FZ(7,I)-1764.0*FZ(6,I)+3675.0*FZ(5,I)
1       -4900.0*FZ(4,I)+4410.0*FZ(3,I)-2940.0*FZ(2,I)+1089.0*FZ(1,I)
2       )/(420.0*DELTA)
   X3(I)=DACX*DEL3/6.0
   Y3(I)=DACY*DEL3/6.0
   Z3(I)=DACZ*DEL3/6.0
   D2AX=(-126.0*FX(8,I)+1019.0*FX(7,I)-3618.0*FX(6,I)+7380.0*FX(5,I)
1       -9490.0*FX(4,I)+7911.0*FX(3,I)-4014.0*FX(2,I)+938.0*FX(1,I))
2       /(180.0*DELSQ)
   D2AY=(-126.0*FY(8,I)+1019.0*FY(7,I)-3618.0*FY(6,I)+7380.0*FY(5,I)
1       -9490.0*FY(4,I)+7911.0*FY(3,I)-4014.0*FY(2,I)+938.0*FY(1,I))
2       /(180.0*DELSQ)
   D2AZ=(-126.0*FZ(8,I)+1019.0*FZ(7,I)-3618.0*FZ(6,I)+7380.0*FZ(5,I)
1       -9490.0*FZ(4,I)+7911.0*FZ(3,I)-4014.0*FZ(2,I)+938.0*FZ(1,I))
2       /(180.0*DELSQ)
   X4(I)=D2AX*DEL4/24.0
   Y4(I)=D2AY*DEL4/24.0
   Z4(I)=D2AZ*DEL4/24.0
   D3AX=(-232.0*FX(8,I)+1849.0*FX(7,I)-6432.0*FX(6,I)+12725.0*FX(5,I)
1       -15560.0*FX(4,I)+11787.0*FX(3,I)-5104.0*FX(2,I)+967.0*FX(1,I)
2       )/(120.0*DEL3)
   D3AY=(-232.0*FY(8,I)+1849.0*FY(7,I)-6432.0*FY(6,I)+12725.0*FY(5,I)
1       -15560.0*FY(4,I)+11787.0*FY(3,I)-5104.0*FY(2,I)+967.0*FY(1,I)
2       )/(120.0*DEL3)
   D3AZ=(-232.0*FZ(8,I)+1849.0*FZ(7,I)-6432.0*FZ(6,I)+12725.0*FZ(5,I)
1       -15560.0*FZ(4,I)+11787.0*FZ(3,I)-5104.0*FZ(2,I)+967.0*FZ(1,I)
2       )/(120.0*DEL3)
   X5(I)=D3AX*DEL5/120.0
   Y5(I)=D3AY*DEL5/120.0
   Z5(I)=D3AZ*DEL5/120.0
   D4AX=(-21.0*FX(8,I)+164.0*FX(7,I)-555.0*FX(6,I)+1056.0*FX(5,I)
1       -1219.0*FX(4,I)+852.0*FX(3,I)-333.0*FX(2,I)+56.0*FX(1,I))
2       /(6.0*DEL4)
   D4AY=(-21.0*FY(8,I)+164.0*FY(7,I)-555.0*FY(6,I)+1056.0*FY(5,I)
1       -1219.0*FY(4,I)+852.0*FY(3,I)-333.0*FY(2,I)+56.0*FY(1,I))
2       /(6.0*DEL4)
   D4AZ=(-21.0*FZ(8,I)+164.0*FZ(7,I)-555.0*FZ(6,I)+1056.0*FZ(5,I)
1       -1219.0*FZ(4,I)+852.0*FZ(3,I)-333.0*FZ(2,I)+56.0*FZ(1,I))
2       /(6.0*DEL4)
C
   FX(8,I)=FX(7,I)
   FY(8,I)=FY(7,I)
   FZ(8,I)=FZ(7,I)
   FX(7,I)=FX(6,I)
   FY(7,I)=FY(6,I)
   FZ(7,I)=FZ(6,I)
   FX(6,I)=FX(5,I)
   FY(6,I)=FY(5,I)
   FZ(6,I)=FZ(5,I)
   FX(5,I)=FX(4,I)
   FY(5,I)=FY(4,I)
   FZ(5,I)=FZ(4,I)
   FX(4,I)=FX(3,I)
   FY(4,I)=FY(3,I)
   FZ(4,I)=FZ(3,I)
   FX(3,I)=FX(2,I)
   FY(3,I)=FY(2,I)
   FZ(3,I)=FZ(2,I)
```

```
      FX(2,I)=FX(1,I)
      FY(2,I)=FY(1,I)
      FZ(2,I)=FZ(1,I)
C
C     SUMATION OF ACC,DAC,D2A,D3A,D4A
      SUMACC=SUMACC +ACX**2+ACY**2+ACZ**2
      SUMDAC=SUMDAC +DACX**2+DACY**2+DACZ**2
      SUMD2=SUMD2 +D2AX**2+D2AY**2+D2AZ**2
      SUMD3=SUMD3 +D3AX**2+D3AY**2+D3AZ**2
      SUMD4=SUMD4 +D4AX**2+D4AY**2+D4AZ**2
C
C
C === DISPLACEMENTS
      DAX(I)=DAX(I)-X0(I)+X0L(I)
      DAY(I)=DAY(I)-Y0(I)+Y0L(I)
      DAZ(I)=DAZ(I)-Z0(I)+Z0L(I)
C
C === CHECK FOR CONSERVATION OF PARTICLES IN PRIMARY CELL
      XX=X0(I)
      YY=Y0(I)
      ZZ=Z0(I)
      IF(XX.LE.0.) XX=XX+CUBE
      IF(XX.GE.CUBE) XX=XX-CUBE
      IF(YY.LE.0.) YY=YY+CUBE
      IF(YY.GE.CUBE) YY=YY-CUBE
      IF(ZZ.LE.0.) ZZ=ZZ+CUBE
      IF(ZZ.GE.CUBE) ZZ=ZZ-CUBE
C
C === STORE NEW POSITIONS
      X0(I)=XX
      Y0(I)=YY
      Z0(I)=ZZ
      X0L(I)=X0(I)
      Y0L(I)=Y0(I)
      Z0L(I)=Z0(I)
  690 CONTINUE
      RETURN
      END
      SUBROUTINE EQBRAT(SUMVEL,AHEAT,TDIST,XDIST,NP,IFLG,NY,KB)
C
C === SCALE VELOCITIES DURING INTIAL TIME-STEPS
C
      DOUBLE PRECISION D4AX,D4AY,D4AZ,D3AX,D3AY,D3AZ,D2AX,D2AY,D2AZ,
     1    DACX,DACY,DACZ,SUMD4,SUMD3,SUMD2,SUMDAC,SUMACC,SUMD4A,SUMD3A,
     2    SUMD2A,SUMDA,SUMA,RK12,RK3,RK123,RK4,RK14,RK1234,RKS4,RK5
     3    ,X5(256),Y5(256),Z5(256),X4(256),Y4(256),Z4(256),X3(256),
     4    Y3(256),Z3(256),X2(256),Y2(256),Z2(256),X1(256),Y1(256),
     5    Z1(256),DEL5,DEL4,DEL3,DELSQ,DELTA
      COMMON /VEL/ X1,Y1,Z1
      COMMON /PROP/ IDIST(300),ISUM
      COMMON /SS/ SUMD4A,SUMD3A,SUMD2A,SUMDA,SUMA,XSUM,SUMV,SUME
      COMMON /KAB/ MAXKA,MAXKB
      IF(KB.EQ.600) GO TO 750
      IF(KB.GT.400) GO TO 760
      HEAT=SQRT(AHEAT/SUMVEL)
   10 FORMAT(5X,'HEAT= ',G12.5)
      DO 730 I=1,NP
      X1(I)=X1(I)*HEAT
      Y1(I)=Y1(I)*HEAT
      Z1(I)=Z1(I)*HEAT
```

```fortran
      730  CONTINUE
      760 RETURN
C
C === AT END OF EQUILIBRATION STAGE, SET PROPERTY SUMS TO ZERO
      750  IFLG=1
           MAXKA=MAXKB-KB
           ISUM=0
           SUME=0.
           SUMV=0.
           XSUM=0.
           SUMA=0.
           SUMDA=0.
           SUMD2A=0.
           SUMD3A=0.
           SUMD4A=0.
           KB=0
           DO 740 I=1,NY
      740  IDIST(I)=0
           RETURN
           END
           SUBROUTINE RDF(DR,RDEL,NY)
C
C === NORMALIZE COUNTERS FOR RADIAL DISTRIBUTION FUNCTION
C
           COMMON /PROP/ IDIST(300),ISUM
           Y=ISUM
           PDEN=DR*(RDEL**3)
           WRITE(6,968)
      968  FORMAT(1H1////15X,'I',6X,'R',6X,'TDIST',3X,'G(F)'/)
           DO 780 J=1,NY
           IF(IDIST(J).EQ.0) GO TO 780
           V=(4.*3.14159/3.)*(3.*FLOAT(J)**2+.25)
           X=IDIST(J)
           GR=X/V/Y/PDEN
           RRR=RDEL*FLOAT(J)
           WRITE(6,972) J,RRR,IDIST(J),GR
      972  FORMAT(14X,I3,3X,F5.3,3X,I6,F8.3)
      780  CONTINUE
           WRITE(6,970)
      970  FORMAT(1H1///)
           RETURN
           END
           SUBROUTINE VELOCO(NP,NVELO,DELTA,MAXKB,TMP)
           COMMON /VELO/ VX(32,2000),VY(32,2000),VZ(32,2000)
           DIMENSION F(2000)
       60  FORMAT(1X,'F=',10G10.4)
       61  FORMAT(8G10.4)
           NAVG=NP/NVELO
           DO 490 L=1,MAXKB
           F(L)=0.
           NDAN1=MAXKB +1-L
           DO 41 K=1,NDAN1
           P=0.
           M=K+L-1
           DO 42 I=1,NAVG
       42  P= P +(VX(I,K)*VX(I,M)+VY(I,K)*VY(I,M)+VZ(I,K)*VZ(I,M))/3.0
           P=P/NAVG
           F(L)=F(L)+P
       41  CONTINUE
           F(L)=F(L)/(NDAN1*TMP)
```

```
490  CONTINUE
     WRITE(6,60) (F(L),L=1,1000)
     WRITE(7,61) (F(L),L=1,400)
     RETURN
     END
```

# APPENDIX D

# BIBLIOGRAPHY

This appendix gives some of the common reference books in equilibrium and nonequilibrium statistical theories. The citations on transport and time-dependent processes are for general interest.

(1) Akcasu, A.Z. and Linnebur, E.J., *Neutron inelastic scattering* (International Atomic Energy Agency, Vienna, 1972)

(2) Bak, Th., *Statistical mechanics* (Benjamin, New York, 1967)

(3) Balescu, R., *Statistical mechanics of charged particles* (Interscience, New York, 1964)

(4) Balescu, R., *Equilibrium and nonequilibrium statistical mechanics* (Wiley, New York, 1975)

(5) Barker, J.A., *Lattice theory of the liquid state* (MacMillan, New York, 1963)

(6) Baxter, R.J., *Exactly solved models in statistical mechanics* (Academic Press, New York, 1982)

(7) Ben-Naim, A., *Water and Aqueous Solutions* (Plenum, New York, 1974)

(8) Berne, B.J., ed., *Statistical mechanics: Equilibrium techniques Part A*, Mod. Theor. Chem. v.6 (Plenum, New York, 1977); *Statistical mechanics: Time dependent processes, Part B*, v.6 (Plenum, New York, 1977)

(9) Biel, J. and Rae, J., *Irreversibility and many body problem* (Plenum, New York, 1970)

(10) Böttcher, C.J.F., *Theory of electric polarization* (Elsevier, Amsterdam, 1973)

(11) Buckingham, A.D., *Molecular structure and properties*, MTP International Review, Series 2 (Butterworth, London, 1975).

(12) Cohen, E.G.D. ed., *The Boltzmann equation* (Springer, Berlin, 1973)

(13) Cohen, E.G.D. ed., *Fundamental problems in statistical mechanics* Volume 2 (North Holland, Amsterdam, 1968), Volume 3 (1975)

(14) Cole, G.H.A., *An introduction to the statistical theory of classical simple dense fluids* (Pergamon N.Y. 1967)

(15) Croxton, C.A., *Liquid state physics--A statistical introduction* (Cambridge University Press, Cambridge, 1974)

(16) Croxton, C.A., ed. *Progress in liquid physics* (Wiley, New York, 1979)

(17) Croxton, C.A., *Fluid interfacial phenomena*, (Wiley, New York, 1986)

(18) De Boer, J. and Uhlenbeck, G.E., ed. *Studies in statistical mechanics* (North Holland, Amsterdam, 1962)

(19) De Gennes, P.G., *The physics of liquid crystals* (Oxford University Press, 1974)

(20) Domb, C. and Green, M.S., ed. *Phase transitions and critical phenomena* (Academic Press, New York, 1972)

(21) Egelstaff, P.A., *An introduction to the liquid state* (Academic Press, New York, 1967)

(22) Eyring, H., Henderson, D. and Yost, W., ed. *Physical chemistry: an advanced treatise* (Academic Press, New York, 1971)

(23) Falkenhage, H., *Electrolytes* (Clarendon Press, Oxford, 1934)

(24) Fisher, I.Z., *Statistical theory of liquids* (University of Chicago Press, Chicago, 1964)

(25) Flory, P.J., *Statistical mechanics of chain molecules* (Wiley, New York, 1969)

(26) Forster, D., *Fluctuations, broken symmetry and time correlation functions* (Addison-Wesley, Reading, 1976)

(27) Franck, F. ed. *Water: A comprehensive treatise* (Plenum, New York, 1973)

(28) Freed, K. and Light, J. ed. *Proceedings of IUPAP conference on statistical mechanics* (Chicago University Press, Chicago, 1972)

(29) Friedman, H.L., *Ionic solution theory* (Wiley, New York, 1962)

(30) Friedman, H.L., *A course in statistical mechanics* (Prentice-Hall, Englewood Cliffs, New Jersey, 1985)

(31) Frisch, H.L. and Lebowitz, J.L., ed. *The equilibrium theory of classical fluids* (Benjamin, New York, 1964)

(32) Frisch, H.L. and Salsburg, Z.W., ed. *Simple dense fluids* (North Holland, Amsterdam, 1952)

(33) Gray, C.G. and Gubbins, K.E., *Theory of molecular fluids*, Volume 1. Fundamentals, (Clarendon Press, Oxford, 1984)

(34) Green, H.S., *The molecular theory of fluids* (North Holland, Amsterdam, 1952)

(35) Haile, J.M. and Mansoori, G.A., ed. *Molecular-based study of fluids*, Advance in Chemistry Series #204 (American Chemical Society, Washington D.C., 1983)

(36) Hansen, J.-P. and McDonald, I.R., *Theory of simple liquids* (Academic Press, New York, 1977)

(37) Hill, T.L., *Statistical mechanics* (McGraw-Hill, New York, 1956)

(38) Hill, T.L., *An introduction to statistical thermodynamics* (Addison-Wesley, Reading, 1960)

(39) Hirschfelder, J.O., Curtiss, C.F. and Bird, R.B., *Molecular theory of Gases and liquids* (Wiley, New York, 1954)

(40) Horne, R.A. ed., *Structure and transport processes: water and aqueous solutions* (Wiley, New York, 1972)

(41) Huang, K., *Statistical mechanics* (Wiley, New York, 1963)

(42) Kohler, F., *The liquid state* (Crane Russak, New York, 1972)

(43) Lovesey, S.W. and Springer, T., *Dynamics of solids and liquids by neutron scattering* (Springer, Berlin, 1977)

(44) March, N.H., *Liquid metals* (Pergamon, Oxford, 1968)

(45) March, N.H. and Tosi, M.P., *Atomic dynamics in liquids* (Wiley, New York, 1976)

(46) March, N.H. and Tosi, M.P., *Coulomb liquids* (Academic Press, New York, 1984)

(47) Martin, P.C., *Measurements and correlation functions* (Gordon and Breach, New York, 1968)

(48) Mason, E.A. and Spurling, T.H. *The virial equation of state* (Pergamon, Oxford, 1969)

(49) Massignon, D., *Mécanique statistique des fluides* (Dunod, Paris, 1957)

(50) Mazo, R.M., *Statistical mechanical theory of transport processes* (Pergamon, New York, 1967)

(51) McQuarrie, D.A., *Statistical Mechanics* (Harper and Row, New York, 1976)

(52) Meeron, E., ed. *Physics of many particle systems* (Gordon, New York, 1967)

(53) Meixner, J., *Statistical mechanics of equilibrium and nonequilibrium* (North Holland, Amsterdam, 1965)

(54) Montroll, E.W. and Lebowitz, J.L. ed., *The Liquid state of matter: fluids, simple and complex* (North Holland, Amsterdam, 1982)

(55) Münster, A., *Statistische Thermodynamik* (Springer, Berlin, 1956)

(56) Nicholson, D. and Parsonage, N.G., *Computer simulation and the statistical mechanics of adsorption* (Academic Press, New York, 1982)

(57) Percus, J., ed. *The many body problem* (Interscience, New York, 1963)

(58) Reed, T.M. and Gubbins, K.E., *Applied statistical mechanics* (McGraw-Hill, New York, 1973)

(59) Resibois, P., *Electrolyte theory* (Harper and Row, New York, 1968)

(60) Resibois, P. and DeLeener, M., *Classical kinetic theory of fluids* (Wiley, New York, 1977)

(61) Rice, S.A. and Gray, P. *The statistical mechanics of simple liquids* (Wiley, New York, 1965)

(62) Rothschild, W.G., *Dynamics of molecular liquids* (Wiley, New York, 1984)

(63) Rowlinson, J.S. and Widom, B., *Molecular theory of capillarity* (Clarendon Press, Oxford, 1982)

(64) Schieve, B., ed. *Advanced course on thermodynamics and statistical mechanics* (Springer, Berlin, 1971)

(65) Singer, K., *A specialist periodic report: Statistical Mechanics* (Chemical Society, Burlington House, London, 1973)

(66) Stanley, E., *Phase transitions and critical phenomena* (Clarendon Press, Cambridge, 1971)

(67) Steele, W.A. , *Interaction of gases with solid surfaces* (Pergamon, New York, 1974)

(68) Temperley, H. Rowlinson, J. and Rushbrooke, G., ed. *Physics of simple liquids* (North Holland, Amsterdam, 1968)

(69) Uhlenbeck, G.E. and Ford, G.W., *Studies in statistical mechanics*, Volume 1 (North Holland, Amsterdam, 1962)

(70) van Hove, L., *The theory of neutral and ionised gases* (Wiley, New York, 1960)

(71) Watts, R.O and McGee, I.J., *Liquid state chemical physics* (Wiley, New York, 1976)

(72) Yamakawa, H., *Modern theory of polymer solutions* (Harper and Row, New York, 1971)

(73) Yvon, J., *La théorie statistique des fluides et l'équation d'état*, Actualités scientifiques et industrielles, Volume 203 (Hermann et Cie, Paris, 1935)

# INDEX